Chemical Process Control 2

Proceedings of the
Engineering Foundation Conference

January 18-23, 1981

The Cloister
Sea Island, Georgia

EDITORS
Thomas F. Edgar
and
Dale E. Seborg

DIRECTOR OF CONFERENCES
Sandford S. Cole

This conference was co-sponsored by the Engineering Foundation and the National Science Foundation, and by the Computing and Systems Technology Division of the American Institute of Chemical Engineers. Any findings, opinions, conclusions, or recommendations expressed herein are those of the individuals, and do not necessarily reflect the position of, nor imply endorsement by, the aforementioned cosponsors. This book is available from the Publications Department of the American Institute of Chemical Engineers under a distribution agreement between it and the Engineering Foundation.

Copyright 1982 by United Engineering Trustees, Inc.
ISBN 0-8169-0203-8
Library of Congress Catalogue Card Number: 81-71594

INTRODUCTION

Dale E. Seborg
University of California, Santa Barbara

Thomas F. Edgar
University of Texas, Austin

This volume contains the proceedings of an Engineering Foundation Conference, Chemical Process Control II, which was held at Sea Island, Georgia from January 18 to 23, 1981. The conference was co-sponsored by the Computing and Systems Technology Division of the American Institute of Chemical Engineers and received financial support from the National Science Foundation. The conference organizing committee consisted of Dale E. Seborg and Thomas F. Edgar, co-chairmen; Edgar H. Bristol; Magne Fjeld; W. Harmon Ray; and George Stephanopoulos.

This intensive, week long conference provided an excellent opportunity for industrial control engineers and their academic counterparts to participate in a stimulating dialogue. The 113 conference attendees had a broad range of experience in the process control field; 48 were from the process industries, 47 were from universities, and 17 were from control system and consulting firms. The conference had an international flavor since ten of the participants were from Britain and Europe, two were from Japan, and one was from China.

The Sea Island Conference emphasized new developments that have occurred since 1976 when the first Engineering Foundation Conference on Chemical Process Control was held.[1] The Sea Island Conference featured the following sessions:

---

[1] The conference proceedings appeared as Number 159, Volume 72 of the AIChE Symposium Series; A. S. Foss and M. M. Denn, Editors.

Systems Software for Process Control
(Organizer: M. Fjeld, AccuRay Corporation)
Human Factors in Process Control
(Organizer: M. Fjeld, AccuRay Corporation)
Advanced Strategies for Process Control and Estimation
(Organizer: W. H. Ray, University of Wisconsin, Madison)
Distillation Column Control
(Organizer: T. F. Edgar, University of Texas, Austin)
Control in Energy Management and Production
(Organizer: T. F. Edgar, University of Texas, Austin)
Design of Control Systems for Integrated Plants
(Organizer: G. Stephanopoulos, University of Minnesota)
Closing Session
(Organizer: D. E. Seborg, University of California, Santa Barbara)

This conference demonstrated that there are several reasons why process control continues to be both an important and exciting discipline:

1. Substantial economic incentives exist for improved control of industrial processes.
2. Inexpensive, reliable control equipment is readily available to implement virtually any control strategy that a process engineer can suggest.
3. Academic research in process control has been invigorated by recent emphasis on challenging new problems which are motivated by industrial needs rather than by theoretical developments in other fields.

As Alan Foss has noted in his summary (pg. 627), "Circumstances are auspicious for significant advances in all of chemical process control."

TABLE OF CONTENTS

INTRODUCTION .... Dale E. Seborg & Thomas F. Edgar   iii

SYSTEMS SOFTWARE FOR PROCESS CONTROL
Systems Software for Process Control: Today's State of the Art
.... D. Grant Fisher   3
Some Principles in Structuring Software for Process Control ...
.... Th. Lalive d'Epinay   33
Operational Needs in Software for Complex Process Control Systems ....
.... Robert G. Wilhelm, Jr.   51
Application of the ADA Language to Process Control ....
.... William C. M. Vaughan   71

HUMAN FACTORS IN PROCESS CONTROL
Important Problems and Challenges in Human Factors and Man-Machine Engineering for Process Control Systems ....
.... John E. Rijnsdorp   93
Human Error and Human Reliability in Process Control ....
.... John W. Senders   111
The Systems Engineering Context of Process Control and Man-Machine System Design .... William L. Livingston   117
Training and Selection of Operators in Complex Process Control Systems .... Lawrence E. Kanous   137
New Results and the Status of Computer-Aided Process Control Systems Design in Japan .... Iori Hashimoto   147
New Results and the Status of Computer-Aided Process Control Systems Design in Europe .... Arne Tyssø   187
New Results and the Status of Computer-Aided Process Control Systems Design in North America .... Thomas F. Edgar   213

On Closing "The Gap"—With Modern Control Theory that Works
.... Magne Fjeld   229

New Approaches to the Dynamics of Nonlinear Systems with Implications for Process and Control System Design...........
.... W. Harmon Ray   245

A Review of Some Adaptive Control Schemes for Process Control
.... Pierre R. Bélanger   269

Model Algorithmic Control (MAC); Review and Recent Developments .... Raman K. Mehra & Ramine Rouhani   287

Application of Modelling, Estimation and Control to Chemical Processes .... E. Dieter Gilles   311

Roundtable Discussion...................... W. Harmon Ray   339

DISTILLATION COLUMN CONTROL

Control of Heat-Integrated Columns .... Page S. Buckley   347

Explicit vs. Implicit Decoupling in Distillation Control ....
.... Carroll J. Ryskamp   361

On the Difference Between Distillation Column Control Problems
.... Thomas J. McAvoy   377

University Research on Dual Composition Control of Distillation: A Review .... Kurt V. Waller   395

CONTROL IN ENERGY MANAGEMENT AND PRODUCTION

Control Analysis of Gasification/Combined Cycle (GCC) Power Plants..................................... George H. Quentin   415

A General Industrial Energy Management Systems—Functionality and Architecture .... C. Ronald Simpkins   433

Plant Utility Management .... Richard A. Hanson   449

DESIGN OF CONTROL SYSTEMS FOR INTEGRATED CHEMICAL PLANTS

Integrated Plant Control: A Solution at Hand or a Research Topic for the Next Decade? .... Manfred Morari   467

Process Operability and Control of Preliminary Designs ....
.... James M. Douglas   497

Design and Analysis of Process Performance Monitoring Systems
.... Richard S. H. Mah   525

Panel Discussion: Control Systems for Complete Plants—The Industrial View .... Irven Rinard 541
.... Joseph P. Shunta 547
.... Thomas E. Marlin 555

DISTRIBUTED COMPUTER PROCESS CONTROL
The Structuring of Distributed Intelligence Computer Control Systems .... Charles W. Rose 565

An Approach to Digital Process Control .... Mart Masak 581

Panel Discussion: Important Design Issues and Recommendations for Prospective Users of Distributed Control ....
.... Remsi R. Messare 599
.... Duncan A. Mellichamp 603
.... Edgar Bristol 605

CONFERENCE SUMMARY
Industrial Theme Problems in Process Control Morton M. Denn 609

Systems Software .... D. Grant Fisher 613

Human Factors .... John E. Rijnsdorp 617

Advanced Strategies for Process Control and Estimation ....
.... John H. Seinfeld 619

Distillation Column Control/Energy Management ....
.... J. Patrick Kennedy 623

Control Systems for Integrated Plants .... Alan S. Foss 625

Distributed Computer Process Control .. Duncan A. Mellichamp 629

Author Index 631

Subject Index 633

Systems Software For Process Control

SYSTEMS SOFTWARE FOR PROCESS CONTROL:
Today's State of the Art

Professor D. Grant Fisher
Department of Chemical Engineering
University of Alberta
Edmonton, Alberta
Canada, T6G 2G6

ABSTRACT

The major, current problem in computer control of chemical processes is software and system design; not hardware, not interfaces, not control theory, but the design and implementation of a software system capable of handling the total plant application. Now that computing power and memory no longer present serious limitations, there appears to be increasing use of vendor software and high level languages such as extended FORTRAN and ADA. As a result users are focusing more on specific solutions to their particular operational needs and putting more of their effort at the supervisory and management levels rather than on "direct-digital-control" (DDC). Network communication support is becoming more powerful, more widely available and more transparent so that corporate-wide computer networks are a reality. The immediate future should see the pendulum swing-back a bit from distributed, "do-it-yourself" networks of mini plus micro-computers to more centralized systems, plus the development of database management systems which guarantee the integrity and availability of information at all levels from management to local display of a process variable. These database and display developments will closely parallel development of in-plant communication systems plus man-machine interface standards. This paper will attempt to stimulate discussion by presenting one man's observations and opinions on views such as those expressed above.

INTRODUCTION

The objective of this *paper* is to present a broad, "top-down" description of computer control systems as they are applied in the process industries. More particularly, the objective is to define the *scope* and examine the *structure* and *requirements* of the application rather than to deal with specifics. The hope is that this paper will provide a basis for integrating the different views and more detailed presentations by others into a consensus that truly reflects the current "state of the art" and identifies directions for future growth.

The objective of the computer control *system* discussed in this paper is to meet all the management and process control requirements of a typical, large industrial plant.

Stated in *control* terminology the objective of the computer control system is to support data acquisition and display, computer-assisted manual control, conventional automatic control, modern multivariable control, estimation, plus a wide range of supervisory and optimization algorithms. Although primarily oriented towards continuous processes, the system must support discrete (batch) operations and meet the needs of plant personnel at the process operator, supervisory, service and management levels.

In *computer* terminology the objective is a real-time, sensor-based, hardware and software system that will effectively carry out the combined tasks of process control and plant management. This requires data input/output via the process interface, information processing, storage, database management, communication support plus an effective man-machine interface and a powerful operating system. None of these requirements is unique to process control applications but there are a number of characteristics such as the large number and variety of I/O devices, the need for real-time processing and "guaranteed response times," the potentially disastrous consequences of system errors or failures, the highly correlated and variable demands of a process during upsets, etc., all of which demand and deserve careful consideration.

In the past, for very valid reasons related to the limited capability of computer hardware and the limited experience of design teams, most computer control installations exemplified a "bottom up" approach to design and implementation. For example, stage #1 was usually data acquisition and/or "direct digital control" (DDC) that closely paralleled conventional analog practice. This was followed by stage #2 in which "supervisory control" techniques were added to provide the objectives (setpoints) for the DDC level. Similarly, problems related to process I/O interfaces, operator communications, "backup" control systems, etc. were solved by adapting existing practice rather than by long-term, top-down design studies. During the same time period other groups were developing management planning support (eg. production planning using linear programming), generalized database management and display support, plus powerful new computing capabilities involving virtual systems, distributed (parallel) computing and powerful, high-level communications support. Unfortunately, as frequently happens with bottom-up design approaches, the design of most existing *process* control systems does not integrate well with available *management* functions and the current means of implementation are not best suited to the broader objectives.

A current requirement is therefore a top-down specification of the requirements and design of a computer control system that integrates both management and automatic control requirements. This is not to imply that all installations will be total packages that meet all these requirements, but rather that *IF* the *system design* defines: *modules* with clearly defined functions, *interfaces* that are complete but minimize "coupling" (interaction) between modules, segmented *databases* with hidden structures, *communication protocols* for both internal and external (eg. process and operator) data transfers, etc. *THEN* subsystems can be designed, built, implemented and operated in a modular, incremental manner but with the assurance that they will function together in the future and that the integrated system will meet the total control requirement.

The following three sections approach the problem of computer control of industrial process from three different perspectives:

i) Section 2 presents a bottom-up description of process control as it is currently practiced in many process plants.

ii) Section 3 describes the distributed network of micro and mini-computers operating in the Department of Chemical Engineering at the University of Alberta. This system illustrates many features and requirements of current industrial systems.

iii) Section 4 attempts to define the system architecture and to identify the main components of a large computer control system.

Commercially available systems are not discussed specifically nor critically evaluated. However, they are important because many users have no practical alternative to purchasing an available system.

The scope of this paper is obviously broad and precludes in-depth treatment of any one subject. It is also recognized that there is a natural tendency for each reader to focus on his own area of expertise. However, the hope is that each reader will not only evaluate the description of his own area of expertise but also look at the *interface* and the *interactions* and/or *common characteristics* of related facets of the total process control problem, i.e. the intended focus is on the structure of the total system rather than on defining the state of the art in each separate area.

PROCESS CONTROL

The following discussion is keyed to the process control system shown in Figure 1.

Figure 1a: Process

In this paper "process" refers to industrial plants such as oil refineries, chemical plants, pulp and paper mills and certain other operations characterized by mainly continuous operation about a set of steady-state design and operating conditions. These process plants typically handle millions of kilograms of raw materials per year, represent a multi-million or multi-billion dollar investment, are physically spread over several hundred hectares, utilize high technology equipment, are not highly labor intensive and typically form part of a multi-national corporation. These plants are sub-divided into *areas,* production *units,* and individual control *loops*. For example, an oil refinery may include crude oil, catalytic cracker, light ends and utility *areas*. The crude oil *area* may contain sulphur removal, atmospheric and vacuum distillation column *units*. These units consists of *equipment* such as distillation columns, heat exchangers, etc. which must be controlled by a feedback (or feedforward etc.) control *loop*.

The size of the process requires that the control problem be subdivided and the nature of the process suggests successive subdivision associated with areas, units, equipment and loops. (However, later sections will suggest other bases for the modularization of the control software.)

Trends

Current trends in process design include larger, often single-stream plants, tighter operating and environmental-impact specifications; greater

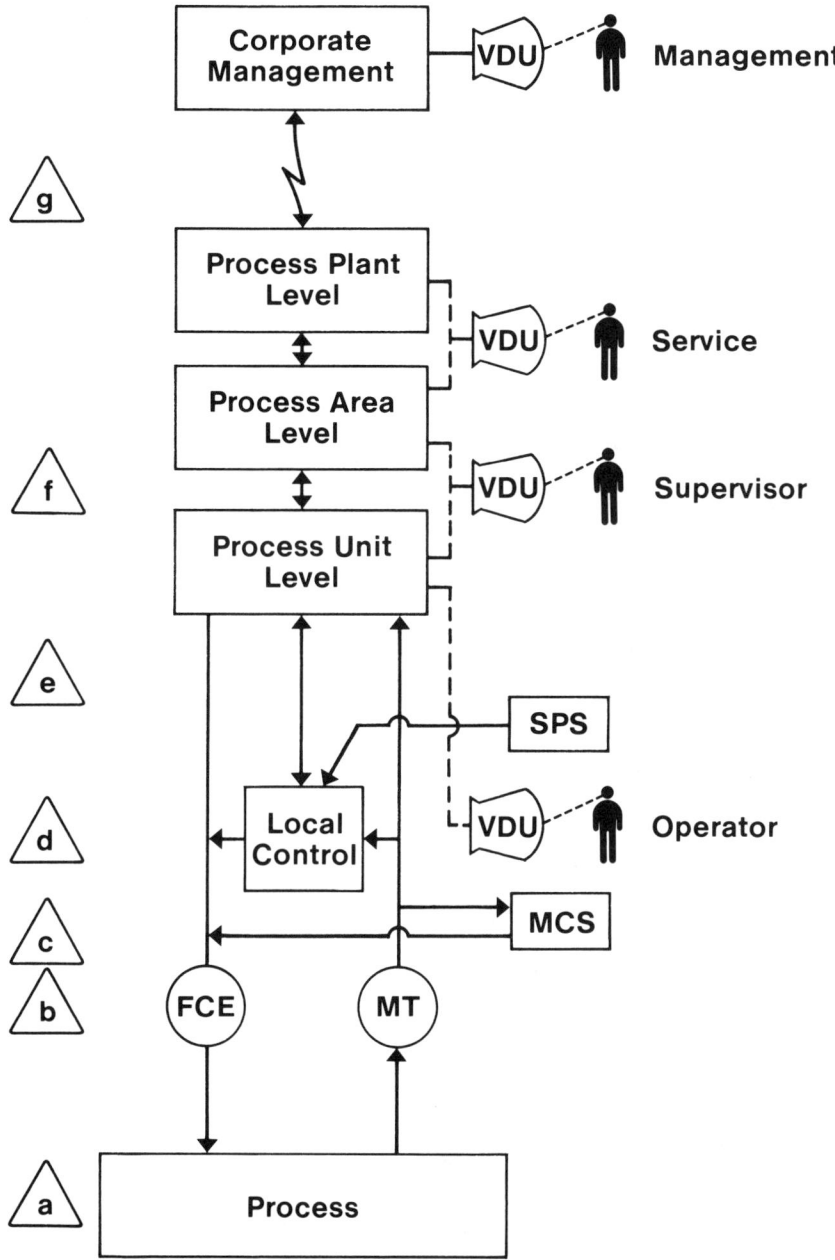

Figure 1. Distributed Industrial Computer Control System.

Table 1. Main Components of a Process Control System.

Process instruments
- measurement transducers
- final control elements
- special instruments: e.g. analysers

Interfaces
- process instrument to computer
- computer to computer
- computer to personnel (man-machine)
  - process operator level
  - supervisory and support level
  - management level

Computer hardware
- main frames and memory
- conventional peripherals
- process I/O interface: AIO, DIO, busses
- operator displays and terminals
- communication support

Computer software
- embedded, e.g. ROM's in microprocessors
- computer operating systems
- computer communications support
- data base management
- program development support
- application
  - loop basis (DDC)
  - process unit controllers
  - plant level
  - corporate or management level

---

vertical-integration, eg. raw material to multi-products in one plant with reduced intermediate storage; greater interaction between units due, for example, to energy saving measures which require exchange of heat between the product from one unit and the feed to the same or a different unit; and faster process dynamics due to smaller pieces of equipment operating closer to design constraints. These trends demand more sophisticated management plus automatic control and require that it be done on a broader basis than a single loop.

Figure 1b: Process measurement and control (MT and FCE)

Process control is almost a perfect area for the application of "input-process-output" or "source-transform-sink" decomposition techniques in software design. *Input* information is obtained from measurement transducers (MT) via the process interface, *processed* by a control algorithm, and then *output* to a final control element(s) (FCE). The general objective of the control system is: (1) *regulatory* control, i.e. keeping the controlled variable at its desired (setpoint) value despite disturbances in other input variables and/or (2) *servo* control, i.e. make the controlled variable follow changes in the setpoint value.

It is obvious that the quality of control possible depends on the amount, accuracy, precision, reproducibility, signal to noise ratio, range, sampling time, etc. of the input/output information. To use an overworked expression: "Garbage-in, garbage out." However, it must be noted that in many applications the limiting factor on process control is the availability of suitable measurement and/or control instruments. For example, the development of a scanning moisture-content analyser for paper resulted in hundreds of computer control installations on paper machines. Surveys of operating people still show that better measurement instrumentation is one of their prime requirements. The area of process I/O instrumentation is a broad one (cf. the Instrument Society of America compendiums which list tens of thousands of different transducers [1]). All that will be done here is to acknowledge a primary need and point out, as Figure 1 implies, that process I/O is the foundation on which the whole computer control scheme is built.

Trends

The trend is to more "solid state devices" eg. pressure sensitive electronic devices rather than mechanical force balance systems, and to more inherently digital devices such as position encoders, eg. for level, and voltage to frequency converters, eg. for temperature measurements. The inclusion of microprocessors directly in the transducers is providing features such as improved linearity, self-checking diagnostics, compensated measurements (eg. temperature compensation), computed values (e.g. mass rate of flow based on temperature, pressure, composition and flow measurements), noise filtering, signal averaging, etc.

Perhaps the greatest impact of microprocessors in process instrumentation will be on the computer interface and communication alternatives. Present practice is to hardwire the process instruments to the nearest process controller station and/or to a local computer interface unit. Future options include "multi-drop, shared data highways" plus various "broadcast techniques' where each instrument contains a send and/or receive broadcast unit (IR, FM etc.) which can communicate with one or more local devices (e.g. a controller located in one of the transducers) as well as the main computer unit. Such "wireless" interfaces are already provided in many new buildings for computer terminals and some process instruments are already being marketed with "terminal (RS-232) interfaces" so the possible, extension is obvious (but to my knowledge not implemented on a broad basis). Micro-processors affect not only the physical communications interface but also the design basis for the computer control system, eg. consider provision for backup hardware, reconfiguration of control strategies, data security, protection against outside interference, etc.

Figure 1c: Manual Control

When necessary the operator uses a manual control station (MCS) connected to the same MT and FCE used by the automatic control scheme. In some cases there may be additional local measurement indicators and manually operated valves that can be used in the event of MT and FCE failures. Historically, this was the first means of process control. Automatic control is now so widespread that there are seldom enough operators on duty to operate a complete process manually and in many cases they do not know how or, at least, have not had much practice in doing so.

Trends

Manual control is only used under emergency conditions and even then only for critical variables. The MCS may be local, remote or a portable plug in unit.

Question: Will we continue to rely on operators for emergency backup or will this function be incorporated into the computer control scheme? (cf. applications such as the Boeing 767 which has an all digital control system with no "analog backup" or complete manual control capability! [2]).

Note that a trend to smaller, higher-performance, integrated, process plants could result in plants that are "impossible" to control manually.

Figure 1d: Local Loop Control

Most current process control loops use proportional-integral-derivative (PID), feedback control of a single variable implemented by either digital or analog controllers. The controller may also include provision for: displaying the measurement, output, setpoint and alarm status; all the functions of a manual control station (MCS); entry of controller constants, sampling time, filter options, etc. Historically, controllers have evolved from field-mounted, usually pneumatic instruments, to "large-case" panel mounted units, to panel mounted operator-control stations with the control functions implemented by separate racks of electronic circuit boards, to computer-based implementations with digital displays shared among several loops. Two observations can be made:

i) change tends to be evolutionary rather than revolutionary
ii) the time constant for change in several years (but possibly decreasing)

Trends

The trend is definitely towards digital-electronic implementations with N ($1 \leq N \leq 8$) loops per microprocessor and a shared process operator's console. Digital technology is providing a broader selection of control algorithms, easier implementation of "advanced conventional strategies such as cascade, feedforward, ratio and preprogrammed adaptive" control, and, most importantly, a simpler (digital) interface to larger general-purpose computer systems.

Database

In digital systems there is a "database" associated with each "loop" which may contain from 10 to 100 fields of information, eg. controller constants, status indications, etc. One fixed database design (perhaps with a few basic variations) is used for all loops. The database is typically only partially accessible to the user, cannot be extended, and is referred to internally by what amounts to direct, absolute memory-addressing. The user-database interface varies from a simple keyboard plus LED display to an elaborate computer console that can display/change the database in a large number of controllers.

Figure 1e: Computer to process instrument communication

Many process plants have a digital computer that is connected to all the single loop controllers in a given process unit, or process area, and sometimes to individual MT and FCE for direct I/O to the process. Of the three basic types of communication in a process control system, computer/people, computer/computer and computer/process instrument, it is the latter that is most unique. Hardware and software development for communication at this level is therefore one of the most active and important areas of development. The hardware connections have involved direct, dedicated wiring, and more recently coaxial cable and fibre optic busses. Protocols include bit-serial international communication standards such as RS232 and RS449, special busses such as the IEEE-488 byte-parallel bus plus a large number of proprietary busses developed by different vendors. The current need is for a hardware/software *standard* for in plant communications. Groups such as the International Purdue Workshops are involved in developing standards such as "PROWAY" but until a satisfactory standard is established, people will be forced to 'go their own way' [4].

Most of the current development in computer communications is directed towards point to point communication between relatively large general purpose computer installations for business communication, electronic mail, integrated corporate computer networks, etc. IBM's message switching network, for example, has 406 processor nodes in 78 cities, in 15 countries and supports thousands of terminals [3]. The market for in-plant communication systems for process control is small by comparison to the business market and the danger is that protocol specification, LSI chip development etc. may overlook the special needs of process control systems.

Trends

The trend is again definitely towards microprocessor or special single-chip, bus-interfaces but beyond that little can be discerned. The problem is complicated because two different groups of vendors are involved (computers and process control instruments) and both have long traditions, international user communities and, to some extent, different criteria for performance evaluation.

Question: Will this problem eventually be solved by vertical integration of process instrument vendors into larger computer systems or by the development of an effective, standard interface between process plant information systems and large, main-frame computer vendors? (In a later section I will argue the advantages of the latter approach. Until this question is resolved process users will continue to be faced by an either/or choice; general purpose process I/O interfaces from mini-computer vendors or proprietary systems from instrument vendors.)

Figure 1f: Computers for process units and/or areas

The most common application for unit or area computers is "supervisory control" where the computer adjusts the setpoints but the actual feedback control is done by the local loop controllers. The setpoint changes can be based on pre-stored "recipes", *ad hoc* decision rules based on previous operating experience, constraint identification, or sophisticated optimization schemes.

There has also been a lot of activity involving preprogrammed computers for specific process units such as distillation columns, furnaces, boilers, etc. The intent is that the user selects a control strategy from the available options and then supplies the necessary parameters using a fill-in-the-blanks, computer assisted procedure. Ideally, the installation could be done by control-oriented personnel since no programming is required and since most approaches are based on conventional multiloop control techniques. Every vendor or user with multiple applications uses some degree of modular design and implementation. However, standardized unit controllers do not seem to be a runaway market success which suggests that in practice, most applications require a significant degree of custom tailoring and tuning.

The most active and most profitable area for *process* control applications is probably computer systems with a sphere of influence that covers a single process unit or an entire process area. Such systems are being installed, for example, to identify active constraints and keep the process operating against the most limiting constraint. Systems have been installed to increase production rates by operating closer to equipment capacity eg. distillation column flooding; to minimize energy consumption, eg. by reducing excess air in a furnace; or the temperature of effluent streams; to improve efficiencies by dynamically altering operating conditions on a minute by minute basis to reflect raw material feed variations; to improve control by estimating and controlling the real objective, eg. product concentration, rather than related variables such as temperature; to improve overall unit operation by providing a dynamic simulator which allows operators to evaluate alternative operating conditions; and in batch-continuous plants, to improve throughput by proper scheduling of the batch units. The most significant features of these applications is the amount of *process* understanding that is required and the fact that many of the key considerations are unique to a single installation. (This is in contrast to other areas, such as feedback control loops, where tuning procedures, stability guarantees etc. can be developed for a broad class of problems and do not depend on the internal details of process mechanisms.) Inspite of the importance of this area, most conference programs and journal indexes do not include details of applications in these areas. Some obvious reasons are the proprietary nature of process data; the fact that many problems are unique (eg. a compressor was sized too small) and hence not of interest to a broad audience; the solution algorithms may be simple ad hoc procedures based on observation of "best operator" practice and hence have no proofs of stability, convergence, etc.; they don't appeal to academically oriented editors and review committees.

Trends

The trend is definitely to more applications at this level since direct digital control systems for implementing feedback control loops are now pretty standard and available on a package basis. By their very nature these problems are process oriented and are the logical area for users to apply their process control expertise, i.e. leave the instrumentation and computer work to others.

Some good work is being done on the development of general algorithms (eg. for the solution of sets of equations) and on the theory and practice of multilevel decision making and optimization, which are basic to many applications in this area. There will probably continue to be few detailed publications in this area but probably an increase in the number of proprietary software packages sold separately or as part of turnkey process engineering packages (eg. $100,000 for software to optimize the operation of cracking furnaces to produce ethylene). The user community is just beginning to accept software as a marketable commodity like computers or process equipment.

More attention should be paid in the design of computer control systems to facilitating the development and implementation of user application programs in this area.

Software aspects

It is hard to define a clear division between the direct digital (DDC) and area control systems. However, in general, most DDC systems are implemented by table-driven processors, which require little, if any, user programming, and which include a system executive which is transparent to the user. Area control on the other hand, is more typically implemented by user-written programs that are written in a high-level language and execute under a general purpose operating system. DDC functions are normally executed cyclically every few seconds, are memory resident and execute to completion at a fairly high priority. Area-level programs are typically larger, more diverse, have longer execution times and may be executed at irregular intervals measured in terms of minutes or hours. There is no reason why DDC functions can not be on an area-computer, but it is obvious that DDC and area-level functions require different treatment in the overall system design. The availability of new languages such as PASCAL and ADA will definitely facilitate program development but the "computer system" should provide better, easier to use support to the user-programmer for database operations, program synchronization, resource allocation, man-machine interfaces, etc.

Figure 1g: Plant and Corporate Level Computing

Most people are aware of the extensive use of computers by the business community for production planning and control, sales forecasting, inventory control, accounting etc. so little needs to be said here except to point out the obvious advantage of integrating the process control and management systems. At the very least, process control systems supply input data for the management systems. More realistically, management can be regarded as simply another level of control to be added to the area supervisory and direct digital (local loop) control discussed previously. It is my view that what is required is not just an *interface* between the two different computer systems but *integration* into one total system.

At the plant-wide level, computers can be used to integrate the various process areas into a single operating system. Many plants, such as oil refineries and paper mills, consist of a series of production areas. Significant advantages can be gained by such obvious means as coordinating a grade (product) change; "following" a disturbance (eg. type of wood chip fed to a paper mill) as it moves through the sequence of process areas; taking advantage of process "capacitance" to minimize the interaction between areas and minimize the effect of large changes in feedback streams; during plant startup and major upsets, use auxilliary resources (eg. steam or recycled product) to meet demands (eg. for heat) that will later be met by other process streams.

Trends

There has been an obvious progression in industrial applications from direct digital control, to area supervisory control, to plant-wide control and finally to corporate management. Since more companies are completing the lower levels, there will be increasing activity at the higher levels with corresponding demands for service and support.

COMPUTER CONTROL AT THE UNIVERSITY OF ALBERTA

The distributed computer control system installed in the Department of Chemical Engineering at the University of Alberta is similar to many industrial process control systems and is shown in Figure 2. As shown by the legend on the right hand side of the Figure, the system can be divided into four levels: maxi, mini and micro computers plus the application. The following sections look at process control with subdivisions corresponding to the type of computer involved.

16 Chemical Process Control II

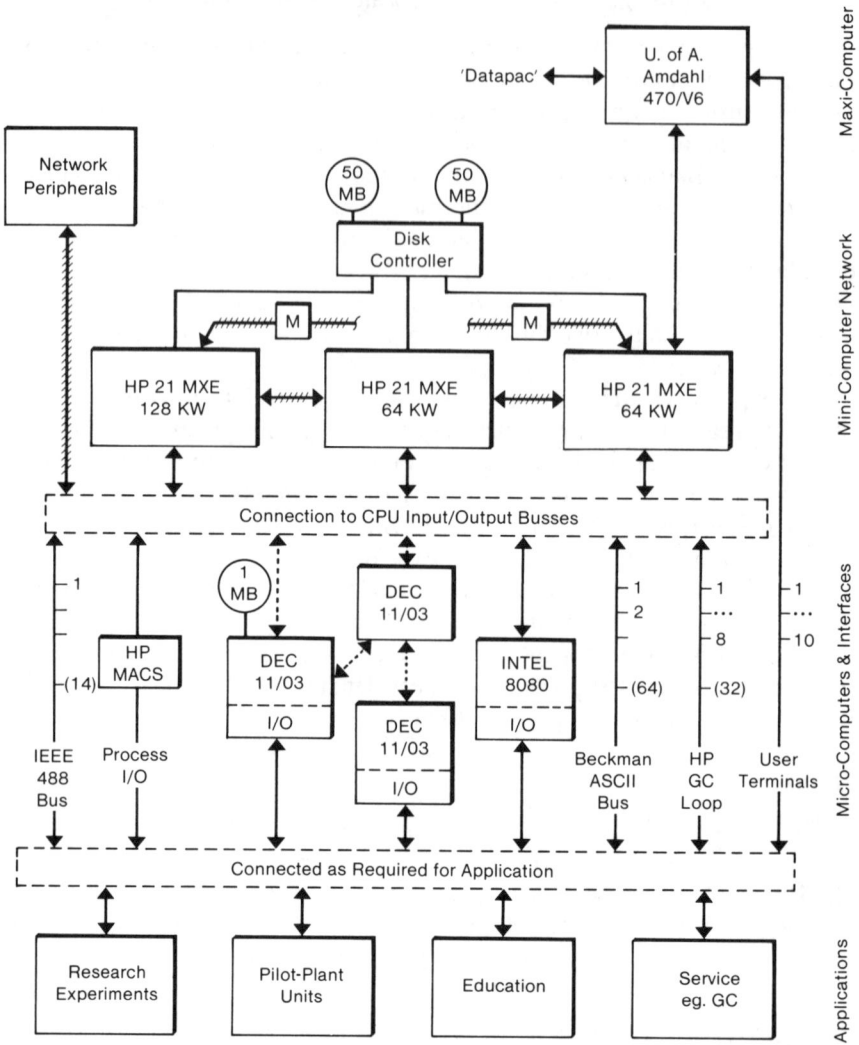

Figure 2. Distributed Computer Network for Process Control and Research in Department of Chemical Engineering at the University of Alberta.

Maxi Computer Level

The main university computer, an Amdahl 470/V7, provides batch and terminal oriented services to the entire university. It is connected to the process control computers by a bit-serial link which is used primarily for the transfer of data files. At the present time the link allows the Amdahl to be used for:

1) executing large programs, e.g. linear programs, to determine optimal conditions, or control parameters to be used on HP/1000 network;
2) cross assembly and downline loading of programs for the DEC PDP 11/03 micro-computers;
3) utility functions, such as high speed printing, tape files, 4 color plots, etc., utilizing data transmitted from the HP/1000 network.

In industry, this level would include the management information system plus corporate planning and control functions. Better performance could be obtained if the link to the large computer supported remote job entry, interrupt service requests, real-time communication, task synchronization and resource allocation semaphores, program to program communication, etc. Note that although it seems "natural" to equate the maxi, mini and microcomputer levels to the process management, supervisory and DDC levels, respectively, there is no *functional* requirement for three levels of user programmable computers. This aspect of system design warrants further study.

Mini Computers

The three HP/1000 (21MXE) mini-computers form the heart of the process control network. The CPU's run under the RTE IVB real-time, multiprogramming operating system and use HP's DS/1000 distributed system software support for network communication support. In the university environment application development is very imporant so the system is designed to include:

- convenient hard copy and CRT terminals
- support for interactive program development in high level languages such as FORTRAN, BASIC, etc.
- peripherals such as card reader, magnetic tape, line printer (with graphics capability), plotter, TV video output, etc.
- micro-processor controlled subsystems, such as the HP2240 measurement and control system, that tend to off-load the HP/1000 CPU's and simplify user programming
- two 50 MB disks that are connected to all three computers via a single multiported disk controller

At the present time minicomputers represent a very cost effective way of meeting the control requirements of a process unit or small process area. The communication support and capacity of communication systems such as DS/1000 or DECNET are adequate for coordinating the control of two semiautonomous process areas but care should be taken to avoid communication overloads.

Micro Computers

The network includes four DEC PDP 11/03 computers. These micro computers are used either as small, single-user, stand alone systems or are dedicated to tasks such as handling the process I/O data required by the mini computers. Communication is by 19.2 k baud, bit-serial communication lines and is controlled by programs written by the DACS Centre staff (4 people). The 11/03's can be down-line loaded from the Amdahl or HP computers. Two of the systems have floppy disks and hence can support single-user assemblers, PASCAL or operating systems such as RT-11.

In industry, micro computers would be most cost effective when several identical controllers were required for a specific dedicated application(s) such as local control of a single piece of equipment, process I/O interfaces, etc. I believe that a single, larger computer should be considered rather than several, general-purpose, user-programmed, interconnected micro computers each running under its own operating system.

Process Interface

For process control and research applications it is essential to have proper interfaces between the computer hardware and the process or experiment. Many of these interface 'buses' and I/O subsystems are critical to the operation of a distributed computer system. The network includes general purpose analog I/O, digital I/O, stepper/counter I/O, thermocouple inputs, interrupts, general purpose interfaces (e.g. IEEE 488 bus) and proprietary interfaces, e.g. for GC's and process controllers.

The process interface is a critical component of process control systems and should not be taken for granted. Power supply failures, ground loops, impedance mismatches, lack of resolution, long term stability problems, etc. can bring the most complex control system to its knees.

Computer Control Software

No computer control system is complete without application-oriented software written by the user or by a third party. Software costs will almost always exceed hardware costs and except for turnkey systems that remain almost static, the costs after the initial startup will usually exceed the original development costs. Thus software is critical! Unfortunately it is usually the least understood and most poorly controlled component of the computer system.

To give some idea of the software required for customized industrial process control the *di*stibuted, *s*upervisory and *c*ontrol (DISCO) system being developed at the University is outlined briefly.

From a software point of view DISCO is a sequential *file* processor. The entries in the data *file* specify the program options to be executed and the parameters required by each program.

As shown in Figure 3 the major components of the DISCO software are:

- DOPE, the *d*ata base *op*erations *e*xecutive, provides interactive, high-level, off-line support to help the user prepare his sequential file (i.e. data base or table).
- ACCESS provides a convenient means of accessing the data base on-line to display or change items, e.g. when requested by the process operator or by other programs executing on the network. It is synchronized with the data base processing part of the DISCO program so that problems do not arise with "contention" or with partially changed data records.

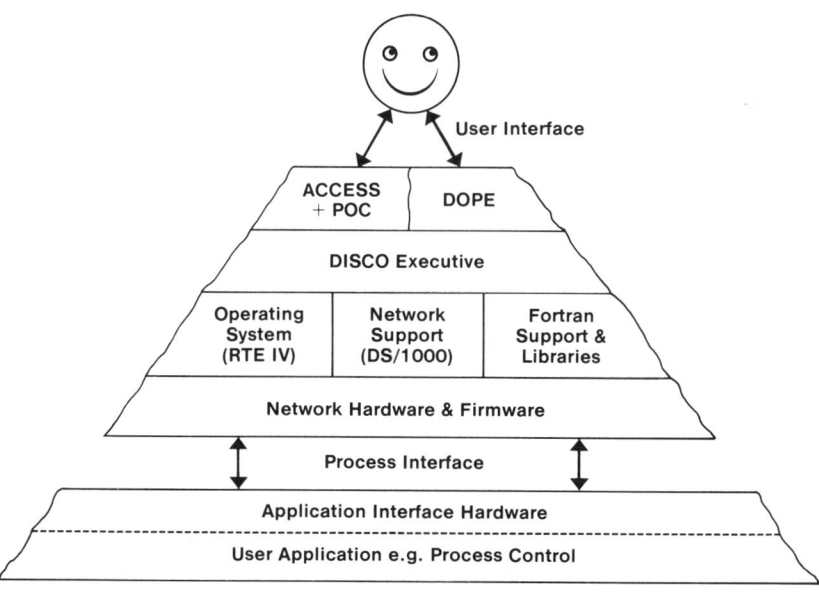

Figure 3. A schematic representation of the *Di*stributed *S*upervisory and *C*ontrol (DISCO) Software System.

- DISCO implements the actual process control functions.
- RTE IV provides I/O and real-time clock support, program sequencing, memory management, interrupt handling, system error detection and recovery, etc.
- DS/1000 supports resource sharing; program to program communication; centralized program development followed by downline loading and "boot-up" of remote nodes; and remote executive calls to RTE IV for program initiation, etc.
- DISCO is written in FORTRAN and wherever possible makes use of system library programs. Extensive use is also made of subroutines to improve modularity.

Some of these software systems such as the operating system and high level language support can be purchased but the remaining, more application-oriented programs still represent several dozen man-years of work and require highly-trained, experienced staff for development and maintenance. The lower level (direct digital control) functions can be handled by standard software and/or hardware packages such as DISCO. However, control programs at the process unit and process area levels are much more application dependent and will probably continue to be written by the user or by specialized consultants.

A THIRD, MORE GENERAL, DESCRIPTION OF THE PROCESS CONTROL APPLICATION

The problem of computer control of industrial processes is a large one and must be subdivided into more manageable pieces before it can be solved. Consider the following characteristics of process control applications:

1. The location of the measurement transducers (MT) and final control elements (FCE) that are the source/destination of process input/output is determined by process rather than computer considerations. They are broadly distributed geographically down to the individual variable (loop) level.
2. Man-machine communication also takes place at several levels but is not tied as closely to the process layout. It is often centralized at the process area level.
3. An industrial plant requires large data processing computers for production planning, accounting, inventory control etc.
4. There will always be a significant amount of application-program development that is unique to a particular process and will have to be done by the user or a specialist consultant.

5. Most users would prefer not to be in the business of computer system development and would be satisfied to purchase process instruments and computer systems on the basis of functional performance specifications, i.e. do not care about computer architecture, internal protocols, etc.
6. The computer load is characterized by a large amount of I/O to process and man-machine interfaces but a relatively small computational load comprised mainly of repetitive operation of a small library of programs.

If these characteristics are true, then they can be used as a basis for the design of a computer control system. Consider the system in Figure 4 which can be compared directly with Figure 1.

Figure 4 a&b: Process & Instrumentation

It is assumed that the process, and the number and type of measurement transducers and final control elements is the same as in the cases discussed previously. However, it is assumed that the MT and FCE are each equipped with a microprocessor that will do self-diagnostic checks on the instrument and provide a standard digital interface to the instrument communication system discussed below.

Figure 4: Manual and Local-loop Control

These capabilities can be included if judged necessary for backup during process emergencies but control will normally be implemented via the computer system. The computer system will be designed to provide the availability and reliability required by the process.

Figure 4e: Computer to process-instrument communication

The *function* of this communication system is to "map" the states of the MT and FCE into designated areas of computer memory, i.e. all measurements are automatically transferred into memory and all values placed in the output buffers of memory are automatically sent to the FCE. Sampling rates would be several times that required by the fastest process dynamics or interrupt specifications and would be done asynchronous to, and independent of any user program in the computer system. Communication rates equivalent to several thousand points per second would be required but are easily obtained with existing technology.

The communication system would consist of high band-width multi-drop busses and/or broadcast hardware. Reliability would be provided by multiple paths for each instrument-computer communication channel and/or parallel, redundant communication channels to be used in case of

22   Chemical Process Control II

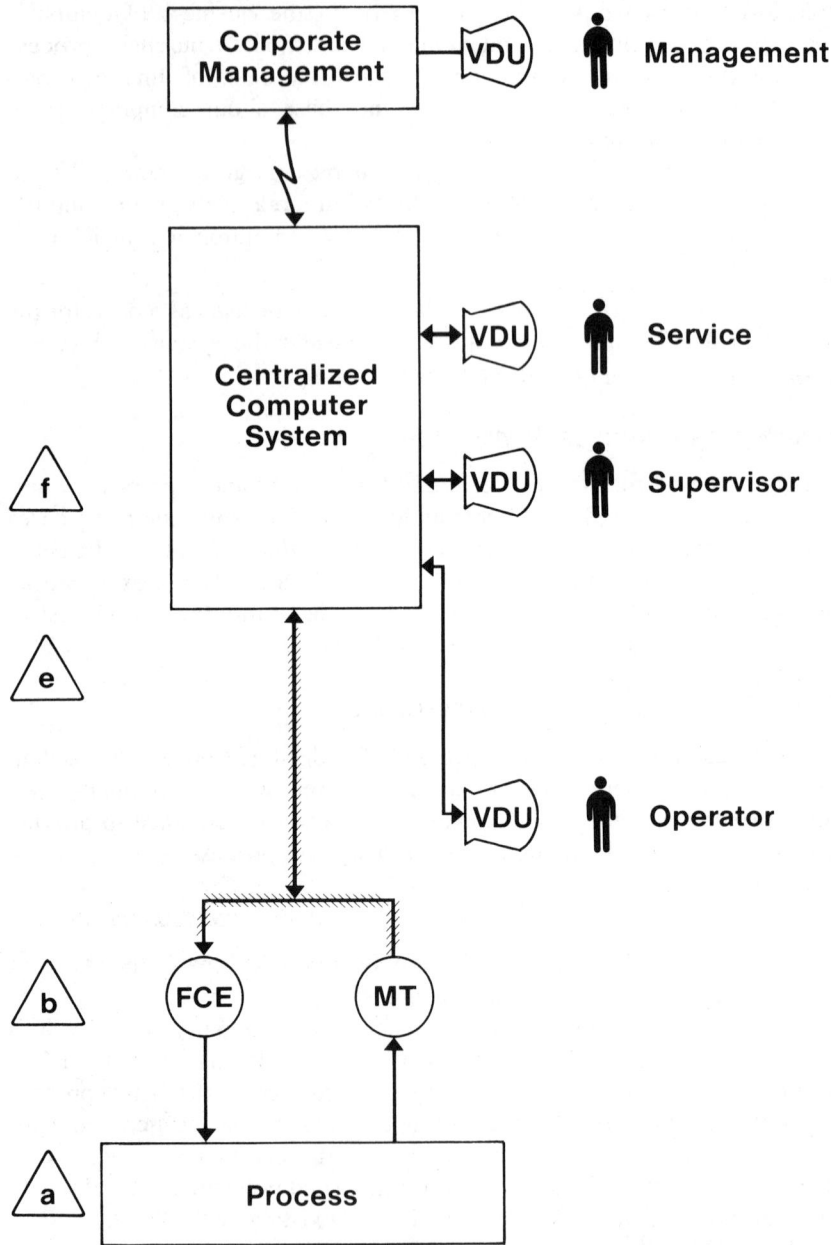

Figure 4. "Centralized" Industrial Computer Control System.

failure of the primary channel. Information concentrators or nodes may or may not be present in the system. All error checking and correction would be managed by the communications controller. Total failure of the communication system would be improbable, but would force a total, emergency plant shutdown.

Design Philosophy

The purpose of the communication system described above is to divide this part of the process control system into three distinct parts:

i) process instrumentation
ii) communication
iii) computer I/O memory buffers

Note that each of these parts has:

i) a simple, well-defined function
ii) a clearly specified, buffered interface

and hence can be designed, implemented, maintained and upgraded independently of the other parts and if desired by a different group of people or company.

Figure 4f: Computer System

It is proposed to purchase a single hardware plus software computer *system* from a vendor of large information processing systems. Note that the computer system is to *function* as a single system but in practice, to guarantee the necessary reliability and availability, would undoubtedly consist of multiple parallel processors, with segmented, but common, data storage capability. The processors and memory could be distributed geographically but would probably be centralized in an area that had the proper operating environment, electrical classification, and easy access for maintenance and computer operating personnel. The computer would undoubtedly have virtual memory and perhaps support virtual processors. The important feature is that it could be accessed by the user as a single system. Note that the only price paid for centralizing the computing system is slightly longer communication lines to the process I/O interface and the man-machine interfaces.

There is no doubt that many of the currently available commercial computer systems can handle the computational and I/O load associated with a process control system. (Some studies would have to be done to ensure that the system could meet the response time specifications but these could probably be met by adjusting the total computer load and carefully assigning execution priorities.) The main question or objection probably concerns cost. At the present time the computer I/O and computational capabilities required for a process control application can probably be obtained more cheaply using a combination of mini and micro-computers such as described in Section 3. However, if the criterion is total system cost over the lifetime of the project then I believe that in the near future the two alternatives will be very competitive and perhaps even favor the single system. For example, I do not believe that the full impact of the latest electronic technology, which is readily apparent in the mini and micro-computers, has reached the market place in large computer systems. The next couple of years should show significant improvements in the price/performance ratios of large systems. A second factor is the highly skilled and specialized personnel required to implement a complete computer control system. If the system can be modularized then process users may prefer to purchase complete functional modules rather than assemble the resources necessary to do the whole job themselves. Finally, careful cost accounting on a *system-wide, project lifetime basis* should favor the large-system approach. Consider the following:

  i) large systems can provide a higher level of user support and a better environment for program development. Software costs are becoming dominant and in many applications over 75% of the user's expenditures on software occur after system startup, i.e. are for system maintenance and upgrading. Therefore development support is important.
  ii) a large part of the process control application is database management and interactive query/display. The transition from instrument panel board to CRT display is now established so we should take advantage of standard "business" packages which are obviously the driving force in this area.
  iii) if the computer system is treated as a single module that must meet only certain *functional* specifications and has a relatively stable, high level interface to the user (i.e. operating system, libraries and programming languages), then the user has the best protection he can get against technology changes, system obsolescence and limitations on future expansions.
  iv) because the large computers are 'standard business systems' then there is a degree of commonality with the plant management computer systems and readily available interfaces to computer networks.
  v) It is difficult to provide data and communication integrity in a distributed system. This is a difficult, specialized area best left to the computer system designer rather than the control engineer.

The advantages of centralized computer *systems* discussed above may be overstated but I am trying to fight the swing of the pendulum from the single centralized *computer* of a few years ago to the distributed, do-it-yourself networks of user-programmed computers that are the current trend.

Database design and management

If the process I/O requirements can be modularized and separated from the main computer system then I would like to argue that perhaps the major factor affecting the design of the computer hardware and software system is the database (including management and communication). I would like to argue in favor of a single, global database maintained and accessed through a single database manager. In practice the database could be centralized or distributed and there could be multiple entry points to the same database system. This is a specialized field and enormous advances have been made in the management, access and display of large commercial databases.

Process control engineers should develop the functional performance requirements and then work closely with database specialists and software engineers to develop the overall design of the computer control system. They should not treat the database as an incidental part of a control problem!

A process database can be divided into two parts: the current or active part and the historical files. The main difference is that historical data is not accessed as often and a longer response time is usually acceptable. On-line access to the current database is typically by:

    i) process control programs (run every control interval)
    ii) supervisory programs (large, scheduled programs)
    iii) process operator consoles
    iv) enquiry/display stations for management, service and engineering personnel.

The response time requirements, the amount of data handled and the frequency of access all decrease as one moves down the list. The size of the database is 'small to medium' by commercial standards but, mainly because of communication with the DDC programs which execute every second or so, the data access rates are relatively high and must be handled on a priority basis. Never-the-less, commercial database management systems could probably be modified to support process control applications.

The principles and advantages of database management systems are not well known outside of computing science and computer system design groups so let me try to illustrate them by use of an analogy. Consider a combined industrial "stores" operation and "tool crib." Plant personnel come to the store to obtain particular items or to borrow and return tools. The analogy is obviously between data in the computer database versus items and tools in the stores. Most industrial people would agree with the following statements:

1) there should be one manager in charge of the whole stores operation
2) the user does not care about the internal organization, i.e. what is stored where
3) the user may use different terminology in referring to an item than is used internally by the stores system. This presents no problem as long as the storekeeper can translate the user's request.
4) items requested frequently can be stored so they are more easily and quickly retrieved.
5) items usually requested together can be handled as "packages" with a single stock number.
6) unauthorized, direct, access to the stores area, even in the event of an emergency, can destroy the integrity of the operation.
7) a system can be devised to facilitate additions and deletions to stock but it should be noted that items related by physical characteristics or functions do not have to be stored near each other. (e.g. all the parameters for one control loop do not have to be stored together).
8) special service can be arranged for high volume or priority users (e.g. process control records can be preassembled based on a specification defined for each loop record).
9) the store manager can be made responsible for quality control and error correction.
10) because of 'overhead' it is usually best to group all related functions into one stores operation.
11) records can be kept or a search can be done for all users of a particular item so they can be notified if special circumstances arise e.g. item no longer available (cf. failure of a measurement transducer).
12) records can be kept of all items used by a single customer so they can be checked for availability or deleted if required. (cf. deletion of one control function and substitution by another).
13) substitutions can be easily made by item, or by customer plus item (e.g. a program under test can get its data from a prepared file rather than directly from the process).
14) the store acts as a buffer between source and end user so that both can be more flexible in their timing.
15) the impact of a proposed change can be evaluated by "simulation" before it is implemented.

16) rarely called for items can be supplied on a "special-order" rather than on "in-stock" basis. (cf. an occasional call from a supervisory or management program for a variable not under active control).

It would be a major undertaking to define and evaluate all the factors that must be considered regarding the use of a database management system as part of a process control application. However, I hope the above analogy indicates the potential advantages to a user who does a significant amount of his own application programming, e.g. reliable data is available on demand, from a single source, through the use of a standard procedure that is defined in user terms. All the details regarding what is stored where, and how the data is obtained are "hidden from" the user.

From a *system design* point of view the major advantage of a database management system is that it provides a *buffer* and a clean *interface* for the other components of a process control system. This is illustrated by Figure 5 and discussed in the next section.

Overall Structure of the Computer System

The major components of the computer system shown in Figure 5 are:
1. Process & instrumentation
2. Instrumentation/database communication system
3. Database and manager
4. Process control programs e.g. DDC level
5. Supervisory control programs e.g. process area level
6. Network & corporate systems.
7. Query/display system (man-machine interface)
8. Computer System Executive
9. Utility/Service Programs
10. Plant personnel

Additional details about some of the above components are given in Figure 5a, however, it is the overall structure and the interfaces that are of concern here. Note that from a communication point of view the design is essentially a "star network" with the database at the centre. The database provides buffering and the manager provides a well defined interface.

The overall operation of the system shown in Figure 5 should follow from the previous discussion in sections 2, 3 and 4. However, Figure 5 should make it clearer how modular the proposed structure is. Each component can be designed or purchased, implemented, tested, maintained and replaced on a modular basis. Although presented as a total, plant-wide system, many of the principles incorporated in the above design can be used to advantage in a much smaller application e.g. a single

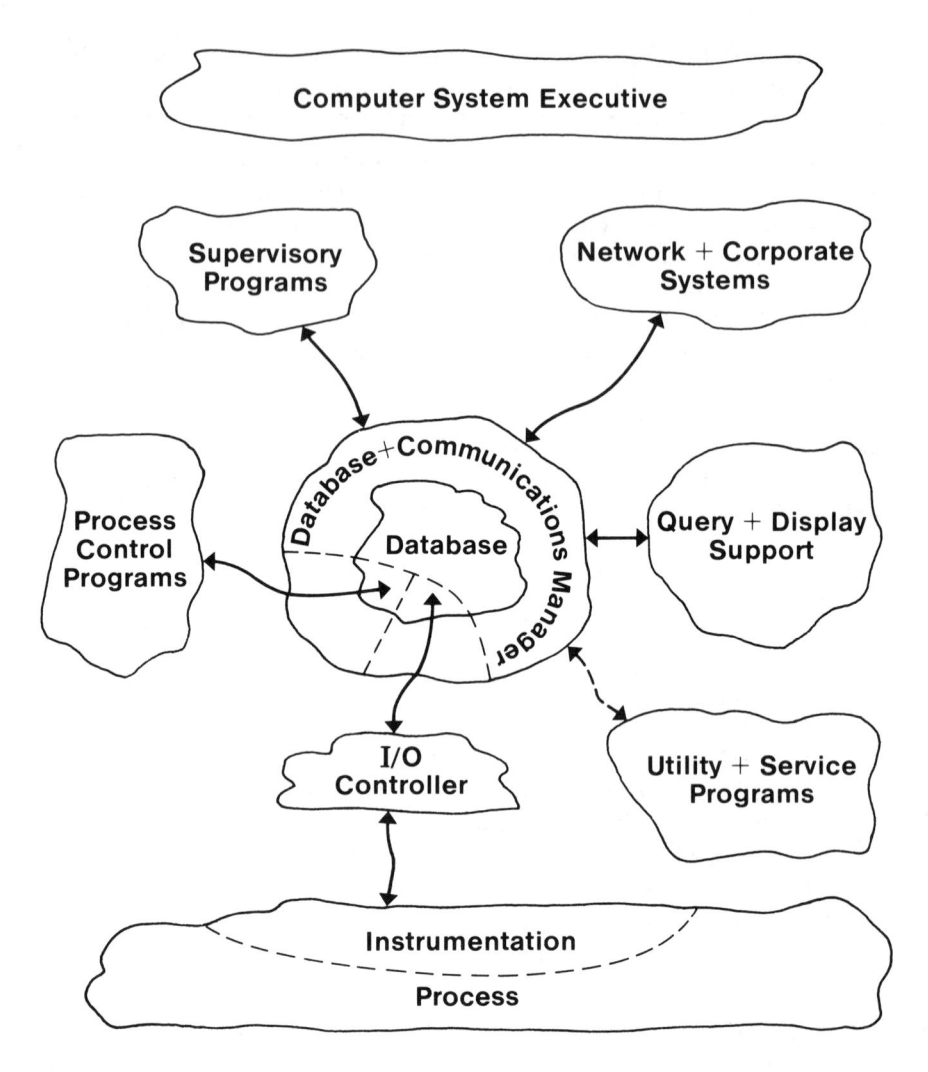

Figure 5. Proposed System Architecture for Industrial Process Control.

minicomputer used to implement a specialized control strategy on a single process unit.

Database Query/Display and Communication Support

No reliable statistics are available, and it is obvious that there would be significant variations from one installation to another, but a realistic

Computer Systems Executive
- program scheduling
- interrupt servicing
- resource management
- system integrity monitor
- real-time clock support

Supervisory programs (system scheduled)
- process unit control
- area optimization
- plant coordination

Process Control Programs (clock/interrupt driven)
- direct digital control
- table-driven sequences

Query/display Support
- process operator
- control engineer
- service personnel
- management
- computer staff

Utility and Service Programs
- program development
  - high level languages
  - debug packages
  - optimizers
- system maintenance
- database maintenance
- display generation
- historical data files

Figure 5a. Details Concerning the Components Shown in Figure 5.

estimate of the total database/query/display/communication load would probably be significantly greater than 50% of the total system load. Thus it is a key factor in the overall design.

Our experience at the University of Alberta has confirmed that once the process control application goes beyond direct digital control (with its strongly structured and segmented database) into higher level supervisory and management programs the database and access problems become much more difficult. It is hard to describe the problem except to say that I think it is severe enough to justify a significantly different system design than is used in most DDC systems.

One of the main objectives in modularizing the database management and query/display systems was to make it possible to use as much of the commercially available systems and software as possible. Consider the following comments and questions:

1. I am convined that totally customized process operator interfaces will be replaced by the most powerful and flexible of the commercial products.
2. Users will require the ability to quickly and conveniently format their own alphanumeric and graphic color displays. This level of support is just beginning to become available.
3. A wide variety of terminals should be supported to meet the needs of management, engineers, maintenance and production record personnel. Will the plant manager's, secretary's word processing system be able to interface to the process control system?
4. Should the "communication subsystem" be expanded to support message switching (electronic mail), verbal message buffering and delivery, video teleconferencing and facsimile transmission, "beepers" and portable communication devices? (my answer is yes)

These features referred to above, plus more are under active development for the business community. They are too complicated to develop separately or even to modify significantly. Therefore we should design the process control system so these standard "business" features can be added when desired.

CONCLUSIONS

1) In the process industries it is not a question of *if* computer control but *when* and *how much*. The industry is now in the process of switching from "traditional approaches" to the full potential of computer control.
2) The process control engineer should continue to contribute in his particular area of expertise but also be prepared to work with computer hardware, software, database, communication and system design specialists to produce a complete computer control system for industrial processes.
3) The computer control area is in a state of rapid change and this fact must be seriously considered when deciding what level of technology to adopt (e.g. "tried and true" or "new but promising"?)
4) The process control system must be modularized and greater consideration given to using commercial modules rather than adopting a totally 'do-it-yourself' approach. Process control engineers can best be

used to write process control programs (not database managers or display formatters!)
5) Database management/query/display and communication are key, perhaps dominant factors to consider.

ACKNOWLEDGEMENT

I acknowledge the contributions of dozens of authors, conference participants, students and professional colleagues who are the source of the work described in this paper. I apologize for not citing them individually and make no claim for originality of individual concepts or practices included in this paper. I have simply collated and integrated a number of factors and added my personal value judgements. I hope I have not distorted the original ideas nor unfairly judged them.

REFERENCES

[1] Berger, J. A. (Editor), "ISA Control Valve Compendium." Instrument Society of America (1973) cf. similar compendium on Measurement Transducers, etc.
[2] Bernhard, Robert, "All digital jets on the horizon" IEEE Spectrum, Vol. 17, No. 10, pp. 38–41, Oct. (1980).
[3] Branscomb, L. M., "What's Ahead in Technology," IBM Data Processor, Vol. 10, No. 2, p. 20 (1980).
[4] IEC SC65A/WG6, "PROWAY, Process Data Highway for Distributed Process Control Systems" (cf. following discussion on "Standards Organizations" IEC and IPW/TC5)

STANDARDS ORGANIZATIONS

The International Organization for Standardization (ISO) is the top international standards organization with 80 member countries, 150 Technical Committees (TC), 1600 Working Groups (WG) and publishes new or revised standards at the rate of about 500 per year. The other principal international organization is the IEC which is affiliated with the ISO and has 42 member committees. The Canadian member of the ISO is the Standards Council of Canada (SCC) and the Canadian Standards Association (CSA) is the national standards writing organization. The principal standards organization in the US is the American National Standards Institute (ANSI) which is the American member of ISO and is an umbrella organization for standards work by the ISA, IEEE, ASTM,

NEMA etc. There is often significant overlap in membership and interchange of information between the different standards groups.

Of particular interest, because of its focus on "industrial computer systems," is the International Purdue Workshop (IPW) and its technical committees, some of which are listed below:

TC1: Industrial, real-time FORTRAN Committee
TC5: Interfaces and Data Transmission Committee
TC6: Man/Machine Interface Communications Committee
TC7: Systems Reliability, Safety and Security Commitee
TC8: Real-time Operating Systems Committee
TC10: Distributed Industrial Computer Systems

Contact with the IPW or its committees can be made via

Purdue University
Laboratory for Applied Industrial Control
WEST LAFAYETTE, Indiana. 47907.

SOME PRINCIPLES IN STRUCTURING SOFTWARE FOR PROCESS CONTROL

Th. Lalive d'Epinay
Brown Boveri Research Center
CH-5405 Badan, Switzerland

ABSTRACT

The design of a process control system and the resulting structure of its software has to reflect the different views of such a system. Each view has associated with it a specific category of problems and requires appropriate design tools.

A layered design model with the levels requirements specification—application and control system structure—global computer system (network)—computer nodes—sequential processors is used to separate the solution of different problems and to combine these solutions to a consistent overall system design.

This approach not only reduces the complexity of the control system, it also limits the influence of changes in the solution to a particular problem.

In addition the layered model represents a consistent framework and precise rules, how process control systems have to be built. This allows to use a maximum of common design effort for different applications and different hardware tools.

INTRODUCTION

The design and structure of software for process control is a problem which appears on different levels; there is software on the level of the application–and control engineer and of the application programmer. There is software to control distributed computer systems and there are problems generally related to real time operating systems.

In the first part the paper shows that the computer ideally leads to an approach using levels; well established tools like compilers are nothing else than translators from one level to another.

The paper then describes a layered system design model which separates different levels of software, where different problems are solved separately. Thus the amount of software which depends on a specific application or on a specific implementation or technology is minimized.

The complexity of the whole problem is broken down into parts which can be treated by individuals and for which well established rules and methods exist.

It is the intention of this paper to provide a general idea of the usefulness of a layered approach rather than to explain the many problems which had to be solved in order to find a consistent model. It is in the nature of such a layered model that the details for the different levels would have to be addressed to different groups of people.

The model itself as well as it application to solve problems of process computer control is still in the state of development and research, although many ideas of it are already used.

SOFTWARE

The term "software" is widely used, but probably interpreted differently by different people and under different circumstances. This is even more the case for 'Structure of Software.' In order to develop principles in structuring software we have to find a suitable definition for these terms.

Software can be understood as a set of rules, conventions or instructions how to use a set of tools or machines in order to accomplish a certain goal.

Some specific properties of the tool envisaged in connection with software, namely a *computer system* executing a program, are responsible for the fact that the structure of software is a multi-layered problem:

- a computer system can automatically execute its program

- a computer system executing its program (micro-program, interpreter) realizes a new tool
- a program for a conceptual computer (e.g. in a high level language) can be translated by a computer to a program (in machine language) for a real computer (compilation).

The two methods of interpretation and compilation may be combined in many different ways to form a more or less efficient method of software and system design.

In addition computer systems allow on all levels to structure the software by procedure- and macro-facilities. Often only this last possibility is considered in design methods; this means that only the definition and use of *one* language or *one* formalism is considered. This is an insufficient approach.

If principles in structuring software have to be found we have to define a number of *levels* with their respective capabilities, languages and structures. Then different levels can be *mapped* to each other using compilation, interpretation and procedures.

LAYERED MODEL

It is important to realize that the main conclusion so far is the general possibility to map one level to another, to *realize new capabilities* and to *hide other ones* in such a mapping. Whether the actual mapping is done with interpretation/microcode, with compilation or by procedures may be a secondary decision.

A model, where different levels are mapped to form an overall design method is called a hierarchy of (abstract) machines [1]. The relations between two machines in such a model can be described as follows (see Fig. 1).

A program P_1 (operations and data) on a machine (computer system) M_1 may solve the same problem as a program P_2 on machine M_2. In this case P_1 on M_1 may be mapped to P_2 on M_2.

A compiler (Fig. 2) maps both P_1 and the specific properties of M_1 into P_2 in the construction phase of the system. Both P_1 and M_1 are only conceptual, they are not existent in the execution phase and may not be referenced for on-line system modifications or exceptions. On the other hand a compiler can make optimal use of M_2 and produce optimized code; compilation time can be neglected for real-time considerations.

An interpreter (Fig. 3) maps only M_1 to P_2; in other words M_1 is realized by P_2 (running on M_2). P_1 remains and M_1 still exists at execution time. This has many advantages, but it is generally less time-efficient, because

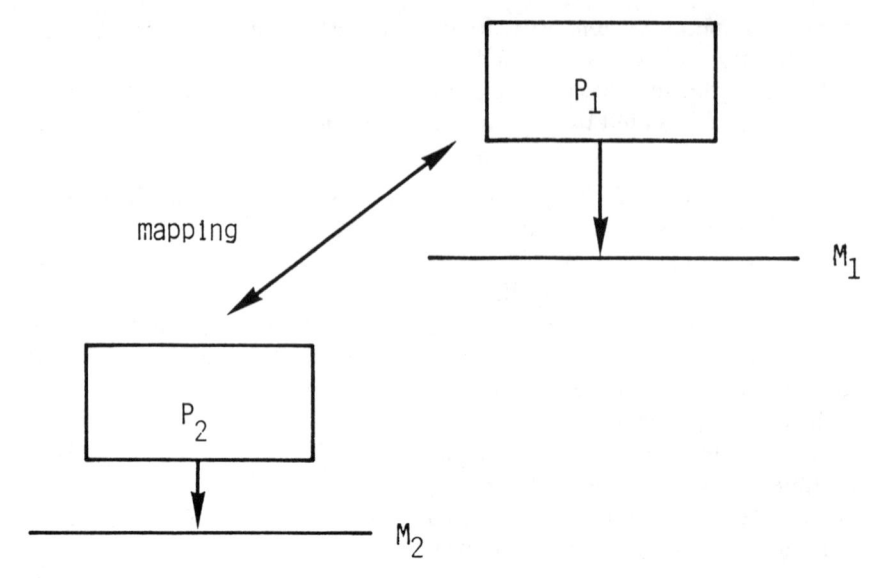

Figure 1. Mapping. The program P_1 (operations + data) running on the machine M_1 solves the same problem as the program P_2 running on M_2.

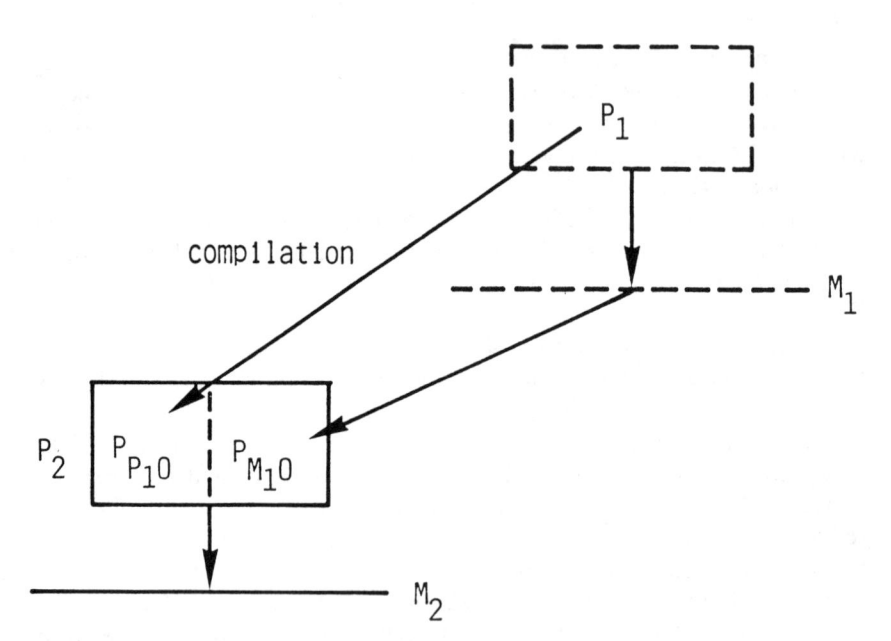

Figure 2. Compilation. Both program P_1 and the capabilities of M_1 are translated to the program $P_2 = P_{P_10} + P_{M_10}$. P_1 and M_1 are only conceptual.

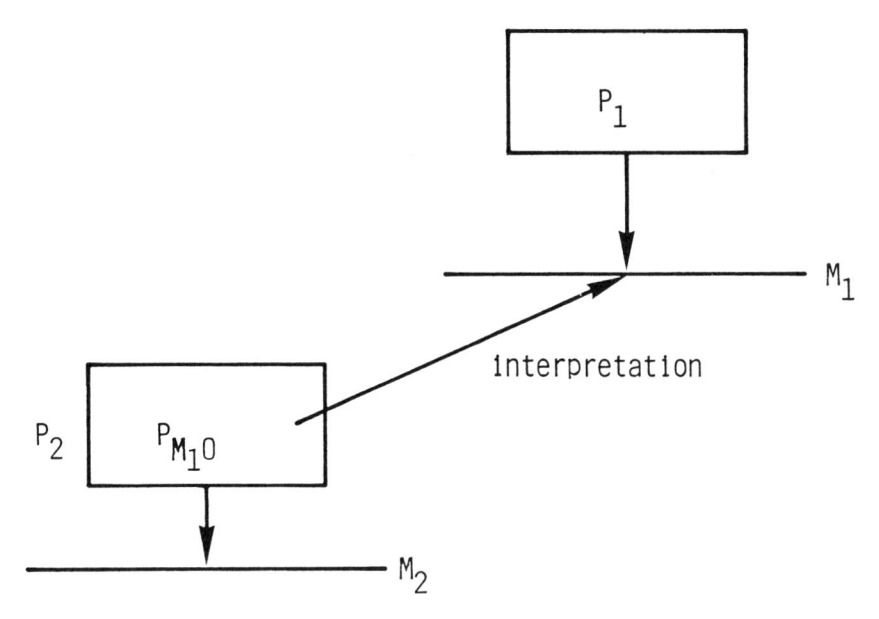

Figure 3. Interpretation. The machine M_1 is realized by a program $P_2 = P_{M_1 0}$. P_1 remains and is executed by M_1.

the interpretation is done in the execution phase of the system. The costs in time for an interpretation depend very much on the machine M_2. If M_2 is specially designed to act as interpreter, these costs are minimized. This is generally the case on the level of the microprogram.

DEFINITION OF LEVELS

There are two levels which act as boundary conditions:
- requirements specifications
- existing set of tools to solve these requirements

Requirements specification level

At this time there are no established methods for requirements specifications, but there is great effort done in this direction at different places, and we can assume that such tools will become more generally available in future.

Being a boundary level, it is basically impossible to create requirements specifications automatically. Nevertheless it is possible to provide formal notation for requirements, to assist a writer of requirements and to detect inconsistencies.

Requirements specifications have to be totally independent from the design of the solution, in order to allow the verification of *all* design steps.

The requirements specifications are the starting point of any system design.

Sequential processors levels

On the other end our design finally has to converge to a system using known elements with well established rules. We assume that a process control system will be a distributed system consisting of computer nodes and an interconnecting network. Both terms 'node' and 'network' will be defined later in this paper. There is no restriction making this assumption, because a centralized computer system is a special case of distributed system.

We further assume that the distribution, the capabilities of the links and of the nodes is influenced by the type of application. Thus a design method has to include the realization of a computer network, of the links and of the nodes. Therefore we set the lower boundary level to a level which corresponds to the building blocks of computer nodes and links. In a model of (abstract) machines the relevant element of this level is a programmable, sequential processor. The capabilities of memories, input/output, etc. are implicitly expressed by the capabilities of the processors.

Of course the regions below the level of sequential processors (e.g. hardware circuits, LSI's, etc.) are not neglected in this approach; their use and development is implicitly defined by this lowest level.

Our model for software and system design so far has three regions (see Fig. 4).

The next step is to define additional levels and the corresponding mapping rules between the two boundary levels.

As a mapping from one level to another allows us to realize new capabilities and to hide other ones we use this mechanism to hide computer dependent properties and to create application specific ones in going from lower to higher levels or vice-versa.

In determining the necessary number of levels we have to consider the following points:

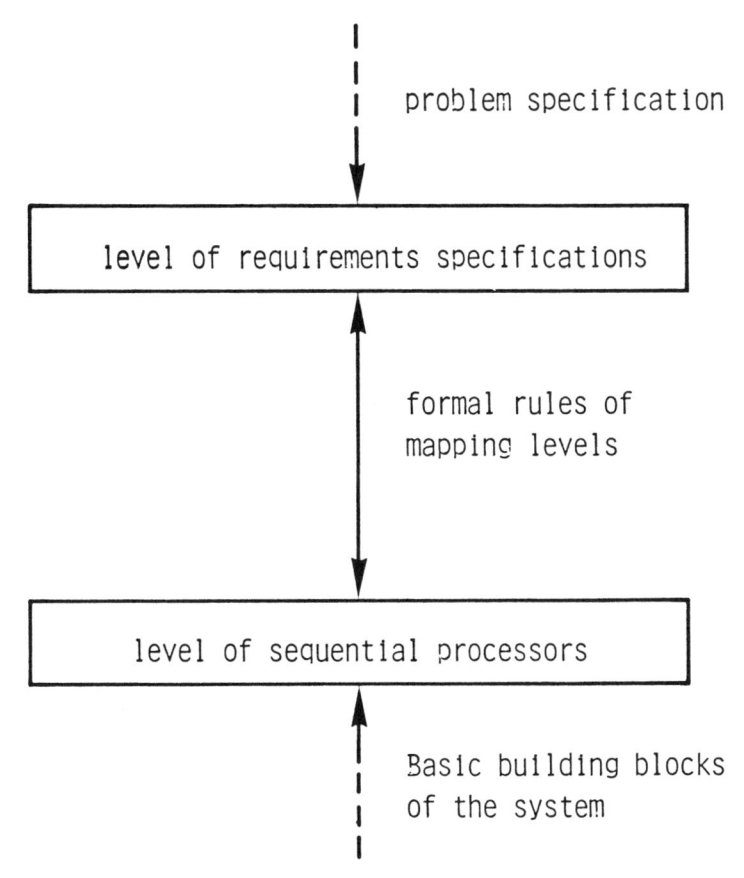

```
                    |
                    | problem specification
                    ↓
┌─────────────────────────────────────────┐
│   level of requirements specifications  │
└─────────────────────────────────────────┘
                    ↑
                    │   formal rules of
                    │   mapping levels
                    ↓
┌─────────────────────────────────────────┐
│     level of sequential processors      │
└─────────────────────────────────────────┘
                    ↑
                    |   Basic building blocks
                    |   of the system
                    |
```

Figure 4. Boundary conditions for a layered model. In a layered model well defined design rules in the form of mapping or translating levels can only exist inbetween two boundary levels; the realization of these boundary levels has to be done intuitively, using state of the art, etc.

(1) Every level introduces certain costs in the design and possibly in the execution phase of a system. Thus, the number of levels should be minimal.
(2) The solution of two independent problems (e.g. how to build a computer network and how to realize a certain control algorithm) should never be included in the same mapping, because the change in the solution of *one* problem would also imply a redesign of the other one (see Fig. 5 and 6).

40   Chemical Process Control II

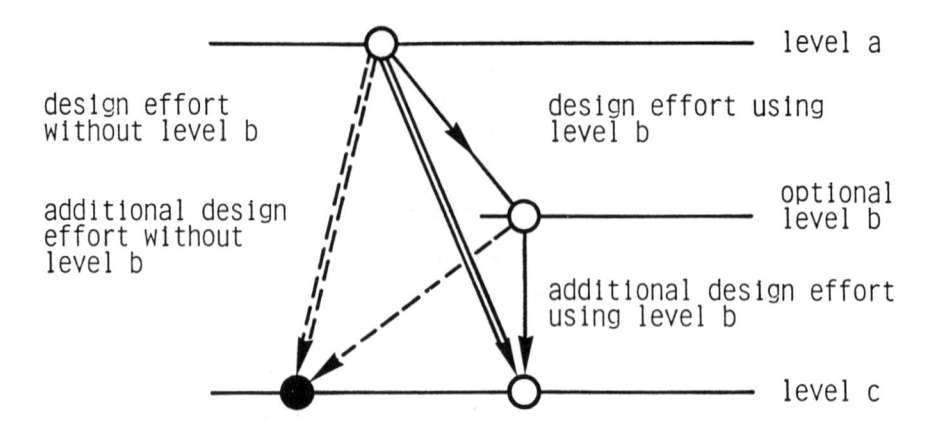

Figure 5. Design effort to solve a problem. The problem is defined on level a and solved using two possible realizations on level c and an optimal intermediate level b. If the solution of the problem can be divided into different independent steps (using level b) development costs due to a change on one level are minimized.

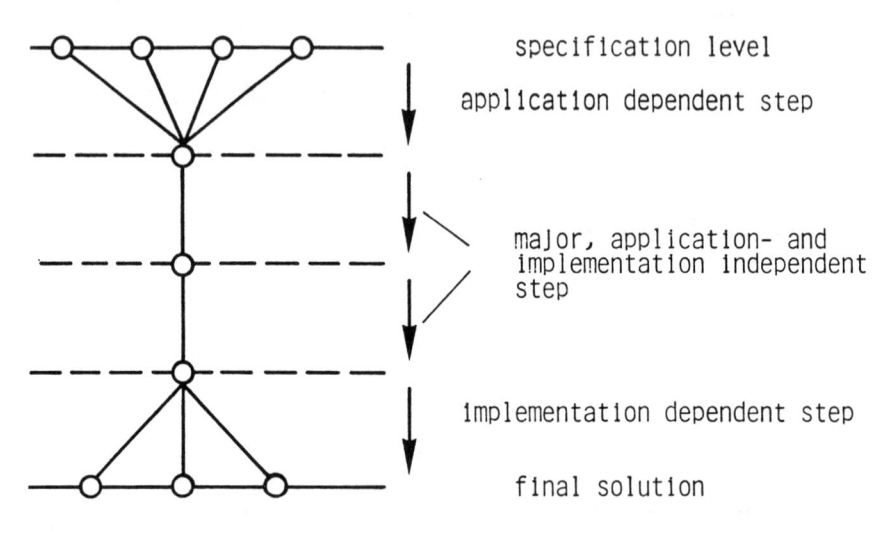

Figure 6. Design effort to solve a problem from specification to final solution. In a well defined layer design model a family of control systems may be used to solve different problems. If intermediate levels are used, only the first design step has to be redone for a new application and only the last step is redone if a new hardware is used.

(3) If a desired capability cannot be realized efficiently on top of an existing level, additional levels are necessary. We will see that a computer network cannot be implemented directly on top of the sequential processor level.

We will now fill in the necessary levels between the two boundary levels. First we will step down from the requirements specifications level and define the next two lower levels (Fig. 7).

Control system level

In the first step the requirements specifications are translated into an application oriented representation of the solution. A direct step from the specifications to solutions in terms of computer systems is not recommended, as it would require the solution of a control problem *and* of a computer problem in one step.

The control system level shows the control structure of the system. This structure contains hierarchically organized control levels, which must not be mixed up with the levels of the design model (see Fig. 8).

The control system level has to be defined by application specialists, control engineers and computer scientists. Whereas the contribution of the first two groups is obvious, the computer scientist is responsible for the formal representation (using the valuable expertise which has been reached in high level languages).

There is no established standard for the control system level representation or language; but most of the computer control equipment

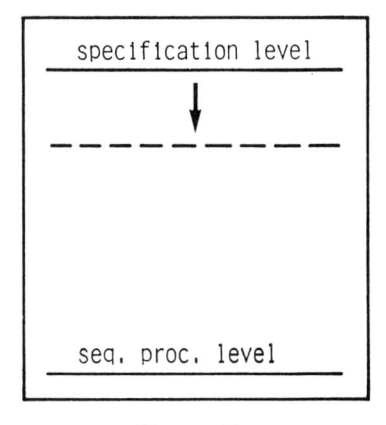

Figure 7.

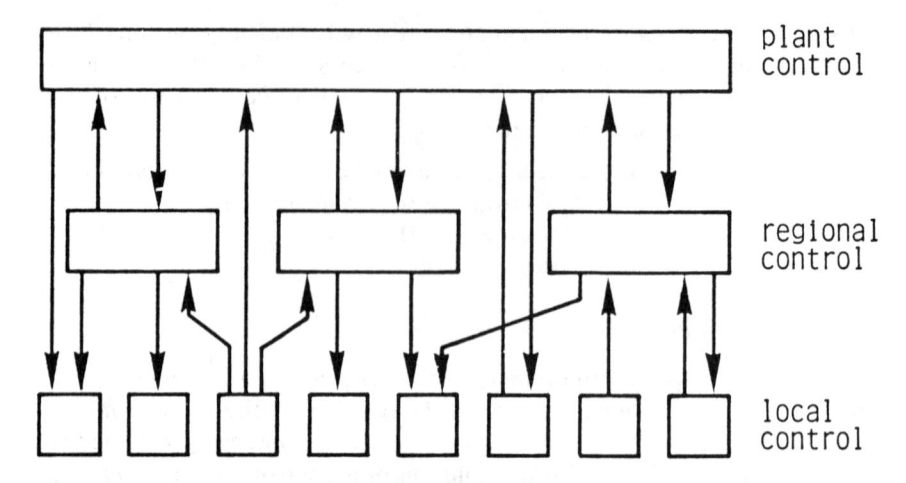

Figure 8. Example of control levels. The control system level of our design model is itself structured in control levels (which have nothing to do with levels of the design model). In general, the relations do not form a tree structure. Attempts to organize the control level tree-like will fail or lead to inconsistent design models (problems are shifted to the "implementor"). There may be fewer or more than three control levels.

manufacturers offer reasonable tools, mostly based on block diagrams, which may be programmed using graphical terminals or textual representations. But these representations have generally the following deficiencies:

- The formalism has by far not reached a state comparable to modern high level computer languages. The languages are at most "control system assemblers."
- There exist no clear statements to represent relations between control elements. It is a common requirement that a local controller should be able to maintain its function if his higher level controller fails. But these requirements and the way they are fulfilled are in most cases based on a intuitive, unspecified behavior of the higher level system in the case of failure and how this situation is detected by the lower level system. Similar situations exist for all kinds of relations in control systems; and statements expressing such relations should explicitly exist and let nothing to intuition.

Realization of the requirements specification level using the control system level

As long as there is no formal definition of requirements specifications there is no formal mapping procedure from the requirements level to the control systems level. Even a formal method will desire the creativity of a control engineer. The advantage of a formal method will be to support this design step and to allow formal proofs for its correctness. Hopefully we will be able to introduce formal methods, starting with specific, small classes of applications, then broadening the scopes of different solutions and combining them to methods at least valuable for related classes of applications. The tendency should be that even different methods should lead to the same formal control system level or to a family of compatible representations of the control system level.

Global computer system level

Going one further step down from the control system level we have to find a level representing general tools of computer systems which allow the construction of the control system level (Fig. 9).

The general rules how to realize certain elements of the control system level using a computer system are independent of the internal structure of the computer system; consequently this first computer system level must not include elements which refer to such an internal structure. There may be no nodes, no communication between nodes and of course no elements of nodes, etc.

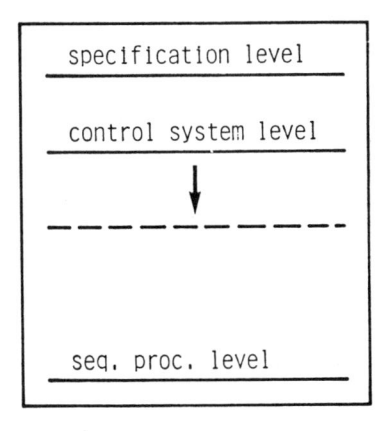

Figure 9.

The logic which is left of a computer system is a machine able to execute any desired number of parallel, communicating tasks. Neither synchronization nor data exchange between tasks may reflect any assumption on the underlying system, e.g. common storage, execution by the same processor, etc.

Levels below the global system level

We have now defined two levels going down from the requirements specification level. We will now start at the other end and try to realize the global computer level on top of the sequential processor level (see Fig. 10).

We can assume that all "conventional" capabilities of a computer system (arithmetic, logic, move data, etc.) are already present in the sequential processor level. The key problem which remains is the problem organizing a virtually unlimited capability to execute parallel tasks and appropriate synchronization functions. These capabilities must be realized in a distributed computer system, i.e. in a network with nodes, where the possibility to have directly executable instructions and directly accessible data is limited to single nodes.

One could try to build the global computer system directly on top of the sequential processor level. This would imply that the design model would lead to a hardware system, where each task is executed by a specific processor.

Besides the obvious inefficiency of such a design it would lead to the following basic drawbacks:

- dynamic creation or deletion of tasks is impossible

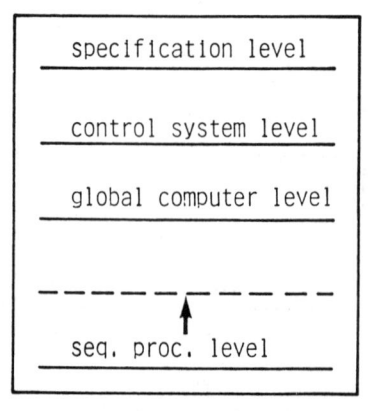

Figure 10.

- the number of tasks is limited; on the other hand modern software design methods, modern languages have a tendency to have many tasks, which may be very short
- a hardware failure is directly related to a task; all possibilities of redundancy and reconfiguration are excluded.

A strict approach to determine the minimally necessary levels to build a global computer system is an analysis of the synchronization problem:

- Synchronization is a restriction in time in the execution of two or more parallel tasks. The only two basic synchronization restrictions are mutual exclusion (specific parts of different tasks may not be executed at the same time) and cross stimulation (specific parts of tasks must be executed in a specified order).
- Synchronization restrictions are expressed and enforced by synchronization operations which are part of the operations executed sequentially within the different tasks. The only way in which a synchronization operation can enforce a restriction in time is in delaying the completion of its execution.
- In order to synchronize parallel tasks these must have well defined or exclusive access to some data, which they use for coordination. Exclusive access is a form of synchronization, which leads to the astonishing fact that synchronization can only be built if it exists already. This is true until we reach a level, where there is no more parallelism. As we allow several sequential processors at the sequential processor level, the level without parallelism is below; it is the level of arbitration (bus- or memory-arbiters). Although these arbiters seem to have parallel parts their innermost method to have deterministic behavior is sequential.

Using these facts we can conclude the following:

- Synchronization at the sequential processor level is based on arbiters, i.e. on access to common busses or common memories.
- Synchronization restrictions on the sequential processor level lead necessarily to delaying or busy waiting of processors. (A sequential processor has *one* task; if a synchronization conflict occurs the processor cannot be switched to another task, because such a task switch would also require synchronization.) As these delays are responsible for the reduction of system efficiency, they have to be minimized. This excludes sophisticated synchronization functions; it allows only primitive operations for mutual exclusion (delays due to cross stimulation are basically unpredictable).
- General synchronization functions including cross stimulation are therefore only available on a next higher level above the sequential

processor level; this level may at most refer to single nodes in a computer system, because communication within a network requires cross stimulation on a lower level.
- A global computer system representing a network can therefore be realized two levels above the sequential processor level.

The local computer level

This is the first level built on top of the sequential processor level. At this level we see the nodes of a network. These nodes support basically unlimited parallelism and explicit communication between nodes. The parallelism of the tasks running on a node is completely decoupled from a potential parallelism within the node, i.e. of the number of sequential processor within a node. A certain number of sequential processors share the load of any number of parallel tasks. The nodes may be realized such that failures of single sequential processors only reduce the throughput of the system; additional processors may be added off-line or even on-line without the need to reprogram anything.

The synchronization functions on this level have to be chosen very carefully, because they are realized using the primitive synchronization of the sequential processor level which possibly keeps processors busywaiting. The correct selection and implementation of the synchronization functions reduces the probability and the duration of synchronization conflicts of the lower level.

As the tasks of the local computer level are decoupled from the processors of the sequential processor level, delayed tasks do not occupy processors and thus do not affect system efficiency.

Exact definitions of the local computer level and of its realization using sequential processors exist [2] and experimental systems realized according to this model have proven its efficiency.

The most important functions of the local computer level are two synchronization functions which send or receive messages to a data structure called synchronization element (mailbox). Messages are stored in the mailbox until they are fetched by the receiving function; a receiver is possibly delayed until a message arrives.

This pair of synchronization functions is used for mutual exclusion and cross-stimulation.

In addition there are task-management functions to deactivate or freeze tasks and to (re-)activate them. These functions are particularly important in a multiprocessor environment in order to handle time-outs, exceptions and on-line system reconfigurations.

Building the global computer level (Fig. 11)

The local computer level has *one* restriction: Exchanging data and synchronization is restricted to tasks having access to common storage. Logically a communication system corresponds to a (small) common storage, so that tasks of different nodes having access to a communication port may also exchange data and synchronize.

Thus we have overlapping regions, in which communication and synchronization is possible (see Fig. 12).

With a concept of communication tasks [2,3,4] it is possible to logically connect all regions in a fully transparent way. This allows to extend the synchronization- and process management functions over the whole network.

The concept of communication tasks is rather simple for send-functions to remote nodes; in this case the sending task (which does not know whether the mailbox is remote or not) sends its message as usual, a communication task (sender) forwards the message to a partner task (receiver) in the node of the mailbox. This receiver task then acts as a normal sender of information in the node of the mailbox.

Receiving a message from a mailbox in a remote node is somewhat more complex, because the possible target of a message is not known in advance (there can be different potential receivers in different nodes). The message may only be sent to another node when a receiving operation has been executed there. In addition no communication task

```
┌─────────────────────────────┐
│    specification level      │
├─────────────────────────────┤
│    control system level     │
├─────────────────────────────┤
│    global computer level    │
│             ↑               │
│    local computer level     │
├─────────────────────────────┤
│     seq. proc. level        │
└─────────────────────────────┘
```

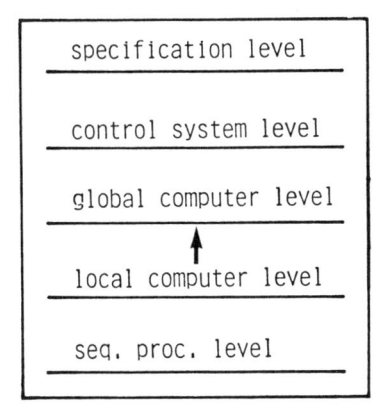

Figure 11.

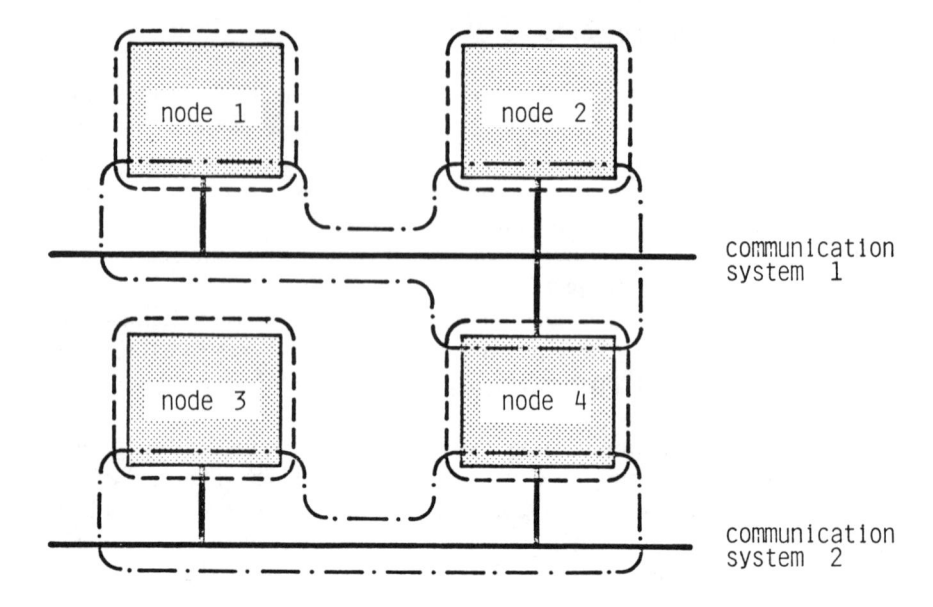

Figure 12. Overlapping regions of synchronization and communication in a computer network.

may be used directly to execute a receive function in order to get the message and then transmit it to the receiving task in another node. Such a communication task could be delayed, if no message is available, and thus leave the communication link unattended. The problem can be solved with auxiliary tasks in a efficient way.

Experimental systems have shown that the overall costs for communication tasks and auxiliary tasks are small. These tasks may consist of fewer than 10 instructions and use common code.

CONCLUSION

We have developed a system design model where we can start on one side, the side of the application, with requirements specifications and translate these specifications first by an application specialist into a control system representation and second by an application-oriented software specialist into a general form of a global computer system representation. Only the first of these two steps depends on a specific application; the second one is a formalization of the accumulated application oriented know-how of the control system designer.

On the other side, the side of the implementation tool, we start with a very simple view the computer system, namely the view of single sequential processors each running its own sequential program. Here in the first step we build computer-nodes where the capability to execute software consisting of parallel, cooperating tasks is decoupled from the basic hardware. This step is the field of activity of the operating system specialist; typical problems are scheduling, processor allocation and problems of redundancy and graceful degradation. Only this first step is dependent of a specific hardware and/or technology.

In a second step we build the global computer system, which was also the target of our approach from the application side. This last step is done by a specialist of distributed intelligence, high level communication mechanisms and data base management systems. Problems of redundancy on the node level are treated here.

Our design model allows the separation of four major problem areas in building and using control computer systems (Fig. 13). It provides rules how to solve these problems. These rules have reached a more precise and detailed form at the computer system end, there still have to be done major research efforts on the requirements specification end.

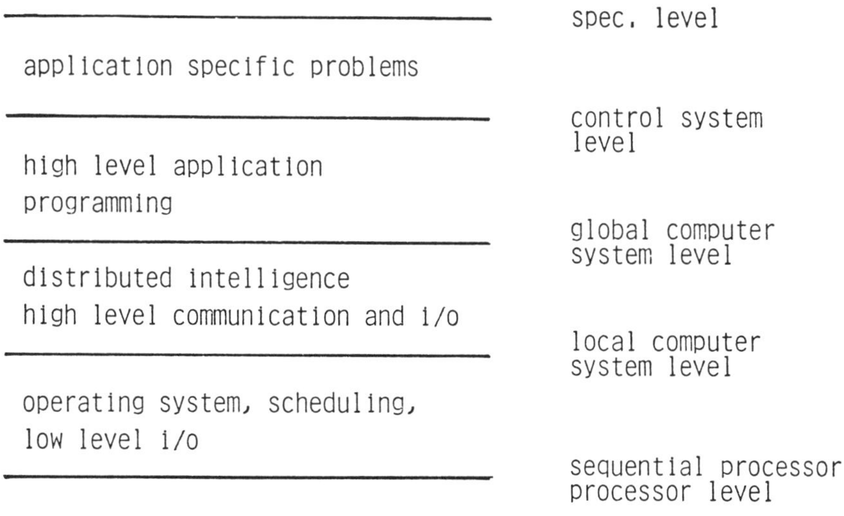

	spec. level
application specific problems	
	control system level
high level application programming	
	global computer system level
distributed intelligence high level communication and i/o	
	local computer system level
operating system, scheduling, low level i/o	
	sequential processor processor level

Figure 13. Separation of different problem areas in the design model.

REFERENCES

[1] Lalive d'Epinay, Thierry, "Structure of an Ideal Distributed Computer Control System," *Distributed Computer Control Systems,* Proc. IFAC Workshop 1979, Pergamon Press, p. 65–70 (1980).
[2] IWICS, TC 8 "Up to date report", ed. by Th. Lalive d'Epinay, BBC Research Center, Baden, Switzerland (June 1980).
[3] Muhlemann, Kurt, "Ein Beitrag zur Synchronisation in Mehrprozessorsystemen und Computer Netzwerken," PhD thesis, Swiss Federal Inst. of Technol., Zurich (1980).
[4] Muhlemann, Kurt "Towards a Model of Communication," IWICS OP/SYS II-14-1, Swiss Federal Inst. of Technol., Zurich (1979).

OPERATIONAL NEEDS IN SOFTWARE FOR COMPLEX PROCESS CONTROL SYSTEMS

R. G. Wilhelm, Jr.
Principal Engineer
AccuRay Corporation
Columbus, Ohio 43202

ABSTRACT

As distributed computer architectures become more popular for process control, modular design practices are elevated from a position of desirability to one of necessity. Simultaneously, it is highly desirable to make system architecture as transparent as possible to the application modules (e.g., the control algorithms). This implies the need for flexible and efficient communication protocols among application modules. While much attention is being paid in the literature to protocols for the lower layers of the distributed system, relatively little has been said about application layer protocols.

This paper discusses some software utilities which promote efficient process control system design, while supporting additional capabilities such as easy on-line control system reconfiguration. In particular, we discuss several functions analogous to the utilities provided by a real-time operating system, but specifically oriented toward process control requirements. These include tools for coordinating the auto/manual status of many interrelated control loops, and efficient methods for obtaining measurements and other essential information from remote nodes in a distributed network.

INTRODUCTION

When designing distributed computer control systems, there is a strong temptation to allow the hardware structure and communication techniques to dictate the structure of the control software—a classic violation of the venerable dictum that form should follow function. To some extent, our infatuation with the design of data highways, communication protocols, and distributed data bases have diverted our attention from the objective: process control. On the other hand, it is equally unrealistic to try to force the hardware configuration to assume the form of the control functional block diagram. There are many other practical forces which influence the hardware configuration, such as the location of actuators and sensors, reliability, suitability of processors for particular tasks, and many others.

In fact, this is a duality problem in designing distributed control systems which a properly designed, layered communication protocol structure can help to solve. Our premise is as follows:

1. The application software should assume a structure dictated by a functional block diagram of the control and information system design (see Figure 1).
2. Communication between the functional (application) modules should follow protocols specifically designed for their class of applications, and these protocols should not require the modules to have any knowledge of how they are distributed in the hardware.
3. The hardware should assume a structure which mets the geographical requirements of process sensors, actuators, and operator interfaces, and provides adequate processing power and communication speed for the application software (see Figure 1).

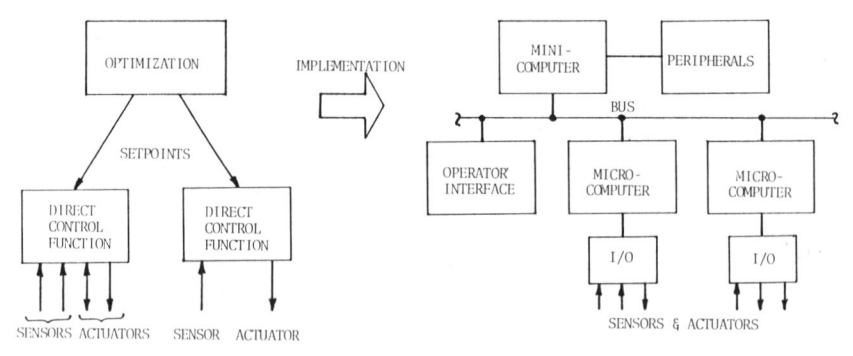

Figure 1. Functional vs. Physical Structure.

4. System utilities (which themselves are distributed) provide the translation from application layer protocols into the message formats required by the hardware and vice versa. These utilities, in effect, allow us to "map" the desired functional structure into a hardware structure which looks quite different.

The ISO model of a layered communication protocol (Figure 2) forms a sound basis for the construction of the utilities in premise 4. Much excellent work has been done on defining standards for the lower protocol

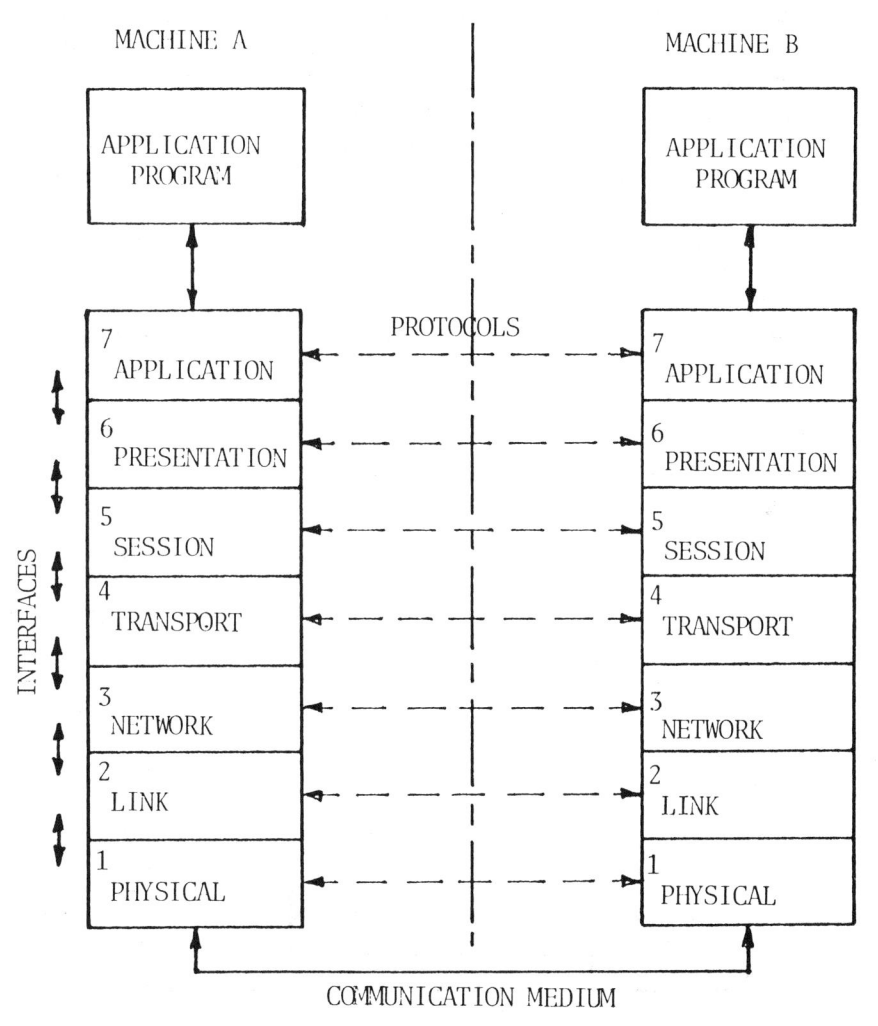

Figure 2. ISO Model for Layered Protocols.

layers [1,2]. This is quite appropriate, since we hope to be able to use similar hardware and operating systems for many different applications. It would be counter-productive to attempt to generate universal standards for the higher layer protocols. This paper will present several examples, however, of protocols and utilities of particular value for process control applications.

TRANSACTION MODELS FOR COMMUNICATIONS

For convenience, let us use the word "element" to represent a distinct functional entity in the control package. Examples of elements could be a PID control loop, an optimization function, or a sequencing package for plant start-up. We may refer to each individual act of communication between elements as a "transaction." Each transaction must be initiated by one element (which we call the "subject" element) and directed to at least one other element (which we call the "object" element(s)). There may or may not be a reply expected from the object, depending upon the nature of the particular transaction. This simple view of communication is illustrated in Figure 3. It implies not only that the two elements share a common language for communication, but that each has enough intelligence to interpret and respond to the transaction.

Character of Control Communications

Transactions between two elements are likely to fall into one of three categories:

1. Petition—One element requests another element to take some action.
2. Inquire—One element requests some information from another element.
3. Inform—One element transmits unsolicited information to another element. (By unsolicited we mean only that the information is not the response to a specific inquiry).

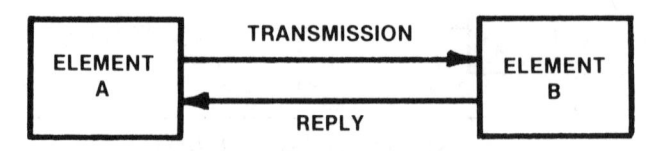

Figure 3. Simple View of Communication.

Examples of a petition would include one element requesting that another element be placed in auto or in manual, or requesting that a specific chain of actions occur on a one-time basis as in a start-up or shut-down sequence. Examples of an inquiry may include asking for the value of a measurement or setpoint, or inquiring as to the operational (e.g., auto/manual) status of another element. An example of the inform type of transaction is when an optimization function sends a new setpoint to a regulating element.

We should note that while some information is requested or transmitted by an element in the process of performing its function, the availability or sudden appearance of new information actually causes some functions to be performed. This reversibility of the causal role between information and task execution is a fundamental factor in dictating appropriate communication techniques.

Structure of an Element

The simple view of communication shown in Figure 3 implies that each element has the intelligence to interpret and respond to the necessary transactions. This would seem to place an undesirable burden on the application designer who wishes to confine himself to the design of clever and effective control algorithms. We may refine our simple view, however, by recognizing that the intelligence to respond to transactions can be separated from the control algorithm itself. For example, a DDC algorithm executing once per second need not concern itself with how its setpoint is updated, or how it is switched on or off. These functions may be implemented as separate command processors which execute only in response to a transaction initiated by another element. Note that one of the other elements which may initiate such actions is the operator interface. The point is illustrated in Figure 4. The algorithm and command

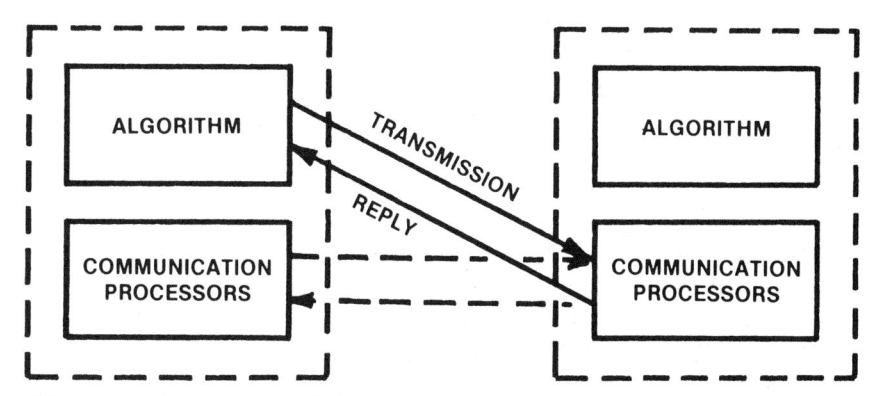

Figure 4. Separation of Communication Processors within Elements.

processors for a given element share the same data base, hence making the communication between them direct and straightforward.

The way in which a command processor responds depends upon the nature of the transaction. For a simple inquiry it may only be necessary to fetch and return the value of a measurement or setpoint. For a petition, such as a request to place the element in auto, the response might involve calling some subroutines which perform some initialization, or even invoking another transaction to petition another element to go into auto. Thus it is important that we draw the distinction that while the transaction itself is uniformly understood by all elements, their individual responses to a given transaction may differ.

An individual element, then, is comprised of an algorithmic portion, a command processor portion, and a private data base. For identical elements (e.g., identical PID controllers) only the data base needs to be replicated. Even elements which are not identical may be able to share parts of the command processors and algorithms.

The Role of System Utilities

Further refinement of the view of communication follows from recognition that a transaction consists of three distinct phases:

1. The handling of the actual communication, involving locating the logical and/or physical address of the object element and, if necessary, translating the message through the various levels of protocol for transmission.
2. The response to the transaction carried out by the object element's own command processors.
3. The handling of the reply, if any, to the subject element, including the translation back through the various layers of protocol.

It is wise to provide system utilities for phases 1 and 3, since they may be designed to be common to all users of the transactions.

Figure 5 shows how the system utilities, collectively called the Transaction Processor, act as a buffer between the elements. The Transaction Processor may simply be a collection of subroutines, one for each type of transaction, which can be called by any element.

Distributing the Utilities

If the elements are distributed among different intelligent devices, the communication utilities must be also. Figure 6 illustrates the fact that this distribution can be rendered invisible to the functional elements themselves by virtue of the buffering service performed by the transaction processors. The elements themselves refer to each other by logical

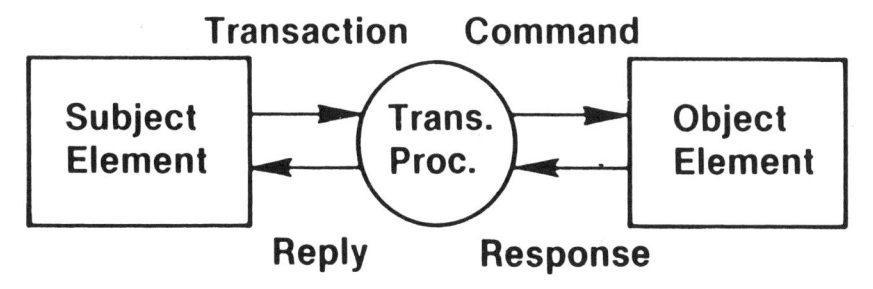

Figure 5. Transaction Processor.

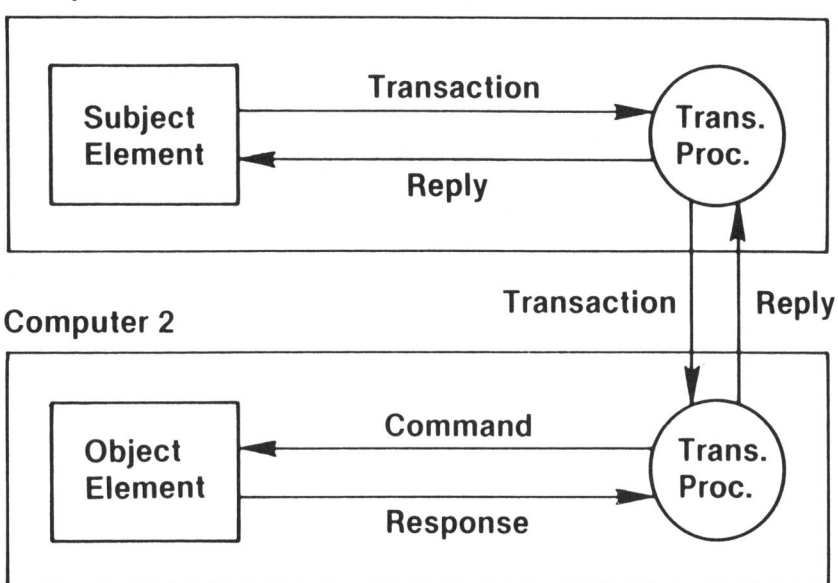

Figure 6. Transactions between Distributed Elements.

names, independent of their actual location. The structure of the system must, of course, be known by the transaction processors in order to properly route the communications. The fact that physical location is invisible to the functional elements does not alter the unfortunate fact that it may severely impact the efficiency and timeliness of communication between elements. For this reason it is imperative that we define transactions which minimize the overhead of communication yet are well suited to the applications being addressed, and that the designer of the functional elements knows how to use them intelligently. The following section discusses two classes of transactions which help to satisfy these objectives.

SUBSCRIPTION AND NOTIFICATION MECHANISMS

In a distributed system, if one frequently executing element requires several pieces of information from other elements, it is grossly inefficient to inquire for each piece of information at the point where it is used in a program. Conversely, if some information is used only rarely by a foreign element, it is inefficient to routinely or periodically transmit it. Two useful techniques for meeting such needs are "subscriptions" and "notifications."

Subscriptions

Unnecessary communication can be reduced if elements are able to enter and cancel "subscriptions" for information at will. By this we mean that the element which needs information (subscriber) enters into an agreement with the element which has the information (source). Subscriptions for data on a periodic basis have obvious application in real-time control. Depending upon the degree of sophistication of the system, the subscriber might be able to specify the frequency at which he wishes to receive the data,. The responsibility then lies with the source element to keep an up-to-date subscription list and send data to the subscribers at the appropriate times. The ability to enter and cancel subscriptions in real-time permits the most efficient use of communication lines, and is especially useful if the control objectives change during normal operation.

Notification

Another valuable use for subscriptions is to request receipt of notification upon the occurrence of a random event. For example, if a high level control element requires the services of a DDC loop, the higher level element should receive notification if the DDC loop turns off for any reason. The notification mechanism eliminates the need for continual transmission of information which changes only infrequently. Status information frequently falls into this category.

The notification mechanism requires that the subscriber's command processors include the intelligence to interpret the notification. However, by cancelling a subscription the subscriber can usually avoid the requirement for complex logic to determine whether a notification is of any interest at present.

EXAMPLE APPLICATION: AUTO/MANUAL LOGIC

The applicability and power of the communication structure discussed thus far can be illustrated by presenting an actual application. The transaction processing approach has been applied to the problem of coordinating the status of individual but interdependent control functions in a complex process control system.

For many process control applications, an element's status may be adequately defined (at least insofar as other elements are concerned) by three binary state variables: ON/OFF, ACTIVE/INACTIVE, and AVAILABLE/UNAVAILABLE. ON/OFF is rather self explanatory. When an element is ON, however, we may wish to temporarily suspend it at times (make it INACTIVE). When an element is OFF, it may not be AVAILABLE to be turned ON, depending upon such external factors as local/remote safety switches, interlocks with other elements, or even the AVAILABILITY of other elements upon which it depends.

In order to coordinate the status of the elements, we can employ the techniques discussed earlier in the following strategy:

1. The auto/manual logic is separated from the control algorithm for each element, and is executed only on demand as a set of command processors.
2. The command processors for each element determine the status of that element, which governs the behavior of the algorithm.
3. Each element must respond to a standard set of commands for state transitions, although the way in which each element responds may be unique.
4. The transactions provided as system utilities provide a cooperative environment which helps to avoid conflicts between elements.
5. Subscription/notification techniques are used to inform elements of status changes in other elements.

Auto/Manual Command Processors

Although the procedures used by each element to change status may be unique, the results are uniform, and may be summarized by the state transition diagram shown in Figure 7. The names of the commands have been chosen to logically relate to the state variables: START (turn ON), STOP (Turn OFF); ACTIVATE (become ACTIVE), DEACTIVATE

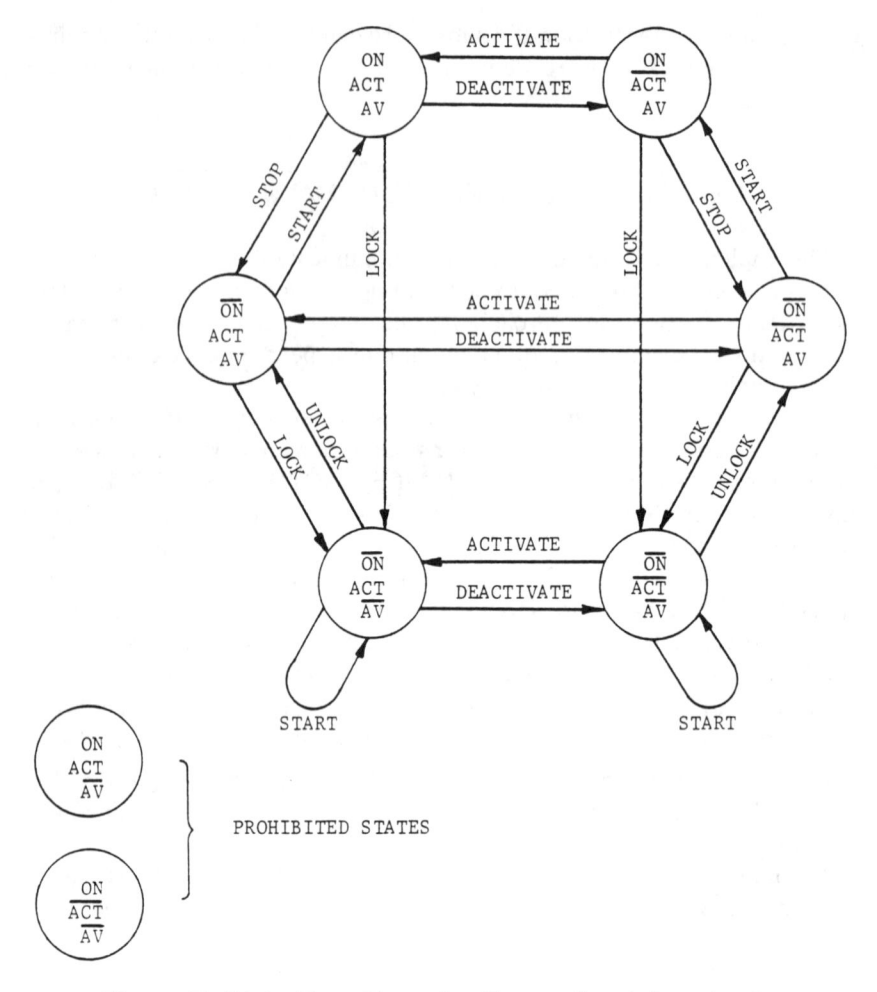

Figure 7. State Transitions for Transactional Auto Logic.

(become INACTIVE); LOCK (become UNAVAILABLE), and UNLOCK (become AVAILABLE). In addition, we provide a NOTICE command processor to respond to notification of state transitions from other elements.

Auto/Manual Transaction Processor

The choice of transactions available to the elements determines the ability to coordinate the behavior of a highly interactive system. Note the emphasis on cooperation and conflict avoidance in the following set.

Transactions for subscription and notification. To support the subscription and notification mechanisms we define two transactions: SUBSCRIBE and CANCEL SUBSCRIPTION. These allow each element to request notification of status changes from any other element, or to terminate such an arrangement. It is not necessary to define a NOTIFY transaction. Since state transitions can only be caused by a transaction, the transaction processor (recall that it is a system utility) can easily be designed to handle the notification function automatically.

Transitions between ON and OFF. If one element wishes another to be on, it may issue a REQUEST. If the object element is already on, the transaction processor does not even bother the object element with any commands; if not, a START command is issued. The REQUEST transaction processor automatically enters a subscription for the subject element as well, so that it will be notified should the object element's status change later. When the subject element no longer requires the object to be on, it may withdraw its request by issuing the WITHDRAW transaction.

The WITHDRAW transaction processor will *not* issue a STOP command unless all subscribers have withdrawn their requests. If the subject element should CANCEL its subscription to the object, its request will automatically be withdrawn by the transaction processor. Should it be necessary to *insist* that the object be turned off, an ABORT transaction can be issued. This always results in a STOP command and subsequent notification of all subscribers unless the object was already OFF.

Transitions between ACTIVE and INACTIVE. If an element wants another to become INACTIVE, it may issue a SUSPEND transaction. As with the REQUEST, the SUSPEND automatically provides a subscription if none existed before. When the subject no longer requires the object to be INACTIVE, it may issue an ALLOW transaction. The transaction processor will not issue an ACTIVATE command as long as any subscriber has suspended and not subsequently allowed the object. Cancellation of a subscription is automatically accompanied by an ALLOW transaction.

Transitions between AVAILABLE and UNAVAILABLE. If one element wishes to lock another, it may issue a DISABLE transaction. When the subject no longer requires the object to be UNAVAILABLE, it may issue an ENABLE. The transaction processor will not issue an UNLOCK command until all subscribers which have disabled the object subsequently enable it. Cancellation of a subscription is automatically accompanied by an ENABLE transaction.

Miscellaneous. Table 1 summarizes the relationships among the transactions, commands, state variables, and forms of notification. An additional transaction not discussed above is the CHECK, which simply

Table 1. Summary of Transactions, Commands, States, and Notices for Auto/Manual Logic.

STATE	COMMANDS	NOTIFICATION	TRANSACTIONS
ON	START	STARTED	REQUEST
OFF	STOP	STOPPED	WITHDRAW REQUEST
			ABORT
ACTIVE	ACTIVATE	ACTIVATED	ALLOW
INACTIVE	DEACTIVATE	DEACTIVATED	SUSPEND
AVAILABLE	UNLOCK	UNLOCK	ENABLE
UNAVAILABLE	LOCK	LOCKED	DISABLE
			CHECK
	NOTICE		SUBSCRIBE
			CANCEL SUBSCRIPTION

inquires as to the current status of the object element. As indicated earlier, the objective of this approach to auto/manual logic is to enable the application designer to construct complex, highly interactive control systems with minimal risk of conflicts, and without allowing the interaction between elements to destroy the modularity of the design. Such modularity assures that future changes or additions to the control system will be able to interact with existing elements without requiring modifications to them.

The methodology and rationale are discussed in more detail in reference [3].

FLEXIBLE PROCESSES IMPLY FLEXIBLE CONTROLS

We are inclined to think of a control system as a fixed structure. Although we often talk of user programmability and on-line modification, we are generally referring to capabilities which are used primarily during a development phase, or in an experimental environment. Since most unit processes are themselves of fixed structure, this view is justified. There are certain types of processes, however, whose structure can change during normal operation. By this we mean that the relationships between certain process elements change. Examples include:

1. Processes in which flowstreams are redirected according to the particular product being produced.

2. Processes in which some elements can be combined in various parallel/series combinations to accomodate changes in load or throughput.
3. Processes in which some elements are interchangeable to allow for maintenance, backflushing, etc.
4. Cases where there are alternate upstream sources of material or energy, and/or alternate downstream destinations for the product.

Such processes require more than the ability to reprogram the control system at will. They require controls which reconfigure themselves automatically or in response to simple operator commands when the process changes. Although control engineers have long been designing such systems both with analog and digital equipment, the task has seldom been easy and the end product has rarely been truly modular.

Independent of whether the process requires the controls to reconfigure, implementation of the controls in a distributed computer system affords the opportunity to use similar techniques to enhance reliability in the event of component failure. Of course, even in a centralized system redundant sensors and actuators can be employed in reconfiguration schemes on a smaller scale for reliability. In any event, it is evident that there is merit in designing tools which render the implementation of reconfigurable controls more straightforward.

Defining a Configuration

Even for a reconfigurable process, the control designer knows the requirements of the individual process elements. What is not known at design time is the particular way in which the elements are interconnected. Obviously the implications for the control design are that it should assume a structure similar to that of the process. That is, for each fixed process element there can be a fixed control element. The interconnection of these control elements can then be modified on line to match that of the process. If we define the reconfigurable aspect of the control system to be the ability to link fixed elements together in different arrangements, the primary requirements unique to such systems are a means of establishing and modifying these relationships on-line, and a record keeping system.

Even the most complex interconnection of multiple elements can be described as a combination of linkages between pairs of elements. A convenient and modular means of recording the entire configuration of the system is to provide each element with an "association list" in which can be recorded the identities of all other elements to which it is presently linked. The system utilities for establishing and changing the configuration can then be implemented as transaction processors which add and delete entries to the association lists of the elements affected by a reconfiguration, much as the SUBSCRIBE and CANCEL SUBSCRIPTION transactions update an element's subscription list.

Defining a relationship. Having established the existence of a linkage between two elements, we must also define the nature of the relationship. Often there are clear "leader/follower" relationships between elements, as in a simple cascade control or ratio control arrangement. Clearly the relationship can also be recorded in the association list. Consider, for example, a simple PID controller as an element. If it is used in a single loop, its association list might contain only the identity of the actuator driver to which it is connected. If, instead it is the primary controller in a cascade loop, its association list would contain the identity of the secondary controller, properly designated as a follower. If the same PID controller is used as the secondary controller in a cascade loop, its association list would designate both the leader (primary controller) and follower (actuator driver). Note that the nature of the relationship is generally a property of the linkage itself, and not an inherent property of an element. Furthermore, the relationships as seen by the two elements in a linked pair are generally complementary, and are designated in both places strictly as a matter of convenience. Thus each element is able to determine its relationship to the rest of the system without ever accessing data outside its own private data base.

Roles. In a structure more complex than a simple cascade control, one element may be linked to several others which bear the same relationship to it, but which nonetheless perform unique "roles." Consider, for example, the multi-fuel combustion control system illustrated in Figure 8. If fuel A is a less expensive substitute for B, but is limited in supply, the fuel apportioning controller may use only fuel A until the demand exceeds the supply of A, then may supplement the flowrate of A with the necessary amount of fuel B. Although both fuel flow controllers bear a "follower" relationship to the fuel apportioning control, each plays a different "role", and is treated differently by the leader. Since fuel prices may change, the roles should be reassignable. Obviously, then, the role must be a property of the linkage and not an inherent property of an element itself. Thus, the role should also be recorded along with the relationship in the association lists.

Implications for the Application Designer

In our earlier discussion of communication between elements the subject element referred to the object element by some form of name. The primary distinction between application programs for systems with fixed configuration and those which are reconfigurable on-line is that, for the

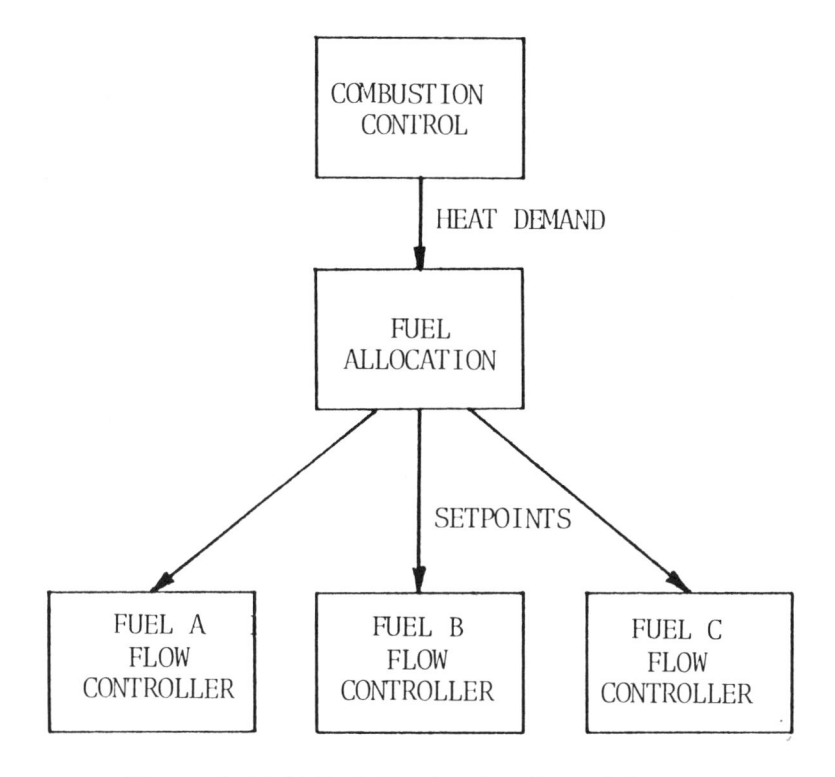

```
                    ┌─────────────┐
                    │  COMBUSTION │
                    │   CONTROL   │
                    └──────┬──────┘
                           │
                        HEAT DEMAND
                           ↓
                    ┌─────────────┐
                    │    FUEL     │
                    │ ALLOCATION  │
                    └──────┬──────┘
```

Figure 8. Multi-Fuel Combustion System (boxes: FUEL A FLOW CONTROLLER, FUEL B FLOW CONTROLLER, FUEL C FLOW CONTROLLER; arrows labeled SETPOINTS).

Figure 8. Multi-Fuel Combustion Control System.

latter, the designer will not know the names of the object elements for a given transaction at design time. Instead, it is necessary to look up the names of the object elements in the subject element's association list at execution time, selecting the appropriate member according to its relationship and role. Once the name of the object element has been obtained, communication transactions can be employed as usual to petition, inquire, or inform.

Where the particular object element is not known at design time, a transaction must be accomplished in two steps in the application program:

1. Obtain from the subject's association list the identity of the object element matching a given relationship and role specification.
2. Use the obtained identity in the usual way in any desired transaction or set of transactions.

In many applications such as continuous blending of several components or the multi-fuel combustion control mentioned above, it may not even be possible to specify *a priori* how many elements may be linked to the subject element with a given relationship and role. In such cases it is often desirable to create a DO loop in the application program to perform such tasks as summing the flowrates of all components being blended (i.e., summing the setpoints or measurements of all flow controllers attached to the blend controller). This requirement leads to the conclusion that a convenient system utility for accomplishing step 1 above is a DO statement which, instead of supplying a series of indices (as in DO 100 i = 1,n), supplies a series of element identities (from the subject's association list. The syntax of a "selective DO" would be:

DO LABEL RELATION-SPEC ROLE-SPEC ID RELATION ROLE

followed by any arbitrary program statements to be executed in the DO Loop, the last of which has the specified LABEL. During each pass through the DO loop and DO statement will provide a new element ID, RELATION, and ROLE which meet the RELATION-SPEC and ROLE-SPEC given in its invocation. It is often useful for specifications to be logical combinations of relationships or roles such as ROLE-1 or ROLE-2, or ALL. As in an ordinary FORTRAN DO loop, when all applicable elements have been found, execution continues with the statement following LABEL. Of course, the selective DO will work just as well if only one element is to be found in the association list.

Meaningful relationships and suitable roles. It is both undesirable and unnecessary to define a limited set or a universal set of relationships and roles. Although the system utilities need to be able to distinguish elements in an association list by relationship and role, the utilities need only to pass this information along to the application programs without applying any interpretation. Therefore, arbitrary symbols may be used to represent a relationship or role, and the application designer is free to choose his own definitions according to his needs.

EXAMPLE APPLICATION: A FLEXIBLE BLENDING SYSTEM

Figures 9 and 10 illustrate a continuous blending control system for the preparation of the "stock" slurry used for paper manufacture which can be reconfigured not only to produce different products, but also to accomodate different ways of specifying the same blend. While it would certainly be possible to link all of the required flow controllers as followers to the main blend controller, this would imply the necessity to specify the amount of each individual component of the blend as a percentage of the total. It may be considerably more convenient to specify the final product as being comprised of certain percentages of some "blend groups", such

R. G. Wilhelm, Jr.   67

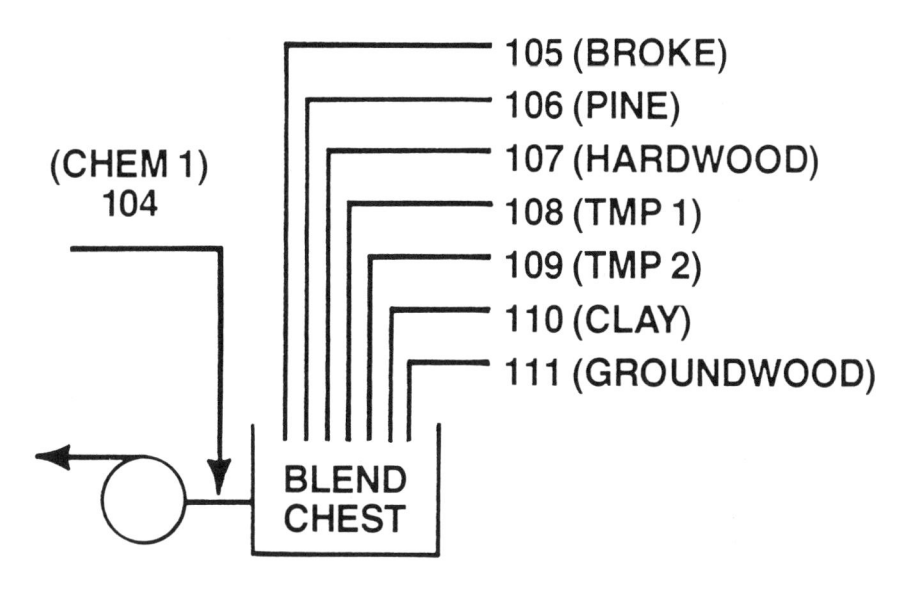

Figure 9. An Example Stock and Additive Blending Process.

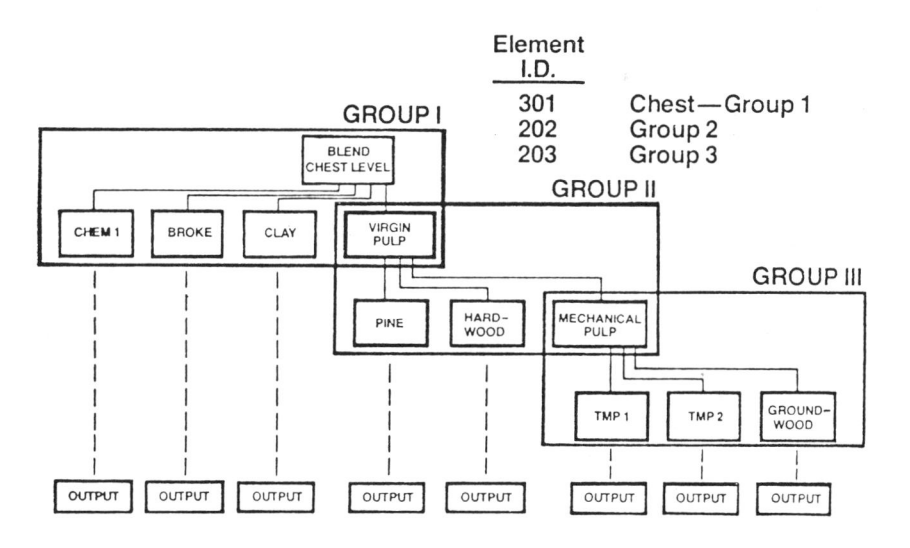

Figure 10. Blend Groups for the Example Process.

68 Chemical Process Control II

as the "virgin fiber" and "mechanical pulp" indicated in the Figures. These blend groups may then be further comprised of given percentages of specific components or other blend groups. Thus the percentages of components which made up a blend group may be altered without changing the relative contribution of that group to the final product.

Figure 11 illustrates one of the video displays which could form the operator interface with the blend control system. The use of "preliminary" setpoint or ratio entries allows the system to indicate to the operator the influence of a single component ratio upon the ratios of other components in the same blend group before the change is actually implemented. By changing the configuration of the control system to include only those components being blended for the present product, rather than always including all possible components and setting some ratios to zero, we realize distinct advantages in clarity of presentation to the operator. The clarity is further enhanced by the artificial introduction of blend groups which serve no concrete control purpose.

In a continuous blending system, three roles suggest themselves. The flowrate of most of the blend components will be increased or decreased to meet the demand, while maintaining specified ratios to each other. These may be called "participating" components. In some cases, however, there may be some component whose flowrate is to remain constant or is determined independently of the demand. Usually such a component already has the correct composition. Since its flowrate contributes to the total but does not respond to the changes in demand, we may call such a

```
MACHINE     3              GROUP STATUS              ENTER I.D.
01-01-79  10:32                                      TRANSFER
                                   ACTUAL   PRELIM
I.D.  LEADER   STATUS    TARGET    VALUE    OUTPUT       OUTPUT

301   BLEVEL   AUTO       150       150      2500         2500
                                 PRELIMINARY      CONTROL       ACTUAL
      FOLLOWER ROLE    STATUS  PERCENT  TARGET  PERCENT  TARGET  PERCENT VALUE

105   BROKE    NPRT    AUTO    17.0     | 425 |  12.0    300     11.8    295
110   CLAY     PART    AUTO    | 8.0 |  200      8.0     200      7.9    198
202   VIRFR    PART    AUTO    75.0     1875    80.0     2000    80.3    2007
104   CHEM 1   NCTR    AUTO     4.0      100     4.0      100     4.0    100
```

Roles:

NPRT = Non-Participating
NCTR = Non-Contributing
PART = Participating

Figure 11. Video Report Layout for Blend Group 1.

component "contributing but non-participating." The third role arises when it is desired to control some flow or other parameter in proportion to the flow of the blended product stream, but that parameter is not added to the blend at this stage of the process. Such an element may be said to be "non-contributing," but can still be adjusted by the total blend flow controller.

The algorithms for the control of such a blending system are readily implemented using the techniques outlined in the preceeding sections, as are the algorithms for the display and operator entry functions. The details of some of these algorithms are given in reference [4].

CONCLUDING REMARKS

This paper has discussed formal methods for communicating between modular functional elements of a process control system, and extensions for the case where the interconnection of functional elements must be modified on-line. Although any formal methodology imposes some overhead, it is argued that the techniques discussed herein encourage modular design of the control functions and help render the hardware architecture less constraining to the application designer. Of course, the application designer must still be aware of the potential for communication delays and their effect on control performance, but the subscription and notification mechanisms can serve to automatically synchronize control functions.

The techniques discussed for auto/manual logic and on-line reconfiguration have been implemented by the author and his colleagues at AccuRay Corporation, and are routinely used in control systems for the paper and other sheet processing industries.

REFERENCES

[1] Graube, Maris, Miscellaneous unpublished reports and communications of the IEEE Local Networks Standards Committee (1980).
[2] Williams, T. J. (ed.), "Minutes of the Seventh Annual Meeting of the International Purdue Workshop on Industrial Computer Systems," Purdue University, West Lafayette, Indiana (1979).
[3] Wilhelm, R. G., *IEEE Transactions on Automatic Control*, 24(1):27–33 (1979).
[4] Wilhelm, R. G., "Flexible Control Techniques for Configurable Processes in the Paper Mill," in the proceedings of the Symposium on Paper and Board Industries Measurement and Control—Toward the Eighties, The Institute of Measurement and Control, London (1979).

APPLICATION OF THE ADA LANGUAGE TO PROCESS CONTROL

W. C. M. Vaughan
Honeywell Inc.
Process Management Systems Division

ABSTRACT

There is a distinct need in process control for the benefits of high-level language programming. Past and present high-level languages have usually been inadequate to meet the specialized needs of process control. In particular, few languages have handled multi-tasking, error recovery, or user control of data representation in a reasonable way.

The programming language Ada has been developed for the U.S. Department of Defense for use in military systems with embedded computers. Such systems are similar to process control systems and require similar language features: therefore Ada should be quite suitable for process control use. Investigation shows that this is indeed the case.

High reliability and high efficiency are important requirements of embedded computer systems, whether military command-control-communication systems or industrial control systems. Most languages have traded off one of these requirements against the other; Ada meets them both. Ada provides a reasonable approach to multi-tasking and error handling. It facilitates programming in a problem-dependent and machine-independent manner, while allowing direct programmer control of machine-dependent constructs such as data representation where necessary.

Ada is the result of an extensive engineering effort directed at analyzing the accumulated computer language experience of the last twenty-five years and synthesizing a better tool for producing and maintaining the software of embedded computer systems. Many observers have concluded that it meets these goals.

INTRODUCTION

The following history of Ada is necessarily brief; readers interested in more detail are referred to [7].

History of Ada

In the early seventies, the United States Department of Defense (DoD) found itself faced with a severe cost problem in its embedded-computer systems.

DoD had been a heavy user of computers since their invention; and, driven by a desire for sophisticated and "intelligent" weapons systems, had caused computers to become embedded in such things as torpedoes, bombs and missiles.

These "embedded-computer" systems usually had to perform complex tasks such as inertial navigation, environment sensing, and pattern recognition; all performed in real time. Embedded-computer systems were further constrained by packaging requirements to operate with a minimum of computational resources such as main memory. Because of this scarcity of resources, operating systems were small or nonexistent.

Such systems were singularly difficult to program in the general-purpose programming languages available in the sixties; yet cost factors made it impossible to program them in assembly language. Each armed service had therefore created one or more of its own programming languages in which to solve the problems of embedded-computer systems.

These languages (e.g. JOVIAL, TACPOL, and CMS-2) had been developed independently and for differing reasons. Consequently, they bore no resemblance to each other. In some cases, incompatible dialects of the same language had developed. To this incipient babel was added the tower of specialized languages supplied by the computer vendors, various defense contractors, and academia. (As early as 1968, over 100 programming languages were cited in [15]).

HOLWG

In 1975, DoD was spending about three billion dollars per annum on software alone, and software costs were projected to increase. The use of a common programming language would help to hold this increase to a reasonable level. To define this common language, DoD formed a High Order Language Working Group (HOLWG).

HOLWG began its task by requiring each future DoD project to use one of only seven programming languages: FORTRAN, COBOL, TACPOL, CMS-2, SPL-1, JOVIAL J3 and JOVIAL J73. The use of assembly language was forbidden unless all seven of these languages could be shown to be impractical or not cost-effective over the project's entire system life cycle.

HOLWG's plan was to define a new language for embedded systems, add it to the list of approved languages, and eventually phase out the earlier languages by attrition (or perhaps by decree).

Strawman, Woodenman, Tinman and Ironman

A preliminary set of requirements for such a language was compiled and sent out for review to the military, other government agencies, industry, academia, and various technical groups inside and outside the United States. This requirements document was called Strawman, since it was intended as a target to draw comment rather than as a realistic specification.

The Strawman requirements were revised following receipt of comments; this revision was whimsically called Woodenman. Woodenman was then circulated for another review cycle.

Meanwhile, HOLWG had begun a massive study of some 23 existing languages, both to determine whether any were suited to become the new standard language and to discover the state of the art. Among the languages studied were the process control oriented languages RTL/2 and PEARL.

The consensus reached from this study and from the Woodenman review was that:

- the idea of a common language was feasible;
- no existing language was suitable;
- the common language should be based on Pascal, PL/I or ALGOL 68.

In June 1976, a set of specifications for the language was released. This specification was called Tinman. Where its predecessors had been discursive and exploratory, Tinman was terse and prescriptive. Along with it was issued a Request for Proposal (RFP) to select competitors for the initial language design.

Responses to the RFP were not long in coming. While these were being evaluated, DoD fed the results of another review cycle into Tinman. The result, Ironman, was widely disseminated.

Rainbow Languages

Of the firms responding to the RFP, four were selected to develop prototype languages compatible with Ironman. Each of these firms had chosen to use Pascal as its base language. In an attempt at secrecy (to prevent reviewers of a language from being biased by corporate affiliations) the four languages were code named Red, Blue, Green and Yellow.

These "rainbow" languages were released for review in early 1978. As a result of the review, two languages (Red and Green) were selected for further development, and Ironman was upgraded and revised, becoming Steelman, the final stage in the series [2].

The winner (Green) was chosen in April 1979 and renamed Preliminary Ada, after Ada Augusta, Countess Lovelace, the daughter of Lord Byron and the world's first programmer. (In assisting Charles Babbage in his pioneering work over a century ago, she wrote programs for the Analytical Engine and Difference Engine, the world's first "computers".)

Preliminary Ada now began a period of extensive review. Many thousands of copies of its defining documents [9, 10] were printed and circulated; it was taught to over one thousand programmers in government and industry; and a panel of "distinguished reviewers", with representatives of industry and academia, was established. Comments and criticism were solicited from these sources and from the public.

In July 1980, Preliminary Ada gave way to Ada, released as a DoD Proposed Standard [8]. Only small changes to the language had resulted from over a year of intense scrutiny.

REQUIREMENTS OF PROCESS CONTROL

Process control systems are similar in many ways to the military embedded-computer systems at which Ada is targeted.

Such systems tend to be both *dedicated* and *coherent*; real-time responsiveness is a key requirement; and reliability and predictability are critical.

Often (but not in every case) there is a need for extreme efficiency at execution time.

Dedicated Systems

A dedicated system is one aimed at a single problem, whose nature is known *a priori*.

Dedicated systems usually require less run-time support than general-purpose systems, since the entire set of run-time needs can be determined at design time; but individual run-time elements are often complex and idiosyncratic.

There is often a need to design, as part of the application system, specialized support elements that would normally be considered operating system components. A language properly supporting the implementation of dedicated systems should, therefore, support the implementation of such components.

Coherent Systems

Process control systems, like military systems, are usually coherent.

A coherent system is, loosely, a set of programs that explicitly interact with one another, either through invocation or through manipulation of shared resources. (A coherent system may be more rigorously defined as a set of modules whose dependency or "uses" graph [14, 16] is not disjoint.)

Every coherent system has an overall design responsibility, to which global issues may be referred for arbitration. Trade-offs between execution space and time are often resolved at this level, as are scheduling conflicts.

Coherence is an important and distinguishing characteristic of real-time systems. Only the existence of the global design authority makes it possible to resolve trade-offs within the application system, rather than delegating their resolution to systems software (operating systems and language processors).

A language supporting coherent systems must therefore support the implementation of routines that detect errors and take action to correct them. To do this, errors must be detectable at execution time, and there must be some mechanism to intercept the reaction to an error and to direct it to a user-written error handler.

To be able to resolve multi-tasking conflicts, it must be possible to write tasks that can schedule, suspend and abort other tasks; to handle resource conflicts, one must be able to write resource allocators.

Real-Time Response

The term "real-time" has been loosely used in the computing industry to refer to any problem which has to be solved in an amount of time corresponding to real-world conditions. By such standards, "real-time" does not always mean "fast," as any astrophysicist modelling galactic evolution could confirm.

This paper is concerned with a more stringent use of the term. Real-time response, with respect to process control systems and military embedded-computer systems, means response to the changing parameters of a physical process under control within an amount of time adequate to close the control loop.

Practical requirements are generally for response to external stimuli within a time ranging from a few milliseconds to a few seconds.

Reliability

The need for reliability in both military and process control systems is clear.

In these systems, the usual notion of reliability (being available when needed, and producing the correct computation) must be extended to include the notion of *robustness,* or fault-tolerance.

In particular, these systems must be able to handle *exceptions,* or erroneous conditions, in the environment, in input data, or (sometimes) in their own computations.

Exception handlers must, as a rule, be application-defined. The range of possible responses to exceptional conditions is too broad for a general-purpose handler to be of much use.

A language supporting reliable systems should not contain many constructs known to be error-prone; it should emphasize error detection, both at compilation time and at execution time. To enhance maintainability and communicability, programs in the language should be easy to read.

Predictability

Both military and process control systems must have predictable performance. In particular, their behavior must be characterized in the worst case as well as in the average case.

This concern with worst-case performance is, like coherence, a distinguishing characteristic of real-time systems. Its impact is felt principally in the area of multi-tasking.

To be able to predict the behavior of a multi-tasking application system, one must know a great deal about the underlying operating system; particularly, its scheduler. The operating system must provide a multi-tasking model that is consistent and complete, that supports priority scheduling, and whose latency characteristics are well-defined.

General-purpose operating systems seldom meet these requirements; consequently, many modern real-time application systems specify means for multi-tasking in their own design.

A language supporting predictable systems should contain a multi-tasking model whose scheduling philosophy is well-defined under all conditions.

Efficiency

In military embedded-computer systems, there are often severe packaging constraints, either upon volume, weight, or some critical dimension (the diameter of a torpedo tube, for example). Regardless, the embedded system often has to do a job that would be difficult to do in real time on a much larger machine.

Not all process control systems share this problem; it is not common in large centralized systems where there are plenty of processing resources, and where the job to be done is primarily mathematical in nature (e.g. simulations and process optimization).

However, in mini- and microcomputer-based process control systems, where resources are scarce and time constraints are severe, the problem of efficiency is real.

Some efficiency accrues naturally from the dedicated and coherent nature of the system (run-time overhead, for example, is reduced), but this effect alone is usually insufficient.

To achieve the required performance in the available space, system implementors may use certain techniques that are normally available only in assembly language:
- data representations are modified to occupy less space or to be accessible in less time;
- error checking is removed from, or reduced in, carefully chosen parts of the system;
- closed subroutines are replaced by open macros;
- properties of a datum's representation, independent of that datum's logical properties, are used (for example, one can change an ASCII numeric character to BCD by logically AND'ing it with the number 15);
- machine dependencies, and occasionally actual machine code, are included in programs.

As a rule, only an expert programmer can use techniques like these with any degree of safety. Because of the added cost of testing and maintenance, the use of such methods is frowned upon as "tricky code" by most practicioners. However, because of efficiency requirements a few tricks like these are found in most real-time systems.

Each such technique is used sparingly, as they all tend to compromise system reliability and because they invariably require more stringent testing than other parts of the system.

No compiler for any language can automatically apply such techniques as optimizations. Consequently, if they are to be used, and if the systems implementation language does not explicitly support them somehow, the programmer is forced either to leave that language or to subvert the language by bypassing its integrity checks or using built-in loopholes.

Specific Requirements

The discussion above leads to the following list of requirements for a language in which to implement process control systems:

- the ability to express specialized run-time routines; this is likely to involve manipulation of pointers and complex data structures, and may involve the writing of trap and interrupt handlers;
- the ability to detect and handle exceptional conditions in the environment and in the execution of system code;
- the ability to control the multi-tasking nature of the system; this requires facilities to communicate, coordinate and synchronize multiple tasks, to declare shared variables and exempt them from certain optimizations;
- the ability to control the representation of data, to write "tricky code" when necessary, to bypass certain error checks and make certain unnatural type conversions in a relatively controllable fashion.

PRIOR LANGUAGES

In the past, process control systems have been wholly or partially implemented in several high-order languages as well as in assembly language. In addition, a number of languages are claimed to be suitable for the writing of real-time or process control systems. It is impossible to discuss each such language here, or in fact to deal with any in the depth which it deserves.

Nevertheless, in this section we shall attempt to give an idea of the applicability of a number of prior languages to process control, and in particular to discuss their "interesting" features with regard to this application.

General-Purpose Languages

By "general-purpose languages" is meant those languages which were not designed strictly to serve the real-time or process control application areas.

The best known of these are FORTRAN, PL/I and Pascal.

FORTRAN

FORTRAN is the most widespread of the languages used to implement process control systems. This is partially due to reasons unrelated to the application, such as the fact that until recently, it was the only programming language included in the average undergraduate engineering curriculum. Nevertheless, it is widely used, and for some good reasons.

FORTRAN was originally designed to handle numerical scientific and engineering problems. It does this well. Since many of the computations required in a control system are numerical, FORTRAN offers a good notation in which to express them.

FORTRAN is ubiquitous. Nearly every mainframe, minicomputer and microprocessor has a FORTRAN compiler.

With respect to the specific requirements of process control, FORTRAN performs as follows:

- Run-time routine writing: this is not one of FORTRAN's strong points. It cannot handle pointers or heterogeneous data structures. Interrupt and trap handlers are difficult to write because it is not possible either to declare data structures compatible with machine status words or to bind them to hardware-defined addresses.

  For this reason, most FORTRAN-based real-time systems contain a large assembly-language run-time package.
- Robustness and exception handling: most FORTRAN compilers emit little or no error-detection code. What errors are detected must be handled by underlying system software, since FORTRAN has no facilities for exception handling.
- Multi-tasking: no multi-tasking facilities are built into FORTRAN. This lack is usually overcome by means of a run-time package, which differs depending upon the underlying operating system. Attempts have been made to specify model run-time packages (see [11] for an example), but these packages are seldom implemented in their entirety, due to incompatibilities with vendor operating systems.

  Even assuming the existence of such a package, a serious problem exists with respect to shared data. Most FORTRAN compilers generate optimized code. For various technical reasons, many optimizations cannot be performed on variables shared between tasks. Attempts to optimize such variables lead to incorrect code which can be a source of lurking timing errors.

  In Standard FORTRAN, there is no way to inform the compiler that certain variables may be shared.
- Extreme efficiency issues: FORTRAN allows some control of data storage, but not of data representation. It is not difficult to write tricky code in FORTRAN, but it is very difficult to control its use.

  In Standard FORTRAN, safety checks are easy to defeat (by EQUIVALENCE, for example), but there is no good way to tell where they have been defeated, except through laborious manual inspection of code.

PL/I

It is not common to find real-time systems written in PL/I; yet it was the first language to try to solve most of the problems discussed above. By inspecting PL/I, therefore, one may discover some of the pitfalls involved in designing languages for embedded systems, and thus determine how well Ada has avoided these pitfalls.

PL/I's design intent was to provide a language suitable to solve all programming problems. Among the problem domains specifically targeted were scientific numerical processing (the FORTRAN domain), commercial EDP (the COBOL domain), and systems programming (then considered the domain of assembly language).

To achieve these goals, many various facilities were embedded in the language, including exception handling, multi-tasking, control over storage allocation, and complex data structuring facilities—in short, most of the facilities required for real-time systems.

Yet PL/I has three major flaws.

- The first is its size. It is a very large language and extremely difficult to learn *in toto*. As a result both of its size and of its design philosophy, there are usually many ways to solve any given programming problem. This renders PL/I programs highly idiosyncratic, making them hard to maintain.
- PL/I's second, and more serious, flaw is its lack of safety. The language provides many data types, but it is not strongly typed: implicit coercions allow automatic transformations between any two types. Also, in the interest of making PL/I programs easier to write, numerous shortcuts were introduced, such as allowing multiple blocks to be closed by a single END.

    These facilities, though well-intentioned, effectively inhibit the detection of many simple errors, both at compilation time and at execution time. As a result, PL/I programs tend to be more expensive to test and to maintain than programs written in a language that permits a higher level of compile-time error detection.
- The third flaw, and perhaps the most serious from the point of view of real-time systems, is the inherent inefficiency of PL/I.

    Folklore has it that all high-level languages are inherently inefficient when compared to assembly language. This notion proves fallacious if highly optimizing compilers are considered. For over twenty years, optimization techniques have been known that can achieve efficiency near assembly-language levels (see [1] for details).

    However, there are features of PL/I that inhibit the use of many such optimizations in real-time programs. In particular, the PL/I style of exception handling and its style of pointer manipulation can cause serious problems.

This is particularly troublesome with a language of the size of PL/I, since in the absence of optimization it is difficult to generate even reasonably good code.

When checked against the specific requirements listed above, PL/I appears very good for run-time routine writing and exception handling and has good multi-tasking facilities but does poorly from the standpoint of extreme efficiency, even though it is good for writing "tricky code." Its effective lack of type checking reduces its robustness.

Pascal

Pascal was designed as a teaching language but has become widespread in a broad range of applications. Some dialects are claimed to be useful for real-time systems.

However, only Standard Pascal as defined by Jensen and Wirth [12] is discussed here, since nearly every proprietary dialect is unique; and Standard Pascal is not particularly suitable for real-time applications.

The reasons can be discovered by comparing Pascal to the list of requirements described above.

- Run-time routine writing: Pascal allows the use of pointers and of quite complex structures. The main class of structures disallowed by Pascal is varying-length strings. However, Pascal has no facilities for trap or interrupt handling. A data structure cannot be associated with a physical address; consequently, device drivers must be written in another language.

   The most important problem in this area (which most vendors have remedied by extending the language) is that Standard Pascal has no provision for external procedure linkage.
- Robustness and exception handling: Pascal is a strongly typed language, and consequently makes many error checks at compilation time and at execution time. However, it makes no provision for exception handling; consequently, any run-time error must abort a Pascal program. This philosophy reduces the robustness of a system by giving trivial errors a catastrophic impact.
- Multi-tasking: Pascal has no multi-tasking facilities.
- Extreme efficiency issues: Pascal allows packing of data structures (although the packing method is chosen by the compiler rather than by the programmer). Beyond this, no user control over data structures is offered. Further, the user cannot create a pointer to a local (stack) object.

Pascal puts great obstacles in the path of the programmer who needs to "cheat"; but by use of two loopholes deliberately built into the language (tagless variant records and the *dispose* function) one can write arbitrarily tricky code.

Pascal is a small language for which it is easy to compile good code. Therefore it must be highly rated for efficiency.

Special-Purpose Languages

Numerous special-purpose languages have been written which claim to be oriented towards real-time systems. Two typical ones have been chosen for examination: PEARL and Modula.

PEARL

PEARL, which is an acronym for Process and Experiment Automation Real-time Language, was developed in West Germany in the middle 1970's. It is explicitly aimed at industrial process control and at laboratory automation, and has found use in Europe. See [13] for a full description of its features.

PEARL is a large language, based on ALGOL 68 but with a PL/I-like syntax. It contains a comprehensive set of multi-tasking operations, with emphasis on flexibility of scheduling. It has some provision for interrupt and exception handling. Complex data structures may be defined.

PEARL is singularly I/O-oriented. A central concept is the data station or *dation*, a complex virtual channel used for most data transmission. PEARL is a good language, though large and of very high level. It suffers needlessly from a cryptic notation consisting almost entirely of idiosyncratic abbreviations.

It is most appropriate for large systems. A small dedicated system would not be likely to contain the large-scale operating system facilities needed to handle PEARL's multi-tasking model and I/O system.

With respect to specific issues, PEARL would be only fair at writing run-time routines; its exception handling facilities are good; its multi-tasking facilities are comprehensive; but where extreme efficiency is required, PEARL is poor.

Modula

Modula was designed by Niklaus Wirth (the developer of Pascal) at ETH Zurich in the late 1970's. It is aimed, according to its author [18], at "programming dedicated computer systems, including process control systems on smaller machines."

Modula is quite the opposite of PEARL. It is extremely small and has no built-in I/O facilities. Its multi-tasking facilities are primitive; and, although permitting arbitrarily complex scheduling through clock calls, permits only the "main program" to create new tasks.

No facilities for exception handling are provided. Data manipulation is a problem since pointers are not provided; neither are floating-point numbers. On the other hand, it is possible to get arbitrarily close to the target machine; Modula contains facilities for encapsulating specialized machine-dependent language extensions. (Wirth's example in [18] uses the PDP-11.) Therefore memory managers can be written, as can trap and interrupt handlers for any machine-detectable exception condition.

Modula would be a good tool for writing the operating system portion of a small, dedicated real-time system. However, it could not be seriously considered for writing the application stratum of such systems.

Following are Modula's characteristics with respect to specific requirements:

- Run-time routine writing: Modula is good in this area because of its ability to get close to the target machine; but interpreting pointers passed by an application, though possible, is messy.
- Robustness and exception handling: Modula has no explicit facility for exception handling. Consequently, execution-time errors, if detected by the language, are not easy to handle. Further, Modula presupposes no run-time support whatsoever, so it is not clear how the compiler could emit error-checking code. Not all underlying hardware can be expected to fault when an error occurs. It appears that Modula is too simple to support robust programming.
- Multi-tasking: the multi-tasking facilities of Modula are sound but not comprehensive. The asymmetric relation between the "main program" and other tasks may force designers away from peer protocols (considered desirable for fault-tolerance) and into master-slave protocols. Insufficient attention is paid to multi-tasking failures, and the primitive synchronization mechanism provided (signals) is known to be error-prone.
- Extreme efficiency issues: Modula provides poor control of data representation. Its ability to support tricky code is excellent. Of the languages described here (including Ada), Modula is the easiest to compile optimally, as its lack of a *GOTO* statement makes all program flow graphs reducible.

OPERATIONAL ADVANTAGES OF ADA

Because of the prior languages' poor fit to the requirements of embedded systems, there was considerable skepticism when the process of defining Ada began. But as straw metamorphosed into iron, the clean and orderly progression of the project turned skepticism into optimism.

Hopes were dashed at the rainbow languages' review. Creating a language suitable for embedded systems was indeed difficult. None of the four languages was perfect; there were those who felt, indeed, that none were any good (see, for example, [3, 4, 5, 6]).

The following year saw Green's multi-tasking model undergo such heavy revision that it seemed almost to have been discarded entirely and reinvented. This came about partly as a result of criticism from the user community. This change to the language was typical of the responsiveness to user needs that characterized the rest of Ada's development.

It must be emphasized that no other language has been subject to so long and so thorough a process of public review and debugging while still in its infancy. This procedure, like the "burning in" of an electronic circuit board, promises Ada, though brand new, the stability of a mature language.

One expects to see, as a result of this artificial maturity, better or more comprehensive solutions to embedded systems' problems. Therefore we examine Ada as we have the prior languages.

Run-time Routine Writing

Ada's ability to express run-time routines and operating system components is superior to all languages discussed above including Modula. Ada achieves this level of expressive power without requiring machine-dependent language extensions.

Hardware-defined data structures of arbitrary complexity (such as Program Status Words and machine instruction formats) can be exactly matched via representation specifications (to be further discussed below).

Any data object or executable program unit may be arbitrarily located within memory. This allows one to write trap handlers for systems in which traps vector through specific addresses or use data structures in "dedicated memory."

In Ada, interrupts act as multi-tasking entry calls. A mechanism is provided to attach an interrupt entry to a hardware interrupt number or address.

It is important to note that such machine semantics are handled within the language; it is necessary neither to leave Ada nor subvert it to handle many machine dependencies. Consequently, the compiler is able to cooperate with the programmer, both in detection of errors and in emission of clean, optimized code.

Robustness and Exception Handling

Like Pascal and Modula, Ada is a strongly typed language. This feature enhances error detection and therefore reliability. Unlike Pascal or Modula, Ada provides means to handle errors detected at execution time. This feature enhances robustness and fault-tolerance.

Ada does not distinguish semantically between errors detected by hardware, by compiler-generated error checking code, or by the application program; each is an exception and each is handled like the others. This feature enhances portability of programs between systems with hardware error detection and those in which the same errors are detected by compiler-generated code or by the application.

PL/I's style of exception handling has the unfortunate side effect of inhibiting code optimization. Ada's style does not. The price paid for this is that a routine in which an exception occurred may not be resumed. Most of the time, this is desirable; but it can occasionally create minor design difficulties.

(It is interesting that, although the Ada exception handling mechanism promotes efficiency, its design was chosen for other reasons, including reliability and program correctness [10, 2].)

Multi-tasking

Ada contains a simple but comprehensive multi-tasking facility.

Like Modula, Ada requires no operating system substrate to support its multi-tasking model; unlike Modula, no master or "main program" is required. Dynamic task creation is supported.

Synchronization is based on the *rendezvous* concept. Two communicating tasks must each arrive at a rendezvous point. The first task arriving at the rendezvous awaits the second task's arrival. When both have arrived, information may be exchanged; thereafter the two tasks may again execute in parallel.

This synchronization mechanism is easy to use and very simple to implement. In particular, there is no need to buffer data being transmitted from task to task, as it may remain in the caller's address space while the rendezvous is pending. In addition, intriguing possibilities for optimization are opened in cases where both tasks potentially share an address space: the caller, arriving at the rendezvous, may conceivably execute the rendezvous code on behalf of the called task, saving a pair of time-consuming context switches.

The possibility of failure is not overlooked. A rendezvous attempt may be timed out from either end. Failure of the called task before or during rendezvous raises an exception in the caller. A task may explicitly raise the FAILURE exception in another task, or may abort another task completely.

Tasks in Ada may have priorities; these are fixed at compilation time and may not thereafter be changed.

Sophisticated scheduling facilities as in PEARL are not supported, but may be written as packages in Ada itself.

Extreme Efficiency Issues

In Ada, extreme efficiency issues are resolved through representation specifications, exception suppression, inline procedure insertion, unchecked programming, and machine code insertion.

Representation Specifications

In Ada, the logical properties of data are separate from their physical properties. Ordinary data declarations specify only the logical properties of the data they declare; the compiler chooses a default representation.

Representation specifications allow the user to specify a representation that may differ from the compiler's choice.

Storage space may be specified for all data. Alignment and offset may be specified for fields of records. Code assignments may be specified for enumeration types.

Exception Suppression

A serious problem with languages that generate run-time error checking code is that the compiler often generates too much of it. Specifically, one would like to inhibit the generation of such code:

- in very time-critical routines;
- or where it is known to be unnecessary.

Surprisingly often, optimizing compilers can discover cases where error checking code may not be required (for an example applied to Pascal, see [17]); but even the best optimizing compiler can neither discover all such cases, nor cover time-critical routines.

By use of the *suppress* pragma (a *pragma* is a compiler directive), error detection code may be suppressed. Each check may be individually controlled, even to the extent of suppressing checks on some data while retaining them on others.

Inline Procedure Insertion

The pragma *inline* is provided by Ada to reconcile the conflict between modularity and efficiency.

In most prior languages, a highly modular program necessarily involves a high degree of subprogram calling. This hinders efficiency in two ways: first, the execution of many calling sequences increases overhead; and second, the compiler is usually unable to optimize code across a procedure call.

In Ada, the same highly modular program can be compiled with inline insertion of small procedures. This removes the calling sequence and its overhead, and allows the optimizer free rein.

Unchecked Programming

There are two facilities of prior programming languages that are known to be highly error-prone, yet are ubiquitous due to the great need for them. These are: deallocation of dynamically acquired storage, and type conversion between logically unrelated objects.

In Ada, these functions may be performed by calling on predefined generic procedures for arbitrary type conversions and for storage deallocation. These are called UNCHECKED-CONVERSION and UNCHECKED-DEALLOCATION respectively.

Because of Ada's importation rules, any procedure using such techniques must name the generic procedure being used in its header. This is easily visible on code review.

Machine Code Insertion

It is possible to define a procedure consisting of machine code statements. Via the *inline* pragma, this code may be inserted inline at the calling site.

Machine code insertions are likely to have an adverse effect on code optimization. The compiler, ignorant of the semantics of the inserted code, must turn off all optimizations surrounding the call. Therefore the efficiency impact of any given machine code insertion will depend on the specific compiler and machine involved.

SUMMARY

Ada was designed to help solve the problems of military embedded-computer systems. In this paper, we have tried to explore its applicability to process control. We have discovered that process control systems have much in common with military embedded-computer systems, and, to the extent to which the applications are similar, Ada seems a good choice.

Not all process control systems are small or dedicated. Some run on large computers, where the presence of a sophisticated operating system, and its overhead, are taken for granted.

Distributed process control systems are becoming common. Often, in these systems, there are small dedicated components (which perform process control *per se*) and large components which are not always coherent or dedicated (performing, for example, process optimization).

The latter parts of these systems can certainly be programmed in Ada; but they could equally well be programmed in any other language supporting numerical scientific applications.

No programming language can ever be perfect for any given application. Ada certainly is not perfect for industrial process control. Nevertheless, it seems to be significantly better than anything heretofore available.

In contemplating the design or implementation of a new process control system, Ada must be given serious consideration. In particular, for the small dedicated system, Ada may present the only viable alternative to the expense and risk of implementation in machine language.

ACKNOWLEDGEMENT

The author wishes gratefully to acknowledge the able assistance of Thomas L. Phinney, without whom this paper would never have been written; and the helpful comments of Magne Fjeld and Dennis Cornhill.

REFERENCES

[1] Aho, A. V., and Ullman, J. D. *Principles of Compiler Design.* Addison-Wesley, 1977.
[2] Department of Defense Requirements for High Order Computer Programming Languages, "Steelman," June 1978.
[3] Dijkstra, E. W. On the BLUE Language submitted to the DoD. ACM SIGPLAN *Notices 13,* 10 (1978).
[4] Dijkstra, E. W. On the GREEN Language submitted to the DoD. ACM SIGPLAN *Notices 13,* 10 (1978).
[5] Dijkstra, E. W. On the YELLOW Language submitted to the DoD. ACM SIGPLAN *Notices 13,* 10 (1978).
[6] Dijkstra, E. W. On the RED Language submitted to the DoD. ACM SIGPLAN *Notices 13,* 10 (1978).
[7] Glass, R. L. From Pascal to Pebbleman . . . and Beyond. *Datamation,* July 1979.
[8] Ichbiah, J. D. *et al. Reference Manual for the Ada Programming Language.* Proposed standard document, United States Department of Defense, 1980.

[9] Ichbiah, J. D. *et al. Preliminary Ada Reference Manual.* Published as ACM SIGPLAN *Notices 14,* 6a (1979).
[10] Ichbiah, J. D. *et al. Rationale for the Design of the Ada Programming Language.* Published as ACM SIGPLAN *Notices 14,* 6b (1979).
[11] Instrument Society of America, *Industrial Computer System FORTRAN Procedures for the Management of Independent Interrelated Tasks.* Draft Standard S61.3, 1978.
[12] Jensen, K. and Wirth, N. *Pascal User Manual and Report,* Second edition. Springer-Verlag, 1976.
[13] Kappatsch, A. *Full PEARL Language Description.* Gesellschaft fuer Kernforschung mbH, Karlsruhe, KFK-PDV 130, 1977.
[14] Parnas, D. Designing Software for Ease of Expansion and Contraction. *Proc. Third International Conf. on Software Engineering,* 1978.
[15] Sammet, J. E. *Programming Languages: History and Fundamentals.* Prentice-Hall, 1969.
[16] Schroeder, m., Clark, D. and Saltzer, J. The MULTICS Kernel Design Project. *Proc. Sixth Symp. on Operating Systems Principles,* 1977.
[17] Welsh, J. Economic Range Checks in Pascal. *Software—Practice and Experience 8,* 1 (1978).
[18] Wirth, N. Modula: a Programming Language for Modular Multiprogramming. *Software—Practice and Experience 7,* 1 (1977).

Human Factors in Process Control

IMPORTANT PROBLEMS AND CHALLENGES IN HUMAN FACTORS AND MAN-MACHINE ENGINEERING FOR PROCESS CONTROL SYSTEMS

John E. Rijnsdorp
Twente University of Technology
Enschede, Netherlands

INTRODUCTION

The gap between control theory and practice has been the subject of extensive discussion (See, e.g. Foss [15], Sargent [47], Foss and Denn [16]). In fact, much work is being done to push advanced control concepts into application, pulled by the availability of powerful and inexpensive computer hardware. Of course, the adaptation of theoretical ideas to practical requirements, and their conversion to application software will keep us busy for a long time.

However, another gap has received much less attention, at least until the Three Mile Island incident occurred in 1979: the non-application of human factors (ergonomics) theory and know-how to process automation systems. Much information is available (Edwards and Lees [12, 13], Stockbridge [53], Kompass and Williams [22], Kraiss and Moraal [26]), and in many countries the educational system is producing experts in the field. What are then the causes behind the human factors/automation gap?

Firstly, attention was mainly focussed on the aerospace sector, particularly in the U.S. Consequently much work has been done on man-machine interaction in fast systems, and relatively little on manual control and supervision of slow processes.

Secondly, the application of human factors data requires special care (Lenior and Rijnsdorp [29]). All too often they have been generated in brief runs during laboratory experiments, with young bright people (students) serving as human subjects in simple tasks. It is tricky to use data obtained in this way in an industrial environment, with operators of different age and experience, carrying real responsibilities in many simultaneous tasks during extended periods of time. Is this the reason why so many automation engineers do not have a human factors handbook at their desk, or are they aiming at doing without the human operator altogether?

One thing is quite certain: In practice, the latter is not feasible. Even in highly automated systems, such as modern electric power plants and space probes, people have to monitor and to cope with malfunctioning.

Experience has taught us that consideration of human factors requires a wide context. For instance, a man-machine interface can only correctly be designed if human tasks and responsibilities are thoroughly analyzed, and even introduced as design variables. Caring for the "machine" subsystem only leads to suboptimalization at the very best; possibly it even results in non-acceptance by the human subsystem. Therefore we advocate an integrated approach, aiming at harmonization of psychological/social and technical/economical factors (Rijnsdorp [36]).

FUNCTIONS OF THE OPERATION SYSTEM

An integrated approach requires a broad term for everything related to operating and controlling a process or a complete plant: the operation system. In terms of human factors theory this is a man-machine, or better a men-machines system.

Table 1 shows the functions the system has to perform (Rijnsdorp [36, 37]), together with the people taking care of the tasks for the human subsystem, and commonly used methods and means:

Almost all plant departments are involved in planning and scheduling the modes of operation for the various processes. For this function, a computer network seems an attractive possibility.

The control functions are arranged in a hierarchy: Mode setting yields set points for process variables. Typical for quality control is the measurement problem: accurate data are only available from time to time and with long delays. Regulatory control (continuous operation) is presently highly automated. The same applies to sequence control for

Table 1. Operational Functions in Relation to Men and Means.

function	men	commonly used means
planning and scheduling	all depts; operation supervisors	oper. res. techniques off-line computer
mode setting	technologists; oper. supervisors or operators	objective function process models on-line computer
quality control	analysts; operators	quality analyzers
regulatory control	operators	control algorithms process instrumentation
sequence control	operators	sequential algorithms process instrumentation
general inspection	operators	operator information systems
off-normal handling	operators; poss. oper. supervisors	alarms; trip-outs
in-situ maintenance	technicians; possibly operators	tools
reporting	management; oper. supervisors; finance & administration	management information system
system improvement	all	

frequently recurring batch-type operations. Less frequent operations are done manually, either from a central control room or locally in the plant.

During a substantial part of the time, operators are involved in general inspection, both in the control room and in the plant. Off-normal handling is shared between operators and instrumentation in several "lines of defence." Correction of off-normal conditions often is a matter of maintenance. First-line maintenance activities are sometimes performed by the operators, for instance the replacement of printed-circuit boards.

Reporting is the information output from operation to other plant departments; it includes the management information system. Finally, modern computerized instrumentation has a high degree of flexibility to allow system improvement.

THE STATE OF THE PROCESS

Also from the point of view of human factors there are good reasons for concentrating most of the operational functions in a central control room. Several studies have been made of the central operator, or, more precisely, of operator behavior in the control room.

In a number of observations, frequencies and duration were studied of the various operator activities (Dallimonti [7], van Droffelaar [11]). Results show a relatively constant fraction of the time spent on general monitoring of displays, irrespective of the time of day or night, or conditions in the plant (See Table 2). The idea that operators are just passively waiting for the alarm buzzer evidently is wrong.

Of course, observations of operator activities do not say very much about mental activities. These have been probed, amongst others, by a technique developed by Bainbridge [2]. She asked the operators to "think aloud" while they are doing their work. The resulting "trains of thought", which tend to be rather fragmentary and chaotic, are arranged into procedures. It is also possible to obtain similar information from interviews (Kragt and Landeweerd [24]). Figure 1 gives an example of results.

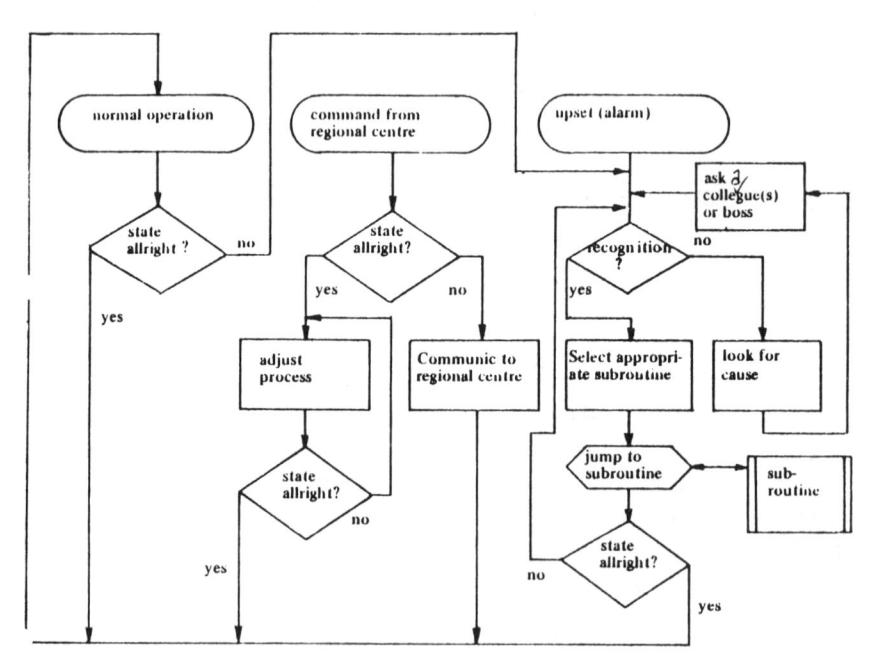

Figure 1. Operator's mental model (Electric Power Generating Plant) in the form of a Computer Program Flow Diagram.

Van Droffellaar (1975)

The results clearly show that the operator returns time and again to the question: "Is the process state O.K.?" This links up with the observed attention to monitoring of displays (Table 2).

What the operator means by "process state" is not completely clear. It certainly encompasses the process line-up if this is frequently changed, as in oil movement and storage in oil refineries, and in batch processes. In addition, the state probably includes a prediction of future values of process variables; the operators wish to anticipate off-normal conditions (Iosif [19], De Keyser [9]). Moreover, they have to divide their attention over hundreds of process variables, so unavoidably intervals between looking at a given variable are rather long, at least in the average.

For slow processes, prediction is greatly aided by displaying historical values. This is in line with the operators' wish to have trend recording for important variables, and not just only indicators.

Thus far, we have only obtained an overall picture of operator activities related to general inspection (see Table 1). Is it possible to develop more detailed models of operator behavior?

This usually requires laboratory experiments with human subjects, under controlled conditions (Sheridan c.s. [50–52], Rouse [45]). Senders c.s. [48, 5] have investigated how human beings divide their attention over several parallel displays. They have used queueing theory for modelling this scanning behavior. Further, Rouse [46] and van Heusden [18] have studied human prediction of time series shown on visual displays. Autoregressive-moving average type models are found to be useful here.

Of course, much work still remains to be done. In particular, it seems worth-while to integrate scanning and prediction behavior into an overall

Table 2. Operator Activities in Electric Power Plant (van Droffelaar [11]).

activity	morning shift	evening shift	night sift	start of shift	end of shift
monitoring	37	39	34	34	32
adjustment	6	5	2	4	5
external communication	2	4	2	1	1
internal communication	29	16	6	23	16
reporting	9	10	11	10	12
other	17	26	45	28	34
	100	100	100	100	100

model of process state observation. Questions like: Can the process state as perceived by the operator be expressed in terms of dynamic state equations, and how does the number of elements in the state vector depend on the number of variables, type of displays, mental stress, are of interest.

DISPLAYING THE "PROCESS STATE"

The great attention operators pay to the process state leads to the design problem: How to display the process state?

Conventional instrumentation, e.g. miniature indicators on consoles, offers an easily surveyable display of, at least, the present values of process variables. Human factor (ergonomic) data about the advantage of lining-up pointers have led to easily scannable configurations, e.g. rows of vertical indicators with aligned setpoints.

Now we have entered the era of the computerized visual display unit (V.D.U.), which offers great flexibility to the control engineer, if, at least, the manufacturer has provided a good software package for picture generation.

Thus far, V.D.U.'s usually are equipped with C.R.T.-screens. Much is known about the visual characteristics of these devices (see, e.g. Kraiss [25], Rijnsdorp and Rouse [41]), such as optimal foreground/background/ environmental luminances (Leebeek [28]), size and shape of characters, limitations in using color (Kraiss [25], Bindewald [4]). However, not much research has been done thus far on the structuring of information presentation, and the relevant man-computer dialogues. Consequently, the designer is left to his own intuition.

Dallimonti [7] has worked out a process overview, which shows the present values of up to 288 process variables on the same screen picture. More details can be seen by selecting process unit, group, or detail displays. Other possibilities for distribution of information over various hierarchies of screen pictures are offered by other manufacturers.

This situation leads to a number of questions for research, such as: what is the optimum number of levels in the display hierarchy? How should pictures be selected? What is the maximum amount of information to be shown in one screen picture? (certainly less than presented by many manufacturers at exhibitions!) Can trend information be integrated into process overviews? How should color be used?

We are engaged in a research program for finding answers to these types of questions. Relatively simple experiments are done in the laboratory, with human subjects sitting in front of a C.R.T., on which various alternative pictures are shown. They have to perform elementary tasks, such as: what is the value of variable X? How many variables are off-normal? Which variable is off-normal? Their performance is expressed in terms of error frequencies, and, in the case of man-paced experiments, also exposure times (Umbers [54] has used a similar method for comparing some commercially applied picture lay-outs).

Verhagen [57] has found that bar graphs are inferior to stroke graphs (see Fig. 2), although with the latter somewhat more errors were made in variable identification.

Further, he has compared flow diagrams and tables for overviewing process variables. Flow diagrams proved to be better for identifying the variable, and tables are more suitable for finding off-normal values. This is typical of many results from these types of experiments: a given structure is superior for one task, but can be inferior for another one. Hence it is imperative to make a careful analysis of the operators' tasks (see previous sections) to be able to make the best choice in a given application.

In another project, Verhagen [56] has compared two ways of presenting the process line-up: in a geographic and in a systematic lay-out. Here the human subjects were the process operators who have to work with the system. To their surprise they discovered that they did equally well with both alternatives, although they had much experience with the geographic lay-out and no experience with the systematic one. An interesting case of user participation in design! (see also Mumford [32]).

HANDLING OFF-NORMAL CONDITIONS

In practice, the handling of off-normal conditions usually takes the form of four "lines of defence" against emergencies:

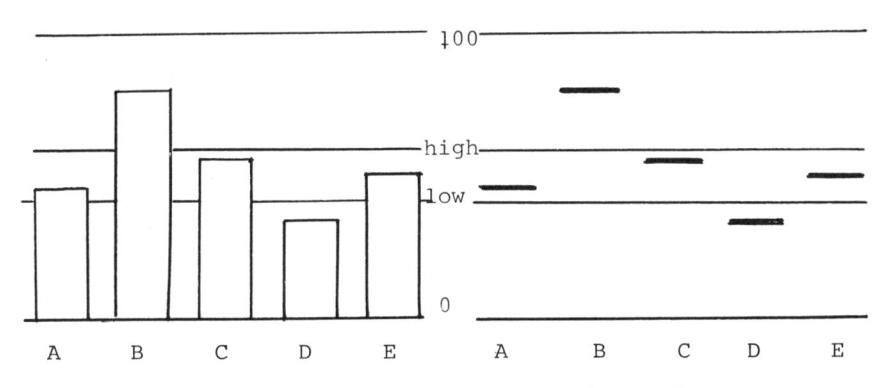

Figure 2. Bar Graphs versus stroke graphs.

- the automatic control system tries to keep process variables near to desired values and away from critical limit values.
- If this is not successful, or when equipment stops functioning, an auditive/visual alarm system warns the operator.
- If the operator does not take adequate action, a trip-out puts the process section out of operation, or brings it to a safe stand-by condition.
- If the trip-out does not function, the operator has a last chance to avoid an emergency condition.

Note that "machines" and "men" alternate; process availability and safety critically depend on human skill and reliability! At first glance the set-up seems to offer reasonable safeguards against emergency conditions. However, investigations have shown a number of difficulties and problems.

Kortlandt and Kragt [23] have analyzed the operators' attitude towards alarms in a chemical plant. They found that the majority of alarms are completely expected, in fact many alarms are the *result* of operator actions instead of being the *cause*. They therefore propose to speak about message systems instead of alarm systems.

During process upsets, elaborate alarm systems generate alarm messages at such a rate that the operator is unable to cope (a dramatic example is the incident at Three Mile Island). Evidently, a clear distinction should be made between serious alarms and "Boolean" messages (see also De Keyser [9]). For instance: different sounds (unpleasant-pleasant), different colors (red-yellow). Further, the maximum frequency of serious alarms is to be kept below the human mental capacity for detection, identification and interpretation, e.g. by reducing alarm clusters to single messages. It goes without saying that these ideas require human factors research.

Not much will be said here about modelling human errors because this is the subject of Senders' paper [49]. We only refer to the investigation by van Gelder [17], who has analyzed operator errors in running a (simulated) batch-type process. He did not find a significant influence of instruction thoroughness on ultimate performance, but noticed a significant increase in human error rate after the human subjects were told that equipment breakdown *could* occur. Stress induces human error, a well-known result from other investigations, e.g. into the effects of poor compatibility of displays and controls.

After an off-normal condition has been identified, the cause has to be determined. In practice, the operator is faced with a number of possibilities:

- the cause is external to the process, e.g. an upset in the feed or in a utility;
- there is an internal cause, e.g. an overload condition;
- a piece of process equipment has broken down;
- there really is no off-normal condition; the alarm is a "false" one.

Good operators try to obtain redundant information (De Keyser [9]) in order to avoid making the wrong choice. The latter can be very dangerous, as it is difficult for human beings to restart from scratch, especially when they are under stress.

Detailed fault diagnosis has been investigated by Rouse c.s. [45] via rather abstract problem-solving experiments in the laboratory. Somewhat surprisingly, they have found that the results of these experiments are representative of actual trouble-shooting skills, e.g. in repairing automobiles.

SEQUENCE CONTROL

In the previous section, reference has been made to operator errors in running a batch process. A design problem is: how to show complicated process line-ups on low-resolution standard C.R.T.-screens (high-resolution screens are too expensive for many small-scale plants).

In practice, many switching operations are still being done by roving operators in the plant. Riedel [34] has noticed that experienced and responsible operators skip steps in procedures, if these steps take much effort and time (For instance: checking an indicator high-up during a shut-down sequence mostly run at ground floor level). Evidently it is important to allocate each task cluster to a coherent work-place with sufficient surveyability and reachability.

REGULATORY CONTROL

There is a great difference between human control of fast and of slow processes. The former is done largely at a subconscious level, and it can nicely be represented by the linear/quadratic optimal observer-controller (Kleinman c.s. [20]). In the latter case behavior is more of a cognitive nature, hence it can be represented by decision diagrams, at least qualitatively (Bainbridge [2]).

Veldhuyzen [55] has tried to find a control model for manual steering of large ships. White [58] is working on the verification of the separation theorem (in linear/quadratic optimal control theory) to human control of a (simulated) utility plant.

In processing plants, operators often have to interfere with automatic regulatory control. This is not too cumbersome with single-loop controllers, as it simply means putting one on manual during a certain period of time. However, advanced multivariable control systems are wrecked if the operator decides to take over one or more of their degrees of freedom for control. This could be circumvented by implementation of more flexible algorithms, and by computer messages to the operator in order to explain the consequences of his (her) action (Bibby c.s. [3]). An additional requirement is to explain what the algorithm actually is doing (multivariable algorithms are too complicated for human understanding under real-time conditions).

QUALITY CONTROL

Quality control in the process industries is a rather neglected area for research. In practice, operators have great difficulty to run close to specification, hence they prefer to maintain broad margins between the *average* quality and the specification limit. The economic consequences can be quite substantial!

The problem is that accurate quality information often has to come from the laboratory, hence it is slow and expensive. There is much to be said for locating the laboratory next to the central control room, and to install sample loops from the process to this central location.

Kok and van Wijk [21] have investigated the problem of manual quality control for a strongly simplified process model. This consists of two partially correlated first-order systems. One is continuously on display, the other one (representing the actual product quality) can only be sampled from time to time, and produces a new value with delay.

They have modelled human behavior by the optimal state estimator-controller. Each time a new quality value becomes available, the state-estimator is backtracked to the time when the sample was taken, corrected, and run forwards to real-time.

A problem with these types of models is the description of the control actions: a wide variety of control patterns yield near-optimal results.

De Keyser [9] has studied the actual performance of operators in a pelletizing process for the steel industry. From interviews she discovered a difference between "good" and "bad" operators: "good" operators sample a broad range of process conditions, both in the control room and in the process equipment. Moreover they pay most attention to the variables in the initial process section, in order to correct upsets before they influence final product quality. By means of a simulation she also discovered that "good" operators have a rich repertoire of subroutines (See Fig. 1) for process correction.

MODE SETTING

In the preceeding section we have already touched upon differences between operators in coping with slow multivariable processes: they have a preference for certain control variables and they choose different control patterns.

On top of this, there seems to be a fundamental difference between computer algorithms and human behavior for optimizing process operation: Human beings tend to take one control variable at-a-time, and then wait to see what happens. In contrast, computer algorithms generate a stream of small changes in many control variables simultaneously. As a result, the operator tends to refuse the computer output. Either he will not implement it (in the case of open-loop control), or he will not like to have the computer adjust the process automatically. This is especially difficult in cases with critical constraints, where only the *combination* of small corrections keeps the process in good shape. Is constraint control here a way out (Rijnsdorp and Maarleveld [31, 40], Roffel and Fontein [43])?

OPERATOR WORKPLACES

Thus far we have discussed important tasks pertaining to the process operator team. Now a human factors problem is: How to design work places where the operators (and other personnel) can perform their tasks in the best possible way.

We have learnt that this can only be realized together with work organization design. For instance, in a certain food processing plant (Lenior c.s. [30]), each operator had to pay much attention to the process equipment under his responsibility. This favors decentralization, e.g. by the installation of local sound-insulated one-person control rooms. The actual set-up, however, had one control room with one V.D.U.-system serviced by a central operator.

An observation study revealed that the central operator really was nothing more than a "slave" for the other operators, who came to him to ask for the values of temperatures, to let him open a valve, etc. Inevitably, this indirect communication between roving operators and central operator caused delays, errors, frustration, and friction. Consequently, for a new plant of the same type, we advized to do away with the central operator function (sometimes human factors analysis results in direct cost reduction!), and to give each roving operator immediate access to displays and controls in the central control room.

In this exercise, the choice between centralization and decentralization has been broken into several dimensions (see Table 3). Note that a decentralized organization can very well be combined with a centralized workplace.

The lay-out of work places, in particular of control rooms, is also influenced by the organization. For instance, if people work in a few groups, the control room can be arranged in a number of working areas with a careful choice intermediate between dividing walls and no dividing walls, in line with the desired balance between group privacy and possibilities for intergroup communication.

A recent redesign of a control room for the oil movement and storage section of a refinery (Verhagen [56]) has resulted in a central area for the central operators (see Fig. 3), a work place for the roving operators, and a separate section for the barge captains, who have to communicate with one of the central operators but should not be allowed to walk throughout the whole control room. In contrast with the original design, the two operators face in the same direction, so it is easier for them to assist each other.

In this example, as well as in many other cases of lay-out design (Åberg [1], Cornelissen-Engels [6], Piso and Kragt [33]), a simple mock-up on actual scale has been built to obtain the suggestions and acceptance from the users.

On this topic, no further research is needed; it simply is a matter of doing. This is not completely true for the way work spaces affect people,

Table 3. Aspects of Centralization and Decentralization of Process Supervision.

aspects	centralized	decentralized
attention for process equipment		x
physical environment (noise)	x	
social contacts	x	
information display		x
responsibility for process operation		x
possibility for mutual assistance	x	

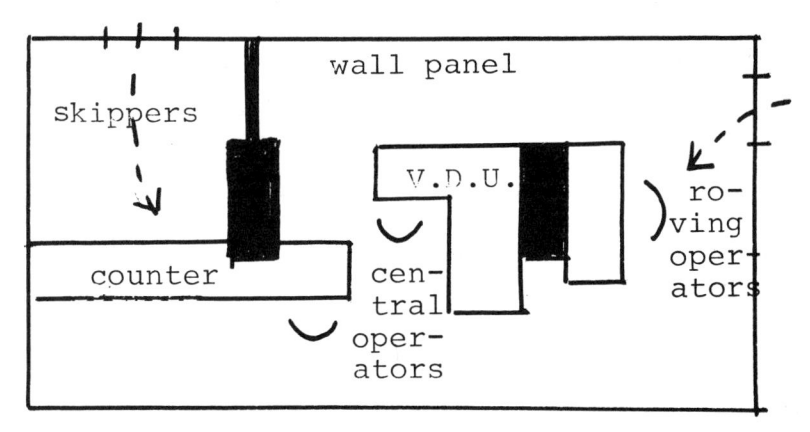

Figure 3. Redesigned Central Control Room for Oil Movement and Storage Facilities in an Oil Refinery.

where more research work is desirable to back up the intuitive way in which lay-outs are now being chosen.

PLANNING, SCHEDULING, AND REPORTING

We have now arrived at both ends of Table 1. They have in common that people outside operations are involved.

It is becoming attractive, both technically and economically, to link individual computers in various departments by digital transmisions. In this way, information has to be coded only once, which reduces (human) errors, saves person/hours, and speeds up communication within the enterprise. However, a number of difficulties can be expected too:

- Relations between people cannot be completely put into digital information streams. For instance, the give-and-take in reaching agreement on a total plant schedule requires a rather informal contact between people. A conference like this one is another example!
- Digital information streams tend to cross departmental boundaries. This can cause many conflicts, unless organization and computer networks are brought into harmony. For instance, in oil refineries there are strong needs for horizontal communication between maintenance, operations, oil movement and storage, buying and marketing. Matrix organizations seem a good solution, but in practice it is difficult to reach a satisfactory balance between the horizontal and vertical dimension (Langstraat and Roggema [27]).

Maybe computer networks can help out, by provided improved horizontal communication between the various departments (Rijnsdorp [38]).

DESIGN METHODOLOGY

For design, the various operational functions discussed in the previous sections have to be arranged into a logical sequence. We shall not go into details here, as this has already been done elsewhere (Rijnsdorp [36]); just a few general remarks:

Most designs are squeezed by tight deadlines, hence the multidisciplinary cooperation required for dealing with psychological/social factors is liable to suffer. Therefore it makes sense to develop general concepts ahead of time, not tied to a given design (Rijnsdorp and Wirstad [42], Rijnsdorp [39]). These concepts enable a "flying start" when a new system has to be designed.

We believe the following concepts are useful:
- automation concept
- work organization concept
- (de)centralization concept
- user participation concept
- training concept.

The interrelations between these concepts are shown in Fig. 4.

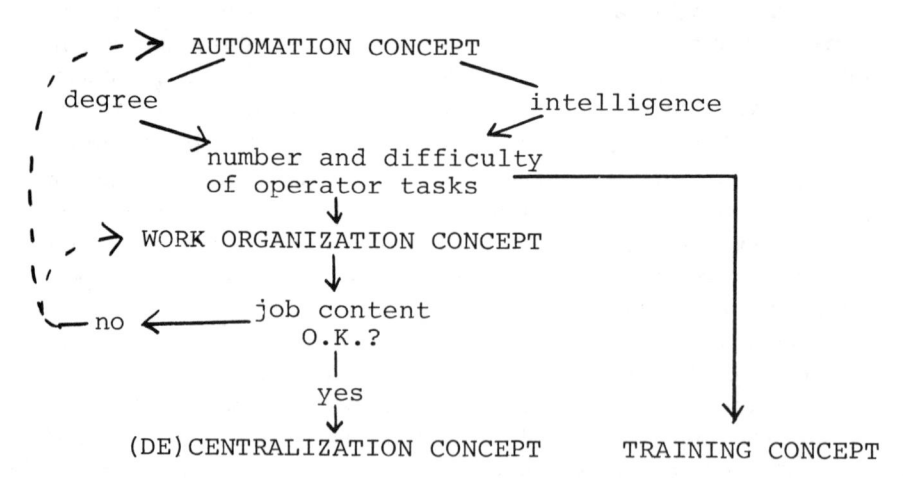

Figure 4. Concepts for dealing with Human Factors for the Design of Operation Systems.

The automation concept deals with the degree and the intelligence of automation. The degree of automation indicates how much of the operational functions (see Table 1) is allocated to the "machine" side, hence how much is *not* allocated to the operators and other groups of personnel. It can be worked out in terms of a profile in Table I (Rijnsdorp [35]).

Process computers offer the possibility to raise the intelligence of automation, e.g. by incorporation of process models, adaptive methods, or self-learning techniques. This will also affect the operators' job, as they have to understand what the computer is doing, and to critically evaluate its output.

Degree and intelligence of automation together determine the number and the difficulty of tasks for operators. Now there are many different ways to distribute these tasks within the operator team. A sensible condition is to ensure that each operator job has sufficient content (see, for instance, Davis and Taylor [8]).

For very highly automated processes, such as electric power plants, the operators' job is just to wait for malfunctioning. The resulting combination of 99% boredom and 1% terror is certainly not satisfying for people, nor suitable for adequate performance (van Droffelaar [11], Ekkers c.s. [14]).

The (de)centralization concept has a number of dimensions (see Table 3). Some of these depend on the previous concepts; other ones, e.g. the geographical one, require decisions.

The participation and training concepts refer to the design and implementation process itself. In some countries, legislation sets conditions for user participation (Docherty [10]; in other countries it is up to management to decide about it (Mumford [32]). Training for modern automation technology is a critical issue, particularly for older people (Rogers [44]).

It is a pity and a waste of investment to lose the skills and experience of older operators and other personnel by paying insufficient attention to training.

Of course, much R&D and application work has still to be done before human factors have been satisfactorily integrated into process automation design and implementation.

CONCLUSIONS

More R & D is needed on:
- Manual control of slow processes, in particular quality control;
- Operator acceptance of multivariable control algorithms;

- The differences between computer algorithms and operator behavior in correcting processes;
- The effects of work spaces on people;
- The meaning of the perceived "process state";
- Structuring of process information on V.D.U.'s, with the corresponding man-computer dialogues.

In automation system design, the following human factors aspects are important:

- 100% automation is a fiction;
- 99% automation results in unsatisfactory jobs and human performance;
- The development of general concepts, not tied to specific designs;
- Work place and man-machine interface design is intimately linked with work organization design;
- Each human task cluster requires a coherent work place;
- (De)centralization has many dimensions;
- Alarms should be used sparingly; they should be clearly distinguishable from "Boolean messages";
- Computer networks and enterprise organization have to be brought into harmony;
- User participation not only favors user acceptance, but it also yields valuable design information;
- Training people, especially older people, for automation technology is a critical issue.

REFERENCES

[1] Åberg, U., Bengtson, L., and Veibäck, T., Appl. Ergonomics, 5, No. 3, 161–166 (1974).
[2] Bainbridge, L., in (13), 146–158.
[3] Bibby, K. S., Margulies, F., Rijnsdorp, J. E., Withers, R. M. J., Man's Role in Control Systems, Proc. Sixth IFAC Congress Boston (1975).
[4] Bindewald, K., Advantages and Limitations of New Displays in Man-Machine Communication (in German), Interkama Congress, Springer, F. R. G. (1980).
[5] Carbonell, J. R., Ward, J. L., Senders, J. W., IEEE Tr. MMS-9, No. 3, 82–87 (1968).
[6] Cornelissen-Engels, M. P. N., Bac. Thesis, Twente Univ. of Technology, Netherl. (1974).
[7] Dallimonti, R., in (22).
[8] Davis, L. E., and Taylor, J. C. (eds.), Design of Jobs, Penguin (1972).
[9] De Keyser, V., Human Behavior in Supervising and Controlling Automated Systems (in French), presentation for IFAP (Rome, Italy) (1980).
[10] Docherty, P., Paper 6121, Ekonomiska Forskningsinstitutet vid Handelshögskolan i Stockholm, Sweden (1978).

[11] Droffelaar, H. van, in Schuh, P., and Sprague, P. A. (eds), Productivity and Man, RKW, F. R. G. (1975).
[12] Edwards, E., and Lees, F. P., Man and Computer in Process Control, Inst. Chem. Engrs. U. K. (1973).
[13] Edwards, E., and Lees, F. P. (eds), The Human Operator in Process Control, Taylor & Francis (1974).
[14] Ekkers, C., c.s., Report Dutch Institute for Preventive Health, Leiden, Netherl. (1980).
[15] Foss, A. S., A. I. Ch. E. J., 19, 209–214 (1973).
[16] Foss, A. S., and Denn, M. M. (eds), Chemical Process Control, A. I. Ch. E. J. Symp. Ser., 72, No. 159, 1-232 (1976).
[17] Gelder, K. van, IEEE Tr. R-29, No. 3, 258–264 (1980).
[18] Heusden, A. R. van, IEEE Tr. SMC-10, No. 1, 38–43 (1980).
[19] Iosif, Gh., Ene, P., Rev. Roum. Sciences Sociales-Serie de Psychologie, 19, 179–197 (1975) (in French).
[20] Kleinman, D. L., Baron, S., Levison, W. H., IEEE Tr., AC-16, No. 12, 824–832 (1971).
[21] Kok, J. J., and Wijk, R. A. van, A Model of the Human Supervisor, Proc. 13th Annual Manual, M.I.T., 210–216 (1977).
[22] Kompass, E. J., Williams, T. J. (eds), Man-Machine Interfaces for Industrial Control, Control Eng. (1980).
[23] Kortlandt, D., and Kragt, H., Process Alarm Systems as a Monitoring Tool for the Operator, 3rd Int'l Symp. on Loss Prev. and Safety Prom. in the Proc. Ind., Basle, Switzerland (1980).
[24] Kragt, H. and Landeweerd, J. A., in (13) 135–145.
[25] Kraiss, K-F, in (26), 85–147.
[26] Kraiss, K-F., and Moraal, J. Introduction to Human Engineering, TÜV, Rheinland, F. R. G. (1976).
[27] Langstraat, A., and Roggema, J., From Vertical to Horizontal, (in Dutch), Intermediair, 9, Feb. 29, 43–51 (1980).
[28] Leebeek, H. J., Journal A (Belgium), 19, No. 3, 153–159 (1978).
[29] Lenior, T. M. J., and Rijnsdorp, J. E., Journal A (Belgium), 19, No. 3, 173–179 (1978).
[30] Lenior, T. M. J., Rijnsdorp, J. E., Verhagen, L. H. J. M., Ergonomics 23, No. 8, 741–749 (1980).
[31] Maarleveld, A., Rijnsdorp, J. E., Automatica, 6, 51–58 (1969).
[32] Mumford, E., and Henshall, D. A., Participative Approach to Computer System Design, Assoc. Business Press (1979).
[33] Piso, E., and Kragt, H., research report Eindhoven Univ. of Technology, Netherl. (1980).
[34] Riedel, personal communication.
[35] Rijnsdorp, J. E., Levels of Automation, a Multidimensional Approach, Symp. on Production Res., Amsterdam (1979).
[36] Rijnsdorp, J. E., in (22), 1–12.
[37] Rijnsdorp, J. E., Journal A (Belgium), 21 No. 4, 177–182 (1980).
[38] Rijnsdorp, J. E., Computer Networks and Organization in the Process Industry, IFAC Workshop on the Impact of Automatic Control and Information Systems on Organization in the '80's, Youngstown, Ohio (1980).
[39] Rijnsdorp, J. E., Ergonomics in the Design of Man/Machine Systems for Industrial Production (in Dutch), report Twente Univ. of Technology (1980).
[40] Rijnsdorp, J. E., and Maarleveld, A., Optimising Distillation by Constraint Control, I. Chem. Eng. Symp. Ser. No. 32 (U.K.), 33–38 (1969).

[41] Rijnsdorp, J. E., and Rouse, W. B., in Van Nauta Lemke, H. R., and Verbruggen, H. B. (eds), Digital Computer Applications to Process Control, North-Holland (1977).
[42] Rijnsdorp, J. E., and Wirstad, J., Management and the Design of Process Supervision, Symp. Design '79, Univ. of Aston, Birmingham, U.K. (1979).
[43] Roffel, B., and Fontein, H. J., Chem. Eng. Sci., 34, 1007–1018, (1979).
[44] Rogers, J., Adults Learning, Milton Keynes, U.K., The Open Univ. Press.
[45] Rouse, W. B., Systems Engineering Models of Human-Machine Interaction, North-Holland (1980).
[46] Rouse, W. B., IEEE Tr. SMC-3, No. 5, 473–477 (1973).
[47] Sargent, R. W. H., Optimal Process Control, in Proc. 6th IFAC Congress, Boston (1975).
[48] Senders, J. W., and Posner, M. J. M., in (52).
[49] Senders, J. W., Human Error and Human Reliability in Process Control, this Conference.
[50] Sheridan, T. B., and Ferrell, W. R., Man-Machine Systems, Information, Control, and Decision Models of Human Performance, MIT-Press (1974).
[51] Sheridan, T. B., in (22), 35–50.
[52] Sheridan, T. B., and Johannsen, G. (eds), Monitoring Behavior and Supervisory Control, Plenum Press (1976).
[53] Stockbridge, H. C. W., Human Engineering Handbooks and Journals: Overview and Use, in (26).
[54] Umbers, I. G., in (22).
[55] Veldhuyzen, W., Ph.D.-Thesis, Delft Univ. of Technology, Netherlands (1976).
[56] Verhagen, L. H. J. M., Ergonomic Aspects of Centralized Supervision and Control of Oil Movement and Storage, ASSOPO '80, Trondheim, Norway (1980).
[57] Verhagen, l. H. J. M., Appl. Ergonomics 12.1, 39–45 (1981).
[58] White, T., Private Communication.

HUMAN ERROR AND HUMAN RELIABILITY IN PROCESS CONTROL

John W. Senders
Dept. of Industrial Engineering
University of Toronto
Toronto M5S 1A4 Canada

ABSTRACT

The fatal flaw in complex systems design is the low level of understanding of human error generating processes and the consequent uncertainties of estimating the reliability of a man-machine system. Human error arises from internal and external sources and the remedies are quite different. Traditionally the processes of selection and training have served to keep the endogenous errors at relatively low rates. As a result, the exogenous errors, those which stem from the physical and informational design of the system now loom larger than ever before. Large stores of data and rules for their application exist which have not generally been applied to complex continuous control process plants. Great improvements in overall reliability and productivity can be expected to follow from such application. However, even with the best selection, training, motivation and man-machine interface design, there will be a residuum of errors which will plague the system designer and operator. Whether there is any hope that these can be eliminated is a complex and intriguing question.

INTRODUCTION

Before one can talk about error, one must be able to define it. There has been much comment in recent years, not only about human error but also about machine error, computer error, system error, and possibly at times even divine error. If one assumes merely that any failure of a system to perform is an error, and if one cares not what the genesis of the error was, then the common notion that an error is anything which isn't what you wanted it to be is acceptable. However, if one thinks more about this idea, one finds that there has to be a difference between an intention and an outcome for an error to exist, and the error must be succeeded by a correct outcome if it is to be distinguished from a "breakdown."

When one deals with the fine structure of behavior and performance, whether of man or machine, it is the moment to moment deviation from what is intended that constitutes the series of errors that characterize the system. From the point of view of the doer of any particular act or event an error is the difference between intention and accomplishment. If I intend to accomplish A, and instead accomplish B, the error exists. Suppose however that I should have intended to accomplish A, but did intend to accomplish B, then the error is not so much one of action as one of intention, and still further, if it were the case that I would have intended to accomplish A instead of B (had I had the information needed) then it is clearly an error of hypothetical intention, rather than of action that we are dealing with. Briefly, one can imagine human error to be one of three kinds: a difference between intention and action—an execution error; a difference between what should have been intended and action—a process error; a difference between what would have been intended and action—an input error. This definition bars non-human, or non-animal, errors unless one can invest a machine with an intention. A machine may fail, but this is not the same thing as making an error. If a system cannot be said to have intention, then it cannot be said to make an error, instead the person who designed the system may very well have made an error. And so it continues. The end of all this, of course, is that *things fail, people err*.

We are generally instructed that the mean time to first failure of human beings is three score and ten years. This kind of failure is too catastrophic to be of much use in trying to assay the reliability of complex man-machine systems. Is it appropriate to attempt a reliability analysis of a human operator and to apply standard reliability concepts?

The notion of mean time to failure, mean time between failures, or mean time to first failure of a human operator is conceptually founded on the idea of hordes of identical clones of the human being in question plugged in to their tasks like light bulbs to a source of current, and performing until they make an error or burn out as a light bulb burns out. The notion would be then that the clones are identical except for small, random variations in their internal structures and wiring diagrams, and that from the data gathered by examining the horde, one can predict the probabilities of error for an individual engaged in the task of being a light bulb. One should note, however, that implicit in this argument is no notion that it is possible to predict when an error will occur or what kind of error it will be, and it is these two characteristics of error which are so important in trying to calculate the adequacy of a large system.

If we look at the gross function of human beings—physiological survival—we find the typical bathtub curve. The curve shows high infant mortality, drops to a fairly steady rate which is reflected, to a degree, in standard actuarial tables, and culminates in an upturn as senility and terminal decline set in. Do we have a parallel in terms of human error to this curve of physiological decay? There is at least a suggestion in the data from 1931 of Bills [1] that relatively error-free human performance suddenly becomes more error-full; that this onset of relatively frequent errors is followed by complete cessation of behavior; that the break in behavior, called "blocking" by Bills, terminates; behavior begins again with frequent errors, and then the errors become infrequent again. Thus between blocks we have the bathtub curve of performance. The difference is that the intervals between blocks are measured in matters of seconds rather than of years. The mechanism underlying this phenomenon is unknown even today, nor has there been adequate analysis of the intervals between blocks to see whether these themselves have any strong periodic component which would suggest an internal clock mechanism driving the error-related blocking phenomenon.

The methodological difficulties which must be faced by anybody who attempts to do work on human error rates are enormous. In particular, one man is not a clone of another, and it is exceedingly unlikely that we will ever be in a situation where we can assume that the ergodic hypothesis holds; that the data taken across an ensemble of human operators will tell us anything significant about any one of them. Furthermore, people are markedly non-stationary, so that even in the very short-term, measured in minutes from one measurement to another, it is unlikely that error probabilities remain constant. So we face a formidable task if what we wish to do is to understand the underlying mechanism of error.

ERROR REDUCTION

We know how to reduce errors, even if we do not understand why they occur. Before exploring this, it is important to classify errors according to their source. *Endogenous errors* arise from factors internal to the human operator; *exogenous errors* arise from factors external to the human operator. The endogenous errors are those that we attempt to eliminate or reduce by processes of selection of personnel, training, motivation, and so on. The exogenous errors we attack in totally different ways: we redesign systems; we automate; we de-automate; we human engineer; in general we manipulate the nature of the man-machine interface in such a way as to reduce the errors that would be performed on the task components in the hopes that the errors that will be performed on the system as a whole will also be reduced.

For the most part these measures work. It is clearly beneficial to select people who can grasp the controls, see the displays, read the numbers, understand the language, and so on. It is clearly beneficial to tell these people what to do, i.e., to train them in the system. It is clearly beneficial to design the displays and controls and the panel layout in such a way that they are within the grasp of one or a team of human beings. Beyond that, however, the simplistic assumption of linearity that tells us that if we make all the parts good, the whole will be good, is unproven. It is not necessarily the case that one can build every system like the "Deacon's Wonderful One-horse Shay" and expect the whole thing to run until all the parts collapse at the same time. (The Deacon, you will recall, understanding that chains broke at the weakest link, resolved to have no weakest link.)

Our ignorance gets in our way again. To reduce production errors you train the people better. It is true that improved training will result in a reduction of errors, but because we do not understand the mechanism which generated the errors in the first place, we do not know whether the training that we engage in, is in fact the most economic, the most nearly optimum of all the possible ways of training that we could imagine.

The same applies for selection. For the most part we have a number of hypothetical, possibly even mythical, relationships between selection criteria and presumed capability on the job. It is rare that such relationships are subjected to statistical validation, and when they are, it is frequently the case that they are found to be wanting.

In the matter of exogenous errors, we feel a little more comfortable because the matters at issue are more of a physical than of a psychological and sociological nature. We know that if we make a display bigger, it will be easier to see. (By easier to see we mean that fewer errors will be made for any given amount of "seeing" time.) Even here, our understanding of the mechanism of error and the process by which it is eliminated is deficient.

In summary, we know how to reduce both endogenous and exogenous errors without understanding why they occur or even whether the error production mechanism is the same for the various kinds of errors. (We have not just two kinds of errors, we have six kinds of errors: the three arising from the classification of intention, and the two which arise from consideration of sources).

Is it possible to understand why errors occurs, to know the mechanism of an error, to predict when an error will occur, and the form it will take? This may be more a philosophic than an engineering or psycho-physiological problem. If errors are quantal in nature, truly indeterminate, then we cannot hope to gain further understanding of exactly when errors will occur and exactly what form they will take. On the other hand, we may become more and more sure of the statistical basis of our predictive efforts. Fortunately, that the error rates are strongly subject to influences like training (if they are endogenous errors) and design (if they are exogenous) shows that there is a hope that prediction of time and form is not theoretically impossible.

HUMAN OPERATOR QUALITY CONTROL

What remedies are there? Endogenous error reduction probably requires the application of a quality-control system applied to human performance and skill: the ability to do the right thing at the right time.

We must select for talent (the innate potential for performance which human beings bring to a situation). We try ordinarily to select people for a particular task, for example, that of process control operator. The selection procedure is, or should be, aimed at finding those with a talent both to perform well ab initio and to absorb training and experience in order to develop high levels of skill. It should also select for characteristics of personality which will result in dedication to the task and, perhaps, pride in accomplishing it. Following selection, there is training, and after that, on-the-job experience. Missing usually are frequent objective assessment and, as needed, maintenance of skill.

Most of the task facing the human operator of a proces control room is routine. The low probability events are, obviously, rarely practiced. Yet it is frequently for these events that the operator is there. Routine can be automated and frequently is. The human operator, therefore, loses his competence in the same way that a pilot or a pianist loses competence in the absence of practice. The only way in which realistic estimates of man-machine system reliability can be had is to have an assessment procedure which provides quantitative data about probabilities of correct performance in critical situations. Such an assessment must involve a whole-task simulation with complete transactional data logging. Everything that flows in either direction through the man-machine interface should be held and be available to analysis. The same level of computer complexity which allows computer control will allow this to be done. Then, given that deficiencies are revealed, skill maintenance programs should be tailored to the needs of the individuals to bring the performance, in all its complexity, up to the desired level. All too often, we depend on supervisors' evaluations and these are unvalidated, at the very least, and have, as a rule, never been assessed for reliability either. The value of the investments placed under human operator control and supervision makes it imperative that improved quality control systems be established. The control component is accomplished by using the information generated by assessment to modify the selection and training systems in a negative feedback way to improve the results of assessment and the success of the training program. In time, then, the predictive capability of the selection 'filter' and of the training 'operator' will be improved as each is modified to achieve a performance criterion.

SYSTEM DESIGN ERRORS—
an unknown quantity

Systems design increases or decreases exogenous errors. We do not know much about the design process itself. That there are both good and bad designers is generally accepted. Yet it is frequently the case that we get camels instead of horses, and it may be that systems design has somewhat the same quality as committee design.

Meticulous investigation of the way in which things are designed may yield nothing, since the process, like many other "cognitive" functions, may very well be beyond analysis. However, like the study of performance error, the closer we can get to an understanding of the underlying mechanisms, the more likely it is that we will be able to deal with it on the higher level to suit our particular needs. The topic is unfortunately virtually unstudied and is beyond the scope of this paper.

REFERENCES

[1] Bills, Arthur G. Blocking: A New Principle of Mental Fatigue. *American Journal of Psychology*, 1931, Vol. 43, pp. 230–245.

THE SYSTEMS ENGINEERING CONTEXT OF PROCESS CONTROL AND MAN MACHINE SYSTEM DESIGN

W. L. Livingston, Chairman
ASME Process Computer Control
Technology Committee U.S.A.

ABSTRACT

Growing energy and regulatory considerations increase the importance of the process control system to safe, efficient, and available production. As process control systems correspondingly get larger and more complex, it is necessary to integrate the control development and design process in a more formalized, objective way that recognizes the contributions of the many established disciplines required for the work. The phased role of the various technologies involved that should precede and surround the basic process control man-machine system design are presented. Several important characteristics of a successful Human Factors man-machine program, as one of the technologies, are reviewed. The conclusion is reached that a step increase in the general field of process control technology effectiveness can be obtained by the timely application of a systems engineering methodology.

INTRODUCTION

Increasing energy costs and concerns for ensuring a safe environment are providing a major driving force to improve process operations in general and process control engineering in particular. The growing significance of control engineering requires more than ever that process controls be designed in context with an integrated systems engineering approach to the larger encompassing complex problem of process operations. As a case in point, the successful development of more sophisticated control algorithms and their implementation in on-line real time computers, as a means to achieve better operation, will occur in no other way.

SYSTEMS ENGINEERING DEFINED [1]

"The engineering management, direction, control, and technical effort applied to a total system to ascertain and maintain overall technical integrity and integration of that specific system as related to design configuration, reliability and performance."

Control engineering will be able to produce a great leap forward with its new contraptions and software when and only when control engineers practice a rational development methodology that systematically deals with content. They can know the control design is good and fits into the larger operations goals when the design, and especially the design development process, is auditable by a multi-disciplined team of peers against the best in available systems engineering [2].

When the technical basis for the control design can be provided and traced back up through a documented hierarchy of tasks, functions, allocations, and objectives; when the integration of work for all of the related inerfaces is documented (Table 0 typical), and when the plans for monitoring and evaluating effectiveness are set, the job is ready to join the community of real contributions to the profession [3].

The constraints to a realignment of typical control engineering practices to integrate with the appropriate systems engineering methodologies appear to be primarily psychological in nature (Table 1) [4]. Control engineering, along with other disciplines, often progresses deeply into design specifics without a commensurate description of the real operational problems, working assumptions, and value criteria to guide the design.

A simple test that can be done by any control engineer to verify the shaky basis on which he is expected to design controls is to bring up an

Table 0. Example Control Engineering Interfaces.

Organizations	Disciplines	Technologies
User-Operator	Mechanical Eng.	Environmental
Equipment Suppliers	Electrical Eng.	Human Factors
A-E	I & C Eng.	Reliability Engineering
Unions	Civil Eng.	Water Treatment
Regulation	Architecture	Structures
Standards	Biology	Licensing
Service/Technology Suppliers	Chemical Eng.	QA–QC
	Nuclear Eng.	Corrosion
	Fire Protection	Construction
	Applied Physics	Health Physics
	Applied Psychology	Purchasing
	Applied Mathematics	Simulation
	Systems Engineering	Software
	Business Admin.	Security
	Legal	Communications
	Computer Science	Management

Table 1. Psychological Constraints to an Attack on Complexity.

1. Instinctive Aversion (Ignore)
2. Lack of rewards for (and general peer understanding of) SE vs plethora of rewards for solution stampede and firefighting heroics.
3. Risk of outlaw label within discipline specialty vs guaranteed shelter for conventional response.
4. Requires extra thinking, effort, and new skills to learn.
5. Lack of qualified, knowledgeable, competent leadership.
6. Unfamiliar, judgemental, and uncomfortable. Perpetual problems of change. Cognitive dissonance.
7. Requires acceptance of SE structure, Participants must experience alienation.
8. Potentially revealing of real discipline and discipline interface limits.
9. Requires commitment to sustained involvement.

issue that requires real knowledge of the dynamics of how the process operates—especially leading to and during off-normal operating scenarios. One sure-fire subject is to propose reducing the information flow to the control room. The usual big secret to safeguard is that the process was engineered in fragments to unit physical design standards with only the slightest notions as to its operation. It is a common but well kept secret and a constraint that can, but not easily, be evaporated through systems engineering.

Systems engineering technology has now been sufficiently developed and has been successfully applied to enough wide-scope problems to establish utility (Table 2) [5]. From the perspective of economics, there is no evidence to suggest that a solution-dominated approach resulting in an endless minuet of backfits and risks to productivity is a less expensive project alternative [6].

Systems Engineering and Control Engineering

It should be recognized at the outset that the proper design for any extensive process operation is clearly a challenge in multi-objective systems engineering. Practically all of the attributes of a wide-scope, interdisciplinary, large-scale systems design problem exist in full measure. The Control Engineering components, as well as the process operations system it serves, rank high on the complexity scales of Table 3

Table 2. Essential Characteristics of the Practice of Systems Engineering.

1. Major Activities and their sequence
   a. Problem Definition
   b. Value System Design
   c. Alternative Solutions Assessment and Selection
   d. Design Development Engineering Testing and Evaluation–iterative.
2. Major SE Activities timely to project. SE starts at outset of new project.
3. Group Consensus—Decisions. System engineer creates structures for group processing.
4. All primary disciplines represented in group—all fully qualified, capable, and involved with continuity.
5. Competent, Experienced, Dedicated Systems Engineering leader conversant with technologies and methodologies.
6. Documentation for traceability.

Table 3. A Measure of Complexity.

Factor	Small	Medium	Large
Number of Simultaneous Objectives	1	7	9
Number of Solutions	1	4	9
Number of Dissimilar Subsystems	1	4	9
Number of Inter Relationships	1	4	9
Number of Simultaneous Constraints	1	4	9
Number of Disciplines, Sciences, Technologies, Professions, Vocations Involved in Problem and Solutions	1	4	9
Number of People Involved	1	4	7
Severity/Frequency of Consequences	1	4	7
Number of Alterables	1	4	7
Required Resources (Costs)	2	6	11
Exposure to Controversy	1	4	7
Time to Complete Project	1	4	7
Totals	13	53	100

Table 4. Elements of a Minimum Value System.

1. Stated Project objectives—Objective tree hierarchy and relationships among the objectives.
2. The constraints impacting on attaining the objectives.
3. Criteria for measuring attainment of the objectives including the sensors and indicators to describe future events.
4. Priorities on the objectives, in accordance with stated criteria, constraints, and consistent with the objectives tree.

[1]. The Nuclear Power Plant operations system design problem to many of us ranks over 100.

Assessment of the span of complexity in the problem at hand is vital. The same techniques that have been proven effective in solving simple problems are a certain driving force for failure in making progress on solving complex problems. The technology and discipline of system

Table 5.

ELEMENTS OF THE USE / USE STAGES	OBJECTIVES	WORK PRODUCT	RELATIONSHIP TO SOLUTIONS	KEY WORDS	TOOLS-MEANS
IA Problem/System Definition Synthesis/ Analysis	Hierarchal structuring of objectives/risks, system/problem elements and their relationships	Work breakdown structure (WBS) Objectives tree structure constraints, scenarios alterables	Solutions forbidden	Multi-objectives Analysis Intent Structure Element I/O Linkages Design functions Element interrelationships Design Criteria Risk assessment System environment	Nominal group technique of consensus evaluation Relevance tree Morpological box Matrix chain Partitioning System model-situation models Relation matrix Normative scenarios
IB Value System Design	Quantitative evaluation criteria, a set of preferedness and overall indices of worth	Performance parameters, their relative values and scoring functions	Solutions forbidden	Performance-consequences merit function Relevance criteria Scalar - valued criteria Priorities-ranking objectives Weighting	Risk Evaluation Nominal group technique of consensus evaluation Worth assessment Sensitivity Analysis Criterion hierarchy
II Alternative Solution Assessment/ Analysis and Selection	Systematically, quantitatively select design solutions	Numerical scores for alternative solutions Merits in hierarchal arrangement Assumptions Allocations	Solutions encouraged - standard - advanced - innovative	Performance measures Merit grading Parametric analysis worth connection Axiological components Value Judgements Consequence set Rationalized traceability	Interaction matrix (chained) Ranking matrix Decision tree Group participation Decision criteria Utility theory Functional allocation Simulation
III Solution Implementation Testing & Evaluation	Design and engineer selected solutions incorporating, performance measures for evaluation	Functioning design selection	Solution incorporated and evaluated	Specific design features Project portrayal Optimization Re-evaluation of value system Resource allocation Program planning	DELTA-Chart - Decisions, Events, Logic elements, Time, Activities Coordination Integration Flow charts Gantt charts CPM PERT Transitive scenarios Task allocation

engineering (SE) specifically addresses complexity—problems that one man cannot solve; problems in which there are only imperfect solutions rather than total solutions; wicked problems; fuzzy problems and so on while simultaneously accomodating man's aversion to complexity, especially under conditions of uncertainty and controversy.

The current status of systems engineering technology provides a reasonably complete methodology, proven tools and a very useful experience base from which to approach complex problems. Good systems are not created by the tool of SE but only by the practitioners using the tools. It can be confidently assumed there now exists several cases of successful system engineering application on processes that essentially are just like yours.

One of the interesting characteristics of the process operations system design problem is that some of the involved disciplines have adopted and adapted a SE approach tailored to their respective functions—primarily because they cannot practice the full applicable scope of their technology without it. An example will be described later [7, 8, 9, 10, 11, 12, 13].

The technology of SE is basically structured to deal with the infinity of relations among the elements of the problem in the context of the finiteness of man's comprehension. At the simplest level, the problem is separated into two parts—the system and its environment with complexity (organized or disorganized) a property of the system.

Complexity has a way of feeding on itself (Fig. 1). As the cmplexity of the problem causes many minds to be applied to it, a new complexity arises—that of sharing among these many minds the products of each other's thoughts and actions through inter-relating and distilling their group efforts into a format where they assume utility [1].

Group Efforts

The SE approach is concerned with dynamic problem analysis and partitioning of the problem into elements and their internal relations. System alterables, linkages, subjective components are all identified to support multi-disciplined group value development for decision making. In this process, language limitations, definitions, and syntax are directly confronted to develop value judgement criteria (Table 4).

Wide-scope problem solving must involve multi-disciplined groups, systematic judgements, and creative actions working within the constraint of some kind of structuring of the problems. There is always a strong need to bring the collective knowledge of specialists to bear on wide-scope problems over relatively brief periods of time. The synergistic effects possible through group interaction can be powerful in reaching or progressing towards solutions. However, group leadership, which cannot be a master of all disciplinary content, is a critical integrating, coordinat-

124   Chemical Process Control II

**THE IMPACT OF COMPLEXITY
ON THE CHANCES OF SUCCESS
OF THE CONVENTIONAL APPROACH TO DESIGN**

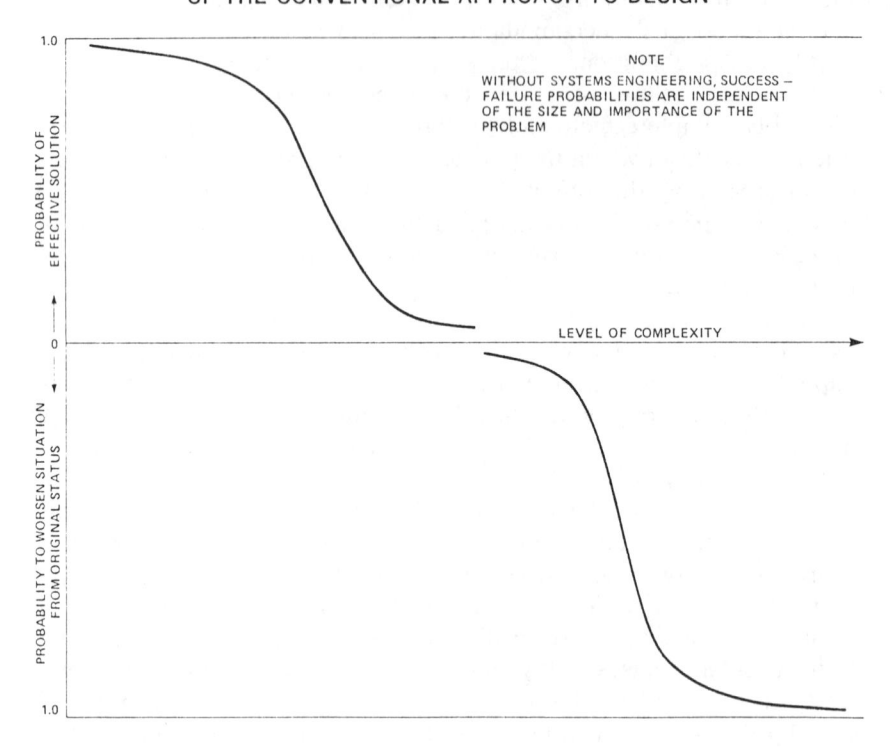

NOTE
WITHOUT SYSTEMS ENGINEERING, SUCCESS –
FAILURE PROBABILITIES ARE INDEPENDENT
OF THE SIZE AND IMPORTANCE OF THE
PROBLEM

PROBABILITY OF EFFECTIVE SOLUTION

PROBABILITY TO WORSEN SITUATION FROM ORIGINAL STATUS

LEVEL OF COMPLEXITY

Figure 1.

ing factor. The leader must know how to gently help the specialists react with some convergency in the midst of their diversity. The acceptance of structure and system discipline is the price specialists must pay in order to become involved in the major effort to solve a significant problem. SE methods help deal with the built-in tyranny to be creative and analytic at the same time. Effective systems engineering counter balances the widespread compulsion towards solutions with a strong organized emphasis on a timely identification of the correct specific problems to be solved [14].

Systems Engineering and Solutions in Stages

The traditional institutional driving force to prematurely bring in solutions is a major, persistent, treacherous inhibitor to SE for wide-scope problems. Once solutions are admitted, there is little chance in slowing the stampede for implementation. Retrofit costs are typically restrictive. If systems engineers are brought in after the design is made and there is only a small set of alterables, there is little useful SE that can be done—and this is precisely how complexity is typically averted. In view of this dominant propensity, the SE methodology is partitioned into three primary sequential stages.

The first stage is dedicated to developing a qualitative and structured understanding of the problems in an environment where solutions are deliberately and expressly prohibited from consideration as improper subjects. Within this first phase are the hierarchial structuring of the problem and its elements (systematics), values (Table 4), objectives, constraints, hazards, risks, consequences, benefits, criteria, system synthesis, system analysis linkages, partitioning, etc., all working with an initial conceptual system configuration framework.

The Unified Systems Engineering (USE) methodology logic is hardly a radical departure from "good sense." This first step is to "make sure the city is worth the siege" spending a little effort to systematically find out what the real problems are without the obscuring clutter of a cacophony of solutions confusing the vital task of incorporating specialized knowledge into a rational structure.

The second stage admits solutions, but in great quantities for evaluation against established criteria as design options. The conversion of identified functions and tasks into hardware/software specifics requires the highest level of interactive competence, knowledge and innovation. The functional allocations to men, machines, and software is of critical and long lasting significance.

There is little question that the persistent pursuit of SE to establish a rationalized traceability from the problem through to the solution generates a lot of documentation. There is no supportable systems approach without it, even disregarding the intrinsically high level of personnel turnover that would readily justify documentation anyway.

The second phase begins when the elements of the right problem, objectives/intent linkages, and the constraints are safely in their respective trees with quantitative assessment criteria. Now the gates of the dammed-up reservoir of solutions (alternatives) can be released for methodical, objective processing and allocation. Once the winning solutions have been identified, quantified and certified, most of the controversial SE work is done.

The second phase is the last realistic opportunity to control the major system design configuration to meet future needs. The third phase involves specifications, implementation, testing, and improving the chosen set of solutions via evaluation and feedback enhancements. Since this phase, except for the increase emphasis on frequent iterative evaluation, is basically the popular conventional engneering process, the related tasks are not discussed further.

Table 5 is an outline of the systems engineering discipline methodology for assaulting complex problems [1, 4, 5, 6].

One current example of ongoing systems engineering work is related to developing the national nuclear waste disposal program—certainly a major complex problem under conditions of controversy. Real progress on what do we stick where, with which, and by whom—was made only after a SE effort was instituted.

The SE Approach in Human Factors Application

Human Factors engineering is one of the professional sciences mentioned earlier intimately involved with process operations, control engineering and the design of man-machine systems. Because most of the Human Factors technology needs to be incorporated before system design details are established, Human Factors has been and is a strong supporter of a system engineering approach. This systems support is embodied in MIL SPEC MIL-H-46855 B (31 Jan 79) broadly describing human engineering requirements for military systems, equipment and facilities. MIL-H-46855 B has been clarified for program managers and users in considerable detail by Human Factors engineers [3, 14].

Table 6 is an outline of this HF established process clearly showing the close alliance with systems engineering. Once again, solutions are held at bay until group participation produces a structured development of the problem. Quantitative measures are then used with much interaction to assess alternatives and allocate functions before the solution stampede is unleashed.

A contrast of these proven system engineering (SE) methodologies with conventional control engineering practices is suggested in Fig. 2 and Table 7. Engineering disciplines such as Mechanical and Chemical engineering are far better equipped with standards and specifications to direct their activities without the benefit of SE than control engineering—which must interact with many more diverse groups depending upon the specifics of the process. It appears to be well within the long term interests of Control Engineering to take a leadership role similar to the Human Factors efforts in adopting a SE methodology to manage the complexity for their own work as well as influencing the project to do likewise. Better process operations will be the result.

SE and Control Engineering

Analysis of SE and HF procedures provides insight and support for establishing a more useful participative role for control engineering

Table 6.

ELEMENTS OF HE / STAGES HE	OBJECTIVES	WORK PRODUCT	RELATIONSHIP TO SOLUTIONS	KEY WORDS	TOOLS-MEANS
IA Problem/System Definition Synthesis/Analysis of the process baseline configuration	· Hierarchal structure of: – objectives/risks – system/problem elements and their relationships · Determine system requirements	Objectives tree structure Work breakdown structure Human Factors design criteria Operating Scenarios Operations Description Functional Flow Diagram	Solution forbidden	Operational goals Operational hazards Regulatory constraints Rationalized traceability Functional indentures Baseline configuration	Program objective – system operational requirement Functional block diagram Group participation System Safety Fault Trees Event Network
IB Value System Design	Human Factors Engineers participate in the group establishing the value system design providing their expertise and knowledge of human limitations in the determination of criteria and relative values. Functional Flow Diagrams show effects of allocation.				
II Alternative Solution Assessment & Analysis Solution Selection	· Convert concepts into system design criteria · Allocation · Systematically select design criteria	Decision/Action diagrams Action/information req. Function allocation Timeline analysis Flow process charts Designs evaluations Task allocation Task description Gross Task Analysis	Solution encouraged · Standard · Advanced · Innovative	Design bases-rationale Time critical tasks Human reliability Evaluation matrix Concept Formulation Baseline Monitoring Configuration Control Maintenance Tasks	Models, mock-ups, manikins Group participation Functional allocation screening worksheets Functional sequence diagram Computer Aids - CAD, CAPE, CAR, CGE Combiman, Cubits, HECAD, HOS "Fitts List" Simulation
III Solution Implementation and evaluation of initial designs Testing and evaluation (iterative)	· Design and engineer allocations to HE design criteria · Evaluation of designs in progressive iterations and performance testing	Operational Sequence Diagrams Task Allocation Task Description Task Analysis Workload Analysis Correlation Matrix Link Analysis Work Area Layout Work Station Design Panel Design Test Plan Performance Plan	Solution Incorporated and evaluated	Information flow Task Statements Hardware Design Workload profiles Task Hierarchy Data transfer adjacency Layout Criticality Error Diagnostics Statistical Analysis Baseline Monitoring Maintenance Tasks Training Requirements	Manual - observation Design Criteria Checklists HF T&E Manual (HFTEMAN) G-2/G-5 Antropameters Environmental Measurements Interviews Questionnairs Physical measurements On-line interactive simulation Contractual documents & specifications Physiological measurements Computer Aids

SE VS. CONVENTIONAL PROJECT DEVELOPMENT ACTIVITIES

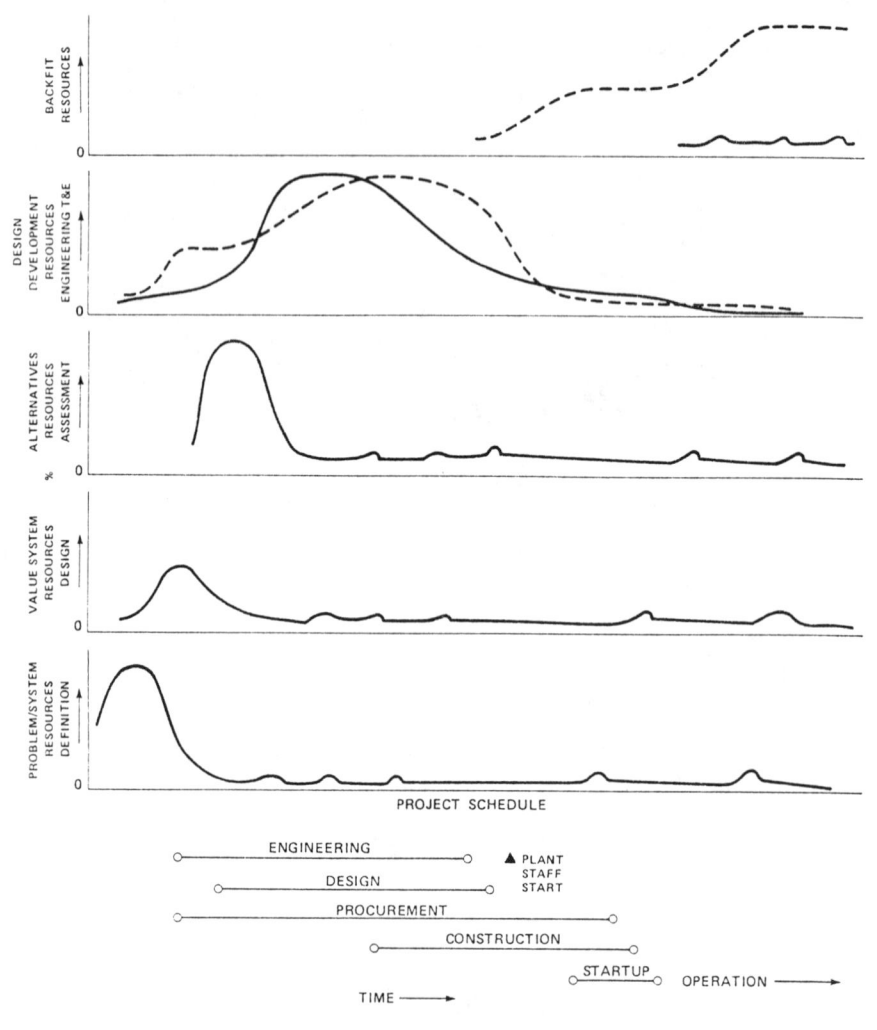

Figure 2.

throughout the project (Fig. 3). In the initial stages, control engineering has particularly important contributions to provide within the multi-disciplined group activities for:

1. Identifying the hierarchal structure, linkages and the information exchange requirements among the work breakdown structure (WBS) elements of the process baseline configuration.

Table 7. A Generalized Comparison of Use and Conventional Control Engineering Practice.

Unified Systems Engineering	Control Engineering
Start early with substantial effort to systematically define problem	Start late and light with solutions
Multi disciplined team evaluation and decision making	Function independently
Integrated quantification of alternatives assessment. Rationalized traceability—documentation	Virtually non-existent
Continually evaluate selected solutions	Engineer, Procure and Install
Monitor performance indicators during operation	Run process, face problems backfit

2. Operating Scenarios—especially those 'off-normal' situations that have serious consequences.
3. Assure that a proper balance is struck for operations and controls in the design of the value system.

Once solutions are admitted, the control engineer has a major responsibility in the joint objective evaluation of alternatives. This phase produces the overall control design objectives and the broad relationships to the other parts of the operations system. Through group interactions, the objectives, the interfaces, and the constraints can be developed in considerable detail and durability at the start of the design phase—a rather good way to begin.

During the design development phase, evaluation is almost constant as various forms of testing and checking are brought to bear on the design [15, 16, 17, 18, 19]. Control Engineers, knowing the importance of keeping the design responsive to evolving project requirements are cognizant of and cooperative with the interface disciplines such as Human Factors and Risk Assessment and assist in documenting for rationalized traceability. In this context, major design changes and backfits are quite unlikely throughout the life of the process.

Systems Software for Process Control

The application of on-line real-time computers to process control is a complex job in itself. The digital world has a way of uncovering all

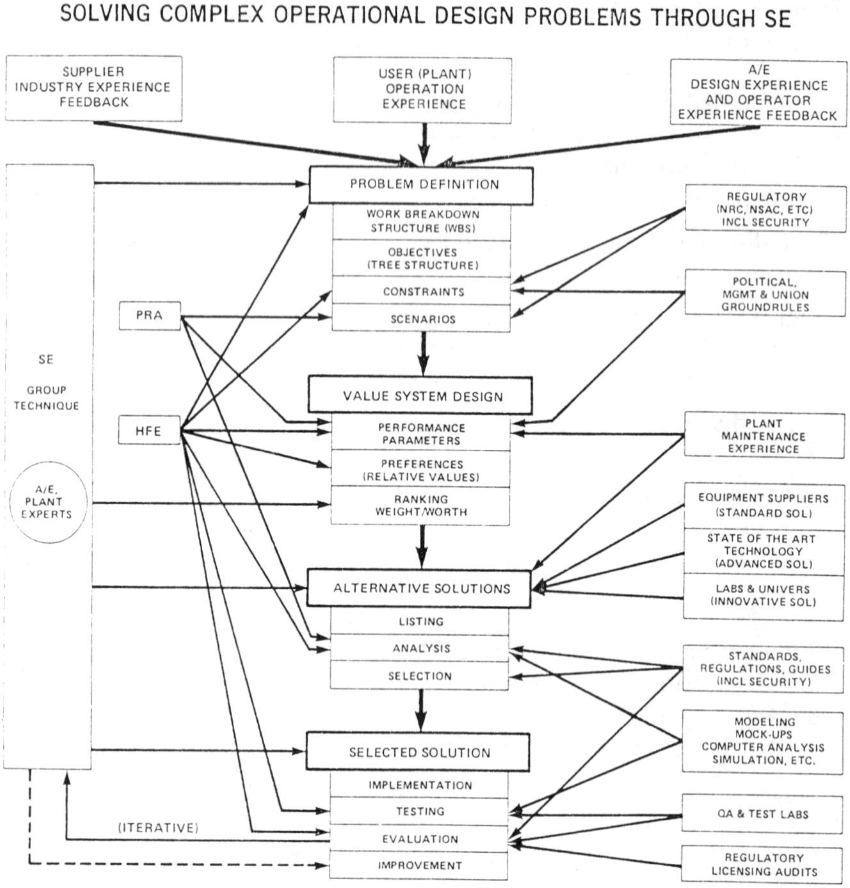

130   Chemical Process Control II

SOLVING COMPLEX OPERATIONAL DESIGN PROBLEMS THROUGH SE

Figure 3.

manner of flaws in preparation for and programming of the information processing to be done. In this regard, few devices have done more to illuminate how little is understood about the real problems and objectives.

Experience with computer applications shows that for every functioning installation, there is one that never is tried and another that is never made to work. This history has created an under-current of skepticism and a fear of making matters worse. Quality assurance of software remains an unsolved dilemma even for batch programs. Rapid obsolescence of computer systems, relative to the process, is another factor often poorly addressed.

The significant shift towards automated computer process control has accentuated rather than alleviated Human Factors considerations in the man-machine interface design. Appendix A provides a summary review of the current technical base for Human Factors in computer systems [17]. There is no universal console design, nor will there be, that eliminates the need for a systems engineering analysis to illuminate the functions to be performed. The responsibility to handle these various issues (and there are many) is acquired at the same time the decision is made to use computers in process operations. Without a systems engineering approach, the chances for success are greatly reduced [19].

SUMMARY

Control engineering is part of a wide-scope complex problem in process operations systems design. When the control system can be developed to a well established set of objectives and constraints integrated with the other parts of the operations systems, the design can proceed with full confidence that process goals will be achieved with problems and backfits at a minimum.

The traditional approach to control engineering as well as many of the other disciplines involved in process operation design, has evolved to establish several conventional practices that are in great conflict with proven systems engineering methodology and seriously inhibit improvements and applications commensurate with the real capabilities of the controls discipline.

If the project was properly equipped with a SE team at the outset, the control engineering design dilemma would not exist. If SE does not exist for the project, a void is created that should be recognized by the affected disciplines and filled. The extended role of controls engineering as system engineering is not inappropriate.

REFERENCES

[1] Warfield, J. N. and Hill, D., "A Unified Systems Engineering Concept," *Battelle Monograph No. 1*, Battelle Institute, Battelle Columbus Laboratories (BCL), June 1972.
[2] Booth, J. N. and Lee, W. H., "Systems and Human Engineering Review of the LOFT Augmented Operator Capability Program," Report No. 2-3755-00JB-062, Battelle Columbus Laboratories (BAC) Idaho National Engineering Laboratory (INEL), 10 June 1980.
[3] Geer, C. W., "Human Engineering Procedures Guide" AMRL-TR-79-, Boeing Aerospace Co. (BAC), Aerospace Medical Research Laboratory, Wright-Patterson Air Force Base, OH, 27 Dec 1979.
[4] Warfield, J. N., "An Assault on Complexity," *Battelle Monograph No. 3*, Battelle Columbus Laboratories (BCL), Apr 1973.

[5] Warfield, J. N., "Structuring Complex Systems," *Battelle Monograph No. 4*, Battelle Columbus Laboratories (BCL), Apr 1974.
[6] Hartway, B. L., "LATA Approach to Systems Engineering for BART System," LATA WP-122, Los Alamos Technical Associates, Inc., Los Alamos, NM, 1 May 1980.
[7] MIL-H-46855B, "Human Engineering Requirements for Military Systems, Equipment and Facilities," Redstone Arsenal, AL, 31 Jan 1979.
[8] Geer, C. W., "Navy Managers Guide for the Analysis Sections of MIL-H-46855," D180-19476-2, Boeing Aerospace Co (BAC). Naval Air Development Center (NADC), Warmingten, Pa, 30 June 1976.
[9] Geer, C. W., "Analyst's Guide for the Analysis Sections of MIL-H-46855," D 1880-19476-1 Boeing Aerospace Co (BAC). Naval Air Development Center (NADC), 30 June 1976.
[10] English, M., "Navy Managers Guide for the Design Sections of MIL-H-46855," NADC-79220-61, BAC, NADC, 26 Sep 1980.
[11] English, M., "User's Guide for the Design Sections of MIL-H-46855," NADC-79220-60, Boeing Aerospace Co (BAC), Naval Air Development Center (NADC), 26 Sep 1980.
[12] Geer, C. W., "Navy Manager's Guide for the Test & Evaluation Section of MIL-H-46855," D194-10006-2, Boeing Aerospace Co, Naval Air Development Center, 30 June 1977.
[13] Geer, C. W., "User's Guide for the Test and Evaluation Section of MIL-H-46855," D194-10006-1, Boeing Aerospace Co (BAC), Naval Air Development Center, 30 June 1977.
[14] Rijnsdorp, J. E., "A General Review of the Field of Man-Machine Interfaces in Terms of the Psychological–Social Effects Involved," *Sixth Annual Advanced Control Conference*, Purdue University, 28 Apr 1980, PP. 1–12.
[15] MIL-STD-1472 C, "Human Engineering Requirements for Military Systems, Equipment and Facilities," Draft 1980.
[16] Mallory, K., et al. "Human Engineering Guide to Control Room Evaluation," NUREG/CR-1580, US Nuclear Regulatory Commission, Washington, DC, Draft Report July 1980.
[17] Ramsey, R. H. and Atwood, M. E., "Human Factors in Computer Systems: A review of the Literature," Office of Naval Research Technical Report NR 196-146, SAI-79-111 DEN, Sept 1979.
[18] Swan, A. D. and Guttman, H. E., "Handbook of Human Reliability Analysis with Emphasis on Nuclear Power Plant Application," NUREG/CR-1278, Sandia Laboratories, US Nuclear Regulatory Commission, Apr 1980.
[19] Anon, "Computer Systems Interface Guidelines for Nuclear Power Plants," TSA-80-361, Macro Corp., Fort Washington, PA June 1980.

APPENDIX A

Summary Review of SAI-79-111-DEN (Ramsey, Atwood)
"Human Factors in Computer Systems: A Review of the Literature"

SUMMARY

1. Relevant Human Factor literature is widely scattered and badly fragmented because of a foundation in several different disciplines. Maintenance of broad Human Factor familiarity is all but impossible for system designers.
2. Existing guidelines concentrate on areas often not under control of system designers.
   a) Most basic design decisions—little Human Factor guidance
   b) Central issues, such as user information requirements—Human Factor guidelines sketchy
   c) Interface device designs—reasonably good and fairly detailed
3. Quantitative data on user performance and user-system interaction are a few and fragmented.
4. The existing Human Factor data base is inadequate to support development of a quantitative reference handbook. There *is* adequate basing for a design guide, procedural in nature, that parallels the major steps of the design process itself—before design decisions are made. After design decisions are detailed, guidelines can be specific and prescriptive. Human Factors then knows the design process well and can help with the specifics when the design is specific.

USERS

1. An interactive Man Machine System acquires its characteristics from the goals adopted by its designer.
2. Behavioral issues are complex, subtle, and in many respects poorly understood.
3. Specific user tasks require a more effective taxonomy than presently exists.
4. Sophisticated users are often unaware of the specialized knowledge, context clues, etc., which transpires in the man-machine interface.
5. Information requirements of managers are more variable than those of most user classes.
6. Very large reductions in error frequency and severity can be achieved if the system design is based on an adequate understanding of the types and causes of errors which occur in the use of the system.

7. Procedures for error analysis during system design may be the most useful information which can be provided.
8. Individual difference effects are often highly task-specific and the relevant data relating these differences to system design considerations is sparse.

TASKS

1. Research results on human behavior often fail to generalize to other tasks even when superficially similar. The lack in characterizing user tasks is probably the greatest existing deficiency of the field.
2. If quantitative human performance data is to be applied by the system designer, a common understanding of user tasks is absolutely necessary.

DETERMINATION OF USER REQUIREMENTS

1. It is necessary for the system designer to have a basic conception of functional requirements before HF data can be applied.
2. User requirements analysis is itself a behavioral research problem.
3. Involving the user in the design process to accurately identify user requirements also will result in greater user acceptance of the system.
4. While users may be quite expert at performing their job, they are rarely expert at analyzing or describing how they do their job, its information requirements, etc.
5. The success of user interviews is heavily dependent on the ability of the user to describe the tasks, information requirements, etc.
6. Tasks analyses must describe behavior in terms of units of activity which are psychologically meaningful with regard to both level of detail and type. Behavioral science is a prerequisite to effective use of task analysis.
7. Optimal aided approaches may differ drastically from the practices of the unaided user.
8. MMS Simulation—Modeling efforts have been successful only when the modeler has a good understanding of a task domain and of user behavior in that domain. No generally applicable models exist.
9. The MMS system designer must explicitly focus on the properties of the user—system interface.
10. A great deal of sophistication is required by the system designer to translate the psychological terms and concepts of the memory and

process limitations of the human information processor into a form suitable to express a system design.
11. No generally applicable approach to developing models of interactive systems exists.
12. Nothing in the traditional task analysis/task allocation method helps the designer to develop a novel MMS approach.
13. The system designers experience is a dominant factor in recognizing the applicability of particular types of operator aids.
14. In data—driven behavior, problem recognition and alternative evaluation are the most crucial problem—solving sub tasks.
15. Knowledge about problem—solving aids useful to the designer is relatively abstract.
16. There is little evidence that existing knowledge of the measurement and effects of information load is being applied and is often not a formal consideration during system design. User performance can be adversely affected by information loads which are either too high or too low with performance effects significantly differing depending whether it is the memory or processing resources that are overloaded. This in turn, is strongly affected by user experience and task complexity.
17. The user's mental model of the MMS may not correspond to the designers or to reality, and has training requirements implications.
18. Graphics design is a major research gap that considerably lags application.

MMIF DEVICES

1. Available empirical data and reasonable guidelines that can be used to support system design is voluminous and often relevant.
2. Little HF research has been done on the overall performance of many types of displays—especially with the selection of one display over another.
3. Few HF studies are available to help the designers decide exactly how to apply chosen coding techniques.
4. It is well established that increasing the number of displayed items (relevant or irrelevant) increases time and errors in search for and extracting information.
5. Operators have subjective preferences in the amount of displayed information.
6. Although not totally satisfactory, several attempts have been made to determine what aspects of displays are of primary psychological importance.

7. A generic framework does not exist which can encompass a significant variety of display or display element types in a way useful to the designer.
8. Display information which is spatially encoded can be recalled with greater speed and accuracy than information which is alphabetically encoded.
9. Research literature related to the graphical—non graphical display question is lacking.
10. Input devices should be selected after considering all user input tasks. Alternate modes of input can be detrimental to performance.
11. The overall body of knowledge of detailed design parameters of non-keyboard interactive input devices in inadequate.

APPENDIX B

SYSTEM ENGINEERING HEXALOGUE*

1. Your client does not understand his problem. You must help him gain this understanding.
2. The problem posed is too specific. You must imbed the specific problem in the next more general question. Generalization of the client's problem is a technique for finding and solving the correct specific problem, not for avoiding the issue.
3. Your client does not understand the concept of an index of performance. You must help him to weight the several desired attributes of the problem solution. The disadvantage to the non-user must be included in your weighted evaluation of each candidate system. You present weighted evaluations of options. The client makes decisions.
4. The goal-centered approach rather than a technology-centered, time-sequential approach is essential.
5. A universal computer simulation model of a complex system cannot exist. You must postulate a priori those specific questions you wish the model to simulate.
6. The role of the "decision-maker" in a socially-relevant, large-scale system is generally unclear. You must expect to engage in building a political consensus if your recommendations are to move to an action phase.

*A truncated, modified version of the system engineers decalogue in Battelle Monograph #3.

TRAINING AND SELECTION OF OPERATORS IN COMPLEX PROCESS CONTROL SYSTEMS

Lawrence E. Kanous
Director of Nuclear Training
The Detroit Edison Company
Detroit, Michigan

Whenever I have the opportunity to discuss my topic of today with an audience like this it seems appropriate to recount a statement attributed to the dean of one of our highly respected graduate schools. He felt that the people of the world could be classified into three groups:

1. A numerically very small group who make things happen;
2. A second group, larger than the first but still numerically small, who watch things happen; and
3. The great mass, all the rest, who wonder "What the devil happened?"

I see all of you here, and all of those involved in the great purpose of these discussions, as members of the first group. If future, more automated, more computerized, more complex man-machine systems are to be economically and societally effective and not to be found seriously wanting, we must not be part of the third group.

Please note that I deliberately used the term man-machine systems. None of the systems which we are concerned about here have any purpose without man. No future system which can be envisioned will function without some interface with man. Man will design and modify them, construct them, operate them, maintain them, and when they have served their purpose dismantle and dispose of them.

Others here will, I'm sure, speak to the need to place man at a central position as complex systems are designed. The traditional design process with its myopic focus on the machine must be turned around and the oft repeated plea to design from the man out heeded. Joe Sublett (2), in his paper in Control Engineering last month, indicates that "for system designers the base line question is: "To what extent will the input capacity, decision making capacity and output of the human component be supplemented by machine components (human aids)?'"

Mr. Sublett earlier in the same paper makes a statement which, as a trainer, I disagree with, i.e., that "humans cannot be engineered."

I contend that through a process quite analogous to the hardware conceive-design-develop-test-implement sequence can be applied to the problem of modification of the behavior of people. The fact that this sequence is not typically applied in the training and education process does not belie its usefulness or applicability.

Since the early 1950's, various associates and I have been conducting research and development programs, the objective of which has been improvement in the methodologies by which employes are selected for and trained to perform various jobs in the power generating stations of The Detroit Edison Company. In the course of these studies, it became clear that both of these problems—selection and training—were more difficult than we judged they needed to be because of the manner in which the control rooms and other operating positions in the plants had been designed and laid out.

In the era during which we began our work, power generating stations tended to be organized utilizing a vertical structure which had three distinct "ladders." Each of these ladders was related to what was perceived to be logical departmentalizing of the work, i.e., a Boiler Room group, a Turbine and Auxiliary (or Condenser) Room group and an Electrical or Switchboard group. In addition there were (and are today) separate groups involved with Plant Maintenance, Fuel Handling and Instruments and Controls Maintenance.

Typically a new employee was hired into the plant as a building cleaner. From this position he bid on a better paying job in one of the "departments" referred to above. If he went into the boiler group, his promotional sequence would be Flue Blower-Ash Handler ⟶ Fan Man ⟶ Water Tender ⟶ Assistant Fireman ⟶ Fireman. In multi furnace stations a few Firemen were advanced to Boiler Room Foreman. It would typically require 8–12 years to move to the top job in one of the departments. Selection was based upon evaluation of background, prior experience and education through the medium of a personal interview. Training consisted of a requirement for self study of equipment descriptive texts either rewritten from manufacturers' literature or the actual equipment data and maintenance manuals frm the same source. In addition, the learner was instructed in his day-to-day operating duties by a senior employee in his classification who "broke him in." There were two voluntary courses which were thought to be helpful: a 180 hour course in Basic Electrical Theory and a 135 hour course in Stationary Engineering.

At that time I believe it is fair to say that operation was essentially "in the hand" rather than in the head. Many operating parameters were directly observed by the man's sensory apparatus. Eyeballs judged the quality of a fuel bed and proper combustion conditions. Hands felt for proper temperatures in components and were directly stimulated when valves or other control devices either "hung up" or "snapped open." Ears listened for the rythmic sound of well operating machinery, and for the shriek of a tube leak. Operators literally "became attuned to the process." At least the good or "fortunate" ones did.

Concurrent with the onset of our personnel research and unbeknown to us, the first of a series of new boiler-turbine-generator units was being designed. Implementation of more advanced remote control systems with the boiler and turbine controls and displays in an air conditioned control room was a key feature of this new concept. Since this control room would be manned by a single operator, the question arose—from which group shall these operators be selected? Is it better to man these rooms with experienced Boiler or Turbine Operators?

Our researches had progressed to the point that we were able to state that although the content of the departmentalized jobs was indeed different in terms of the tasks performed, the jobs were different only in degree in terms of the basic aptitudes or abilities which underlay their performance. Our work had led us to conclude that in most instances it was a fortuitous accident which had led persons with potential to become "good" operators to select or be selected into one or the other department. Reinforced by this information the management of the Power Production Organization decided to implement a new organizational concept for the new unit. This concept did away with the departmentalized vertical ladders and divided the work tasks horizontally instead. At the lowest or entry level which is called Assistant Power Plant Operator are found all of the Boiler, Turbine, and Electrical tasks which formerly were performed at the lowest rungs of the three ladders. The intermediate level tasks were combined into a single job called Power Plant Operator. Finally, the highest level tasks from all three ladders with total responsibility for unit operation including work direction of Power Plant Operators and Assistant Power Plant Operators were combined to form the job of Supervising Operator.

This organizational concept has been very successful and has been used as the basis for staffing 17 new units since and will be the concept used for the three we now have under construction. Two of these are large coal fired fossil units and one is a boiling water nuclear unit. Computers are found in some form in all of the recent units. Performance of control functions via computer is minimal. Data logging and display are the major tasks assigned to these machines. The first of these new "unit system" plants started a trend away from "in the hand" to "in the head" operation which has progressed at an accelerating pace with each successive unit and continues today.

Our research on the selection of personnel to man these stations used two methods. In the first, the researchers acted as participant-observers. Each task performed by operators was identified by the researchers who also learned to perform it. The tasks were subsequently analyzed to identify the set of aptitudes which were judged to underly their performance. In addition, the critical incidents methodology (1) was used to augment and enhance our understanding. Interviews were conducted with more than 200 incumbent operators and their supervisors. These interviews produced $\propto$ 3,000 descriptions of operating behaviors (incidents) which the interviewees felt were examples of significantly "good" or "poor" operating performance. Each of these incidents was evaluated by a panel of qualified psychologists. Each evaluator worked independently of the others. At the conclusion of this exercise, all incidents had been classified in terms of the aptitude or ability which the members of the panel agreed could "explain" the behavior described.

Eleven traits or variables were identified. As a further check on the reasonableness and relevance of the variables, descriptions of each, in layman's terms, were prepared and 100 senior operators and their supervisors were asked to rate the importance of each using the paired-comparison technique. Statistical analysis of these ratings showed very good agreement among the operator-supervisor group and between that group and the panel of psychologists. Reliable means for testing these aptitudes were then selected or developed and then validated using the concurrent validation strategy. The criterion against which the tests were validated was supervisory ratings of operator performance.

The traits which survived the validation process are the following:
- Space Perception—the ability to visualize objects or the relations of objects in three dimensions.
- Decision Time—the ability to make quick, accurate decisions between two or more courses of action especially under stressful conditions.
- Mechanical Experience—general knowledge of mechanical apparatus and principles. (not mechanical ability which includes a phycho-motor component)

- Planning—the ability to consider the steps of a problem well ahead of the step being worked on.
- Perceptual Speed and Accuracy—the ability to perceive differences and similarities even when these may be very slight.
- Perceptive Attention—the ability to apprehend unusual conditions even though these conditions are not the objects of the person's immediate attention.
- Judgment—the ability to make good decisions in situations where not all of the facts are obvious or available.
- Reasoning—the ability to draw valid conclusions from facts known or assumed from a knowledge of general principles.
- Integration—the ability to simultaneously keep in mind and properly account for several conditions, premises or rules in order to correctly solve a problem.
- Flexibility—the ability to quickly and easily change in action or throught when the situation demands such change.
- Carefulness—the ability to perform correctly in spite of working for speed. It is also concerned with the ability to perform correctly even when the task is repetitive or routine.

Use of a test battery assessing the above traits resulted in raising the success ratio of new hires from 7 of 10 prior to putting the selection system in place to 8.5 of 10 following its emplacement.

Concurrent with the conduct of the selection studies work was being done to put together a more systematic, more defensible and more effective training system.

Prior to the work I have been describing the Power Generation organization had on three separate occasions attempted to put into place a formal training process for Operating personnel. Each of these had, after an intial period of enthusiastic use, fallen into disrepute and then into disuse. All of these programs were constructed on the traditional instructor dependent, classroom-lab model. The fatal flaw which characterizes this model in this environment is that it fails to take into account the realities of the system it should serve. During the participant-observer phase of the selection study, the research team had many opportunities to become aware of the realities of the plant operating environment and had listened to a great many inputs from operators and operating supervisors which were critical of the training processes then in use. A compilation of these inputs (gripes?) led to the design of a training system which

1. Provides operators with a useful model of the Power Plant as a system composed of subsystems which must all interact effectively to produce the product.

2. Provides operators with the necessary skills and knowledge to perform each and every task they are called upon to perform.
3. Provides operators with frequent opportunity to assess themselves and be assessed against the standards and objectives set for them by the program.
4. Provides supervisors and managers with frequent information, in a timely manner, on the progress and performance of their trainees in order that they could take whatever actions they needed to to assist in the process.
5. Provides a means to quickly and easily revise or otherwise modify the program to reflect current conditions.
6. Provides a means by which senior operators, especially those who operated in the control room, could maintain their skills and knowledge of the conditions out in the plant as well as the tasks performed there.
7. Consisted of training exercises designed to fit in with the realities of plant operation i.e., rotating shift schedules, plant emergencies, operator selected vacation and other offtime, etc.
8. Finally, the system was to require heavy but not burdensome participation by operators and operations supervisors in its design, development, implementation and modification so that it would ultimately become a functioning "organic" part of the life of each power plant rather than an invasive "cancerous" process.

The system has three phases. In Phase I all new operators are enrolled in a 5 week, 8 hour/day pre-training course. In this course, which is instructor based, the trainee learns to specified criteria: plant administrative procedures, safety procedures, shift organization, and gets his/her first introduction to the systems and cycles which plant operations involves. In addition, the trainee qualifies in First Aid, Cardiopulmonary Resuscitation, fork lift truck operation and the care and use of rescue breathing apparatus. Frequent tests and performance demonstrations are required throughout this Phase.

Trainees who successfully complete Phase I are assigned to a plant and begin Phases II and III which proceed somewhat concurrently.

Phase II or Job Procedures Training follows the Job Instructor Training (JIT) concept. For each operating job task, what is called a JIT Unit is written by an operating supervisor. Such units describe in sequence each step of the operating procedure along with any key or critical point of information pertinent to each step.

These are written down on a formatted form with steps on the left and key points on the right. After one shift writes a unit, it is circulated to the other shifts for critical comment. All comments are returned to the original writer. If there are unresolved disputes the Operations Engineer resolves them in a meeting with the parties. The edited unit is then reviewed for technical accuracy by plant management, especially Technical/Equipment Engineers. Following this the unit is edited for grammar, sentence structure, etc. It is then submitted to the responsible plant Operating Manager for approval and printed and issued. Each Unit is targeted to require only 30 minutes or less to give a complete description and demonstration of the procedure. A rare one will require 45 minutes and frequently the additional time is due to travel inherent in the task or lags built into the procedure. As an example there are 5 units on each boiler feed pump. One on Pre-Ops Inspection; one on Start Up; one on Operating-Normal; one on Emergency Procedures; and the fifth on Shutting Down. Some purely informational units are also included. In one plant, for the first or beginning level job, there may be 3,000 of such units.

Upon entering the plant, the new person is assigned a senior operator as his/her trainer for a period of 6 months. This trainer, using the units, will schedule as many 30 minutes time blocks with the trainee as he requires to be able to show the trainer that he can properly perform the task. (The blocks are usually not consecutive.) When the trainer is satisfied with the proficiency of the trainee he recommends the trainee to the shift supervisor for certification. The shift supervisor then takes the trainee to the work site and asks the trainee to demonstrate the task, actually operating if possible or "walking it thru" while verbally explaining what he expects to occur and what he will do if it behaves other than as expected. The supervisor, if satisfied, certifies the trainee as competent and he is then allowed to perform that task on his own. If not, he is sent back to his trainer. The supervisor makes an appropriate note in his training record. Trainees typically pass these certification tests on the first try.

The training units are sequenced in 6 month increments from easy jobs to more complicated. The program for the beginning job requires 36 months to complete. A comprehensive written test is required at 6 month intervals.

During the time period covered by Phase II the trainee, at his/her option, can enroll in two additional programs which constitute Phase III. There is a mandatory course in Applied Electricity. This program differs in content, design strategy and delivery technique from the old Basic Electrical Theory Course. First the content is specifically job relevant. For example after learning the concept of resistivity of materials at a practical level the trainee learns to identify various insulating materials— liquids, solids (various forms) and gases and also how to preliminarily evaluate a piece of insulating material which has failed in service using his senses. He/she is not required to work mathematical exercises using the dielectric constants. Mathematical problem solving is required and learned at the time the skill or process is needed to solve a given problem *not* as a pre or corequisite.

The program is designed to be individualized, self paced and multimedia. Its basic purpose is to provide a "gut-level" understanding of important electrical concepts. Hazard avoidance and failure symptomatology are emphasized. The program is delivered by a Program Monitor who acts as a learning facilitator, coach, counselor, examiner and sometimes disciplinarian rather than as a lecturer. The content of this program is equivalent in breadth and depth to the former instructor-based course. When the employee completes this program a one step advancement in job grade and pay is awarded. Also the Company pays a one time only bonus of 200 dollars.

The second part of Phase III requires that each employe acquire a Stationary Engineers License appropriate to his job classification. This license is awarded by the City of Detroit. This achievement requires passing of written and oral examinations. The employee must achieve the appropriate license prior to being considered for promotion to the next position. The Company provides an instructor based course which the employee can attend on his own time. There are also similar courses offered in several Community Colleges in our service area which can be used. Again on the employee's own time. Employees can if they choose, and some have, prepare by self study on their own. When a license is received the employee again is rewarded, on a one time only basis, a bonus of $200 for a 3rd Class License, $400 for a 2nd Class License and $800 for the 1st Class License. No employee is eligible for consideration for promotion to Shift Supervisor without a valid 1st Class License.

The system works very well. It was installed in the first plant in 1958–1960 and is humming along today. The units are almost self-correcting since the operators who conduct the training are required to be aware of changes in plant characteristics or procedures by reviewing the previous shift's work before they relieve. They become immediately aware when information in the unit is out of date. They provide the trainee with correct information and submit the unit for correction. Often he supplies a corrected unit with it.

This system functions sufficiently well that personnel have moved from off the street to the top operating position in 5 years. Prior to this it required 8 to 12. Of equal value, the men and managers in the plant believe in the system and have confidence in the competence of operators produced by it. They feel it is necessary, relevant, and theirs. It also produces a side benefit in that all senior operators are constantly up-to-date on the jobs being done by their lower level associates. Since each of them must act as JIT Masters, they are not isolated in the control room.

We are now taking the final step to complete the system. Over the last 4 years we have been using and evaluating the use of the full scope, dynamic Simulator. We have concluded that for the types of units we are now bringing on the line which are of such size—500, 800 and over 1000 MW capacity—complexity and significance to our system and those to which we are interconnected that the capital investment is justified. We are now developing the specifications for the first of these machines and should contract its acquisition this year.

REFERENCES

[1] Flanagan, J. R., "The Critical Incident Technique," *Psych. Bulletin* 51(4):327–358, (1954).
[2] Sublett, J., "The Human Component in Man-Machine Systems," *Control Engineering*, (December, 1980).

NEW RESULTS AND THE STATUS OF COMPUTER-AIDED PROCESS CONTROL SYSTEMS DESIGN IN JAPAN

Iori Hashimoto and **Takeichiro Takamatsu**
Department of Chemical Engineering
Kyoto University
Kyoto, Japan 606

ABSTRACT

The drastic change in the economic situation due to higher energy costs since the oil embargo of 1973 and tighter restrictions on safety and environmental emissions during the last ten years have caused a great change in the philosophy of process system design. This has resulted in more difficult control problems, most of which can be referred to as the control of a highly interacting multivariable system. There now exists strong impetus for the application of advanced control techniques. In order to design a sophisticated control system based on these techniques, the help of a CAD system is indispensable to facilitate the necessary calculation and data handling.

In this context, in this paper an attempt is made to review the present status of chemical process control and the computer-aided design of control systems in Japan.

INTRODUCTION

The chemical industries in Japan attained their high degree of development with the aid of very cheap energy, low labor costs, and imported advanced technology during the country's period of high economic growth in the 1960's. During the early stage of this period, large chemical complexes were constructed along the Seto Inland Sea and in the Pacific coastal zone. These locations facilitated the ocean-borne import of raw materials and the export of products on a large scale. Most of the newly constructed chemical processes were operated continuously with large throughputs. There were no strict environmental emission regulations, and product specifications were also quite lenient.

In the steady state design of process systems, many intermediate storage tanks were installed, and large design margins were usually added to each unit. This was to break the interactions between the subsystems and to absorb unfavorable influences due to unexpected disturbances or uncertainties in order to assure stable steady state operation. Since energy costs were low, energy conservation was not such an important objective in designing process systems and, therefore, less heat integration was implemented.

Under these circumstances, control system design did not have a strong relationship to steady state design. The influence of disturbances on the quantities of the product streams were usually masked by the large amount of design margins in the units. The control system played a relatively minor role in maintaining the whole system at an optimal steady state condition. A single input and single output control strategy was used mainly to design control loops to hold variables, such as flow rate, liquid level, temperature and pressure etc., at desirable set points.

During the last 10 years, the rapid increase in raw material and energy costs—especially the enormous increase in energy costs since the oil embargo of 1973—as well as stricter environmental and safety regulations have caused a great change in the philosophy of process system design. Process engineers have been forced to utilize energy and other natural resources more efficiently and to operate their processes at high performance levels, yet under tighter constraints.

Intermediate storage tanks may be made smaller or eliminated completely to reduce inventory and capital cost. This, however, results in more sophisticated and highly coupled process systems. Design margins are also reduced in order to improve steady state optimality, which causes loss of flexibility and operability in the process. In order to minimize energy consumption, heat integration has to be thoroughly implemented by connecting the heat-consuming units with the heat-producing units, the heat of which was previously wasted. Energy integration can save energy from the static viewpoint. However, from the dynamic viewpoint, it makes the process a highly interconnected, multivariable system which might lose flexibility, operability and controllability. The loss of these favorable dynamic attributes has to be recovered by implementing good quality control. All of these factors combine to pose more difficult process control problems. This offers strong impetus for devising and applying modern control techniques.

The rapid development of computer hardware technology, that is, the emergence of cheap highly capable minicomputers and microprocessors seems to have greatly changed the prospects of implementing sophisticated control strategies. At the same time, advanced control theories have rapidly developed, and efforts have been made to apply them to more realistic problems. For example, in the United Kingdom, several graphics-based computer-aided design packages for multivariable control systems are now commercially available [1, 2, 3].

Under such circumstances, it is very important to survey the real situation with regard to chemical processes control and computer-aided design of control systems in Japan. In the following sections, this topic is introduced.

THE PRESENT STATUS OF COMPUTER AIDED PROCESS CONTROL SYSTEM DESIGN IN JAPAN

In order to obtain the most up-to-date information on this topic, several companies were interviewed, and a questionnaire was sent to about 80 leading companies, most of which are chemical engineering and construction companies, and chemical and petrochemical processing companies. Some 60 companies, the rough classification of which is shown in Table 1, responded to the questionnaire as of October 31, 1980.

The process systems which have been designed or are now operating in these 60 companies can be categorized into three different groups. Processes that are continuously operated at their steady state points without great fluctuations and that typify the basic process industry are classified as Type I processes. Processes which are continuously operated yet are accompanied by large scale time-variant fluctuations are categorized as Type II processes, and batch processes are classified as

Table 1. Classification of Responding Companies.

Engineering and Construction Companies	11
Heavy Industries and Computer Companies	9
Chemical and Other Processing Companies	40
(Chemical and Petrochemical) 34)	
(Paper, Cement, Food etc. 6)	

Type III. The percentage of these three different types of processes is shown in Table 2.

All of these processes are essentially operated in a dynamic manner. In order to assure stable operation of these processes, it is indispensable to implement control systems irregardless of their complexity, because the implementation of control systems is an effective means of giving favorable dynamic properties to processes.

As mentioned earlier, control system design is traditionally separated from the steady state design stage. Ideally, however, there should be an integration of the control system and the steady state design stage. There is a trade-off between high quality control system design and optimal steady state design. By increasing design margins which result in the degradation of steady state optimality, the system often becomes a stable self-regulating system which does not need a high quality control system.

For example, if a distillation column that has a large number of surplus stages is designed, the composition of the distillate would be kept at a high quality and would be insensitive to various disturbances, without any sophisticated control system. If we reduce the surplus stages, the composition of the distillate will be very easily influenced by disturbances, and a good quality control system would be required to keep the composition at a desirable value.

In this way, there is a strong interrelationship between steady state design and control system design. The more complex and interacting the system becomes, the more difficult it is to design the optimal system from both the steady state and dynamic points-of-view. The problem is how to reconcile steady state design with control system design. In other words, a method has to be developed to incorporate control structure design into steady state design optimization. Research relating to this topic has just started, and much effort must be devoted to resolve this problem [4, 5].

Table 2. Classifications of Process Systems.

Type I	
A continuous process operated mainly at a stable steady state	52%
Type II	
A continuous process operated with large scale time variant fluctuations	18%
Type III	
Batch processes	30%

The procedure presently used for control system design can be summarized as shown in Fig. 1. The control system design is performed after the completion of the steady state design. The steady state design is partially amended, only when a control system with sufficient dynamic performance cannot be designed due to the poor steady state design of the process.

The present use of computers in each step of the design procedure shown in Fig. 1 is described below.

Almost all of the control problems are formulated on the premise that the measured and manipulated variables have been selected. But in order to design a control system for the whole process, the most fundamental and difficult problem, namely, that of how to synthesize control structures, has to be solved. In other words, the problem of how to select the measured, manipulated, and controlled variables has to be solved in order to synthesize a control system that can assure the desirable dynamic characteristics of the whole system for the fulfillment of a set of specified objectives.

Very few systematic studies dealing with this important problem have yet been reported [6, 7, 8]. Experience is still playing a dominant role in the synthesis step of control structure. The survey conducted here reveals that empirical approaches in the selection of the variables in Type I processes constitute slightly less than fifty percent. Static Simulation, such as sensitivity analysis and static gain analysis among variables, is utilized in 30% of the cases, while dynamic simulation is utilized in 25%. For Type II processes, dynamic simulations play an important role in the determination of control structure while for Type III processes, empirical and dynamic simulation approaches are used almost equally, as shown in Fig. 2.

Roughly speaking, there are two approaches to the development of dynamic models, i.e., mathematical descriptions of process dynamics. One model is based on the rigorous interpretation of physical and chemical principles, and the other is usually obtained by fitting the empirically assumed model structure, such as first or second order plus dead time, with experimentally measured process dynamics.

As shown in Fig. 3, 61% of the companies that responded to the questionnaire have utilized a rigorously developed model for analysing the dynamic behavior of the processes. Some 21% of them utilize a so-called "general package for dynamic simulation," such as DPS, DOPL, MODYS etc., the detailed explanation of which will be given later. In other cases, simulation programs are made every time the requirement emerges.

Seventy four percent of all the companies have utilized the model developed by the fitting of experimental data, such as the model obtained

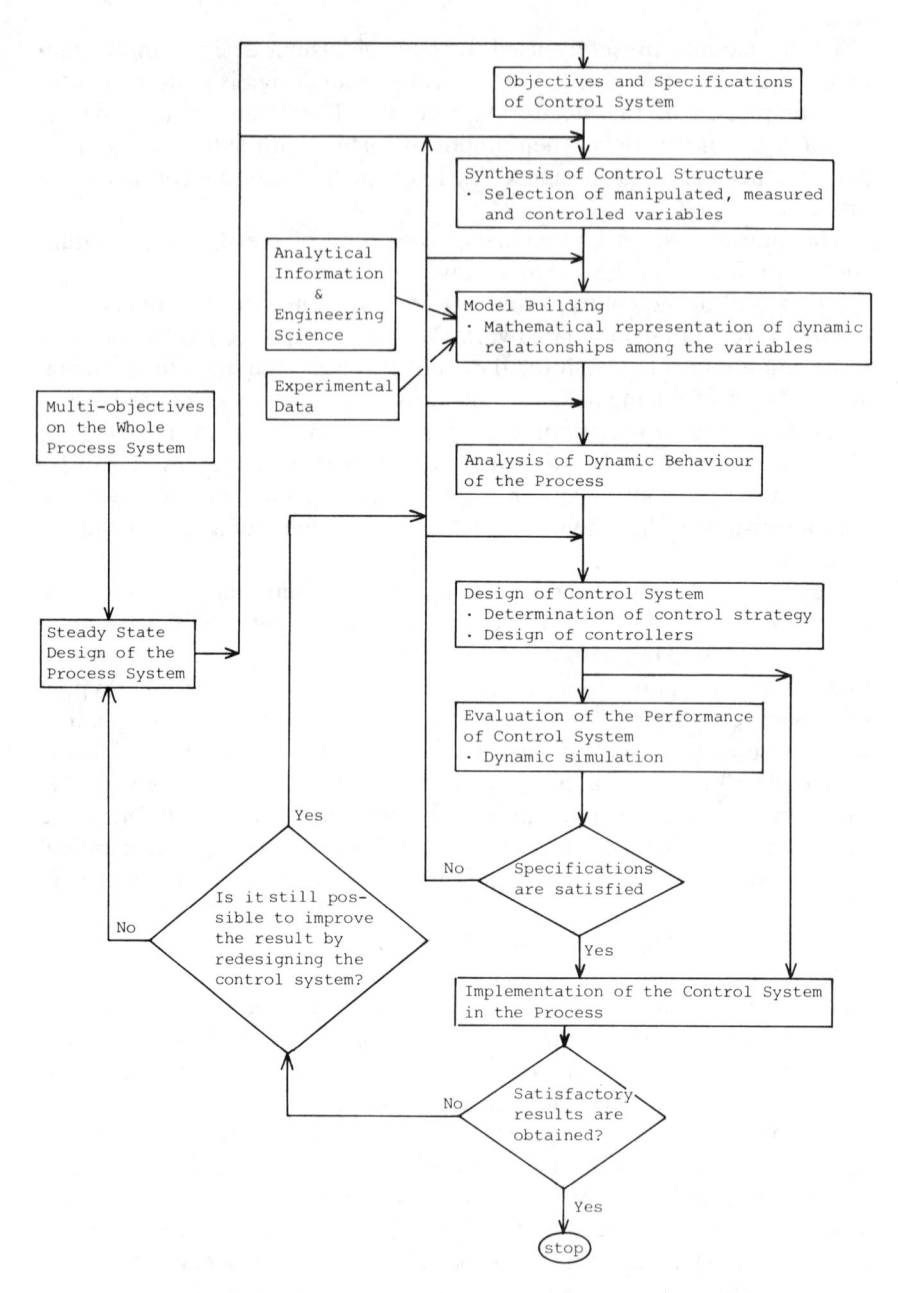

Figure 1. Design Procedure for a Control System.

For Type I Processes
- Empirical 45%
- Static Simulation 30%
- Dynamic Simulation 25%

For Type II Processes
- Empirical 27%
- Dynamic Simulation 73%

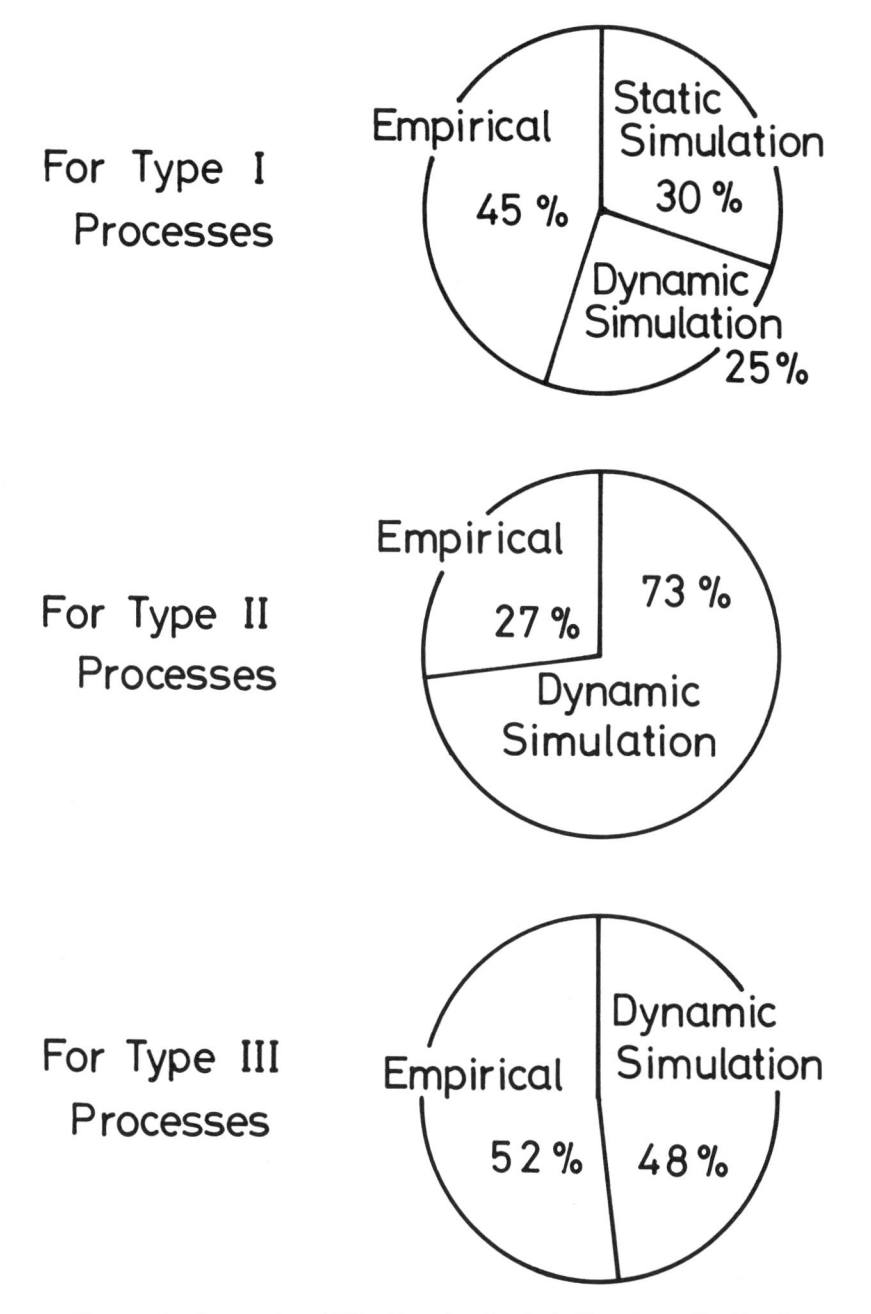

For Type III Processes
- Empirical 52%
- Dynamic Simulation 48%

Figure 2. Computer Utilization in Control Structure Synthesis.

Utilization of a general package for dynamic simulation

21 %

61% : Use of dynamic simulation based on a rigorously developed model

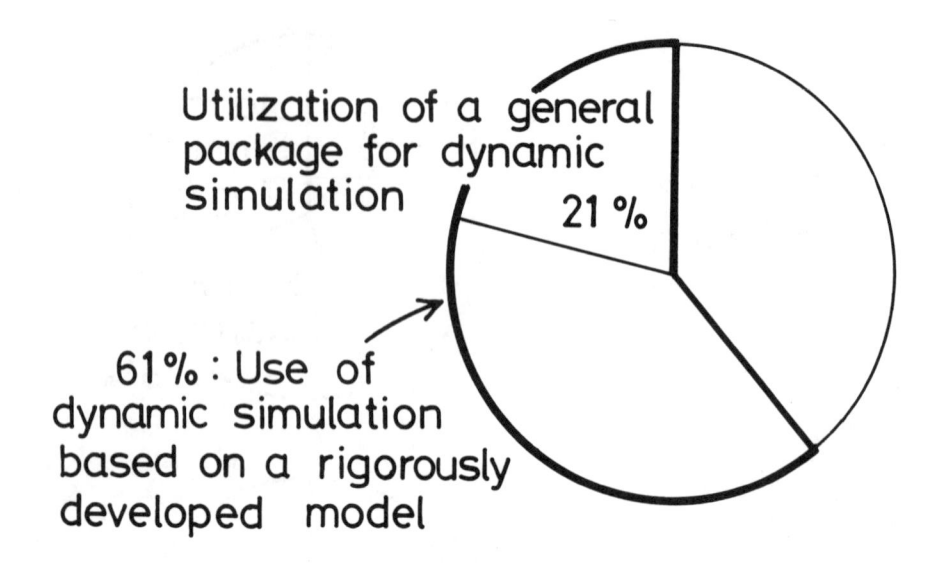

Figure 3. Utilization of a Rigorous Model.

by parameter fitting of the empirical model structure of first or second order plus dead time, as shown in Fig. 4. Some 21% of them have prepared computer packages for parameter fitting or obtaining a model by the handling of input and output data based on regression techniques.

After the dynamic characteristics of the process are fully analysed, the actual design of the control system begins. It is first necessary to determine which control strategy should be used, single input–single output control or multivariable control. Then, a suitable control scheme has to be selected for each controller.

It is often said that most chemical processes are multivariable and highly coupled systems. The problem however, is whether or not there are strong interactions between the various input and output variables. If the system does not have strong coupling between variables, each control loop could be designed separately by classical single input—single output design methods.

Fig. 5 shows how often process control engineers encountered difficulties due to significant coupling between variables. The number of companies which experienced difficult control problems resulting from the detrimental interaction in less than 30 percent frequency, exceeds 71%. That is, in 71% of the companies that responded to the questionnaire, the control engineer encounters less than three difficult cases out of

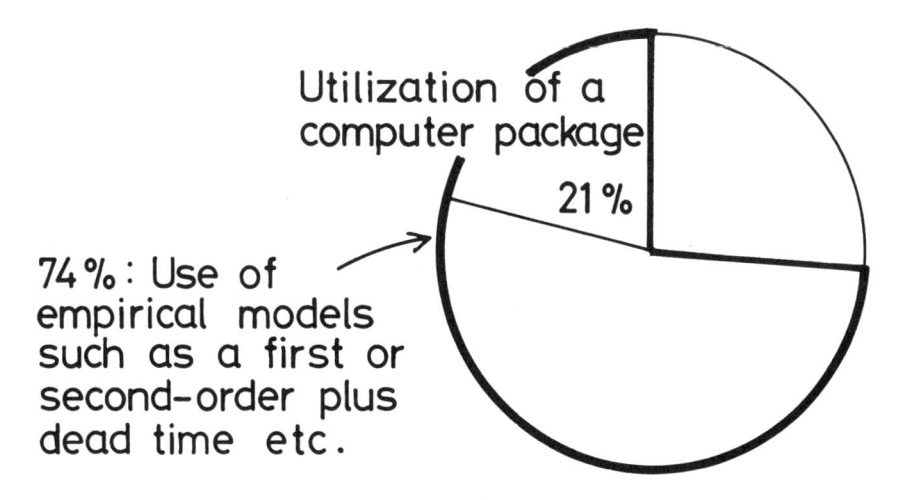

Figure 4. Utilization of Empirical Models.

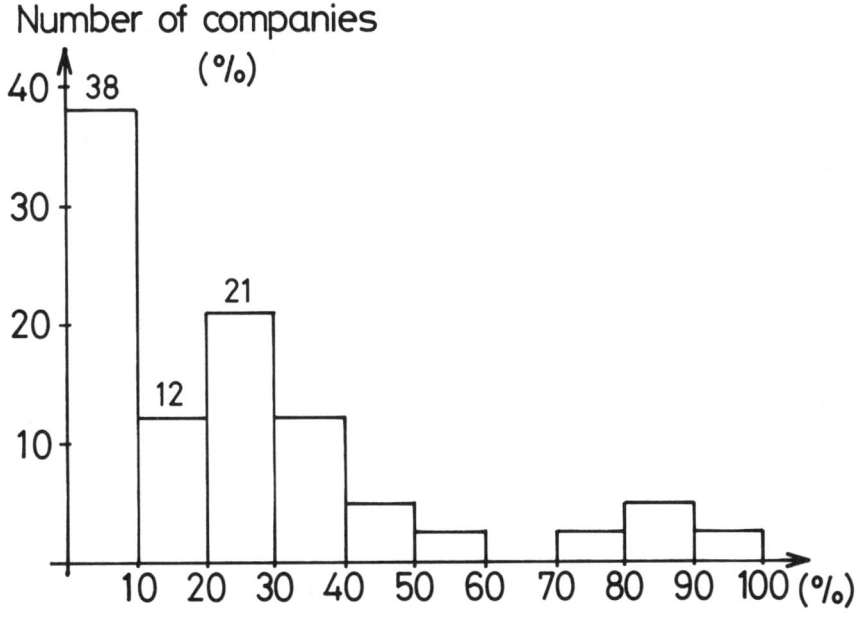

Figure 5. Emergence Rate of Difficult Control Problems.

156   Chemical Process Control II

ten in control system design. This may indicate that most chemical processes are designed by process engineers in order to easily be controlled. In other words, hard-to-control processes were discarded, and consequently safe self-regulatory processes were realized. The control system design, therefore, seldom poses difficult problems and is carried out by applying classical design methods.

Fig. 6 shows the present status of computer utilization in designing single input–single output control systems. In 64% of all the companies, control systems of this kind are designed without the help of any kind of CAD system. Some 13% of the companies use a computer in the form of CAD or sets of program packages based on the frequency domain approach. In 33% of all of the companies, program packages for the

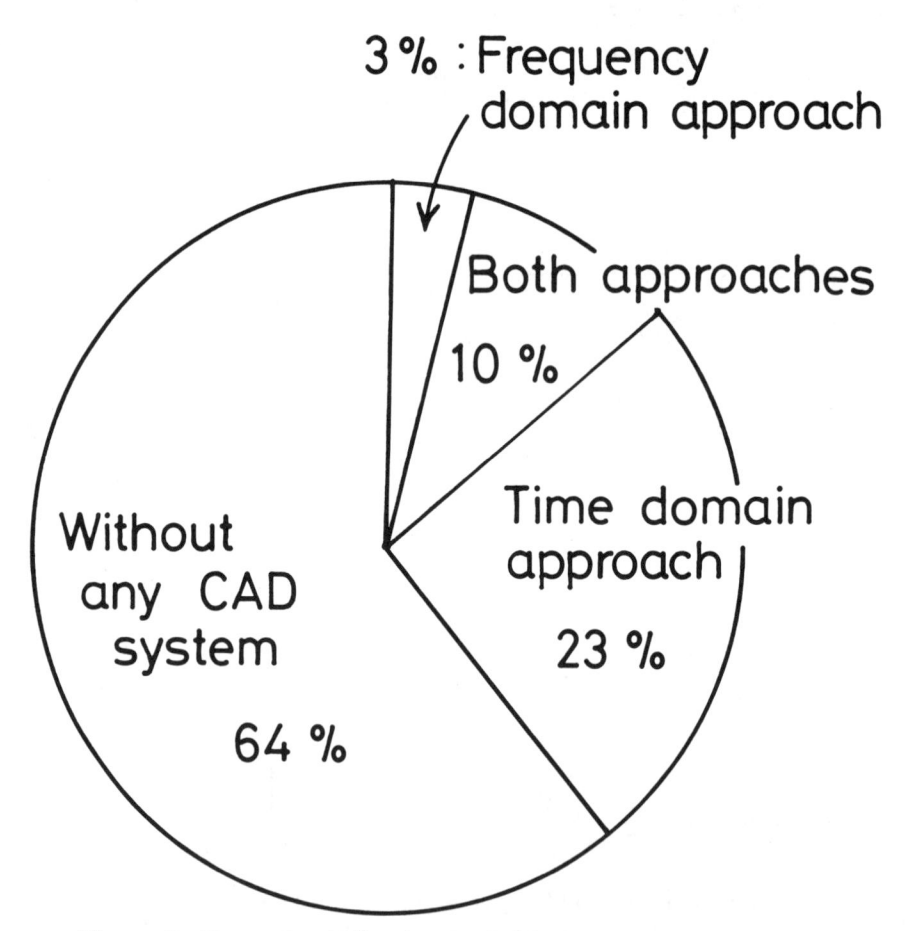

Figure 6. Computer Utilization in SISO Control System Design.

dynamic simulation are utilized. The dynamic simulation of the process with a control system is repeated in the time domain by changing the design parameters of the controller in order to make it optimal.

When difficult control problems emerge from the strong interaction among variables,—though the frequency of encounter of these problems may not be so great, as is clear from the result of the survey as shown in Fig. 5—, it is necessary to remove detrimental interactions by applying some kind of countermeasures.

As practical countermeasures, the following three can be used to overcome the so-called "interaction problem" which is the main cause of difficulty in the application of a single input-single output control strategy.

COUNTERMEASURE I

Change the process system structure by installing intermediate tanks and/or auxilliary equipment at proper locations so that the system results in fewer interactions between variables.

COUNTERMEASURE II

Find proper control loop pairings that have less interaction, and utilize different control schemes for different loops, even though each controller might be designed based on classical methods. For example, if a feedback control scheme is used for one loop, use a feedforward control scheme for the other. Use ratio-control for one loop and feedback control for the other.

COUNTERMEASURE III

Design a multivariable control system by taking into account interactions.

Figure 7 shows the utilization ratio of the three countermeasures mentioned above. Some 49% of all companies that encountered some serious trouble in designing control systems due to strong interactions have recourse to Countermeasure III; i.e., a multivariable control system is designed to overcome the difficulty.

Countermeasure II is used in 37% of all the companies, the single input—single output control strategy is still utilized by devising the selection of control loop pairings and adopting a different control scheme for each different loop. In 14% of all the companies, countermeasure I, i.e., the process system itself, is redesigned.

158   Chemical Process Control II

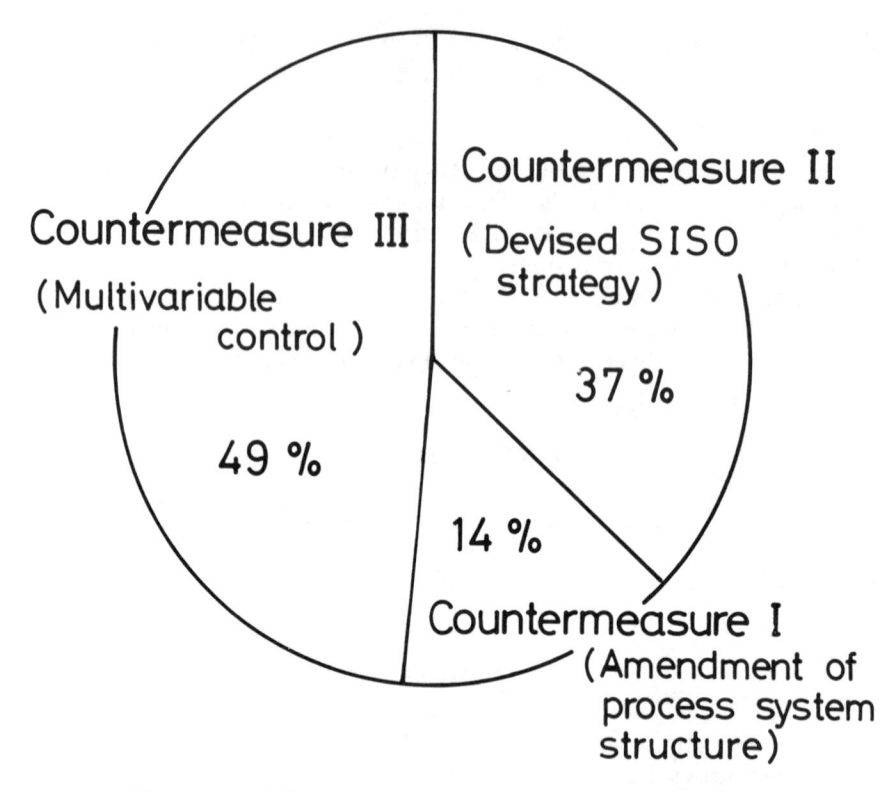

Figure 7. Utilization Ratio of Three Countermeasures.

Figure 8 shows the ratio of control techniques used for the design of multivariable control systems. In 44% of the companies which used countermeasure III, a decoupling control strategy is utilized. The multivariable control system designed based on the frequency domain approach, such as the Inverse Nyquist Array Method etc., was used in 33% of the companies which adopted countermeasure III. Optimal control policy based on the state space approach is also utilized in 27% of the companies.

Needless to say, the design of a multivariable control system with large dimension cannot be achieved without the help of a CAD system. At present, CAD systems called DPACS-F, CONPACK, CSDP1, CSDP2, etc., are developed in Japan based on the "state space approach" or the "frequency domain approach." The detailed explanation of these CAD systems is given later.

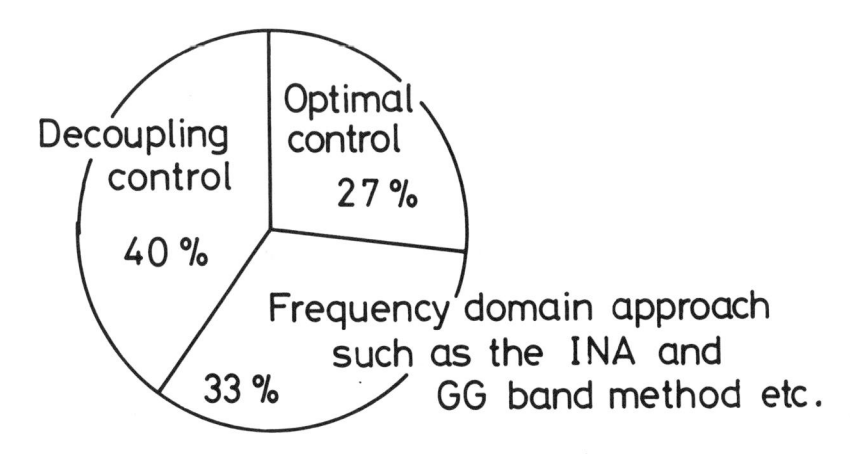

Figure 8. Control Techniques Used for the Design of Multivariable Control Systems.

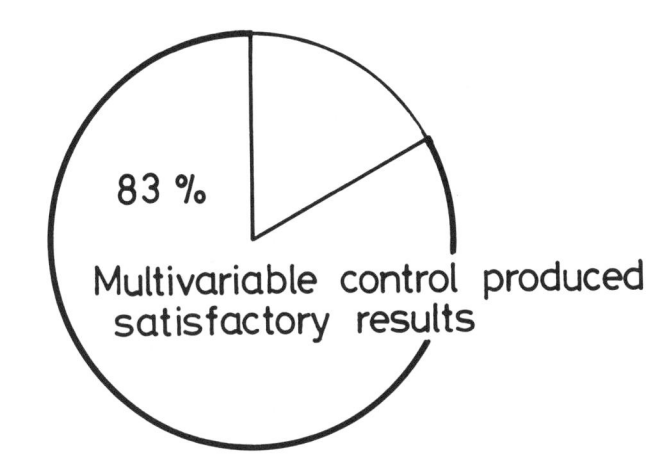

Figure 9. Percentage of Success of Multivariable Control System Design.

Although the multivariable control strategy has not been applied as frequently to actual problems as Fig. 5 and 7 might suggest, Fig. 9 shows that the designed multivariable control system produced satisfactory results in 83% of all applied cases.

As mentioned above, advanced control strategies, including multivariable control policy, have not been utilized fully in most chemical processing companies in Japan. The percentage of the companies that intend to utilize multivariable control techniques in the future, even though they have not yet used any advanced control, is as high as 63%. These companies are also interested in the introduction or development of CAD systems for advanced control. Of the remaining companies, 28% foresee little possibility of further successful application of advanced control theories. Approximately 9% of the companies have no interest in advanced control techniques, as shown in Fig. 10. It can safely be concluded that so far, most companies in Japan do not sufficiently utilize advanced control techniques for the design of high-quality control sys-

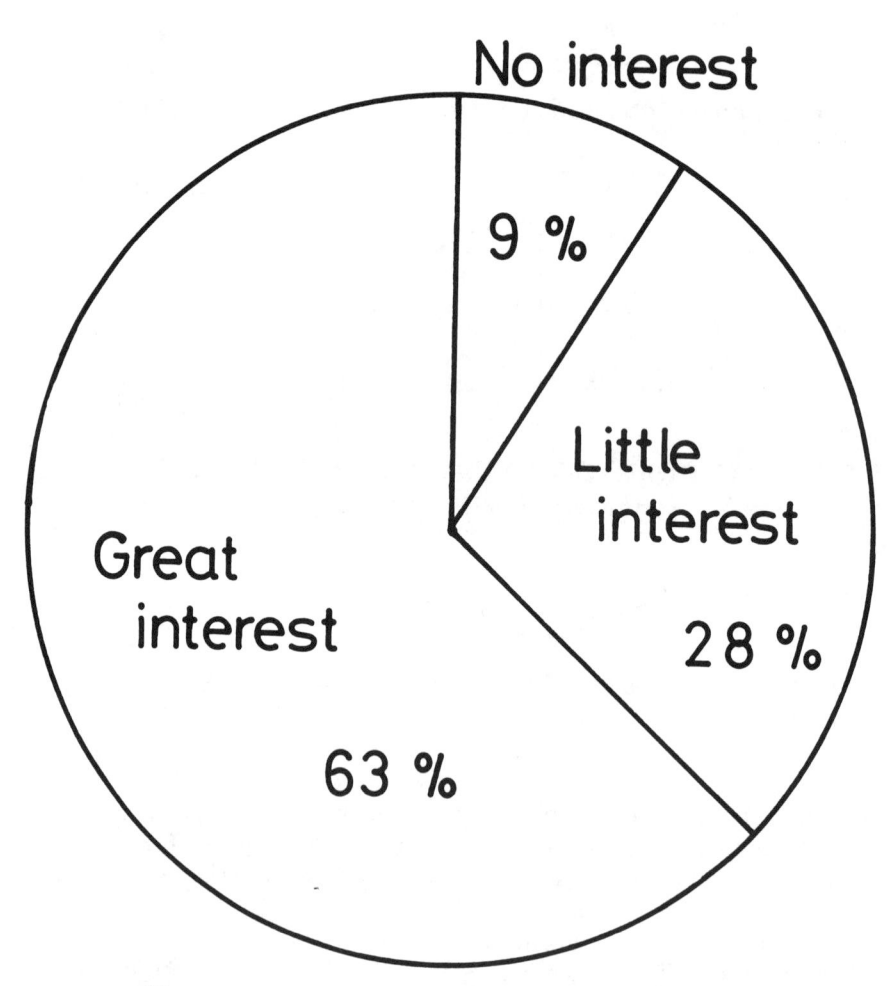

Figure 10. Interest in Advanced Control Techniques.

tems. But they have fully realized that there is a high probability that difficult control problems will arise and that there is an inevitable trend towards the intensive utilization of advanced control techniques in the future.

Fig. 11 shows that the performance of designed control systems is evaluated by performing dynamic simulation calculations in 40% of the companies surveyed. In 88% of the companies, the on-the-spot tuning of controllers is practiced in order to adjust the controller settings so that the process can be controlled in a desirable manner. Some 60% of them used only the method of adjusting and testing controllers on the spot. The remaining 28% used both dynamic simulation and on-the-spot tuning to adjust and evaluate control systems.

The type of controller, i.e., analog or digital, chosen for the implementation of a designed control system was surveyed. The abscissas in Fig. 12 and 13 represent the ratio of the cases where the designed control system is implemented in the digital type controller, to the total cases of

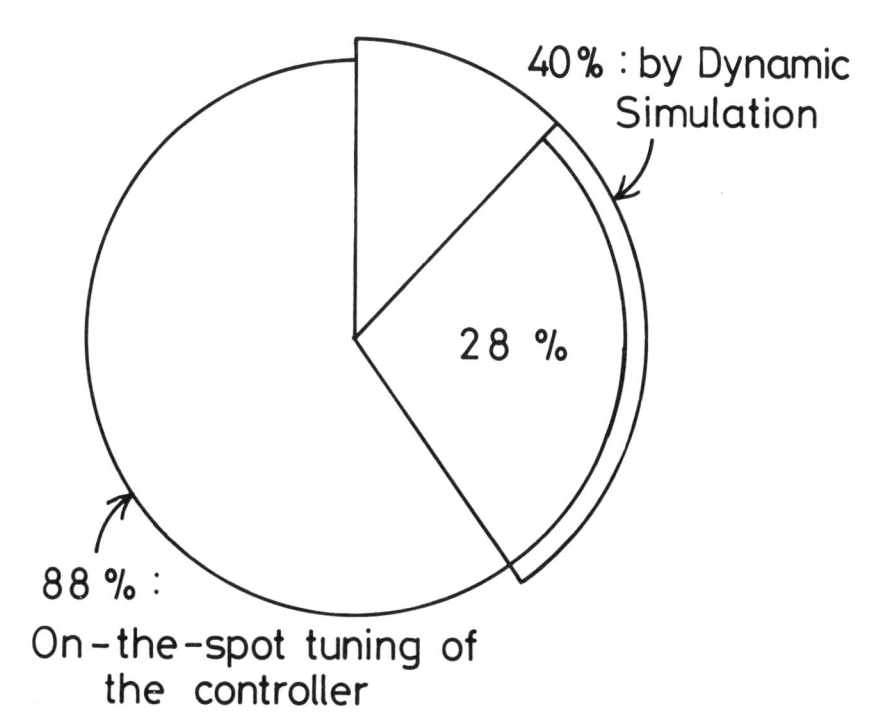

Figure 11. Evaluation of the Performance of the Control System.

162 Chemical Process Control II

frequency (%)

[Histogram with bars: 50 at 0-10, 14 at 10-20, 11 at 20-30, 7 at 30-40, values near 0 at 40-60, small bars at 60-70 and 80-90, axis labeled 10 20 30 40 50 60 70 80 90 100 (%)]

←—— Analog type   Digital type ——→

Ratio of the digital type controllers to the total cases of implementation

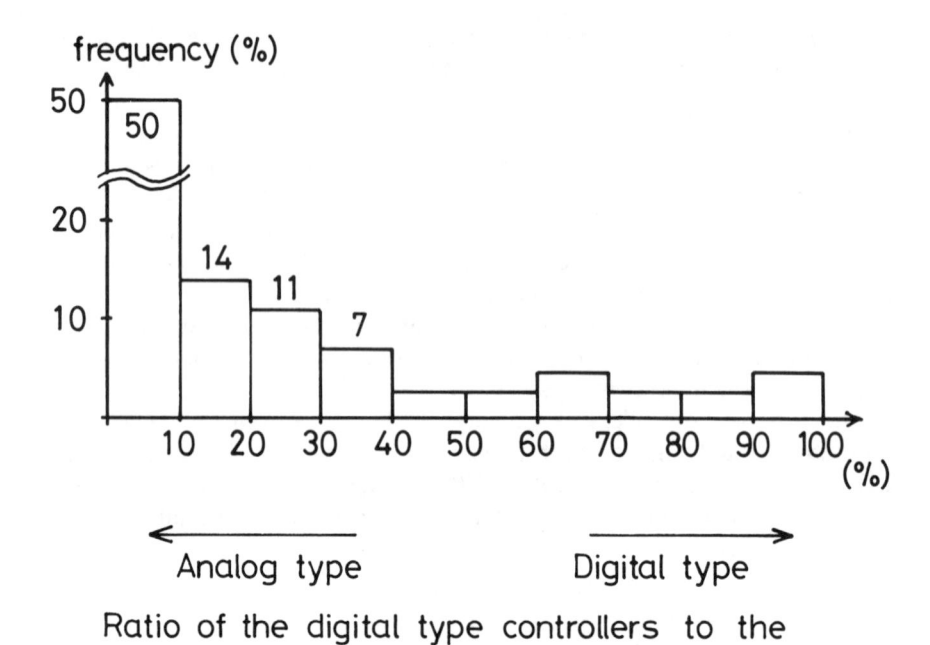

Figure 12. Implementation Ratio of Analog and Digital Type Controller at Present.

implementation. As is clear from Fig. 12, in 50% of all the companies, in 10 out of 100 cases, a control system is implemented in the digital type controllers. This means that the controllers were principally of the analog type.

Figure 13 shows the opinion of process control engineers in Japan with regard to the future of digital type controllers based on microprocessors over the next 5 to 10 years. Most engineers are quite confident that the digital type controller will become more popular. However, some insist that the analog type of controller will be fairly widely used even in the future. So far, the present status of chemical process control design and the computer utilization in Japan have been briefly surveyed. The so-called "general purpose type computer package" for simulation and CAD system for control system design developed in Japan are discussed in the following sections.

frequency (%)

[Histogram with bars: 10-20: 9, 20-30: (small), 30-40: 7, 40-50: (small), 50-60: 21, 60-70: (small), 70-80: 21, 80-90: 12, 90-100: 16]

← Analog type   Digital type →

Ratio of the digital type controllers to the total cases of implementation

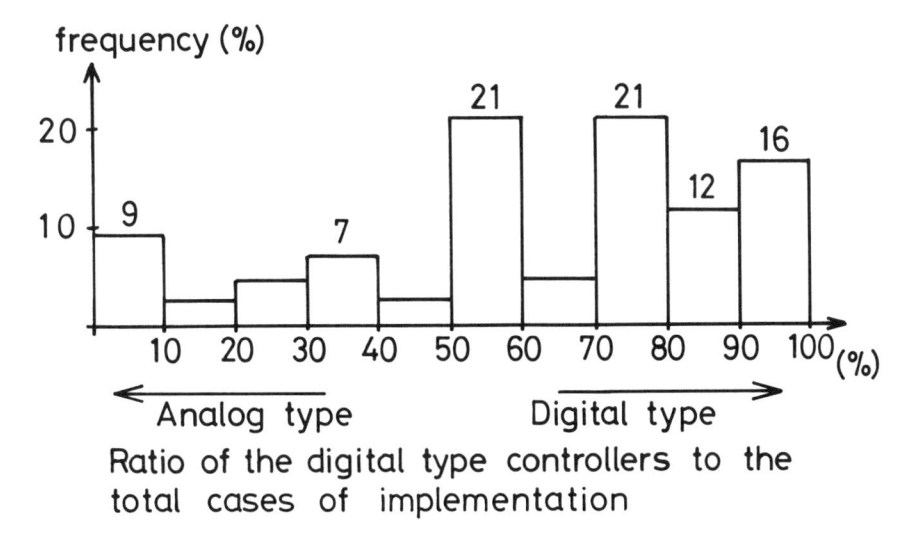

Figure 13. Implementation Ratio of Analog and Digital Type Controller in the Future.

DYNAMIC SIMULATION PACKAGES DEVELOPED IN JAPAN

As shown in Table 3, five dynamic simulation packages have been developed in Japan. These general-purpose type simulators can be used to facilitate;
  i) the analysis of the transient behavior of the process during start-up and shut-down.
  ii) the study of the dynamic responses of the process to various disturbances and to the change-over of various feeds.
  iii) the analysis of the operation of batch processes.
  iv) the design and evaluation of control systems, and
  v) the education and training of plant operators.

Most of the packages are developed according to the module-oriented approach so that the user can easily build the necessary model.

The functions and the structure of each package are briefly described below.

i) MODYS (Module Oriented Dynamic Simulator) [9]

MODYS was developed by Mitsui Toatsu Chemicals, Inc., in 1977 to facilitate many dynamic simulation studies by that company.

Table 3. Unsteady State Simulation Package.

Program Name	Developer	Completed	Size	Mode
MODYS (Module Oriented Dynamic Simulator)	MITUI-TOATSU	1977	250 KB (Minimum)	BATCH
TRASPO (Training Simulator for Plant Operator)	TORAY & FUJITSU	1976	64 KB	INTERACTIVE
TRSP (Transient Response Simulation Program)	IHI	1975	20 KW	BATCH
DOPL (Dynamic Operation Predictor Library)	CHIYODA	1974 (Version 1) 1978 (Version 2)	64 KW	BATCH & INTERACTIVE
DPS (Dynamic Process Simulation)	JUSE	1974	40 KW	BATCH

The user of MODYS can build the model of the total process by connecting "blocks" according to the topological structure of the process. Here, block means a set of "modules" connected in advance in order to represent the mathematical model of a processing unit, such as a reactor, heat exchanger, distillation column or controller, etc. A module is a FORTRAN subprogram consisting of some model equations and does not necessarily correspond to a processing unit.

The MODYS system consists of three parts as shown in Fig. 14, i.e., the Problem Oriented Data Base, MODYS BASIC, and MODYS Subsystem.

The Problem Oriented Data Base is prepared by the user in order to supply necessary information concerning an individual problem, such as the model expression obtained by assembling blocks and modules, the "operations" related to manipulations of process operators, actions of process control computers and disturbances set up by the user, and initial values of process variables and parameters, etc.

MODYS BASIC is the most essential part of this package. This part decodes all of the data given in the previous stage and generates

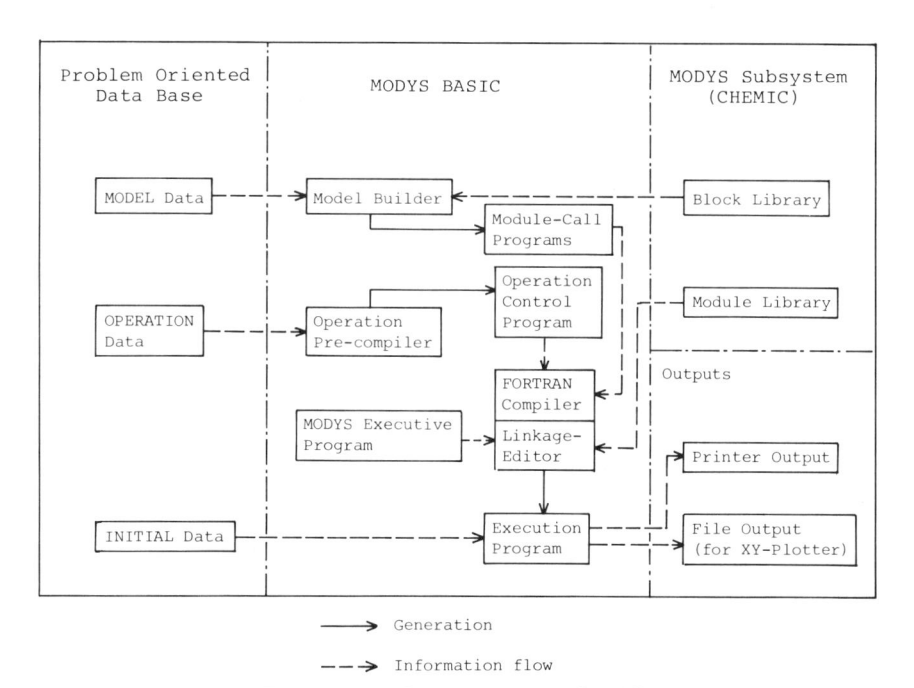

Figure 14. Structure of MODYS.

FORTRAN codes of subprograms which call module programs in the decided order and control the execution of the operations. These generated codes are compiled and linked with module programs in the Module library and MODYS executive program in order to make the execution program for carrying out the actual simulation calculation.

MODYS subsystem is a collection of module programs and blocks. The library in the subsystem CHEMIC contains about 50 blocks and 60 modules for various chemical processing units, control devices, physical property calculations, and utility subprograms for mathematical treatments. The user can easily supplement the library by adding modules and blocks prepared by himself.

This program package was developed under the IBM/370 Operating System. The source programs are coded in FORTRAN IV. Only the IBM version is now commercially available.

ii) TRASPO (Training Simulator for Plant Operator) [10]

TRANSPO was developed through the cooperation of Toray Industries, Inc. and Fujitsu, Ltd. in 1976. This package is designed to be usable in an interactive mode. TRASPO is able to re-create dynamic behavior which is exactly the same as that of the real plant by the execution of a real time simulation. This package, therefore, can be utilized especially to educate and train operators so that they can operate the process properly even in case of an emergency. In order to facilitate the model building of the total process system, this package provides basic subprograms for nine fundamental processing units and control devices, such as a reactor, vapor-liquid separator, heat exchanger, pump, tank, compressor, control valve, etc. Models for any other equipment can be easily added as supplementary subprograms to the library prepared in advance. Information concerning physical properties, such as enthalpies of vapor and liquid, equilibrium relationships and densities, etc. is also prepared in the library. The input data required of the user are, a) the number of components and streams, b) the mathematical models of the processing units and equipment, c) the topological structure of the process, and d) the initial values and parameters. The upper limits on the permissible number of components, streams, and unit operations are 10, 50, and 50, respectively. This package was developed by FACOM U-300 and needs peripherals, such as card reader, disk, line printer, CRT and operator panel, etc. The package is now available commercially.

iii) TRSP (Transient Response Simulation Program) [10]

TRSP was developed by Ishikawajima-Harima Heavy Industries Co., Ltd. in 1975. TRSP is provided with many modules consisting of subprograms for various types of equipment, such as values, tanks and pumps, etc. The digital simulation of the chemical, petrochemical, and LNG plants can be performed easily without the preparation of a transfer function for each unit. The input data required of the user are, a) the number of components, streams, and equipment, b) the models for equipment, c) the topological structure of the process, and d) the initial values and parameters. The maximum permissible number of components, streams, and operating units in TRSP are 200, 200, and 200, respectively. The package was developed for UNIVAC 1110. TRSP is not available commercially.

iv) DOPL (Dynamic Operation and Predictor Library) [11, 12]

DOPL was developed by Chiyoda Chemical Engineering and Construction Co., Ltd. in 1974 and revised in 1978. DOPL was devised as an aid to building models for dynamic simulation and to solving the problems the mathematical expression of which includes a set of differential equations. DOPL system consists of the two subsystems called BASIC and APPLICATIONS. BASIC, which could be referred to as a general purpose simulation language for the continuous system, can perform the three routines listed below. The input-output routine controls data, equations, and output by a printer and by graph plotting. The integration routine executes the numerical integration calculation based on the Euler method and the Runge-Kutta method, etc. The routine for all numerical calculations except for integration treats the convergence calculation of non-linear algebraic equations, stability analysis and local linearization of a non-linear equation, etc.

APPLICATIONS consists of several groups of application subprograms. This library provides models for various processing units and control devices as shown in Fig. 15.

The simulation model can be generated by connecting these modules according to the topological structure of the process system, the same as in CAPES (Computer Aided Process Engineering System developed by Chiyoda as a general purpose type steady state simulator).

DOPL is designed to be used in an interactive mode as well as in a batch mode, and to be supplemented by a user file which provides the user with a set of data, such as initial conditions and computational results.

FORT-mini language is used to describe a user's problem in the form of one of the main programs that control DOPL library modules. The grammar of the language is similar to that of elementary FORTRAN and BASIC, though some extended facilities are available to facilitate model building.

The inputs to be provided by the user are, a) mathematical models for equipment, b) process and control structure, and c) initial values and parameters.

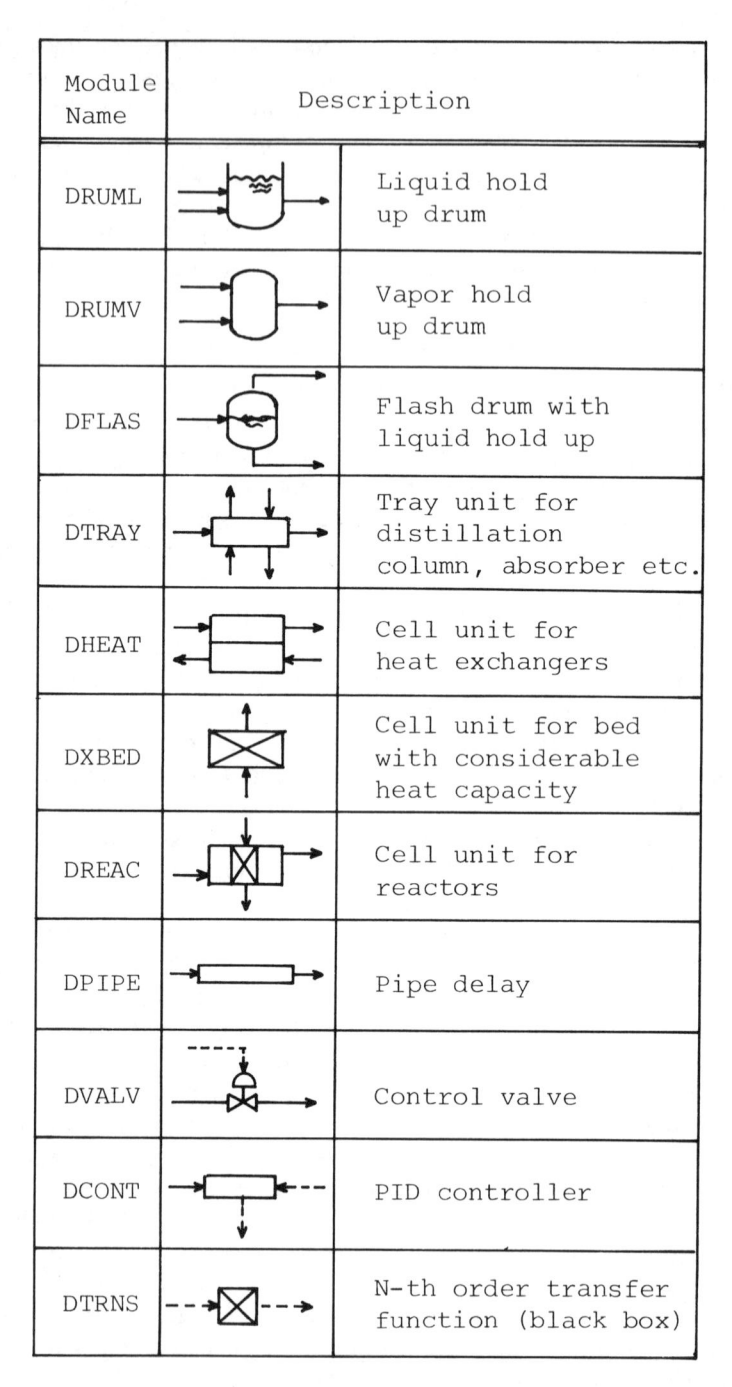

Module Name		Description
DRUML		Liquid hold up drum
DRUMV		Vapor hold up drum
DFLAS		Flash drum with liquid hold up
DTRAY		Tray unit for distillation column, absorber etc.
DHEAT		Cell unit for heat exchangers
DXBED		Cell unit for bed with considerable heat capacity
DREAC		Cell unit for reactors
DPIPE		Pipe delay
DVALV		Control valve
DCONT		PID controller
DTRNS		N-th order transfer function (black box)

Figure 15. Typical Modules Provided in DOPL.

The package was developed by IBM 3033. At present, DOPL is not commercially available.

v) DPS (Dynamic Process Simulation) [13, 14]

DPS was developed by the Institute of the Japanese Union of Scientists and Engineers in 1974. DPS is an equation-oriented, general-purpose simulation package with especially facilitates, i) the analysis of the dynamic perturbation of a process from its steady state condition due to various disturbances, ii) the testing of the performance and local control loops, and iii) the assessment of various operation schemes of both continuous and batch process.

DPS is provided with data bases which contain the model equations for about 80 standard "units" and "elements." A unit, which represents a reactor, distillation column, heat exchanger and PID controller, etc, consists of several elements, which are the minimal constituents of the module expression in DPS and correspond to the parts of chemical processing equipment, such as the control value and the switching valve, etc. The user can add any supplementary models to the library as part of the input data. The mathematical model equations for a whole system are automatically generated by the execution program based on data bases and topological information, which represents the structural connection of the units with mass and heat flows, and information flows. The whole system is finally expressed by a set of ordinary differential equations and non-linear algebraic equations. DPS can automatically determine the Optimal order of calculations of this large dimensional equation system by utilizing a graph-theoretical approach.

The input data required of the user are, a) the number of components, streams, and equipment, b) the models for the equipment, c) the topological structure of the process, d) the initial values and parameters, e) conditions to be altered in the course of integration which correspond to the operational procedures, such as opening or closing valves, f) physical properties, and g) specification of the method for solving equations.

The upper limits on the permissible numbers of components, streams, and unit operations are 20, 50, 100, respectively. This package can be used in the following computers; TOSBAC-5600/160, FACOM 230/55, IBM 370/158, CDC 6600, and UNIVAC 1110. DPS is commercially available, and 9 companies in Japan have already introduced it. ICI and CAD Centers of Cambridge University, U.K., have also introduced it.

THE CAD SYSTEM DEVELOPED IN JAPAN FOR CONTROL SYSTEM DESIGN

The help of computer software is indispensable to facilitate the complicated calculation and data handling necessary for the design of a

170   Chemical Process Control II

sophisticated control system. A CAD system, which can be used in an interactive mode and is based on a digital computer with graphic display terminals, provides a powerful tool for control system design and synthesis studies.

A brief explanation of the structure and functions of CAD systems developed in Japan is given below.

i) DPACS-F (Design Package for Control Systems developed by the Furuta Laboratory)

DPACS-F was developed by the Furuta Laboratory, Department of Control Engineering, Tokyo Institute of Technology, for a minicomputer (NOVA with 96 KB cpu memory and 10 MB disc memory). DPACS-F can perform data management, analysis, design, and simulation. The control systems are designed by choosing appropriate program modules specified by commands from the console.

The CAD program package can deal with linear time—invariant systems of any of the following five types: state space description, transfer function, Markov parameters, differential or difference equations, and input-output data. This system provides design algorithms based on both state-space and frequency domain approaches. The execution of the program can be controlled by a common line.

The structure of the program and the data-files of DPACS-F is shown in Fig. 16. A set of parameter values in any of the five types of the systems

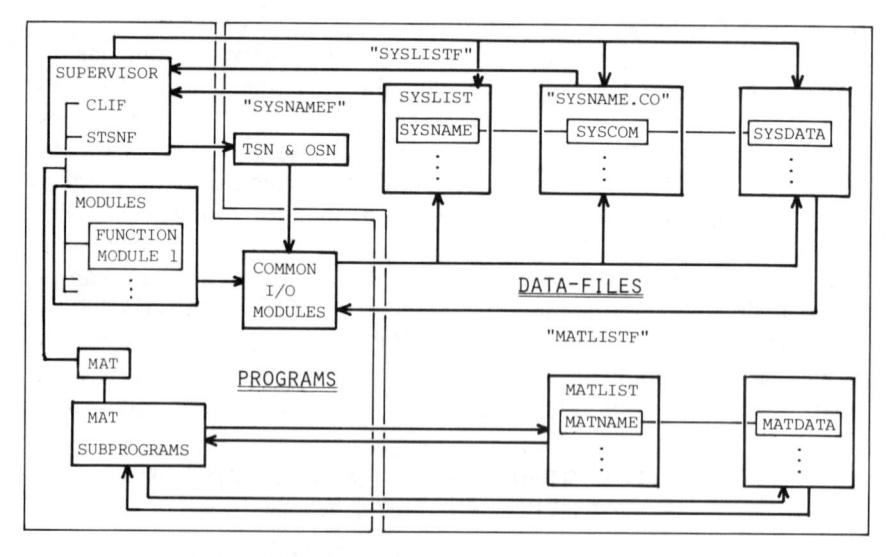

Figure 16. Programs and Data Structure of DPACS-F.

described above is called "system data (SYSDATA)." DPACS-F enables the designer to manage systems and their system-data under the same name. The system registered is managed by using a table SYSLIST which contains the information on, 1) the system name, ii) the numbers of inputs and outputs, iii) the sampling interval, and iv) the number of a set of registered system-data for representing the same input-output relationship. SYSLIST is stored in a file SYSLISTF. A set of system-data representing a system named SYSNAME is controlled through a table SYSCOM stored in a file SYSNAME.CO.

The supervisor of DPACS-F has a Command line Interpreter CLIF as an interface between the designer and DPACS-F, which can interpret the command line in the following form:

CMD/SW SYSNAME NEWSYSNAME/O

where CMD represents the command name to be executed corresponding to one of the function modules. SW specifies a subtask in the function appointed by the command name. SYSNAME is the name of the objective system. NEWSYSNAME with the switch "/O" is the name of the system newly produced as the result of the command. For example, when an observer is designed, the command name "OBS" is used as CMD. As for SW, one of G, M, F, and K can be selected. G, M, F, and K represent the subtask to design the Gopinath's state observer, Miller's state observer, functional observer, and Kalman filter, respectively.

CLIF checks the command line. If there is no error, a subprogram STSNF asks the designer to choose the type of system-data to be used in the execution of the command in case the system-data is stored in several different forms. Then STSNF sets all the information (Tables TSNE & OSN) necessary for data-handling in the corresponding function module in a file called SYSNAMEF, and starts the execution of the function module corresponding to the command. After the execution of the command, the system then enters the "posing state," ready for the next command from the designer. The designer, therefore, can design a control system in an interactive mode by repeating the input of command lines without being bothered by tedious data handling.

All of the functions of DPACS-F are listed in Table 4. Fig. 17 shows the hierarchical structure of these function modules of DPACS-F. Fig. 18 shows how the types of system-data are related by the various transformations under discussion.

After the controller design is finished, the responses of the closed-loop system should be simulated in order to evaluate the performance of the designed controller. For this purpose, DPACS-F has a module called LINK which enables the designer to easily connect subsystems according to an arbitary configuration and to combine the controllers and all the

Table 4. Descriptions of junctions in DPACS-F.

[1] System-Data Management
 1) SYSIN   : to input system-data
 2) TYPE   : to type out system-data
 3) REVISE   : to revise system-data
 4) COPY   : to copy system-data
 5) DELETE   : to delete systems or system-data
 6) RENAME   : to rename system name
 7) LIST   : to list up systems or system-data

[2] System-Data Transformation
 8) CMR   : to obtain minimal realization
 9) TF   : to obtain transfer function
10) MRKV   : to obtain Markov parameters
11) DIF   : to obtain differential equation
12) IO   : to obtain input-output data

[3] System Analysis
13) POLE   : to investigate stability
14) CCCO   : to investigate controllability
15) NYQST   : to draw (Inverse) Nyquest diagrams
16) BODE   : to draw Bode diagrams
17) LOCI   : to draw root loci
18) BAND   : to draw Gershgorin bands
19) RDCT   : to obtain a reduced order system
20) COOD   : to transform a coordinate
21) DUAL   : to obtain a dual system
22) DCMP   : to decompose a system

[4] Controller Synthesis
23) PLOC   : to solve a pole assignment problem
24) OPT   : to solve an optimal regulator problem
25) TRACK   : to solve a tracking problem
26) DCPL   : to solve a decoupling problem
27) OBS   : to obtain observer and Kalman filter

[5] Simulation
28) LINK   : to link systems
29) CLPS   : to obtain a closed-loop system
30) DIGIT   : to digitize a continuous system
31) SMLT   : to plot free responses

Table 4. Cont'd.

[6] MAT	:	to execute the following
1) MAT	:	to input, revise, copy and type
2) LIST	:	to list and delete
3) ADD	:	to execute addition
4) SUB	:	to execute subtraction
5) MUL	:	to execute multiplication
6) SMUL	:	to execute scalar multiplication
7) EIG	:	to obtain eigen-values & vectors
8) GEIG	:	to solve a general eigen-problem
9) MAP	:	to obtain the basis of range & kernel
10) INV	:	to obtain the inverse
11) LSQ	:	to obtain the least square solution
12) NORM	:	to calculate the norm
13) TRACE	:	to calculate a trace
14) VEC	:	to investigate column-independency
15) COOD	:	to transform a coordinate
16) TRNS	:	to obtain a transposition
17) PICK	:	to pick columns
18) EMBED	:	to embed columns
19) PRMT	:	to interchange columns
20) NORMZ	:	to normalize
21) RSLV	:	to obtain a resolvent matrix
22) EXP	:	to obtain a transition matrix
23) IEXP	:	to obtain an integral of trans. mat.
24) BYE	:	to stop MAT
[7] BYE	:	to stop DPACS-F

external signals, such as disturbances, reference signals, etc. Therefore, the simulation of the total system can be carried out without any difficulty, and the testing of the controller performance becomes an easy task.

The upper limits on the permissible number of the pairs of input-output variables, and the dimension of the state variables are 10 and 20, respectively.

ii) CONPAC (Computer Aided Control System Design Package) [17, 18]

CONPAC was developed by the Engineering Research Association of Nuclear Steelmaking as a modified version of DPACS-F for a large computer. The function of CONPAC is almost the same as DPACS-F,

174   Chemical Process Control II

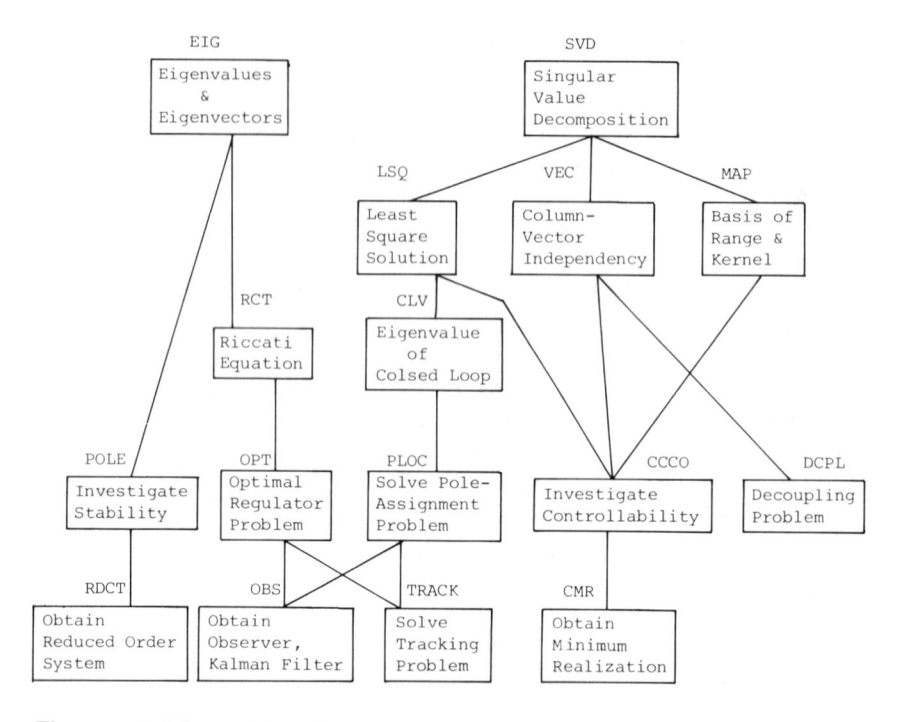

Figure 17. Hierarchical Structure of the Function Modules in DPACS-F.

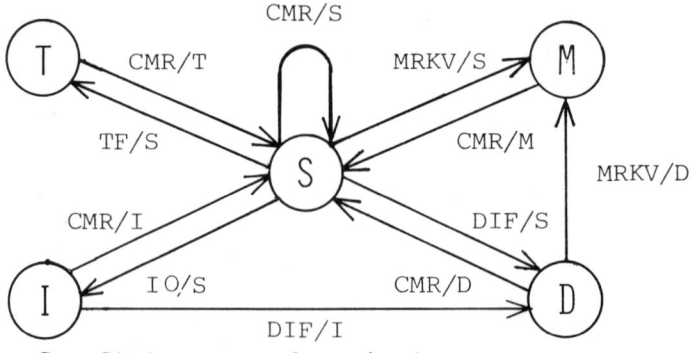

S: State-space description
T: Transfer function
M: Markov parameters
D: Differential or Difference equation
I: Input-output data

Figure 18. System-data Transformation.

except that CONPAC can deal with problems of higher dimensions. The upper limits on the permissible number of pairs of input-output variables and the dimension of the state variable are 20 and 50, respectively.

The hardware configuration of CONPAC is shown in Fig. 19. CONPAC is written in FORTRAN for UNIVAC 1110 with 6000 steps.

iii) CSDP-1 and CSDP-2 (Control Systems Design Program) [19, 20]

CSDP-1 & 2 were developed by Kawasaki Heavy Industries, Ltd. CSDP-1 is a CAD system for control system design based on a minicomputer and used in an interactive mode.

The software structure of CSDP-1 is shown in Fig. 20, and the structure of data-handling is shown in Fig. 21. All of the data handled by CSDP-1 is filed in a disk in the following five forms: 1) a transfer function matrix $G(s)$ (File No. 10–29), 2) a frequency response matrix $G(j\omega)$ (File No. 30–39), 3) a state space expression $(A, B, C, D; \dot{x}=Ax+Bu+Dd, y=cx)$ (File No. 40–49), 4) a constant matrix (M) (File No. 50–69), and 5) the

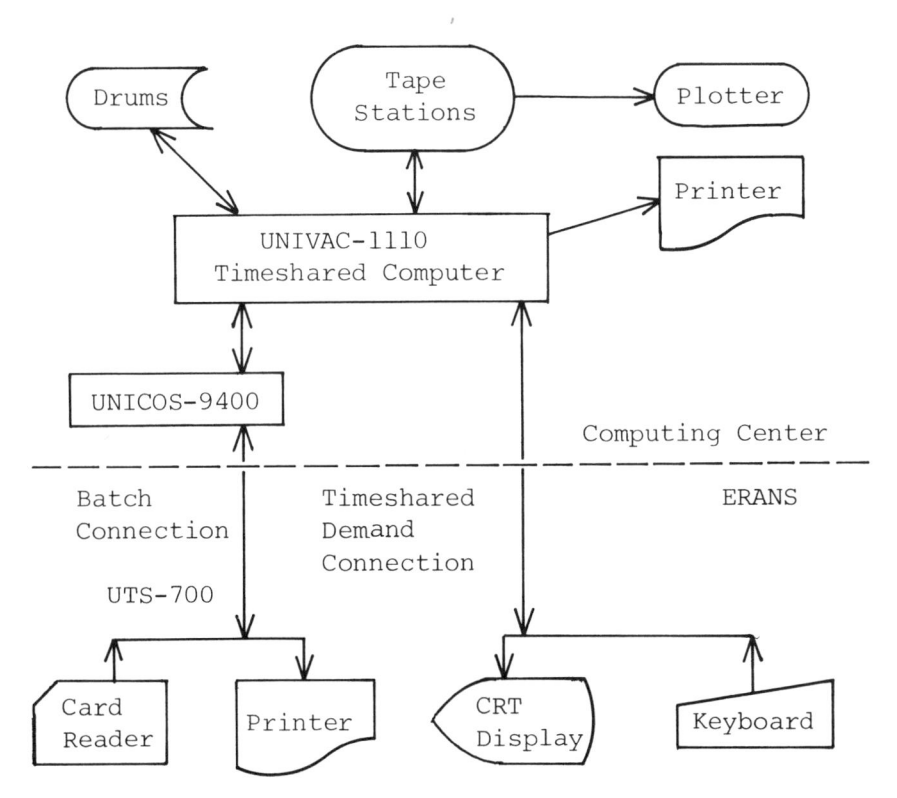

Figure 19. Hardware Configuration of CONPAC.

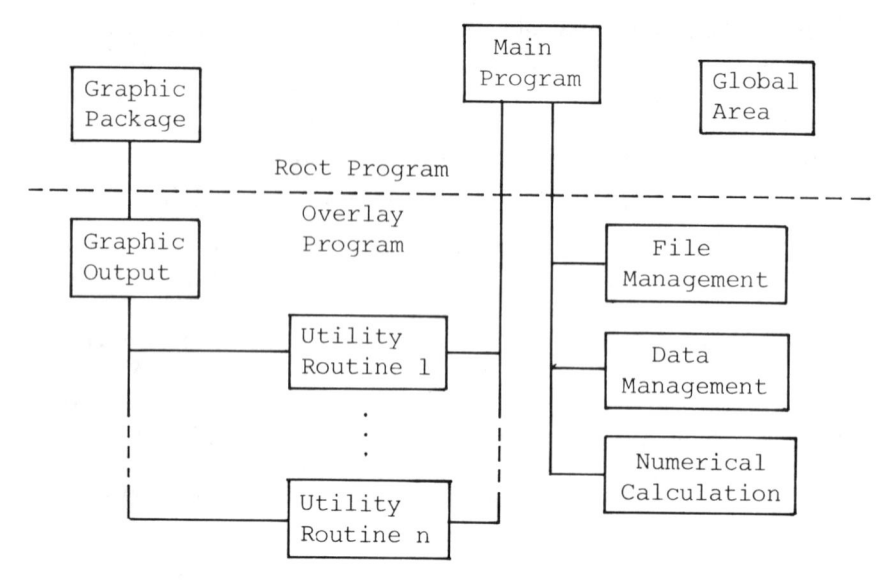

Figure 20. Structure of Software in CSDP-1.

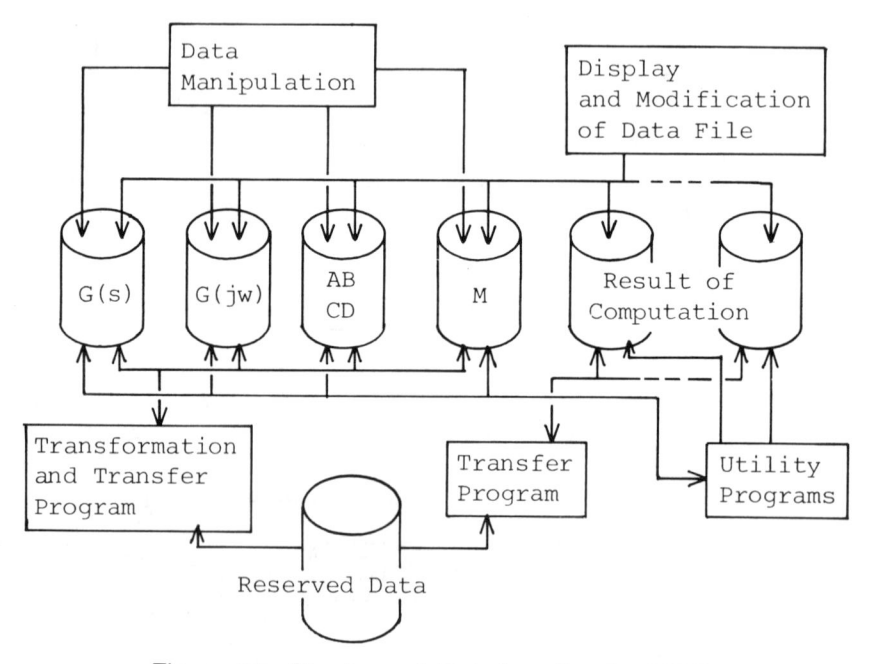

Figure 21. Structure of Data-handling in CSDP-1.

sequence of data of fixed values (the calculated values, the data necessary to be preserved). The user can stipulate the system by using the file number assigned to each system description shown above.

As the user does not need to worry about the difference in the forms of data, he can handle data and files very easily. The program is executed by inputting the commands represented by two letters of the alphabet. When a command related to the file management, utilities, and methods is input, the control is transferred to each program, and the task is operated in an interactive mode. As a communication device, a seven-colors graphic display is used to facilitate the interactive design procedure. CSDP-1 is programmed by FORTRAN.

The function of CSDP-1 can be summarized as shown in Table 5. The hardware configuration is shown in Fig. 22.

CSDP-2 is a modified version of CSDP-1 for a large computer. CSDP-2 can accept the model developed by using a simulation language, such as a CSMP (Continuous Systems Modeling Program by IBM) etc. CSDP-2 can treat problems where the dimension of state variables and the number of pairs of input and output variables are less than 100 and 10, respectively.

ACTIVITIES RELATED TO DEVELOPING CAD SYSTEMS IN JAPANESE ACADEMIA

In Japanese academia, considerable effort has been devoted not only to educate students better, but also to develop advanced control technology itself, both theoretically and practically.

In this context, many staff members in Japanese educational institutions are involved in the development of CAD systems which provide various benefits to education, research and industry. Most CAD systems developed in the academic field are a set of program packages to facilitate one or some combination of the following procedures; modeling, design, and evaluation of control systems. They are not assembled in the form of a complete CAD system, such as DPACS-F, CONPAC, or CSDP-1 & 2 mentioned in the previous section.

For industrial use, a well organized large CAD system with many different functions is the most helpful. However, from the educational viewpoint, it provides a possible disadvantage in that a student may learn only how to use it and thus become a "cookbook engineer" who cannot understand the methods used in the CAD system and is also unable to check and recognize the significance of the answer obtained. This is probably one of the reasons why most computer-aided program packages in academia are not assembed in a complete CAD form.

Table 5. Functions of CSDP-1.

Classification	Functions
Control of the Execution	Initiation of CSDP-1 Termination of CSDP-1 Analysis of commands List and revise of parameters
File-Management	Generation of files Revise of data Transfer of data
Data-Manipulation	Addition of matrices Multiplication of matrices Calculation of inverse matrices Transformation of the type of data
Controller Synthesis based on Classical Theories	Calculation of transient responces Drawing of Bode diagrams Drawing of Nyquist diagrams Drawing of Nicholes diagrams Drawing of Gershgorin bands
Controller Synthesis based on Modern Control Theories	Solution of optimal regulator problem Solution of observer problem Solution of pole assignment problem Solution of decoupling problem
Others	Execution of simulation

As an example, MFSD (Multivariable Feedback System Design) developed in the Kondo-Laboratory, Department of Electrical Engineering, Kyoto University, is now taken up and briefly explained. This program package enables the design of multivariable control systems based on a newly developed design method called the "Generalized-Gershgorin-Pseudo-Band method."

GG-Pseudo-Band Method

This method was proposed by Araki [21] as an extension of the INA

```
   IBM                IBM                       Process
   370-168            SYSTEM/7      (32KW)       I/O
         Channel
```

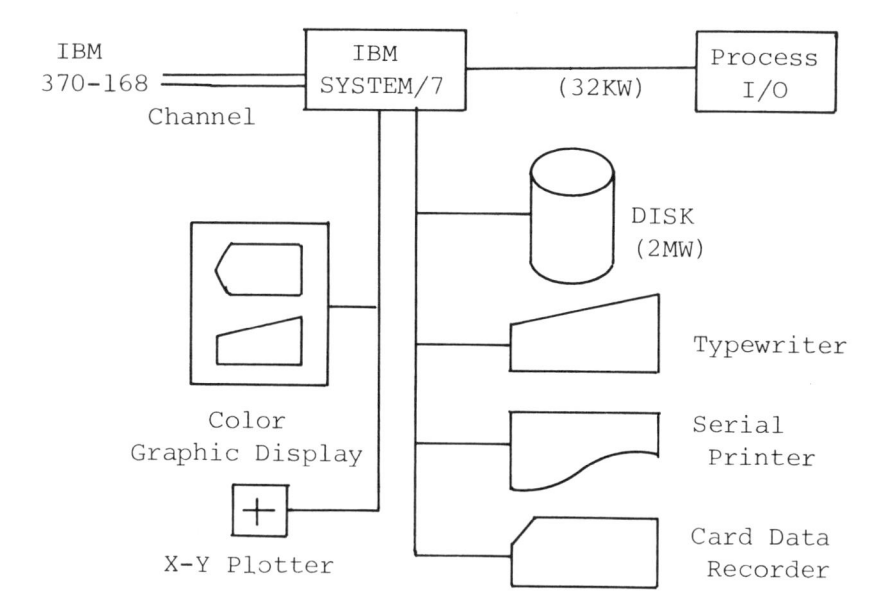

Figure 22. Hardware Configuration of CSDP-1.

method. Here, only a short description is given in order to clarify the main features of the design procedure.

Consider an n-input n-output system as shown in Fig. 23. Here, r, u and y are n-vectors representing the reference, manipulating and output variables, respectively. $G_p(s)$ is the transfer matrix of the plant to be controlled. $G_c(s)$ is the precompensator to make $G(s) \triangleq G_p(s) G_c(s) = [g_{ij}(s)]$ nearly diagonal. $K(s) = [\text{diag.} (k_i(s))]$ is the diagonal main controller.

For an arbitrary square matrix $Q = [q_{ij}]$, here the interaction matrix $C(s|Q) = [c_{ij}(s|Q)]$ is introduced as follows;

$$c_{ij}(s|Q) = \begin{bmatrix} 0 & i = j \\ q_{ij}(s) / q_{ij}(s) & i \neq j \end{bmatrix}$$

Then, the maximum eigenvalue of $C(s|Q)$, $\lambda(s|Q)$, is called the interaction index of Q.

When n=2, $C(s|Q)$ and $\lambda(s|Q)$ are given by

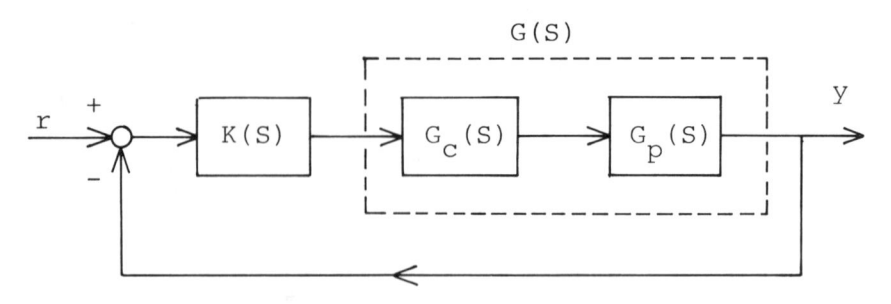

Figure 23. A Controller System.

$$C(s|Q) = \begin{bmatrix} 0 & q_{12}(s)/q_{22}(s) \\ q_{21}(s)/q_{11}(s) & 0 \end{bmatrix}$$

and

$$\lambda(s|Q) = |q_{12}(s) \, q_{21}(s) / q_{11}(s) \, q_{22}(s)|^{1/2}$$

It is clear from the definition of $\lambda(s|Q)$ that $\lambda(s|GK) = \lambda(s|G)$, that is, the interaction index depends only on the non-diagonal matrix $G(s)$ and is not influenced by the diagonal controller $K(s)$.

It is also verified that the relative change of the open-loop transfer function of the i-th loop of the system shown in Fig. 23 when the other loops are closed, is bounded by $\lambda(s|G)$.

The disk with center $k_i(jw)g_{ii}(jw)$ and radius $\lambda(j\omega|G) \mid k_i(j\omega)g_{ii}(j\omega) \mid$ on the complex plane and the band swept out by the disk when ω changes from 0 to $+\infty$, are called "the GG-disk (Generalized Gershgorin Disk)" and "the GG-band (Generalized Gershgorin Band)," respectively. A theorem has been derived which implies that the polar plot of the i-th loop lies inside the i-th GG-band if the GG-bands of the other loops do not include the point $(-1, 0)$.

The disk and the band obtained by mapping the GG-disk and the GG-band on the gain-phase plane are called "the GG-pseudo-disk" and "GG-pseudo-band," respectively (see Fig. 24). Here, it should be noticed that the GG-pseudo-disk is determined completely by $\lambda(j\omega|G)$, and the change of $k_i(j\omega)$ only causes the parallel displacement of the GG-pseudo-disk of the i-th loop. Thus, the theorem mentioned above can be restated as follows: the log-modulus plot of the open-loop transfer function of the

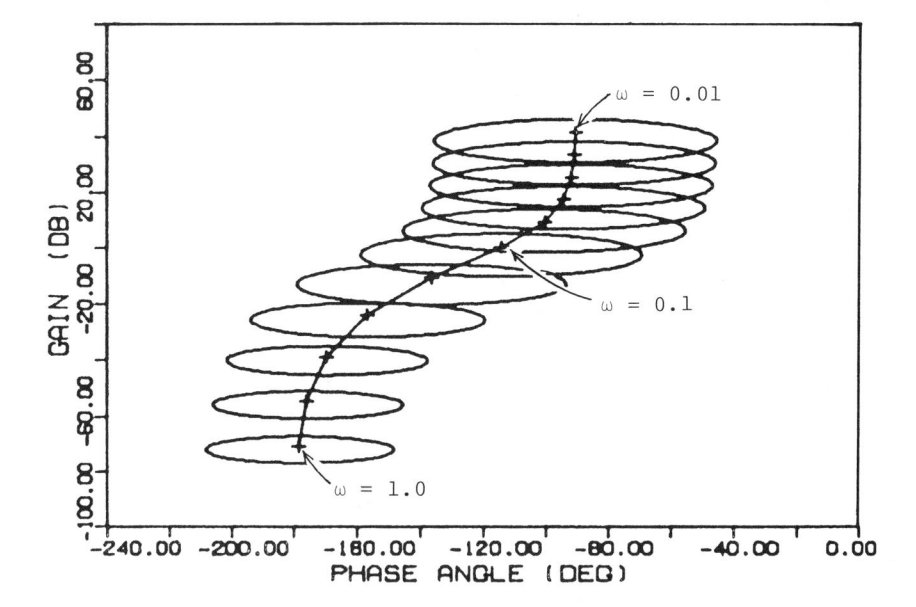

Figure 24. GG-pseudo-band in the Gain-phase Plane.

i-th loop lies inside the i-th GG-pseudo-band if the GG-pseudo-bands of the other loops do not include the point (OdB, −180°).

As for the stability condition, the following theorem has also been derived:

When $G(s) \triangleq G_p(s) G_c(s)$ has no unstable poles, every closed loop in the system of Fig. 1 is stable if each GG-band neither includes nor encircles the point (−1, 0) (equivalently, none of the GG-pseudo-bands include the point (OdB, −180°)).

Based on the theorems derived, Araki proposed the following design procedure:

step i) Find suitable pairing between the input and output variables by taking into account the strength of their physical connection.

step ii) Determine the precompensator $G_c(s)$ as to make $G(s)$ ($=G_p(s) G_c(s)$) diagonal-dominant by checking the interaction index $\lambda(s|G)$.

step iii) Draw the GG-band on the complex plane or equivalently, the GG-pseudo-band on the gain-phase plane.

By regarding the boundaries of these bands as the true response of the controlled object, evaluate the stability characteristics of the controlled process and adjust the main controller by applying some classical design technique.

When the GG-pseudo-band on the gain-plase plane is used, the Nichols chart technique can be very easily applied to adjust the gain constants of the controller because the radius of the GG-pseudo-disk is not changed, and the GG-pseudo-band moves only up and down when the gain parameter of the controller is changed.

step iv) Evaluate the over-all performance of the controller obtained in the above steps by performing simulation calculations. When a desirable response is not obtained for some disturbances, return to step iii) to redesign the main controller. If the interaction between loops is strong, return to step ii) to redesign the precompensator. Repeating these steps, a rational multivariable control system can be designed.

This design method has been successfully applied to the design of multivariable control system for a practical boiler system [22].

In the author's laboratory, a program package was developed based on this method and utilized to design a multivariable control system for a binary distillation column. The performance of the designed control system was evaluated by using an experimental distillation column with 10 plates of 12 cm diameter, and a satisfactory result was obtained [23]. This method is also applied to the multivariable control design problem of a heat integrated distillation columns system [24].

CONCLUSION

The present situation with regard to the control of chemical processes and the utilization of computer-aided design systems were surveyed by means of interviews and questionnaires responded to by 60 leading companies, most of which belong to the chemical and petrochemical industries. General purpose type packages developed in Japan for the dynamic simulation and the control system design were also reviewed.

The results of the survey might be distorted to some extent due to the companies' secretiveness, i.e. they would not make public even the fact that they applied advanced control techniques to some practical problems, irregardless of their success or failure. However, the general tendency with regard to this topic in Japan can be said to be satisfactorily revealed in the survey reported here.

Most chemical process system have been designed by adding generous design margins and installing many intermediate storage tanks and much auxiliary equipment so that designed systems become safe and self-regulatory processes which do not pose any difficult control problems. Control systems, therefore, can be designed very easily by applying classical methods.

According to the present survey, companies have not often encountered difficult control problems that could not be solved without applying more sophisticated and advanced control techniques. Consequently, CAD systems which are necessary for the design of advanced control systems are scarcely utilized at present.

However, most companies are greatly interested in the future application of advanced control techniques. This probably means that they anticipate the emergence of more difficult control problems and the necessity of improvement in the performance of process control systems in the future.

As long as the process system is designed to be stably operated without any sophisticated control systems with high performance, the necessity for advanced control techniques will not drastically increase.

In the future, process system design itself will be integrated with control system design, and process systems which are fully backed-up by a sophisticated control system will be designed and operated under conditions which used to be discarded in the traditional design procedure. In this way, advanced control techniques will play a crucial role in order to back up the stable operation of future chemical process systems which are likely to be highly interconnected and have smaller design margins and fewer intermediate tanks.

The future control system which can secure the optimal operation of such large complex process systems must have a decentralized structure. The control system of each subsystem would be designed by an advanced control theory such as the multivariable control technique by taking into account their existing strong interactions and using Micro- or Minicomputers. The coordination between subsystems could be performed by a large computer so as to optimize the overall objective.

We must endeavour to exploit advances in this direction, such as the development of powerful CAD systems, in order to provide a far better control system for the actual complex and large chemical systems of the future.

ACKNOWLEDGMENT

The authors wish to express their sincere thanks to all of the companies that responded to the questionnaire, and all the engineers who accepted our interview and offered their most recent data. The authors are grateful to professor N. Suda, Department of Control Engineering, Osaka University, the chairman of the research committee of multivariable control of actual systems, sponsored by the Society of Instrument and Control Engineers, Japan, for his permission to send questionnaires to all the committee members from the industrial sector, and to Professor K. Furuta, Department of Control Engineering, Tokyo Institute of Technology, for his offer of the variable information on DPACS-F. Professor M. Araki, Department of Electrical Engineering, Kyoto University is also greatly acknowledged for his helpful discussions and his offer of the information of the GG-pseudo-band method and its CAD system.

REFERENCES

[1] Munro, N., Broland, B. J., and Brown, L. S. "The UMIST Computer-aided Control System Suite—Introduction and User's Guides," National Research Development Corporation, Great Britain (1974).
[2] Rosenbrock, H. H., "Computer-aided Control System Design," Academic Press (1974).
[3] Cripps, M. D., "An Integrated Approach to Computer-aided Design of Control Systems" Imperial College Publication 74-40 (1974).
[4] Morari, M., A. M. Lenhoff, D. F. Marselle and D. F. Rudd, "Design of Resilient Energy Intergrated Processing Systems," Presented at 72nd Annual Meeting, AIChE, San Francisco (1979).
[5] Arkun, Y. and G. Stephanopoulos, "Interaction beween Process Design and Control: A Study in Operability," presented at 72nd Annual Meeting, AIChE, San Francisco, (1979).
[6] Govind, R. and G. J. Powers, "Control System Synthesis Strategies," presented at 82nd National Meeting, AIChE, Atlantic City (1976).
[7] Umeda, T., T. Kuriyama and A. Ichikawa, "A Logical Structure for Process Control System Synthesis," Preprints of 7th IFAC Congress, Helsinki, Finland (1978).
[8] Morari, M., Y. Arkun and G. Stephanopoulos, "Studies in the Synthesis of Control Structures for Chemical Processes, Part I, II, and III," *AIChE J.*, 26(2):220 (1980).
[9] Oguchi, G. and K. Okano, "Dynamic Simulators," Private communication from Mitsui Toatsu Chemicals, Inc.
[10] Oh'shima, E. & H. Shono, "Chemical Engineering Calculation," *bit*, 9(9):975 (1975). (in Japanese)
[11] Shindo, A. and T. Maejima, "Outline and Features of DOPL," Private communication from Chiyado Chemical Engineering & Construction Co. Ltd.
[12] Maejima, T., "Computer-aided Process System Design (VI)—Dynamic Simulation," *Chemical Engineering*, No. 7 (1975). (in Japanese)
[13] Syono, H., "Dynamic Process Simulation—DPS," *Software Circulation* (Saftware Ryutsu), No. 5 (1980). (in Japanese)
[14] Harima, M., "Introduction of DPS Program," Private communication from Inst. of Japanese Union of Scientists and Engineers. (in Japanese)
[15] Furuta, K. and H. Kajiwara, "CAD system for Control System Design," *Jr. of the Society of Instrument and Control Engineers, Japan,* 18(9), (1979). (in Japanese)
[16] Furuta, K., H. Kajiwara and K. Tsuruoka, "Computer Aided Design Program For Linear Multivariable Control Systems," Proceedings of the IFAC symposium on Computer Aided Design of Control Systems, 267, Zürich, Switzerland (1979).

[17] Furuta, K., H. Kajiwara, T. Okutani, K. Tsuruoka, T. Miyasugi and T. Yoshida, "Simulation in Computer Aided Design for Large Complex Systems," Simulation of Control Systems, I. Troch (ed.), IMACS, North-Holland Publishing Company (1978).

[18] Tsuruoka, K., T. Miyasugi and H. Miyamoto, "Computer Aided Design of the Control System of a Nuclear Steelmaking Process," Proceedings of the IFAC Symposium on Computer Aided Design of Control Systems, 535, Zürich, Switzerland (1979).

[19] Ogami, K. and T. Yamauchi, "Development of a CAD Program for Control System Design," presented at the National Meeting of Electrical Engineers, Japan (1979). (in Japanese)

[20] Yamauchi, T., K. Sugimoto, K. Ogami and M. Nakagawa, "CAD system adjusting to the Model Expression by a General-Purpose Simulation Language," presented at 24th Meeting on System and Control, sponsored by Japan Association Automatic Control Engineers, Japan (1980). (in Japanese)

[21] Araki, M., K. Yamamoto and B. Kondo, "GG-Pseudo-Band Method for the Design of Multivariable Control Systems," to appear in 8th IFAC Congress, Kyoto, Japan, (1981).

[22] Shibata, H., K. Yamamoto and M. Araki, "Multivariable Control for a Boiler (1)," presented at Symp. on Multivariable Control Applied to Practical Examples, Kansai Branch of SICE, 133, (1978). (in Japanese)

[23] Takamatsu, T., I. Hashimoto, M. Iwasaki and M. Nakaiwa, "An Experimental Study of the Multivariable Control of a Distillation Column," to appear in 8th IFAC Congress, Kyoto, Japan (1981).

[24] Takamatsu, T., I. Hashimoto, and T. Watanabe, "Multivariable Control System Design for a Heat Integrated Distillation Columns System," to appear in 2nd World Congress of Chemical Engineering, (1981).

NEW RESULTS AND THE STATUS OF COMPUTER-AIDED PROCESS CONTROL SYSTEM DESIGN IN EUROPE

Arne Tyssø
Division of Engineering Cybernetics
The University of Trondheim
Norway

ABSTRACT

This paper surveys design principles and structures of the most important interactive CAD packages in Europe concerning model identification, parameter estimation, simulation and design of control systems. Hardware as well as software problems are discussed and the most frequently implemented methods are shortly reviewed. The presentation will focus upon CAD systems which have been used in industrial applications and that present the state of the art of CAD in Europe.

INTRODUCTION

During the past few years the most significant developments in CAD in Europe have been the increasing numbers of interactive program systems tailored to meet the needs of industrial users and the implementation of complex algorithms to handle problems like nonlinear estimation and adaptive control.

Development and use of CAD packages in Europe started in the early sixties. The packages were based on batch operation and the field of interest was mainly concentrated on simulation and control system design. The design methods were based on classical theory such as Nyquist diagrams, Bode-plots and Evans root-loci.

Late in the sixties interactive operation began to dominate, and multivariable time domain and frequency domain methods were introduced. The CAD systems were mainly used at universities and polytechnics, and the reported applications were characterized by simple low order, linear systems, often representing a pilot plant or a laboratory model of some kind.

The first available program systems were made strongly dependent upon the computer system and minor or no attention was paid towards the quality of software, for instance, standardization, portability, and modularity. The spread of CAD packages was therefore very limited. When the modern minicomputer, with graphic displays and improved computation capabilities emerged in the mid-70s, the field of CAD had an immense growth. The packages, handling the modelling, identification, and parameter estimation problem were developed and at the same time improved control system design methods oriented towards practical applications, were introduced. The CAD systems became powerful tools in control education and research, and engineers from industry turned their attention to the packages. The software quality had improved considerably, and several CAD systems were successfully transferred to computer systems in industry and research institutes, and interesting applications were reported. However, it often turned out that the numerical robustness of the implemented algorithms were poor and therefore failed in operation with larger, industrial and non-technical problems. There also clearly existed a gap between the theoretical description of the CAD packages, for example, "we emphasize efficiency, modularity, good structure, general methods" and the practical experiences from using them.

In order to give CAD the real breakthrough in industry there has been a continued improvement of algorithms and software, and today there are many CAD systems available in Europe, which one may say quite well satisfy the ideal requirements.

There are a large amount of CAD systems available in Europe and it is, of course, impossible to discuss every package. I will try to concentrate on CAD systems which are oriented towards industrial problems and seem to satisfy the general requirements to an applicable CAD system. If some packages are missing, it does not necessarily mean that the packages are of minor quality but, in the author's opinion, that they are too specialized or too academically oriented. It is natural to classify the existing program packages in three categories: Control System design, Identification and Parameter Estimation, and Simulation. Within each category there are also general purpose programs for matrix and time series manipulation, and programs for system analysis and transformation of system representation forms.

A common feature for most of the packages is that many different methods are implemented and various system representation forms are accepted. The more advanced packages deal with multivariable, non-linear systems, and there are also packages that operate on-line for adaptive control operation and on-line parameter estimation.

The most well known and widely used CAD packages, developed in the beginning of the 70s, are concerned with either of the categories above, for instance [1, 2, 3, 4] handle the control system design while [5, 6, 7] take care of the identification and parameter estimation task. However, during the last years, there have been developed several CAD systems that combine the three categories in one general package. Some of these packages have implemented special methods [8, 9] while others are more general, containing a large variety of different control, identification, and estimation methods [10, 11, 12, 13]. Some of the packages [10, 11] combine the process identification and controller design to a systematic conception of adaptive control and implementation, and on-line realization of control algorithms are sometimes also included [10, 12].

The paper is organized as follows: starting with some general remarks on requirements of a CAD system it follows a review of the implemented methods for simulation, analysis, control system design, and identification and parameter estimation. Adaptive methods will also be discussed. Some of the most well known packages in Europe will be described in some detail. The paper ends with some comments on future trends and developments in CAD.

GENERAL REQUIREMENTS OF AN APPLICABLE CAD-SYSTEM

The design of an interactive CAD system is rather complicated, for two primary reasons.

First, the CAD system should contain relevant methods that cover a wide range of control engineering problems, not only analysis and design of control systems but also system identification and parameter estimation. The demands upon the CAD system will reflect different levels of complexity and expertise, and it is therefore important that the implemented methods will be useful to a general-user community.

Second, it is essential that the individual subroutines and programs in the CAD system should qualify as "good" software (modularity, portability, efficiency), and that the interaction organization is sufficiently flexible to meet requirements from the experienced user as well as the temporary user. It must also be emphasized that the CAD system should be implemented on a computer with a reasonable core capacity in order to give an acceptable response time and, that there are available graphic devices such as visual display units and plotters.

Field of Application

The control engineer meets a wide range of tasks when he/she is going to design a control system for a real life operating process. Initially a model of the process must be available. This can be obtained from physical *modelling* (differential and algebraic equations) or by identification. Then it would be necessary to *estimate* the unknown model parameters. Application of linear system theory requires linear models, and as most process models are nonlinear it would be necessary to *linearize* the model equations with respect to a given operating point. Often it would also be necessary to perform a model *reduction* and a *transformation* of representation form. Assuming that an adequate model has been found, the next function of the CAD system is the derivation of a *control strategy*. Then the *simulation* and the *testing* follow and finally the on-line realization of the control algorithms. Figure 1 shows a systematic approach to the control system design problem and tasks that should be implemented in a general CAD system.

The CAD system will contain several special purpose packages (simulation, design, etc.) and within each package there should be some

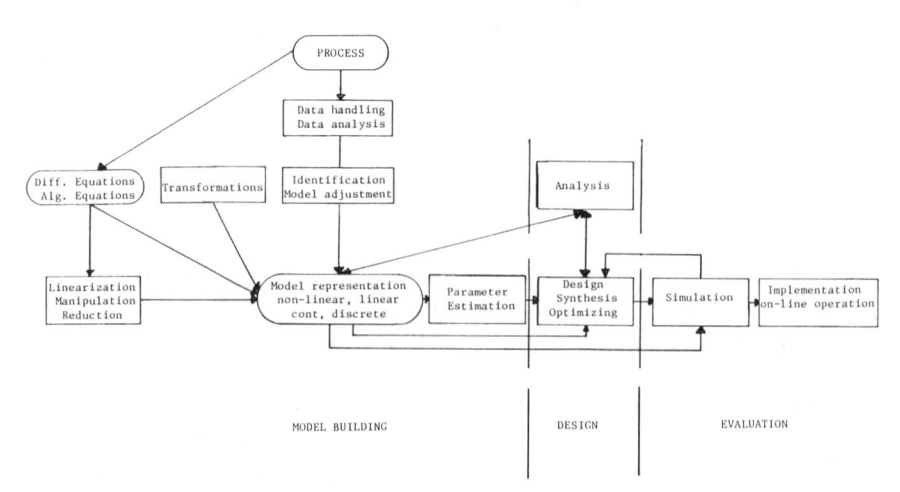

Figure 1.

possibilities to choose among a number of algorithms which have proven to be robust and applicable to industrial problems. Packages which incorporate too many different methods are difficult to use and should thus be avoided.

Hardware

A major factor that influences the rate at which CAD is taken up and accepted by industry, is the performance characteristics of the computer hardware. Running a CAD problem must be economical and therefore the response time must be sufficiently good. As long as the processor, memory and graphic displays are shared, the response obtained by one user depends, to a large extent, on the activities of the other users of the system. This variable response time is a source of irritation to the interactive user and therefore a single-user computer system would be preferable. Concerning real-time operation of a CAD system, for instance performing on-line estimation, a single-user computer is a must.

Cost reduction and improvements in technology has made the minicomputer very attractive for CAD purposes, and the majority of program packages which are presented in the sequel are implemented on minicomputers. Most commonly used is the 16-bit minicomputer with typical core capacity of about 64k and with possibilities to organize overlays.

Computer graphics devices are musts in CAD systems. These include the visual display units and the ink plotter and the electrostatic plotter. The visual displays commonly used in the presented packages are of two kinds. First, there are storage tubes, which can display a picture of any complexity, but, which have only limited picture editing. Second, there are refreshed displays in which the displayed picture can be modified easily, but which are limited in the complexity of the picture which can be displayed.

Plotters and display units represent a rather expensive part of a CAD system and software is of considerable complexity and of various standards. This is often reported as being one of the major problems in connection with transfer of program packages.

Computer peripherals such as magnetic tape and disc systems are mainly used as secondary storage media. I/O devices such as paper tape punches and readers and card punches and readers are today only used to a very limited extent. The floppy disc has taken over most of their functions.

The alphanumeric visual display unit has almost completely replaced the previously ubiquitous teletypewriter and, intelligent terminals, in which microprocessors are used to relieve the host computer of tasks such as editing and display of data, are in use in some packages.

Software

Besides all computational sophistication, it is very important that the CAD packages have a good organization of programs and data, and that the user-machine (program) communication is as flexible as possible. The programs must be structured in order to allow the introduction of new methods, and to allow sufficient flexibility for uses not envisaged by the implementors. Further it is important that the programs are made portable and easy to maintain. This includes a strict documentation and a good user's guide.

Most of the existing program systems use Fortran as the programming language. Obviously Fortran is not a very effective language but, on the other hand, Fortran compilers are available at almost all computer systems. There exist several dialects of Fortran but mainly the Fortran IV, USA ANS X3.9, 1966 is used.

To overcome the fact that software and hardware facilities differ quite a lot from one computer system to another, most of the referred packages are structured in such a way that basic mathematical functions and all input-output handling, e.g. use of graphical displays and files, are taken care of in special computer dependent subroutines, while all general computation is taken care of in high level programmed subroutines.

In order to give the packages a uniform structure there are suggested certain guidelines for the program implementor to follow [14]; for instance show how to deal with the problem of dynamical allocation of matrices in Fortran, and how a subroutine/program should be documented. Regarding maintenance and transfer of programs it is very well accepted that a strict documentation is a necessity.

The interaction organization is either of command dialogue type or conversational (question-answer) dialogue type. Depending on the category of users one will be found more convenient than the other. The experienced user will demand maximum freedom in the choice of solution steps and access to many parameters in the algorithms being used. The temporary user (student) will prefer to have an interaction scheme that gives guidance through a predefined sequence of actions (YES/NO). In certain situations the user would like to have information on alternative strategies and this is obtained by the HELP function. It is, however, to be noted that there is often natural to operate interactive programs in a so-called batch mode. This will, for example, be the case if the user wants to repeat a simulation problem with a given set of parameters. In most existing programs the interaction structure is static (either command or conversational) but there are also examples of program packages where it is possible to dynamically define and alter the interaction structure [15, 10, 16]. This is made possible by the so-called MACRO facility, which is a preprogrammed input sequence stored in the data base. One good example on how to design a communication module to meet the requirements regarding a general interaction structure, is found in the CAD system from Lund, Sweden [15].

The information flow between the user and the program is usually parameter values, data sets, and choices of operation. It is, however, also important to have the possibility to interactively specify computer code, for instance in nonlinear estimation and simulation where not only the parameter values, but also the model equations have to be specified. In most of the existing nonlinear simulation and estimation programs the model equations have to be specified beforehand, commonly by means of programming statements or predefined block functions. Program packages which contain own interpreters (compilers) have, however, the possibility to specify computer code interactively [17].

IMPLEMENTED METHODS

Because of the nature of the control enginering field problems, the design of a CAD system includes decisions that are inherently somewhat subjective. There exist a lot of different approaches to the problems of control system design, identification and parameter estimation, and each approach has its enthusiastic supporters. Generally speaking it is not possible to say that any method is best for all kinds of problems and, therefore the user must have the possibilities to investigate several methods to the same problem.

The objective of this chapter is to present CAD methods for simulation, control system design and identification and parameter estimation, methods which have proven to be useful solving control engineering problems.

Model descriptions

The general, nonlinear state space model is given by

$$\dot{x}(t) = f(x(t), u(t), v(t), \theta(t), t)$$
$$y(t) = g(x(t), u(t), w(t), \theta(t), t) \quad (1)$$

where

$x(t)$ = state vector
$u(t)$ = control vector
$v(t)$ = process disturbance vector
$\theta(t)$ = parameter vector
$w(t)$ = measurement disturbance vector
$y(t)$ = measurement vector

In the discrete case the models have the form:
$$x(k+1) = f(x(k), u(k), v(k), \theta(k)) \\ y(k) = g(x(k), u(k), w(k), \theta(k)) \quad (2)$$

The corresponding linearized versions are
$$\dot{x} = Ax + Bu + Cv \\ y = Dx + Eu + w \quad (3)$$

and
$$x(k+1) = \phi x(k) + \Delta u(k) + \Omega v(k) \\ y(k) = Dx(k) + Eu(k) + w(k) \quad (4)$$

eq. (3) can also be described as a transfer function
$$y(s) = G(s) u(s) \quad (5)$$

where
$$G(s) = D(sI-A)^{-1} B + E$$

or on Rosenbrock's polynomial form [18];
$$\begin{bmatrix} sI-A & -B \\ D & 0 \end{bmatrix} \begin{bmatrix} x(s) \\ u(s) \end{bmatrix} = \begin{bmatrix} 0 \\ y(s) \end{bmatrix} \quad (6)$$

The corresponding matrix polynomial model of Eq. (4) can be represented as:

$$A(q^{-1}) y(k) = B(q^{-1}) u(k) + G(q^{-1}) v(k) \quad (7)$$

where
$$A(q^{-1}) = 1 + A_1 q^{-1} + \ldots A_m q^{-m} \\ B(q^{-1}) = B_1 q^{-1} + \ldots B_m q^{-m} \\ G(q^{-1}) = 1 + G_1 q^{-1} + G_m q^{-m}$$

It is also possible to describe the process as a sample response function (weighting sequence)

$$y(k) = \sum_{i=0}^{k} h(k-i)u(i), \quad k\epsilon(1,N) \quad (8)$$

where k is the sampling instant and N is the period of observations.

Since some of the design methods and estimation methods require state space models, some require transfer function models, and others require matrix polynomial models, it is necessary to have programs which perform transformations between these different descriptions (see for instance [19, 20], and [10]).

Most CAD packages for control system design initially require a process model on the form 3, 4, 5, 6, or 7. If the model is derived as a result of an identification procedure, the model is automatically given on the required form, but if the model is given on the form (1) with an additional set of algebraic equations on the form

$$h(x,u,v) = 0 \qquad (9)$$

it can be difficult to derive at the linearized state space model, Eq. (3). The problems of finding stationary solutions and perform numeric linearizations are taken into consideration in only a few number of the available CAD systems [21, 22].

Control system design

A general control system consists of a feedback strategy and a feedforward strategy. Most CAD systems are concerned with the feedback loop design and minor attention is paid to the feedback control design. The main reason for this is that the feedforward controller does not influence the system stability, and besides, it would always be possible to treat the feedforward loop as a feedback loop by augmenting the statevector to include a dynamic description of the disturbances and the reference variables. In some cases, however, when the disturbance is directly measured, it would be preferable to have different types of feedforward algorithms available, such as realized in the CAFCA package [23].

It is necessary to distinguish between programs based on the time domain approach and the frequency domain approach. Some package designers seem to prefer time domain methods, while others seem to have most confidence in frequency domain methods. In a general design package, however, both approaches should be represented such that the user can have as much freedom as posible in the choice of method.

Frequency domain methods. Most frequency domain design methods in CAD are based on models of continuous system expressed in terms of the Laplace transform variable, s. Some packages, however, which aim at digital computerized control strategies also accept z-domain models. The main advantage of the frequency domain methods is that during the interactive design procedure the designer gains considerable insight into the system structure and is able to interpret most steps in physical terms. The commonly used design methods are in the single-input/single-output case (SISO), the classical methods of Nichols, Nyquist, and Evans (root locus). Concerning multivariable systems (MIMO) the Characteristic Locus Method (CL) [24] and the Inverse Nyquist Array (INA) [18] are most frequently implemented. The Nyquist array of a m × m matrix is the m × m set of Nyquist loci associated with the elements of the matrix. Rosenbrock [18] has shown that the Nyquist array provides considerable guidance in choosing the elements of the compensator matrix to improve diagonal dominance, and the Gershgorin bands associated with the diagonal elements of the open loop transfer matrix, provide a means for checking stability and designing controllers for each loop.

The frequency domain methods have met with computational difficulties when used to design realistically, large industrial control systems and, one of the main computational problems results from the inherent numerical sensitivity of the methods, e.g. the dependence on the specification of "zero." Frequency domain methods require the use of complex z and bilinear transformations to take into account the sampling process, and time-delays cause problems since they cannot be expressed exactly as rational transfer functions.

However, in order to meet the computational difficulties, new improved algorithms are continuously developed and in some programs there are implemented more methods for solving the same problem [25]. Thus, it would be possible to check the result when one suspects numerical errors.

The INA design procedure normally involves three steps:

a) Column and row operation on the open loop transfer function matrix in order to obtain minimal interaction.
b) Design of a dynamic precompensator to make the inverse open loop transfer function matrix and the closed loop system matrix diagonally dominant.
c) Design of controllers for each loop using classical design methods.

Point b) is known to be the most tricky point and there are several methods to choose among.

The CL design procedure involves in most cases four steps: a) Stability investigation, b) Integrity investigation, c) Interaction investigation, and d) Performance investigation. In each phase a basic controller is designed to achieve part of the design objectives for the closed loop system. The final controller is a product of simple controllers, and MacFarlane and Belletrutti [24] have shown how the basic controllers should be chosen in order to obtain the desired design objectives. Each step in the CL procedure involves the display of three plots for each characteristic transfer function (eigenvalue or characteristic value). These plots are the characteristic locus, the magnitude of the characteristic eigenvalues and the corresponding aligment. The characteristic locus must satisfy the CL-Nyquist stability criterion, and the magnitude of the characteristic value and the aligment tell about the interaction as a function of frequency.

Although the multivariable frequency methods require a certain engineering insight and manipulating skill, they are obviously a powerful tool in the hands of a skilled designer. The largest problem up to now tackled with these methods, is the design of a control system for a jet engine with five input, five output, thirty-three states and with open loop poles varying from 0.5 to 500 [25].

An alternative approach is to let the designer after a preliminary CL analysis, choose a suitable specification of the closed loop performance and the controller structure, and then use an optimization algorithm to find the parameter values of the controller that minimize in some sense the difference between the desired closed loop behavior and the actual closed loop behavior [25].

Time domain methods. In time domain the methods have mostly centered on state-space feedback pole allocation [26], decoupling [27], and linear optimal control theory. The design is whether based on an optimal choice of poles to minimize a specified cost function, or a more intuitive choice of poles to satisfy engineering constraints of rise time, settling time, overshoot, etc. An important part of the optimal control strategy is a state estimator which provides an estimate of the state vector. These estimated states are used in computing the control variables. Concepts of controllability (assignment of closed loop system poles) and observability (condition for reconstructing the system state from process measurements) play important roles in the algorithms.

Instead of requiring access to the full set of plant state variables or to a state reconstructor, one can develop alternative suboptimal design strategies. The feedback configuration is prespecified and tailored to suit the process under consideration; for instance, output feedback and dynamic compensation.

Some design programs make use of optimization to find an appropriate controller. The designer selects a controller structure and a cost function, after which the program determines the parameters of the controller which minimize the cost function.

For system models of matrix polynomial form there are algorithms for deadbeat controllers, minimum variance controllers, and decoupling controllers. The deadbeat controller is characterized by a finite time settling of the input and output signals after a step change of the reference variables.

For system models given on a weighting sequence form (Eq. (8)) there are developed CAD methods based on discrete convolution algebra [8]. Terms from the Characteristic Locus method, such as characteristic eigenvalues and alignment [24] are interpreted in the time domain, and the design methodology follows the same ideas as in the frequency domain method.

One of the more recently developed methods, in particular suitable for large multivariable systems, is based on a state space decomposition technique [28]. The idea is to decompose the system into sets of subsystems with well defined interactions which can be controlled individually with local feedback controllers (decentralized control). The decomposition can be performed using simulation in order to study the influence of each control signal on each output variable or by a linear controllability subspace analysis. Other techniques for design, such as the parameter optimized, decentralized control algorithm [29] also fit into the framework of the decomposition technique.

Simulation

The simulation packages are utilized in the modelling phase, the control system design phase, and finally in the verification and testing of the implemented control system. A wide class of process models (Eq. 1–7) are accepted and integration methods like Euler's method and various versions of Runge Kutta methods are most often implemented. This includes methods with fixed integration step size and methods with variable step size. However, only a few of the packages contain methods for simulation of "stiff" systems [30, 58].

The nonlinear simulation packages are either of equation-oriented type or block-oriented type. The equation-oriented packages are more flexible concerning model descriptions; for example, the models can be described as Fortran statements, but unless an interpreter is used [17], it is not possible to modify the model structure during a simulation which is an important feature in an interactive program system. A block-oriented package requires a description of the model that is closely related to the block-diagrams familar to almost every control engineer, but only blocks supplied by the program are accepted [31, 10]. The model structure can be modified during operation and in general the execution speed can be made faster than in equation-oriented programs. To improve the applicability of block-oriented programs it has been developed simulation programs that combine the advantages of block-oriented programs with some of the useful facilities of equation-oriented programs [32, 33].

Some of the equation-oriented packages accept both continuous and discrete models [17, 29], and thus it is very convenient to simulate computer controlled processes. Optimization is also often directly available [17, 32, 33], and this makes it possible to optimize parameters in a system. As each optimization may require quite a lot of successive runs, it is very important that a fast integration algorithm is available; for instance, one of the variable step size methods.

Analysis

It is difficult, concerning CAD operation, to separate the analysis from the synthesis and design operations. Typical analysis tasks are to calculate eigenvalues and eigenvectors, calculate the time response for a given system subjected to a step input, calculate system zeros, and determine observability and controllability. These tasks are implicitly provided by most of the design and synthesis methods, but sometimes it is convenient to have them available explicitly.

It is interesting to see that many packages [13, 20, 34] use the algorithms available in EISPAC [35] for solving standard analysis problems like eigenvalue, eigenvector, zeros, and singular values calculations (QR-algorithm), but there are also a lot of dedicated methods implemented. System zeros, which play an important role in frequency response methods, are often difficult to calculate and it is often necessary to have two or more separate methods available [36, 37, 38]. The singular value of a system is often used when determining the rank of the controllability and observability matrix.

Identification and Parameter Estimation

The field of CAD for identifaction and parameter estimation is characterized by a huge amount of implemented methods. However, almost every method is concerned with linear systems and in many cases, only SISO models are accepted. Some packages use only the knowledge of input-output data for model structure determination, while others are based on a user specified model structure. The packages generally contain programs for generation of test signals, data analysis, model structure identification, parameter estimation, simulation, and validation.

The most implemented methods for off-line and on-line identification are the well-known techniques such as least squares method, the generalized least squares, the maximum likelihood method, instrumental variables, extended matrix method and extended Kalman filtering. A survey of identification methods is given in [39]. Different model representations are accepted (linear, nonlinear, parametric, nonparametric, continuous, discrete, SISO, MIMO) and obviously, not all forms are suitable for design of control systems. However, by means of model transformation algorithms one may end up with an applicable model [10].

In most packages it is possible to choose between different methods and versions of methods, and in methods which solve an optimization problem, like the maximum likelihood method, it is possible to choose among various optimization procedures. The selection of a proper optimization procedure is an important problem, and is closely related to the model description and its parameterization. For linar models, techniques based on sensitivity derivatives and Lagrange multipliers are often applied, while for nonlinear models, techniques based on loss function evaluations are applied [40]. Algorithms which only use function evaluation allow great flexibility in the model description and are therefore sometimes applied for linear models, too [41].

Many packages cannot operate on continuous models, and this is a serious drawback because models based on physical laws and construction data will be in this form. Packages which accept continuous models, also easily handle problems like nonuniform sampling and missing measurements.

It is well known that the accuracy of a parameter estimation depends strongly on the nature of the input signals used in the experiment. In most packages there are available PRBS and random signal generators [8, 9] but it is also an example of a package that generates optimal test signals [42]. This means that the process is excited in such a way that the measurements are most sensitive to changes in the unknown parameters.

Adaptive control systems

By combination of a suitable control algorithm and a recursive estimation technique, it is possible to design an adaptive control system. Another approach to adaptive control is to use the parallel reference model technique which allows the identification of the plant dynamics and the adaption of the controller parameters to achieve a given model behavior for the closed loop.

There are several types of adaptive controllers based on the combination of identification and control, but up until now the only implemented methods are based on the so-called certainty equivalence principle [43]. The certainty equivalence principle holds if it is possible to first solve the deterministic problem with known parameters and then calculate the controller parameters for unknown parameters by substituting the true parameter values with the estimated values. One package which utilizes this approach is the MACS package [11, 44]. Various kinds of control algorithms are implemented, such as pole assignment, optimal controller, minimum variance state controller for state space models (Eq. (4)) and deadbeat, decoupling, and minimum variance controller for matrix polynomial models (Eq. 7)). The recursive estimation methods are based on the least squares method and extended least squares method. The KEDCC package [45, 10] contains adaptive control methods based on both approaches, but mainly SISO systems are regarded. Design methods such as self-tuning regulator, minimum variance controller and methods using the hyperstability concept are implemented.

SURVEY OF CAD-PACKAGES

In the sequel it follows a survey (geographically organized) of the most important CAD packages available in Europe. A brief review of implemented methods programming languages and computer systems is given.

Belgium

The most important development has taken place at the University of Ghent. The Ghent package [9] contains programs for off-line and on-line identification and control system design. Only SISO systems are considered. The off-line identification program is based on least squares, generalized least squares, and the maximum likelihood method. The on-line identification program is based on least squares, instrumental variables, extended matrix, and the maximum likelihood techniques. All methods are available in two or more versions. The control system design method is based on a parameter optimization technique.

Bulgaria

Of special interest in this context are two packages from the University of Sofia [34, 46]. Both contain programs for analysis and design of multivariable control systems. The first package is based on state space models, and the implemented methods are: linear optimal control, pole and zero assignment, output regulator, reduced order observer, and others, more specialized algorithms. The second package contains various versions of pole assignment techniques which also include optimal control and eigenvector assignment. There are also special methods for calculation of zeros available. Fortran IV is used as programming language and the programs are implemented on IBM S/360 and IBM S/370 computers.

England

The most important developments in England have taken place at the Universities of Manchester [20], Cambridge [25, 5, 47, 48], Bradford [8, 49], Imperial College, London [4, 7, 50] and University college North Wales, Bangor [5].

The Manchester package (UMIST) [20] is mainly concerned with the design and analysis of continuous control system, but additionally there are programs for data input-output facilities, data transformation and simulation. The design techniques are based on frequency response methods, Bode, Nyquist, CL, INA but also synthesis procedures based on state space theory such as pole assignment, decoupling and optimal feedback with Luenberger observers and Kalman filters are included. In the data transformation package there are available programs for curve fitting, the Faddeev algorithm, minimal realization, and inversions. The UMIST package is implemented on a DEC-10 computer and the programming language is Fortran IV.

The Cambridge packages are concerned with analysis and design of multivariable systems based on frequency response methods [25] and identification and time-series analysis, CAPTAIN [5]. The CAPTAIN package is available in several different versions [47, 48], but the original version was developed at Cambridge University. The most recent versions of CAPTAIN accept models on multivariable polynomial forms (continous and discrete) and the recursive instrumental variable and approximate maximum likelihood are the most important identification methods implemented. The CAPTAIN packages are written in Fortran IV, and are marketed by Oxford Systems Associates, Ltd., Oxford, England.

The control package accepts models on Laplace transfer function form and state space form (continuous and discrete). Several methods for designing control systems using Nyquist arrays, characteristic gain loci, generalized root loci and closed-loop Nyquist arrays, are provided within the package. Wherever possible, at least two methods have been implemented for calculating any given function in order to provide a way of checking the results produced in those cases when numerical errors are suspected. There is also available a program for matrix manipulation that makes it possible to calculate Kalman filters, system zeros, optimal controller, and bilinear transforms. The language which has been used is Fortran IV, and the programs are available on several different computers: PDP11, PDP10, PRIME400 and GEC4070. The programs are also available commercially through Compeda, Ltd.

The Bradford packages consist of the CAIAD [8] and the CSNCS [49] package. The CAIAD package contains programs for identification and control, and it operates totally in the time domain. All systems are represented in terms of their input-output properties using discrete convolution, Eq (8). The identification program operates with PRBS test signals in conjunction with least squares parameter estimation or model matching. A variety of control system structures are available; decoupling, standard PID controllers, the CL method interpreted in the time domain and optimal control. The discrete weighting sequences are manipulated by means of convolution algebra. CAICAD is written in Fortran IV and implemented on the HP2100, HP21MX, and HP1000F computers. There is also a real-time version of CAICAD, written in CORAL66, and implemented on a Ferranti Argus 700 process computer. The CSNCS package, also written in Fortran IV, is concerned with analysis and simulation of nonlinear control systems. The CSNCS package also uses discrete time sequences.

F.R.G.

The Federal Republic of Germany has played an important role in development of CAD packages in Europe and, in particular, the Universities of Stuttgart [52, 53, 54], Darmstadt [55, 11, 23, 56], Bochum (Ruhr) [10], and Karlsruhe [33] have contributed with advanced packages. Some of the packages from Stuttgart and Darmstadt are closely related since programming activities which originated in Stuttgart have been further developed and improved in Darmstadt [55, 11].

The Darmstadt packages, MACS [11], CADCA-MIMO [55], CAFCA [23], and REDECK [56] cover a rather broad field in control engineering. MACS contains various discrete time control algorithms for linear multivariable systems in state space form or matrix polynomial form, which are assumed to be a result of an on-line system identification procedure. The control algorithms are based on pole assignment, optimal control, decoupling, and extended and general minimum variance methods. As the system operates on-line, it can be used to design adaptive control systems. The CADCA-MIMO package is used to design multivariable control systems based on discrete-time, parametric process models (state space or polynom matrix representing the system transfer matrix). The control algorithms are based on parameter optimized control and state variable control including observers. The CAFCA package is dedicated for design of feedforward controllers. The models are given on polynomial form and may be achieved from previous process identification. Several algorithms are available, parameter optimized feedforward, static feedforward, minimum variance and regular dynamic feedforward, but only SISO systems are accepted. Fortran IV is used as programming language for MACS, CADCA, and CAFCA, and the packages are implemented on a HP 21 MX-E process computer system.

The REDECK package is concerned with analysis and design of control systems. Both SISO and MIMO models are accepted and time domain as well as frequency domain methods are implemented. The time domain design algorithms are based on pole assignment and optimal control (state and output control), while the frequency design algorithms are based on the INA method, decoupling, and parameter optimization. The analysis program contains methods for traditional calculations and model reduction. REDECK is intended for use on small computers such as desk top calculators or minicomputers. Programming language is Fortran IV and BASIC and implemented versions are available at PDP11/34 and HP 9825, respectively.

The Stuttgart packages OLID-SISO [52], OLID-TITO [53], CADCA-SISO [54], are restricted to operate on SISO and TITO (two inputs-two outputs) models. Recursive least squares and recursive instrumental variable methods are implemented in OLID-SISO and recursive correlation analysis with least squares parameter estimation is implemented in OLID-TITO. The control package CADCA contains algorithms for parameter optimized controllers, deadbeat and state controller with observer for external disturbances. The programming language is Fortran IV and the packages are implemented on an HP2100A computer.

The KEDDC package [10] from Ruhr-University Bochum, is a very general package which contains programs for signal analysis, off-line and on-line identification, controller design, simulation, and adaptive control operation. The package accepts model representations in state space, frequency domain, and time domain and a wide variety of transformations are available. The identification programs for multivariable, linear systems are based on the extended maximum likelihood method and the instrumental variable method. The multivariable control system design is based on optimal state feedback controllers with reduced or full-order observers and the INA method. For SISO models a large amount of methods for identification and control are available such as recursive maximum likelihood, hyperstable parameter estimation, optimized PID control, minimum variance, self-tuning, deadbeat controllers, and model reference adaptive control [45]. The KEDDC package is written in Fortran IV and implemented on the HP 2100 and HP 1000 computers.

Hungary

The MERCEDES package [12] developed at the University of Budapest contains programs for control system design, off-line identification and real-time process control. The control system design package is concerned with frequency domain methods like the CL method and classical design techniques, and time domain methods like optimal control and minimum variance, stochastic control. The real-time process control package is concerned with modelling and realization of discrete control algorithms. Parts of the system are programs for on-line identification, test signal generation, data handling, DDC, and self-tuning minimum variance control. The on-line identification module is suitable for linear and non-linear single and multi-input (MISO) processes and the implemented method is based on a least-squares parameter estimation and correlation computation. The control module is suitable for SISO, MISO, and MIMO systems and discrete PID control and deadbeat control algorithms are implemented. The off-line identification module contains methods for identification of SISO and MIMO linear and non-linear systems and time-series analysis. The MERCEDES package is written in Fortran IV and OPAL for off-line and real-time usage, respectively. It is implemented on a TPA computer, similar to the PDP-8 computer, and together with standard peripherals, it makes a portable computer laboratory.

Italy

There exist rather few interactive CAD packages in Italy and the only one to be mentioned here is the control design package developed at Enel, Milano [29]. It is based on a parameter optimization technique. The user specifies the regulator block diagram, consisting of summing points and SISO dynamic or algebraic systems, and the optimization module, based on a conjugate gradient method, determines the optimal setting of the regulator parameters with respect to a given performance index. The transient behavior can then be verified by means of a simulation module. The program is written in Fortran IV and implemented on a H66/60 computer.

The Netherlands

The most important developments in the Netherlands have taken place at the universities of DELFT [32, 19, 22] and Eindhoven [57].

The DELFT packages PSI [32] and TRIP [19] are concerned with simulation and identification, respectively. The PSI program is block-oriented with approximately 30 different block types, like the integrator, function generator, time delay, switches, discrete PID controller, etc. However, it is also extended with facilities like optimization (simplex method), storing of variables in different simulation runs and solving algebraic loops, etc. The TRIP program is restricted to operate on SISO systems. Different model representations are accepted and the program offers a lot of various transformations. The identification methods are based on recursive least squares method and complex curve fitting. The TRIP and PSI packages are written in Fortran IV and implemented on PDP-11/20 and PDP-11/60 computers. The packages have been transferred to research institutes and industries in the Netherlands and many interesting applications have been reported.

Norway

In Norway the development of CAD systems has mainly taken place at the University of Trondheim [13, 3]. The CYPROS package [13] consists of programs for simulation, identification, parameter estimation and control system design. All programs, except for some design and identification programs, are based on state space model descriptions (Eq. 1–4). The identification methods implemented are based on the extended Kalman filter and the maximum likelihood method. The maximum likelihood program accepts nonlinear models, and it is based on a Powell minimization technique. Optimal test signals can be designed. The control programs are concerned with optimal control and the INA and CL methods. There are also available programs for matrix manipulation and time-series handling. The programming language is Fortran IV and the package is running on a Nord-10 computer. CYPROS has been applied successfully to industrial problems of various kinds, and also transferred to research institutes and industry. The DAREK package [3] is in particular concerned with state space models, and handles problems like multivariable control system design and linear and nonlinear simulation. The design methods are based on optimal control with Kalman filters and optimal controllers with prescribed eigenvalues. The programming language is Algol and the system is running on a UNIVAC 1108 computer.

Poland

A very special, but interesting, program package for modelling and simulation based on the bond-graph technique has been developed at the University of Warsaw [58]. A user specified bond-graph is automatically converted into a simulation model, linear or nonlinear, in state space form. The simulation module contains several integration algorithms and special care is taken for stiff equations. The program is written in Fortran IV.

Romania

The SIPAC package [59] developed at the University of Bucharest covers a rather broad field in control engineering. There are available various programs for identification, control system design and simulation. Model representations given by Eqs. 3–6 are accepted and various transformations are available. The identification methods are mostly concerned with SISO systems and nonparametric identification, like numerical deconvolution, correlation, and spectral analysis and parametric identification, like model-adjustment, instrumental matrix method, generalized least squares and maximum likelihood techniques are implemented. The control programs are based on optimal control with additional features like reference tracking, measurable disturbance compensation, steady state error cancelling and state estimators (Kalman filter, Luenberger observers). Dynamic compensator design by a pole assignment technique is also available. The programs are written in Fortran IV and implemented on a Felix C-256 computer.

Suisse

The INTPOS package [60] developed at the University of Zürich is in particular concerned with analysis and design of control systems. Time domain and frequency domain methods are implemented. The time domain programs are based on state space models and methods like optimal control, pole assignment, steady state decoupling, dynamic compensators, observers, and Kalman filters are implemented. The frequency domain methods are only concerned with SISO systems and the most well-known classical design methods are implemented. The programming language is Fortran IV and the package is implemented on the PDP11/60 and IBM-3033 computers.

Sweden

The most interesting developments have taken place at the University of Lund [6, 61, 41, 17]. The Lund packages consist of programs for simulation [17], identification [6, 41] and synthesis of control systems [61]. The simulation program, SIMNON, accepts nonlinear, discrete, and continuous models and the optimization facility is also included. The program operates in an interpretive way. The IDPAC program [6] is designed for identification of SISO sysems. Identification methods like least squares and max likelihood method are implemented, and the program also performs simple statistical analysis. The macro facility is available. The LISPID program [41] is designed for maximum likelihood identification of linear, continuous or discrete, multivariable models. Two different optimization methods are available which allow a great flexibility in the description of the model and its parameterization. SYNPAC [61] is a control system design program which operates in time domain and frequency domain. Optimal controllers and design based on the INA method are implemented. All packages are written in Fortran IV and are available on different computers: IDPAC (PDP-11, PDP-15, UNIVAC 1108), LISPID (IBM 360, UNIVAC 1108), SIMNON (PDP-15, UNIVAC 1108, DEC-10), SYNPAC (PDP-15). A lot of interesting applications have been reported, for more details see [41].

U.S.S.R.

The field of interactive CAD in the USSR is rather new, and only a few examples have been presented. One of the most interesting developments is the control system design package from Tallinn [21]. This package also contains methods for generation of state space models from nonlinear differential and algebraic equations, and methods for model reduction are also available. The design methods are based on optimal control theory and eigenvalue specifications. The programming language is Fortran IV and the package is implemented on an ES computer system.

TRENDS AND FUTURE DEVELOPMENTS

With the increasing availability of computers and graphic displays the field of CAD will continue to grow. The microcomputer and minicomputer will play important roles in the coming CAD systems, and powerful data bases and advanced communication systems will influence the operation and organization of the programs.

Time-shared operations of interactive CAD systems are not particularly well suited when effectiveness is demanded, and when it comes to real-time computation, exclusive use of time-shared computer systems becomes illusive. The tendency exists to use microcomputers in small, stand-alone CAD systems or as stand-alone modules in a large CAD system. Since the core memory available for programming on such computers is usually quite limited, it becomes important to make the CAD programs modular, and let them operate as subprograms (data analysis, DDC, etc.) which communicate data through a common data base.

It is likely that there will be an increasing tendency for CAD systems to become dedicated to a single user. This simplifies many software problems, but it also ensures a good response which is in particular required when large scale problems are to be handled.

Concerning the contents of the CAD systems, it is likely to believe that more general and industrial oriented methods will be developed, and the number of general purpose packages which combine the modelling, estimation, and control problem in a unified form, will increase. More attention will also be paid to the development of numeric robust algorithms.

Much effort will be put into standardization and portability of software and packages will be transferred to research institutes and industry. The field of application will be getting wider, and there exist already today several examples [62] on how the CAD technique can be applied to modelling and control of non-technical systems such as biological, economical, and ecological, often referred to as large scale systems. Studying and analysis of such systems require efficient CAD packages including sophisticated display and drawing programs, powerful data bases and command languages.

As computers and graphic devices become cheaper and the CAD systems prove to be robust and applicable, they will turn out to be engineering tools and design aids rather than a computer system to be kept in a carefully protected environment, surrounded by computer professionals. CAD will become common in engineering workshops, in research laboratories, and above all at the plant.

CONCLUSION

Over the past decade interactive computer aided design of dynamical systems has established as a firm and useful tool in research institutes and education in Europe. There are available various program packages for system identification, parameter estimation, simulation, analysis and design of control systems. Many interesting applications have been reported, but the real breakthrough among industrial users has not taken place. This is mainly due to the fact that many of the implemented methods are not properly oriented towards practical problems, but also insufficient theoretical background among industrial users and problems like software and hardware standardization must be taken into account.

Recent research in the field of CAD in Europe has generated numerous enhancements to the status of CAD, and in the 80s we see clearly that it will be more flexible, economical, and easier to use, and it will ensure its place as a major tool in future development and research in the field of automatic control.

REFERENCES

[1] Munro, N. and R. S. McLeod, "Computer aided design of Control Systems." Report no 130. University of Manchester U.K. (1970).
[2] Dyne, M. D. C. and M. O. Mutch, "Algorithms and System Design Packages," University of Cambridge, U.K. (1973).
[3] Pedersen, J. O., "Darek User's Guide," Report STF 48 A7301 SINTEF department of Automatic Control, Trondheim, Norway (1973).
[4] Shearer, B. R. and A. D. Field, "Multivariable design System; MDS," Publ. NO. 75/29, Department of Computing and Control, Imperial College of Science and Technology, London (1975).
[5] Shellswell, S. H. and P. C. Young, *Computer Aided Control System Design,* IEE Publication Number 96, 149 (1973).
[6] Weislander, J., "IDPAC, Users Guide," revision 1, report 7605, Department of Automatic Control, Lund Institute of Technology, Sweden (1976).
[7] Abaza, B. A., "Interactive Package for the Parameter Estimation of Dynamical System, PEST," Publication 75/2, Department of Computing and Control, Imperial College of Science and Technology, London (1975).

[8] Gough, N. E., Kleftouris, D. N. and R. S. Al-Thiga, *Multivariable Technological Systems,* IFAC, 357 (1977).
[9] DeKeyser, R. M. C. et al., Preprints, IFAC Symposium on CAD of Control Systems, 473 (1979).
[10] Schmid, Chr. and Unbehauen, H. *Software for Computer Control,* IFAC/IFIP, 251 (1979).
[11] Schumann, R., Preprints, IFAC Symposium on CAD of Control Systems, Zürich, 163, (1979).
[12] Fehrer, A. et al., *Software for Computer Control,* IFAC/IFIP, Prague, 261 (1979).
[13] Tyssö, A., Preprints, IFAC Symposium on CAD of Control Systems, Zürich, 383 (1979).
[14] Elmqvist, H., Tyssö, A., and J. Wieslander, "Programming and Documentation Rules for Subroutine Libraries-Designed for the Scandinavian Control Laboratory," Nordforsk, Stockholm (1976).
[15] Wieslander, J., Preprints, IFAC Symposium on CAD of Control Systems, Zürich, 493, (1979).
[16] Edmunds, J. M., Preprints, IFAC Symposium on CAD of Control Systems, Zürich, 253 (1979).
[17] Elmqvist, H., "SIMNON, And Interactive Simulation Program for Nonlinear Systems," Report 7502, Department of Automatic Control, Lund University, Sweden (1975).
[18] Rosenbrock, H. H., *Computer-aided Control System Design,* Academic Press (1974).
[19] Van den Bosch, P. P. J., *Journal A., 16,* 107 (1975).
[20] Munro, N., "The UMIST Computer-aided Control Systems Design Suites," Institute of Science and Technology, University of Manchester, U.K. (1979).
[21] Aarna, O. and Raiend, K., *Software for Computer Control,* IFAC/IFIP, 275 (1979).
[22] Kappetijn, F. K. and Jonkers, H. L., Preprints, IFAC Symposium on CAD of Control Systems, 311 (1979).
[23] Dymschiz, E. and Schumann, R., Preprints, IFAC Symposium on CAD of Control Systems, 241 (1979).
[24] MacFarlane, A. G. J. and Belletrutti, J. J., *Automatica, 9,* 575 (1973).
[25] Edmunds, J. M., Preprints, IFAC Symposium on CAD of Control Systems, 253 (1979).
[26] Wolovich, W. A., *Linear Multivariable Systems,* Springer Verlag, New York (1974).
[27] Falb, P. L. and Wolovich, W. A. *IEEE AC-Trans.* 12, 651 (1967).
[28] Denham, M. J., Preprints, IFAC Symposium on CAD of Control Systems, 25 (1979)
[29] Maffezzoni, C. and F. Parigi, Preprints, IFAC Symposium on CAD of Control Systems, 579 (1979).
[30] Opdal, J., "SIM, and Interactive Program for Dynamic Model Simulation," Report STF48A77026 SINTEF, Trondheim, Norway (1977).
[31] Grepper, P. O. and M. Djordjevitc, IFAC Symposium, Trends in Automatic Control Education, Barcelona (1977).
[32] Van den Bosch, P. P. J., Preprints, IFAC Symposium on CAD of Control Systems, 223, (1979).
[33] Moll, H. and H. Burkhardt, Preprints, IFAC Symposium on CAD of Control Systems, Zürich, 337 (1979).
[34] Konstantinov, M. M., et al., Preprints, IFAC Symposium on CAD of Control Systems, Zürich, 319 (1979).
[35] EISPAC, Release 2, Eigensystem Subroutine Package, Argonne National Laboratory, Argonne, Illinois, USA (1978).
[36] Laub, A. J. and B. C. Moore, "Calculation of Transmission Zeros Using QZ Technique," Report No. 7618, University of Toronto, Control Science and Engineering Department, Toronto, Canada (1976).

A. Tyssø   211

[37] Davison, E. J. and S. H. Wang, *Automatica 10*, 643 (1974).
[38] Konvaritakis, B. and J. M. Edmunds, *Int. Journal of Control, 29*, 293 (1979).
[39] Eykhoff, P., *System Identification*, Wiley (1974).
[40] Jensen, N. A., "MLPROG, and Interactive Program for Parameter Estimation," Report STF48F78017 SINTEF, Department of Automatic control, Trondheim, Norway (1978).
[41] Källström, C. G., et al., *Identification and System Parameter Estimation, Part 2*, IFAC Proceedings, 1315 (1978).
[42] Kristoffersen, K. and S. Sälid, Preprints, IFAC Symposium on CAD of Control Systems, Zürich, 119 (1979).
[43] Bar-Shalom, T. and R. Tse, *IEEE Trans. Autom. Control, 19*, 494 (1974).
[44] Schumann, R., Preprints, IFAC Symposium on Identification and System Parameter Estimation, Darmstadt, 1203 (1979).
[45] Schmid, Chr. and H. Unbehauen, Preprints, IFAC Symposium on Identification and System Parameter Estimation, Darmstadt, 1087 (1979).
[46] Danev, E. and A. Kukleva, Preprints, IFAC Symposium on CAD of Control Systems, Zürich, 235 (1979).
[47] Venn, M. V. and B. Day, "The Captain Package," Rep. No. 39 Institute of Hydrology, Oxon, U.K. (1977).
[48] Young, P. C. and A. J. Jakeman, Preprints, IFAC Symposium on CAD of Control Systems, Zürich, 391 (1979).
[49] Dimirovski, G. M., et al., *Software for Computer Control*, IFAC/IFIP, 265 (1979).
[50] Daly, K. C. and P. Katzeberg, "The Classical Design Suite." Publ. 77/34, Department of Computing and Control, Imperial College of Science and Technology, London (1977).
[51] Flemming, P. J., Preprints, IFAC Symposium on CAD of Control Systems, Zürich 259 (1979).
[52] Baur, V. and R. Isermann, *Identification and System Parameter Estimation Part 1*, IFAC Proc. 263 (1978).
[53] Blessing, P., et al., *Identification and System Parameter Estimation, Part 1*, IFAC Proc. 461 (1978).
[54] Isermann, R. and E. Dymschiz, *Software for Computer Control*, Tallinn, USSR (1976).
[55] Dymschiz, E. and R. Isermann, *Digital computers Applied to Process Control*, IFAC/IFIP, Proc. The Haag (1977).
[56] Asselmeyer, B., et al., Preprints, IFAC Symposium on CAD of Control Systems, Zürich 217 (1979).
[57] Van den Boom, A. J. W., *Journal A, 20*, 92 (1979).
[58] Grabowiecki, K. A., et al., Preprints, IFAC Symposium on CAD of Control Systems, Zürich, 273 (1979).
[59] Popesch, Th., et al., Preprints, IFAC Symposium on CAD of Control Systems, Zürich, Vol. 2, 7 (1979).
[60] Agathoklis, P., "INTOPS, An Interactive Program for Computer Aided Control Systems Design," Report No. 79-05, ETH, Zürich (1979).
[61] Wieslander, J., Computer Aided Control System Design, IEE, Publ. no. 96, 1 (1973).
[62] Balchen, J. G., "The need for Computer-aided Design in Modeling and Control of Nontechnical Systems," Report 79-100-W, NTH, Trondheim, Norway.

NEW RESULTS AND THE STATUS OF COMPUTER-AIDED PROCESS CONTROL SYSTEM DESIGN IN NORTH AMERICA

T. F. Edgar
Department of Chemical Engineering
The University of Texas at Austin
Austin, TX 78712

ABSTRACT

Research activity in North America in developing software for computer aided control system design has been primarily located in academic institutions, while industrial application of such tools has been rather limited. Available domestic software is generally not well-documented nor is it being disseminated to other users. The results of a survey of users of multivariable control and estimation software are reported herein. In addition, new advances by investigators in the U.S. and Canada which bear on future software development are also discussed.

INTRODUCTION

The need to implement advanced process control strategies, often requiring multivariable control and estimation, has become more pressing as world energy costs have accelerated during the past five years. Energy-integrated chemical plants exhibit multivariable interactions and generally require control systems which take such interactions into account. In order to deal effectively with the analysis of such control systems, the designer must have flexible computational tools (often employing computer graphics) at hand.

General purpose control software has been available for state space and optimal controller (linear-quadratic-Gaussian or LQG) synthesis for nearly five to ten years [1, 2]. More recently, design tools incorporating multivariable frequency response [3, 4, 5] have been marketed commercially. While the state space tools were developed in the U.S., these later developments have occurred in Europe, largely due to influence of H. Rosenbrock and A. MacFarlane. The frequency response packages offer more flexibility than their earlier counterparts because they incorporate design techniques both in the time and frequency domain and employ a significant component of computer graphics.

Since such software packages are well-documented and tested and available commercially, it is pertinent to examine how they are being utilized in the process industries in North America. In order to identify industrial and academic applications of the software, a survey of control researchers in the U.S. and Canada has been carried out; the results of the survey are discussed in the next section. In addition, the research efforts in extending the available design tools at a variety of institutions are detailed. In this review, we do not summarize basic concepts in linear control theory, due to space limitations. The reader is referred to several review papers and textbooks [6–9].

SURVEY RESULTS

In order to determine the status of computer-aided control system design in the process industries in North America, approximately 150 active researchers in the U.S. and Canada were contacted. These scientists and engineers represented both academic institutions (60%) and private industry (40%). Respondents were requested to identify only software which is being used for multivariable control and estimation (rather than techniques for single input-single output systems). In addition, respondents were asked if they were acquainted with commercial design packages for controller design, such as those developed in Europe or elsewhere. Finally, it was requested that projects or applications involving multivariable control or estimation software be identified along with estimating the number of professionals who use the software.

Some general observations can be made about the responses received.

(1) Fewer than five industrial respondents in the U.S.A. have used multivariable controller design packages. A somewhat larger number indicated that multivariable interactions were a concern in

control system design, but normally simulation is used to examine the multivariable effects (rather than linear system analysis tools). There is a general lack of familiarity with established software, but a reasonable amount of latent interest in their use exists.
(2) Roughly ten universities are active in multivariable controller software development; there are many "homegrown" packages, some outgrowths of established, well-documented packages and others developed for special purpose use (rather than general application).
(3) Because of the evolutionary nature of the software developed in North America, and the lack of support personnel, only a few packages are well-documented enough to assure portability of the packages. As a result, there are presently very few "commercial" programs being marketed, nor does there appear to be any concerted effort in North America to publicize a given control software package.
(4) Rarely was there a concentration of more than two or three professionals at a given site who were involved in use of such software.

Industrial Applications

A wide variety of industrial applications (few of them documented in the literature) were reported where multivariable control and estimation were being employed; a list of such projects is shown below (excluding pilot scale tests):
   (1) six stand hot rolling gauge control
   (2) continuous casting metal level control
   (3) slab temperature control
   (4) paper machine control
   (5) rubber calender control
   (6) ammonia plant
   (7) biological wastewater treatment
   (8) pH control for wastewater neutralization
   (9) boiler control (steam and gas turbines)
   (10) process heat solar collector
   (11) hydrocracker temperature control
   (12) chemical reactor control
   (13) distillation column control

By way of contrast, in 1976 Rijnsdorp and Seborg [10] performed a literature survey of industrial applications of modern control and found relatively few production-scale applications. It is apparent that at least in North America, there has been very little change in the methodology used in industrial control system design (except for more extensive use of simulation) since 1975, although this author feels substantial changes could occur in the next five years.

SOFTWARE PACKAGES

Software packages for design of multivariable control systems identified by the survey respondents can be classified into two groups:

(1) modeling and parameter estimation
(2) multivariable control and state estimation.

The second group can further be categorized into programs which have a reasonable capability for design using frequency response techniques (as opposed to time domain design only).

Modeling/Parameter Estimation

In the past decade, general purpose dynamic and steady state simulators have achieved respectability in the practice of chemical engineering. Therefore such software can be and is an integral part of testing control strategies, although such programs are quite cumbersome for control system synthesis. Packages for dynamic simulation of individual process units or chemical plants mentioned by the survey respondents include

(1) CSMP
(2) ACSL
(3) FLOWTRAN
(4) Box and Jenkins
(5) IDCOM

The development of general purpose software for dynamic simulation continues to be an active area of research (e.g., Patterson and Rozsa [11]). The emphasis of this paper is on multivariable controller/estimator design, hence we choose not to discuss simulation programs in detail. The last two packages are of special interest because they are concerned with optimizing control as an integral part of the modeling strategy. Later in this paper we shall discuss features of various time-series analysis programs.

Multivariable Control and Estimation

The various software packages for multivariable control and estimation which have been developed in North America are discussed below. Their general features and some assessment of the level of documentation are provided. Naturally we devote more discussion to those programs which are available to the public. The first six packages are oriented towards controller and estimator/observer synthesis in the time domain, although many of these programs have the capability of performing calculations in the frequency domain (but not necessarily controller design). The remaining packages, most of them recently developed, primarily focus on frequency domain design techniques.

1. VASP

The first code for lumped LQG computations, the Automatic Synthesis Program (ASP), was developed by Kalman and Englar [12]; this code consisted of a special purpose executive program, utilizing macro-instructions for implementing the matrix operations necessary for solving the LQ optimal control problem. For instance, the instruction RICAT plus appropriate arguments performs the integration of the Riccati equation. ASP was designed for use with IBM machines, and it was very difficult to adapt ASP to other computers due to its dependence on assembly language programming. A modification of ASP, completed by White and Lee [1], is known as VASP. VASP is a variable dimension version of ASP, and its advantages over ASP include a more versatile programming language, a more convenient input/output format, some new subprograms, and variable dimensioning.

VASP consists of 31 subprograms for analysis, synthesis, and optimization of complicated high-order time-variant problems associated with modern control theory. These subprograms include the basic operations of matrix algebra, computation of the exponential of a matrix and its convolution integral, solution of the matrix Riccati equation, computation of dynamical response of a high-order linear system, and input-output functions.

Since VASP is programmed in FORTRAN the user has at his disposal not only the VASP subprograms, but all FORTRAN built-in functions and any other programs written in the FORTRAN language. All of the storage allocation is controlled by the user so the largest system that the program can handle is limited only by the size of the computer, the complexity of the problem, and the ingenuity of the user. All matrices and scalar variables are assumed to be in double precision. A typical instruction consists of CALL RICAT, which implements the Ricatti equation integration. VASP can be used on any computer with a storage capacity greater than 50 k decimal bytes.

One of the disadvantages of the above computer code is an inefficient subprogram used for the matrix Riccati equation, requiring integration to the steady state. Methods utilizing eigenvector transformations for the continuous time Riccati equation solution have been proposed by several authors (e.g., Marshall and Nicholson [13]). The reduction in computational requirements resulting from these noniterative approaches has been spectacular (approximately a factor of 50).

VASP is available from COSMIC (Computer Software Management and Information Center), Suite 112, Barrow Hall, University of Georgia, Athens, Georgia 30602 at a price of $590 for the program tape and $40 for the documentation.

2. EASY5

The Engineering Analysis System 5 (EASY5) [14] is a proprietary simulation and control system analysis language developed by Boeing Computer Services Company. EASY5 allows the scientist or engineer to simulate dynamic response and perform control system analysis for systems described by nonlinear differential or difference equations.

EASY5 uses a familiar English-like language to define commands required to operate the program. EASY5 assists the user in preparing models and performing analyses with a full set of verification procedures and error messages. EASY5 has complete FORTRAN compatibility, allowing any user or system library to be accessed. The user can operate in either a remote job entry interactive mode or batch mode. Analysis capabilities of EASY5 are nonlinear simulation, nonlinear steady state analysis, linear analysis, linear model generation, frequency response analysis, root locus analysis, and optimal controller design.

3. MIT

Several respondents reported having used this program. The MIT program was developed by Sandell and Athans at MIT to complement their videotaped lectures on "Modern Control Theory," a short course given by the M.I.T. Center for Advanced Engineering Study. The programs are a set of FORTRAN based subroutines employed for design calculations for linear time-invariant systems [2]. The subroutines are divided into "high level" and "low level" routines; an experienced user may prefer to employ only low level routines to maximize program flexibility and to solve extensions of the basic LQG problem. High level routines include optimal linear regulator, steady state Kalman-Bucy filter, simulation of forced and unforced systems (including stochastic systems), elimination of non-diagonal cost matrices, calculation of rms error, eigenvalue calculation, controllability and observability computation, and input-output routines. The emphasis throughout is on continuous time systems. The graphics employed are based on the use of a standard line printer.

The so-called "low level" routines contain the standard matrix operations and well as solutions to linear matrix equations. The Riccati equation is solved using a successive approximation algorithm by Kleinman [15]. The exponential matrix is computed using a matrix series of 36 terms, with the Cayley-Hamilton theorem employed for matrix products.

4. ORACLS

ORACLS (Optimal Regulator Algorithms for the Control of Linear Systems) is a set of FORTRAN subroutines developed at NASA for aircraft control by E. S. Armstrong for designing linear control laws and filters for systems with linear time-invariant multivariable differential and difference equation models [16]. The primary advance over similar packages previously developed is that more reliable and efficient numerical methods for matrix operation are employed, plus there are more "specialty" and alternative calculations included.

Procedures included in this package include computation of eigensystems of real matrices, solution and/or least square approximations to solutions of matrix algebraic equations, and controllability and observability. The same general methods as used in VASP are employed for discrete and continuous optimal regulators, optimal servomechanisms, and Kalman-Bucy filters. However, the steady state Riccati equation (continuous and discrete) is solved using a Newton-Raphson method. Interestingly enough, many of the ORACLS subrountines have the same names as those used in VASP. There are 62 subroutines, requiring 6500 source statements in FORTRAN IV.

The features of ORACLS not included in the first three programs discussed above are as follows:

(1) computation of transfer function matrix
(2) model-following control
(3) pole placement (scalar control) and stabilization operations
(4) supporting programs for matrix calculations (Hessenberg, Schur forms, LU decomposition).

The procedures used in ORACLS are described in a text by Armstrong [16]. The program tape and documentation are available from COSMIC (see VASP description above) for $650 and $21.50, respectively.

5. RPI

Investigators at RPI have developed software for generalized state space operations, including those subroutines in ORACLS. Special features include pole placement subrountines as well as simultaneous eigenvalue/eigenvector assignment. Model data must be entered in the state space (time) domain; conversion to transfer function form is handled by the program. Computation of system properties such as decoupling and transmission zeros can also be performed. The program does not perform minimal realizations in order to convert from the s domain to the time domain. Frequency domain design can only be performed one loop at a time, using the Bode plot. A recent addition to the program is a special algorithm for selection of quadratic weighting matrices [17]. A manual for the RPI program is presently being written.

6. GEMSCOPE (University of Alberta [18])

GEMSCOPE (General Multipurpose, Simulation and Control Package) is a series of separate programs which share common disk storage files and can be executed either singly or sequentially to perform operations such as model definition, control calculations or simulation of time domain responses. The main programs used for simulation are

(1) specification of the programs to be executed
(2) definition of the state-space model
(3) calculation of the state difference equation
(4) calculation of the time domain response.

Other software functions include:

(5) generation of an equivalent state-space form for models or block diagrams entered as transfer functions (including time delays)
(6) reduction of the order of the state-space model
(7) generation of state feedback required for non-interaction
(8) calculation of the control matrices for optimal regulatory control
(9) utilization of the discrete control vector for open-loop or optimal state-driving
(10) limited multivariable frequency domain design
(11) observe, filter design
(12) pole placement

Utility programs for input-output, editing, listing, etc., are not included explicitly in this package but are assumed available through a main computation center. The program was designed to run on an IBM 360/67 computer and is not directly transferable to other machines without some editing.

The GEMSCOPE programs are written in FORTRAN and except for critical calculations, such as determining eigenvalues, are single precision. In the original form they consisted of about 5500 source cards and occupy about 100,000 words of disk storage. It should be emphasized that each program in GEMSCOPE is a separate mainline program and can be executed individually or in any sequence the user specifies. A series of integer code numbers identify which sequence of programs are to be used.

This program has been developed in several stages; an earlier version of the program was discussed in a 1972 article in *Automatica* [18]. Since that time several additions to the program have been implemented, including software for multivariable frequency response (direct and inverse Nyquist arrays and characteristic loci); documentation is available from the University of Alberta, Department of Chemical Engineering.

7. LOPCON (Drexel University)

This multivariable control system package has evolved over the past ten years but a major revision and restructuring of the code is planned, although the basic analytical methods will remain the same. LOPCON itself has the capability of providing standard LQG designs or disturbance rejecting servo-mechanism designs. The latter still selects gains (both control gains and observer gains) on the basis of quadratic penalty function minimization. The flexibility exists to solve problems with 'degree of stability' constraints and singular performance criteria. LOPCON can be used to provide full order or minimal order observers or Kalman filters as appropriate [19, 20].

LOPCON can also provide a complete modal analysis (using EISPAC routines) for the open and/or closed loop system as desired. It can also provide the (multivariable) zeros for the open loop plant, closed loop systems, or compensator. It also allows for the use of estimation for only part of the state. LOPCON's current state is somewhat of inefficiency which is why it will be revised, according to its authors.

LOPCON is used in conjunction with two other major programs. One is a simulation program (either DSL, developed at Drexel, or IBM's CSMP) and the other is a linearization program LINEAR. The procedure almost always is to start with a nonlinear system simulation using DSL (or CSMP). LINEAR interfaces with DSL (or CSMP) and produces a linear model at one or more equilibrium points. LINEAR can also provide a reduced linear model based on modal reduction and/or a complete modal analysis at each equilibrium point. LOPCON uses the models generated by LINEAR to produce a control system design; this controller can be readily incorporated in the DSL simulation of the full order nonlinear plant via automatic program generation. Professor H. Kwatny is the primary developer of LOPCON.

8. Purdue University

Leininger [21] has developed an interactive design suite for multivariable Nyquist array methods. The main advance over the existing CAD suites in Great Britain is an automated diagonal dominance subroutine, designed to eliminate much of the tedious trial and error calculations involved in testing various square pre- and post-compensators. In essence, the program allows the designer to specify the frequency response desired for the controlled variables. The use of an optimization method for diagonal dominance usually takes less than two minutes on the computer, compared to up to several weeks by trial and error. Optimizing both pre- and post-compensator constant matrices seems to be sufficient for obtaining diagonal dominance in most applications (rather than using dynamic compensation).

The dominance algorithm [22] uses a conjugate direction function minimization algorithm to adjust the parameter set until a performance index composed of dominance definitions is minimized. For the inverse Nyquist array method in a row dominance mode, the optimization problem can be separated into m independent optimization efforts, one for each row. Here the performance index by row is

$$J_i(K_{ij}) = \max_{\omega} \sum_{\substack{j=1 \\ i \neq j}}^{m} |\hat{q}_{ij}(s)|/|\hat{q}_{ii}(s)| \quad i=1, 2, \ldots, m \quad (1)$$

where $\hat{q}_{ij}(s)$ is an element of $Q^{-1}(s)$. For each i, the ith row of K^{-1} is adjusted until $J_i(K_{ij})$ is minimized. In practice, the ratio in (1) is computed for each discrete frequency point in the range of interest. This array is then scanned to identify the maximum ratio. Adjusting the elements of row i in K^{-1} yields a set of final dominance levels.

$$\theta_i = \min_{K_{ij}} J_i(K_{ij}) \quad (2)$$

If the dominance levels in (2) are all less than unity, diagonal dominance has been achieved. In the event that some of the dominance levels are greater than one, the designer may initiate a dominance sharing search or restart the program using new starting values of the unspecified compensator parameters.

The concept of dominance sharing, as detailed in [22], consists of rescaling the compensator matrices so that low dominance levels may be intentionally increased. This permits previously non-dominant levels to be shifted to a range of acceptability. This procedure is initiated by the designer after a set of dominance levels has been evaluated. The automated dominance search algorithm as described above evaluates the dominance conditions over a finite prespecified frequency range. The theoretical foundations for the finite range search are established in [23] for strictly proper transfer matrices (all elements of Q(s) tend toward zero as ω increases).

Using conformal mapping (w = 1./z), the dominance condition in one polar plane (z-plane) can be graphically portrayed in the opposite polar plane (W-plane). Thus, the establishment of diagonal dominance for Q(s) can be imaged to the inverse polar plane using the mapping above (and vice versa). Since the mapping is conformal, all phase margin, gain margin, and stability relationships are retained. In this way, diagonal dominance can be obtained in either polar plane with closed loop design characteristics determined using the original polar plane or the image plane. This imaging process is independent of Gershgorin dominance considerations arising in the evaluation of the transfer matrix associated with the image plane [23].

The automated dominance search procedure takes over at this point to obtain system dominance. Data files are created for the Gershgorin, Ostrowski and image bands. When completed, the requested Nyquist, Bode and Nichols plots are displayed for designer interpretation and evaluation. The CAD suite normally requires five to ten minutes of elapsed time with an approximate expenditure of 50 to 100 CPU seconds to reach this point in the design calculations.

The designer may proceed directly to a dynamic compensator design for each loop using classical single input-output techniques or display unit step response curves. To design using Bode or Nichols methods [24], the user may elect to insert lead, lag, lead-lag, etc. compensators in one or more of the control loops. In this case, the interactive commands request the amount of phase advance and/or db attenuation required and the respective crossover frequencies for the desired compensator. Upon entry, the new Bode, Nyquist and/or Nichols plots are displayed.

The step response segment of the CAD suite allows the designer to command input or output set point levels with or without the dynamic compensation. If dynamic compensation is to be included, the program will convert the frequency response compensators to state space form and augment the system state model automatically. System transient responses are then displayed following an interactive command sequence.

At any point during the graphical evaluations the visual cursor can be used to obtain:

a. Specific frequency points on any curve
b. x and y coordinate points
c. Compute phase and/or gain margins

The program by Leininger (Department of Mechanical Engineering, Purdue University) is not yet fully documented; full documentation will not be available for several years, although there are presently several outside users of this software.

9. University of Wisconsin

Ogunnaike and Ray have developed a series of programs for multivariable controller synthesis with special emphasis on time-delay compensation. In addition, they have developed computer control software to accompany a series of laboratory experiments in chemical process control.

The basic program for multivariable dead time compensation is based on recent results for optimal control (time domain as well as frequency response design, as reported by Ogunnaike and Ray [25]. They have shown how a multivariable analog to the conventional Smith predictor can be employed to remove time delays from the characteristic equation. In the frequency domain this permits a two stage design

(1) compensate for the delays in the process transfer function
(2) selection of a precompensator feedback matrix, using any of the standard multivariable controller techniques discussed earlier.

If desired, a parallel path using either pole placement or linear optimal control can be implemented, as discussed by Ogunnaike and Ray [25]. A control system designer can initiate calculations with the Wisconsin program using either a state space or a transfer function model. Minimal realization as well as numerical simulation subroutines are provided to link s and t domains.

10. Princeton University

Kantor and Andres [26, 27] at Princeton University have developed on extended version of the basic Nyquist array approach; the program is designed for local use only. Kantor and Andres have developed a graphics-based scheme for frequency-dependent pre-compensation. Diagonal dominance is achieved through a measure of interaction which is independent of the scaling of inputs and outputs; for 2×2 systems contour plots of a scalar interaction parameter can be analyzed to determine a suitable precompensator [26].

11. Other Institutions

The particular orientation of linear multivariable control research at Notre Dame is to use frequency dependent pre-compensation to obtain digonal dominance. A program called CARDIAD (Complex Acceptability Region for Diagonal Dominance) for column operations makes the Nyquist array diagonally column dominant [28]. The heuristic used is to normalize the compensator matrix to yield all diagonal elements equal to one and to employ a gradient search algorithm to optimize off diagonal elements of the compensator. Schafer and Sain have demonstrated this approach for a sixth order, four input, four output descriptions of a turbofan engine.

Chemical Engineering Departments at two other universities reported having developed some software for multivariable controller turning, mainly through the efforts of graduate students. The University of Colorado (D. Claugh) has Tektronix-based rountines for Nyquist and inverse Nyquist arrays as well as for multivariable self tuning regulation. The University of Texas (T. Edgar) has Tektronix-based software for 2 × 2 systems only, with capability for displaying Nyquist, Bode, and Nichols arrays as well as dead-time compensation.

OPTIMIZING CONTROL

Recently there has been great interest in discrete-time optimal control based on a one-step ahead optimization criterion, also known as minimum variance control. A number of different approaches for minimum variance control has been developed in the last decade. MacGregor [29] and Palmor and Shinnar [30] have provided overviews of these minimum variance controller design techniques. Generalized software for minimum variance control was originally developed by Astrom [31] and Box and Jenkins [32], both of which are marketed from Europe. Several respondents in the survey mentioned using these techniques. A more recent time series statistical package developed in Japan [33] is now available in the U.S. through the Mathematical Sciences Division of the University of Tulsa (TIMSAC-78, costing $25 for the tape).

More recently, techniques based on on-line impulse response models [34, 35] have been used in the U.S.A. for optimizing control. "Dynamic matrix" control [34] has been developed by Shell Development for multivariable control of cracking reactors. IDCOM [35], developed in France, has been licensed by several firms in the U.S. for process applications. Setpoint, Inc. plans to employ IDCOM for a hydrocracker application in 1981.

CONCLUSIONS

There appears to be a growing interest in using advanced control techniques and associated software in the U.S. and Canada. Very little marketable software is indigenous to the U.S. or Canada, probably because of the availability of well-documented packages developed in Europe [36] or Japan [37]. These packages have largely evolved at universities and similar institutions, in spite of the lack of industrial interest in their application. That situation now seems to be changing, based on responses in the survey conducted for this paper. Japanese industry [37] is already showing relatively heavy involvement in using advanced control methodology.

One way to encourage more activity in testing and evaluating multivariable control design software in the process industries may be to develop a theme problem of relevance to process control. The 1977 International Forum on Alternatives for Linear Multivariable Control [38] was a conference organized to compare and evaluate various controller design approaches for a turbofan engine. The theme problem consisted of a 16 state, 5 input, 5 output model; actuator and sensor dynamics were also included. Model reduction was permitted for controller synthesis. This theme problem has continued to be a focus of research interest since the conference.

REFERENCES

[1] White, J. S. and Lee, H. Q., "Users manual for the variable dimension automatic synthesis program (VASP)," NASA TM X×2417, Washington, D.C., October, 1971.

[2] Sandell, N. R. and M. Athans, "Manual of FORTRAN Computer Subroutines for linear-Quadratic Gaussian Designs," Massachusetts Institute of Technology, Cambridge, Massachusetts, 1974.

[3] Munro, N., Broland, L. S., and L. S. Brown, "The UMIST Computer-Aided Control System Suite-Introduction and Users Guide," National Research Development Corporation, Great Britain, 1973.

[4] Rosenbrock, H. H., Computer Aided Control System Design, Academic Press, New York, 1974.

[5] Pedersen, J. O., "darek Users' Guide," Report STF48A7301 SINTEF, Trondheim, Norway.

[6] Edgar, T. F., "Status of design methods for multivariable control," AIChE Symp. Ser., 72(159):99 (1976).

[7] MacFarlane, A. G. J., *Automatica,* 8:455, (1972).

[8] Edgar, T. F., "Advanced control strategies for chemical processes: a review," Computer Applications to Chemical Engineering, ACS Symposium Series, 124, 1980.

[9] Kwakernaak, H., and R. Sivan, linear Optimal Control Systems, Wiley, 1972.

[10] Rijnsdorp, J. E. and D. E. Seborg, "A survey of experimental applications of multivariable control to process control problems," *AIChE Symp. Ser.,* 72(159):112, (1976).

[11] Patterson, G. K. and R. Rozsa, *Computers and Chemical Engineering,* 4:1, (1980).

[12] Kalman, R. E. and T. S. Englar, "An Automatic synthesis program for optimal filters and control systems," RIAS report, 1963.

[13] Marshall, S. A. and H. Nicholson, *Proc. IEE,* 117:1805, (1970).

[14] Boeing, Computer Services Company, EASY5 Product Newsletter, 1980.

[15] Kleinman, D. L., *IEEE Trans. Auto. Cont., AC-13,* 114, (1968).

[16] Armstrong, E. S., *ORACLS-A Design System for linear Multivariable Control*, Marcel Dekker, New York, 1980.
[17] Harvey, C. A., and G. Stein, *IEEE Trans. Auto. Cont., AC-23*, 378, (1978).
[18] Fisher, D. G., R. G. Wilson and W. Agostinis, *Automatica*, 8:737, (1972).
[19] Kwatny, H. G., *IEEE Trans. Auto. Cont., AC-19*, 274, (1974).
[20] Kwatny, H. G., and K. G. Kalnitsky, *IEEE Trans. Auto. Cont., AC-23*, 930, (1978).
[21] Leininger, G. G., "An interactive design suite for the multivariable Nyquist array method," IFAC Computer Aided Design of Control Sysems Symposium, Zurich, Switzerland, 1979.
[22] Leininger, G. G., *Automatica*, 15:339, (1979).
[23] Leininger, G. G., *Int. J. Cont.*, 30(3):459, (1979).
[24] Leininger, G. G., "Multivariable compensator design using Bode diagrams and nichols charts," IFAC Computer Aided Design of Control System Symposium, Zurich, Switzerland, 1979.
[25] Ogunnaike, B. A., and W. H. Ray, *AIChE J.*, 24:1043, (1979).
[26] Kantor, J. C. and R. P. Andres, *Proc. JACC*, 884, (1979).
[27] Kantor, J. C. and R. P. Andres, *Int. J. Cont.*, in press, (1980).
[28] Schafer, R. M. and M. Sain, *Proc. JACC*, 348, (1979).
[29] MacGregor, J. F., *Can. J. Chem. Engr.*, 51:468, (1973).
[30] Palmor, Z. J. and R. Shinnar, *IEC PDD*, 18(1):8, (1979).
[31] Astrom, K. J., Introduction to Stochastic Control Theory, Academic Press, New York, 1970.
[32] Box, G. E. P. and G. M. Jenkins, Time Series Analysis, Forecasting and Control, Holden-Day, San Francisco, 1970.
[33] Akaike, H., G. Kitigawa, E. Arahata, and F. Tada, TIMSAC-78, Computer Science Monograph No. 11, Institute of Statistical Mathematics, Tokyo, Japan.
[34] Cutler, C. R. and Ramaker, B. L., "Dynamic matrix control-A computer control algorithm," AIChE National Meeting, Houston, Texas, April, 1979.
[35] Richalet, J., A. Rault, J. L. Testud, and J. Papon, *Automatica*, 14:413, (1978).
[36] Hashimoto, I. and T. Takamatsu, "New results and the status of computer aided process control systems design in Japan," Engineering Foundation Conference on Chemical Process Control-II, Sea Island, Georgia, 1981.
[37] Tysso, A., "New results and the status of computer aided process control systems design in Europe," Engineering Foundation Conference on Chemical Process Control-II, Sea Island, Georgia, 1981.
[38] Sain, M. K., J. L. Peczkowski, and J. L. Melsa, Eds., Alternatives for Multivariable Control, National Engineering Consortium, Oak Brook, Illinois, 1979.

ON CLOSING "THE GAP"—WITH MODERN CONTROL THEORY THAT WORKS

Magne Fjeld
AccuRay Corp., 650 Ackerman Rd.,
Columbus, Ohio 43202

The title of this paper could appear to be illogical to many of us in this conference. Indeed, what would the implied complement "Modern Control Theory that Does not Work" really mean? By itself, it is nonsense. If a theory "does not work" (implication: in practice), it may not deserve its status as a serious theory, i.e., one which is mathematically based on a stringent model formulation, on the realities of the process objectives of control and proofs of solution. Since most of us think that the main body of what is named modern control theory, is a serious theory, any argument which implies that there is something wrong with the theoretical development, should be easy to refute. However, such arguments are frequently backed by a conglomerate of theoretically unfounded opinions: subjective imagination based on experience of what could possibly be applied practically, or some hard valid points. All of these should be dealt with in a rational manner.

However, it is extremely difficult to open a meaningful discussion when trying to counter an entirely "positivistic" view of control engineering. Although positivism as a popular philosophy belongs to the 19th century, positivism in control engineering is alive and well.

By definition, in positivism, mathematical transformations cannot be considered facts. Therefore, to the positivist, neither formal mathematical representation of conditions, states, nor extrapolation of such facts into a domain which is unknown to experience, can be accepted as a valid basis for control engineering design.

It is of interest to examine on a more rational basis what underlies the infamous gap between theory and practice.

Some of the more important obstacles appear to be:

1. The apparent dichotomy between modern and classical methods.
2. Improper application of methods, and human resources limitations.
3. Tools and utilities.
4. Practical problems not directly treated by the theory.
5. Systems software for process control.

Item 5 has been thoroughly treated elsewhere in this conference, and will not be discussed here.

NEW THEORY VERSUS CLASSICAL THEORY

When talking to newly graduated or graduating Control Engineers, some of you may have discovered that quite a few of them think there exist two disjunct worlds of control engineering theory: design methods based on state space theory, and methods based on theory of frequency analysis. Maybe the same was true when we graduated ourselves, if exposed to state space theory at all. However, we now know that this is much more a dichotomy of mind than something inherent in the theories. Although progress has been made in texts to unify the two, in order to convey the idea that these theories are somehow two different projections of roughly the same "design solution space," still more can be done. I have not yet seen the basic tutorial paper on this topic presented—without being loaded with theoretical subtleties.

Hence, since we have learned about the equivalence between various approaches, why bother with new methods at all? Are not those new theories "old wine in new bottles?" Yes—and No.

We must not forget that state space theory has led to design methods and techniques for implementation that we did not have before, and establishing equivalent results in the frequency domain is in this sense a backwards exercise. Also, state space theory has sparked off some promising advances in adaptive control, which have no direct interpretation in the frequency domain. Therefore, any claim that a solution by modern control theory X does not work—with the understanding that a classical solution of the same problem Y based on frequency analysis does work—is a problem that can in many cases be resolved as an academic exercise, and therefore, is relatively easy to handle. I will therefore not pursue further discussion on the possibility of catastrophic or successful results with modern control theory because of ill-conceived ideas about the apparent dichotomy above.

HUMAN RESOURCES, CHOICE OF PROBLEM, METHOD AND IMPLEMENTATION UTILITIES

As one Control Engineer put it to me in a recent correspondence, the success of a computer control application depends as much upon the ability of the designer as anything else.

Such factors as formulation of the problem to be solved, and the method used to solve it, are dependent upon this ability. Further, what appears to be more related to experience, ability, appreciation of details, and the "art of design," are what we could designate as software "implementation utilities." Experience tells me the implemented application control program may contain as little as 2–20% plain text-book control theory implementations. As we all know, several well-known necessities in industrial control systems contribute to such low figures. If we talk about advanced control algorithms, let me mention only a couple of characteristics that tend to make software voluminous, beyond what the bare algorithm requires:

— The case where very little a priori information is needed to get the control running.
— The case where a lot of a priori information is needed to start the control up (constraints, design parameters, etc.)

Both approaches can involve advanced algorithms, but in different ways. The first one needs built-in adaptivity that tends to increase complexity. The second gets its complexity from the fact that more a priori data tends to increase the number of logical tests, computational paths and algorithm complexity.

When we talk about modern control theory, we mostly think of theory for linear systems, based upon the state space concept. This gives clean design results, free of nonlinear constraints and logical decisions. However, the industrial environment is distinguished from the linear model by requirements of system start-up, computer/local transfer, auto/manual transfer, saturation of actuators, different operating regimes that destroy linearity, failing measurements, varying measurement noise levels, etc. Since we may not want to discard an otherwise powerful control design method for such and other conceivable reasons, the full design will become an exercise in getting the process behaviour as perceived by the algorithm to fit the framework of the theory applied, rather than finding a special control theory which fits the particular situation (which is a nearly insurmountable task). There is no doubt that this is a task which may effectively discriminate the less competent designer from the competent designer.

Another potential failure situation is the newly graduated engineer or Ph.D. who is looking for the first and best opportunity to apply his (or his professor's) favorite theory. I think that the Control Engineer who is looking for the best suited of available theories to the particular situation at hand, is likely to be more successful than the former, not to mention the one who looks at control engineering as a discipline which can be exercised quite independently of the needs of the process and its practical controllability as dictated by the actuator configuration, measurement accuracy and noise, etc.

ORGANIZATIONAL CONSTRAINTS

This constitutes undoubtedly one of the most—if not the most—significant set of factors that determine whether modern control theory will work. Depending on size and the characteristics of the companies' business, there are entirely different ways to organize internal and external activities. Larger companies will certainly have some kind of functional organization, whereas very small companies may only have a project organization or task force to deal with a control system design and delivery from start to finish.

Case histories may be classified into a few (not exhaustive) organizational and user/supplier relationships:

Case 1.

A small process control company gets a contract with a process industry to specify, design and install an advanced process control system as an expansion of one of the customer's existing in-house computer process control systems. The project includes a process study to precede the design. Three to four engineers are forming the project group, and will do the project to its finalization. The customer is open to any well-founded solution that results in a system that is easy to use. Since the value added, the cost of raw materials and their throughput are relatively high, extra efforts to make a good control system can be proved to pay off handsomely. This organization is fairly simple and is close to an ideal situation for possible application of modern control theory. The task force effort can ensure successful completion of the objectives, with full control of all aspects of the system development.

Case 2.

A large process control manufacturing company decides to design and market a new flexible process control system for processes of kind "X," applying some new concepts for control, based on theory "Y."

The control engineers involved in the development may in all probability have to face several requirements, constraints and challenges, as for instance:

— Sufficient generality to be applicable to a certain percent of the market (universality costs too much).
  This means modularity and configurability in order to meet reasonable variations in each customer's requirements.
— Total cost for the customer less than N$.
— Good & flexible man-machine engineering.
— System to be sold by a decentralized field sales organization on a customer contract basis.
— System to be installed, tuned and maintained by a field & customer service organization. Experience shows that it is increasingly difficult to get control engineers to do this kind of job in these types of organizations.

There is no doubt that these requirements may have a good deal of impact upon both overall system design and control engineering design in particular. The requirements make necessary some attributes of the system like:

— High quality of the design (reliability, well-tested, visibility in maintenance and modification situations, good, unambiguous documentation on several levels of detail, etc.)
— Detailed, easy-to-understand tuning or set-up procedures for field installations.
— Expandability and modularity may be important to accomodate future upgrading and enhancement of this product line.

When it comes to applications of modern control theory, experience shows that the following project organization patterns are more likely to be a vehicle for such development:

Case 3.

Contract research & development institution or university gets contract with a process industry, for joint development of a control system for a particular process.

Case 4.

Process control manufacturing company obtains agreement with a customer to try out a new design, being a prototype for a new class of control products.

Case 5.

Contract R&D institution or university carries out a reasonably well-defined development project for a vendor or process control manufacturing company. Unless a third partner is involved—the final user—this tends to result in a less than immediately marketable product.

Case 6.

In-house development group with the user industry organizes its own development for a specific site and/or corporate use.

I will not discuss the possible outcomes of these organizational set-ups, but we may agree that within each of them, modern control theory as a basis for design may thrive with various degrees of difficulty.

THE GROWTH RATE OF APPLICATIONS OF MODERN CONTROL THEORY WILL INCREASE— BUT PROBABLY SLOWLY

Notwithstanding circumstances and organization, a design solution based on modern control theory must lend itself to an evaluation that shows definite advantages for which the user is willing to pay. Initial development costs range typically from moderately higher to very much higher than conventional design solutions. However, in some systems with high capital equipment costs as part of the sales price, this factor may be less important, in particular, if the system is a "high" ($> 10\text{--}20$ units) sales volume item. Installation and tuning costs may be either lower or higher than for a conventional solution. For instance, adaptive control loops may cut costs of otherwise elaborate tuning procedures considerably. In any case, economical development and installation may depend significantly on convenient configuration, design and analysis (as simulations) tools.

In both vendor and user industry, such tools are still fairly limited, but are on the increase, as you will observe from other papers in this conference.

Another factor limiting a rapid increase in the use of modern control theory is that results of one organization are not easily transferrable to another organization, even if these results are well documented. Efficient transfer of knowledge at this level requires a good match in educational and training background between some people in each organization.

Finally, there is at least one more factor involved in the "rate-of-spread" function. For more complex algorithms, nitty-gritty details and the aforementioned control algorithm utility functions will be proprietary information that will not be easily accessible by outsiders, and academia traditionally does not take much interest in those aspects of an implementation. With the know-how built up internally within an organization including those important design details, the organization does not disclose much of value to potential competitors when publishing basic background material, (mathematical development) and results obtained. In fact, a manufacturer of process control systems normally publishes such information for marketing reasons, and there is no doubt that this sharpens its competitive edge in the marketplace.

On the other hand, a development within a user industry may often be of a confidential nature, in particular if it is crucial to the nature and quality of a specific product, which in turn is critical for the product's ability to compete in the market.

Since the spread of practical applications of modern control theory will be slow, innovators may take advantage of this situation by correctly and ingeniously applying modern control theory. When satisfying the technical and business constraints, they may conquer the lion's share of the market they are competing in.

SOME INDUSTRIAL APPLICATIONS

Some of the work that has been carried out, mostly in the Nordic countries, will be briefly reviewed here. The discussion is not intended to be exhaustive in any sense.

Self-Tuning Regulators and Adaptive Control

Self tuning regulators may be defined as follows:

Once initialized to have some regulator coefficient values at start-up time, the regulator automatically tunes these coefficients on-line while controlling, such that eventually a criterion, such as

$$J = E\ [(y_{t+k+1} - y_{ref})^2 + Qu_t^2] \rightarrow Min \quad (1)$$

is satisfied asymptotically. Here, k is the process transport lag in units of sampling time, Q is a polynomial in the shift operator, y is the variable to be controlled, y_{ref} is the setpoint, and u the control (manipulated variable). The STR is composed of an identification part plus a dynamic controller algorithm, such that the computed control at any time is a linear function of a number of previous deviations from setpoint, plus a linear function of another number of the last and previous control actions:

$$u_t = -\frac{1}{b_o}[\sum_{i=1}^{n} a_i y_{t-i+1} + \sum_{i=1}^{m+k-1} b_i u_{t-i}] \qquad (2)$$

Thus the STR algorithm is in principle very simple, but—as with all adaptive control systems—a number of pitfalls exist in practical applications. Unbehauen and Schmid [1] list more than 50 applications of various kinds of adaptive process control systems. However, very few of these applications have been confirmed to be in current operation, as the vast majority of them appear to be experimental.

As with any other control algorithm, the STR should not be applied blindly to any process, and consideration must also be given to points mentioned earlier in this paper. However, good candidates for applications do exist. A well documented application is the one by Westerlund and Toivonen [2] on continuous cement raw material mixing. In this application, the lime saturation factor, the silica ratio and the alumina ratio were all controlled by STR's, which adjusted the feed ratios for the raw materials. The system could be thought of as three SISO (Single Input/Single Output) systems in parallel. Because of the nature of the control output computation and the computed measurements, full decoupling between loops was achieved.

The analysis measurement is taken before the silos, but after the ball mill. Best performance was obtained by using a simple model of silos to produce computed measurements $y(t+k+1)$ *after* the silos on the basis of the real measurements. The STR was then applied by using the computed measurement as the process output; producing a control law of the form

$$\Delta u_t = -\frac{1}{b_o} \cdot [a_1 y_{t+k} + a_2 y_{t+k+1} + b_1 \Delta u_{t-1}] \qquad (3)$$

One interpretation is to consider the STR as a feedback control acting upon a process which is part physical (the ball mill) and part simulation (open loop estimation of the output from the silos), as illustrated in Fig. 1.

This strategy worked so well that it is a standard feature of the computer system which the company markets for cement raw meal mixing (Oy Partek AB).

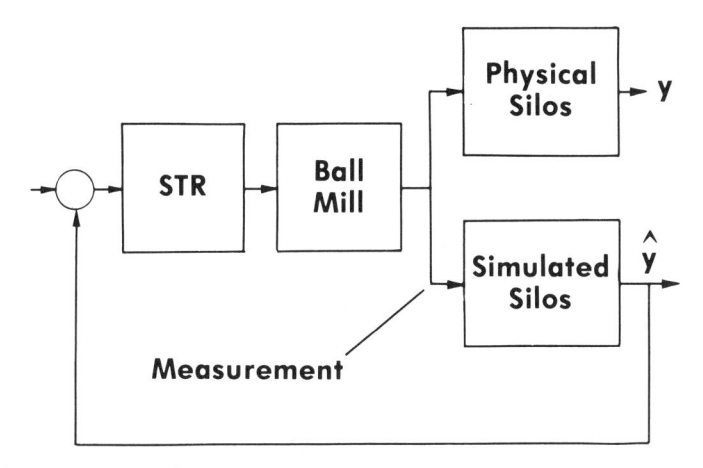

Figure 1. STR application on a raw meal mixer system in a cement plant.

Other good applications were the pioneering installations on paper machine moisture control reported by Cegrell and Hedqvist [3] and Borisson and Wittenmark [4], adaptive autopilots for large tankers by Kallstrom et al. [6], self-tuning control of an ore-crusher by Borisson and Syding [5], and more recently the application to the tension control problem on a cold strip rolling mill [7] by Bengtsson.

In the U.S., there seems to be a reserved attitude toward applying adaptive or self-tuning control to processes. To the author's knowledge, the first running commercial field application is one designed by Fjeld, Kelly, Somerville and Wilhelm [15], which is successfully applied to moisture control on paper machine No. 4 at Fox River Paper Company in Wisconsin.

Moisture control on paper machines is often an ideal candidate for advanced control. Sheet uniformity is of major importance, and the high number of different grades of paper (weights, furnishes, amounts of sizing) together with variations in machine speed and changing conditions in the dryer system, make it practically impossible to make grade-dependent tuning of the controller. On the above mentioned paper machine a self-tuning regulator was applied outputting to the analog steam pressure setpoint of the dryer section. The measurement is obtained from a gauge scanning continuously across the sheet.

To document the improvements, two types of functions are shown:

— The autocorrelation function of the control error (deviation from setpoint).

— A histogram of the standard deviation from setpoint of machine-direction variations in moisture, as measured over a record of data corresponding to a full reel of paper; each sample in the data record is a scan average.

Figure 2a shows a typical autocorrelation plot for the conventional control system, which is a type of model reference system with fixed tuning parameters.

Figure 2b shows the same function, now with the STR applied. Both dynamics and variance have improved significantly. According to theory, the autocorrelation function should for minimum variance control theoretically decrease to zero after k+1 intervals, where k is the apparent transport lag in units of sampling intervals (k=2).

We observe that this is close to being the case. However, since a slight penalization on control amplitude changes is applied, the bandwidth will be slightly lower, and the interval for the autocorrelation function to reach zero will be slightly longer.

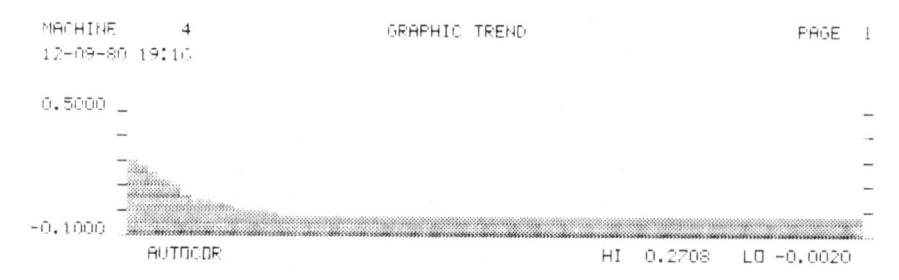

Figure 2a. Autocorrelation function for moisture control error under the conventional control.

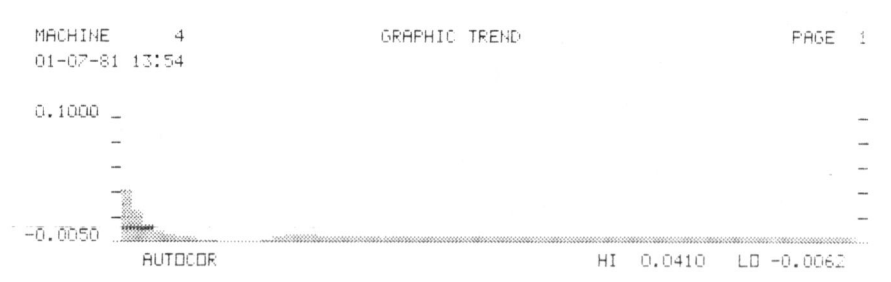

Figure 2b. Autocorrelation function for moisture control error under STR.

Figure 3a and b show histograms of σ in moisture as defined above. Both histograms are normalized to the same total number of samples to ease comparison. We observe that both the variability from reel to reel, and the average σ, have significantly decreased.

LQG and multivariable Approaches

So-called Linear-Quadratic-Gaussian control design approaches have been implemented quite extensively in navigation and aerospace systems. However, their use in process control has been very modest. In the late sixties it was said among quite a few chemical engineers, and in publications, that the theory was not well suited because the quadratic criterion did not fit the optimization criteria for most chemical processes. Since then such a viewpoint probably has prevailed due to confusion between "optimal" feedback control and steady-state optimization. This clearly shows a misunderstanding of LQG theory and its intent. Although the assumption of linearity may make the theory cumbersome or impractical to implement in strongly nonlinear cases, I still believe it could be applied more extensively if the CAD tools were widely available.

A major part of the bag of tricks for such implementations is to come up with a simple cookbook procedure for tuning adjustments in the field,

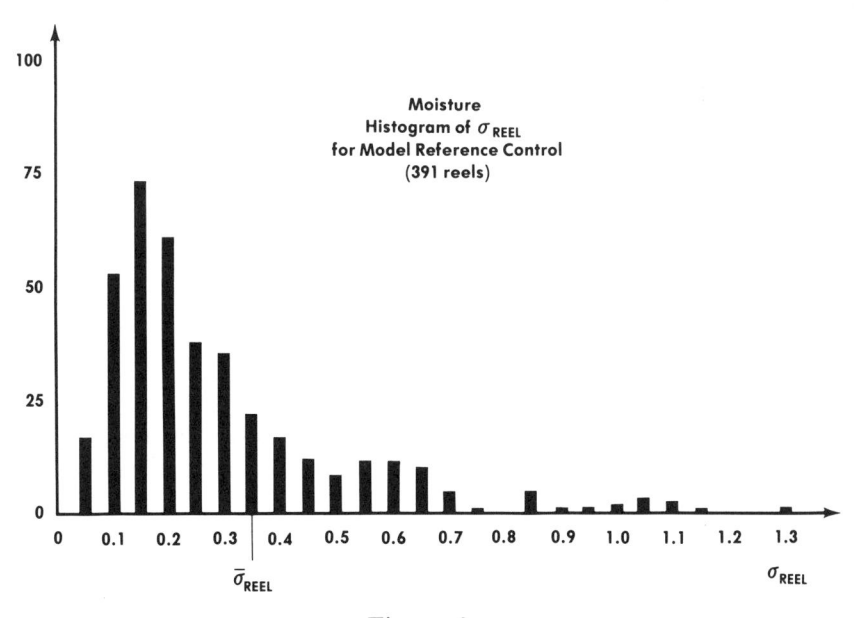

Figure 3a.

Figure 3b.

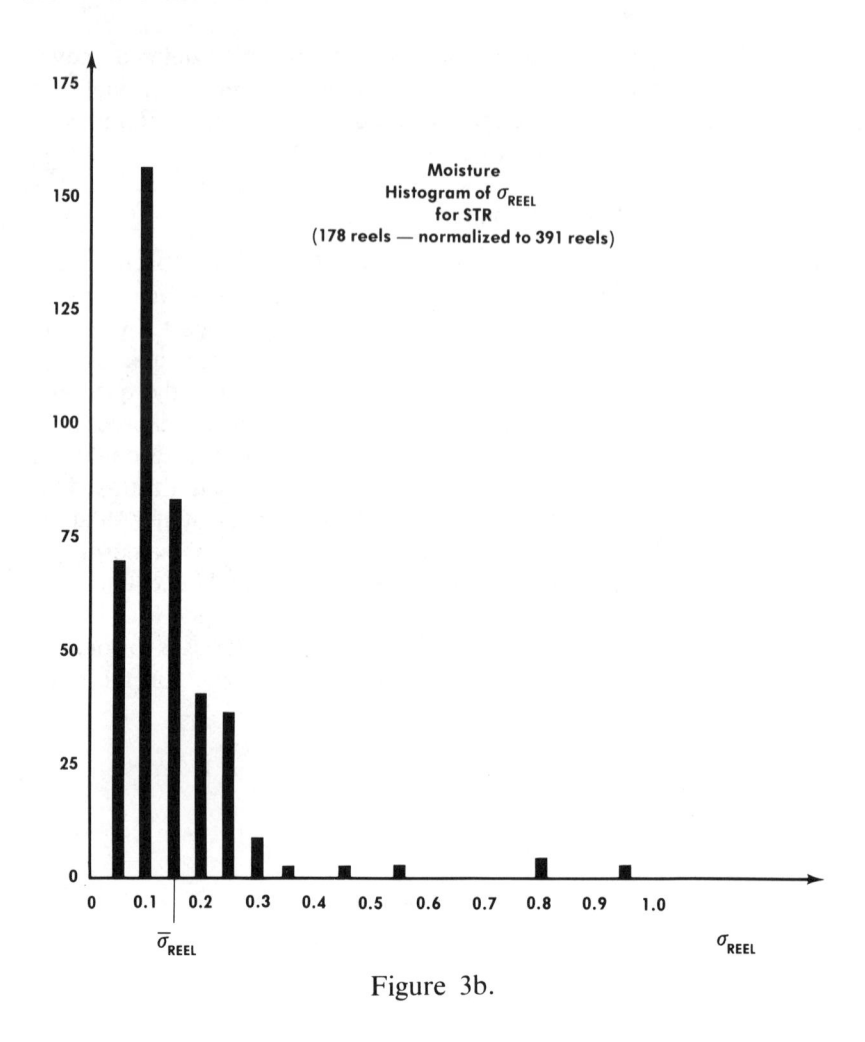

after using key process configuration data in a CAD-based custom control design.

Another concern has been the accuracy of models. Part of this problem can be resolved by using identification packages. However, I also think that little work has been done in order to find out how good models really need to be in order to produce control improvements which are salable at increased price for a given process.

An early industrial process control implementation of Least Squares and LQG control theory was the application by Fjeld, Flatabö and Stai [8], [9] on a paper machine headbox and for thick stock consistency control. For the headbox application, the process was considered second order for design purposes. The quadratic criterion penalized deviations in total head and liquid level. In order to obtain reasonable control amplitudes (limited closed-loop bandwidth), amplitudes of the liquid flow pump and the pressurized air valve were also penalized, in the form

$$J = \frac{1}{N} \sum_{k=1}^{N} [\Delta x^T(k) Q \, \Delta x(k) + \Delta u^T(k-1) P \Delta u(k-1)] \to \text{Min} \quad (4)$$
$$N \gg 1$$

Integral action in control was obtained by extending the process description with integrators.

Compared to the previous control system, the application gave a reduction in standard deviation from setpoint of about 30–40 percent.

The consistency control application had to be handled as a time-delay problem. In this case, the LQG approach was taken in order to filter the measurement and in order to get predictions of "internal states," i.e. consistency not yet transferred to the point of measurement. The estimated internal states (2) plus the environmental state (1) accounting for the varying process disturbance bias, were then used in a feedback control law in the form

$$u(k) = g^T \hat{x}(k) \quad (5)$$

where $\hat{x}$ is the estimated dynamic model state.

In this case the environmental bias state provided for the integral action [16]. The reduction in standard deviation from the setpoint, compared to the best obtainable tuning of a PI analog controller, was in this case 50 percent. Both of these papermaking applications are justified by these loops being important for achieving paper of uniform quality.

Another example of successful application of LQG control strategies is the ship boiler control system as developed by Tyssö and Brembo [10], [11], [12]. The justification for the advanced control in this case is the severe demand on control response in narrow waters, safety, such as to avoid tripping of the boiler under severe process "disturbances," and economy. It is a fact that there are all too many ship boiler tragedies. In particular, the wave induced disturbance in drum level measurement gives an excellent opportunity for demonstrating superior performance of the multivariable control system, using modelling of the wave generated motion. It also diminishes feedwater control valve wear, which also thus contributes to increased safety.

Although not exactly what most chemical engineers would classify as process control, Balchen, Jenssen, Mathisen and Saelid [13] developed a dynamic positioning system for floating vessels. This is one of the most convincing applications of industrial, albeit off-shore, applications of LQG theory, both from technical and business viewpoints. Just as for the previously mentioned implementations, the basic principles used were well-known. The achievement in this case was the mixture of correctly and thoroughly applied theory and the innovative application that resulted in a product with remarkable success in the marketplace. The system uses a Kalman filter for estimation of vessel motions and position, and environmental forces from wind, waves and current. It also automatically tracks the 1st harmonic of the high-frequency movement of the vessel, induced by waves (surge, sway and yaw). This makes it possible to avoid hunting in the actuators while trying to control frequencies that are beyond closed-loop system bandwidth, thus also contributing greatly to maneuvering accuracy. Using the estimator information, the system can also detect probable measurement failures. The advantages of the total system turn out to be more precise dynamic behavior, improved measurement noise suppression, and improved reliability. And, most importantly, the market is willing to pay a higher price for this performance.

CONCLUSION

Any industry, whether it manufactures control systems or is an end user of such systems, should critically analyze its own situation for opportunities in the area of application of modern control theories. All too often, in the face of better knowledge, the assumption is made that such applications are not worth the efforts given the prevailing industrial constraints. In the longer run, it may also be necessary to change some of these constraints in order to create an innovative climate for new systems development.

As an overall assessment, notwithstanding some good demonstrations, process control has missed the "modern control bandwagon." Now it is time to catch it.

REFERENCES

[1] Unbehauen, H., Chr. Schmid: Application of Adaptive Systems in Process Control. Proceedings from the International Workshop on Applications of Adaptive Control, August 23–25, 1979. Yale University, Conn., U.S.A.
[2] Westerlund, T., Toivonen, H., Nyman, K.-E., Modeling, Identification and Control, *1*, (1), 17–37 (1980).

[3] Cegrell, T., Hedqvist, T., Automatica, *11*, 53–59 (1975).
[4] Borrison, U., Wittenmark, B.: An Industrial Application of a Self-Tuning Regulator. 4th IFAC Symposium on Digital Computer Applications to Process Control, Zürich, Switzerland.
[5] Pehrson, B., Automatica, *15*, (6), 631–639 (1979).
[6] Källström, C. G., K. J. Åström, N. E. Thorell, J. Eriksson, L. Sten, Automatica, *15*, (3), 241–254 (1979).
[7] Bengtsson, G.: Industrial Experience With a Microcomputer Based Self-Tuning Regulator. Paper in 18th IEEE Conference on Decision and Control, Ft. Lauderdale, Florida, December 12–14, 1979.
[8] Fjeld, M., J. Flatabö, E. Stai: Headbox Control. (In Norwegian). Report 72-2-T, SINTEF, Division of Automatic Control., Trondheim, Norway. (1972).
[9] Fjeld, M., Automatica, *14*, 107–117 (1978).
[10] Tyssö, A., J. C. Brembo, K. Lind, Automatica, *12*, 211–224 (1976).
[11] Tyssö, A., J. C. Brembo, Automatica *14*,213–221 (1978).
[12] Tyssö, A., Automatica; to appear in 1981.
[13] Balchen, J. G., N. A. Jenssen, E. Mathisen, S. Saelid: Modelling, Identification and Control, *1* (3) 135–163 (1980).
[14] Åström, K. J., B. Wittenmark, Automatica *9*, 185–199 (1973).
[15] Fjeld, M., W. Kelly, G. Somerville, R. G. Wilhelm, Jr.: Application of a Self-Tuning Regulator and a Continuous Performance Monitor to Paper Machine Control. (To appear in 1981)
[16] Balchen, J. G., T. Endresen, M. Fjeld, T. O. Olsen, Automatica, *9*, 295–307 (1973).

NEW APPROACHES TO THE DYNAMICS OF NONLINEAR SYSTEMS WITH IMPLICATIONS FOR PROCESS AND CONTROL SYSTEM DESIGN

W. Harmon Ray
Department of Chemical Engineering
University of Wisconsin
Madison, Wisconsin 53706

ABSTRACT

Some strongly nonlinear processes, such as chemical reactors, have sufficiently perverse dynamic behaviour in the open loop that even single input/single output control system design can be very difficult. In this paper examples of this nonlinear behaviour are presented and some new methods of dynamic analysis discussed. Control system design problems, such as parametric sensitivity, ignition/extinction phenomena, and nonlinear oscillations are illustrated by example. Some outstanding research problems are discussed.

INTRODUCTION

Much of the current control literature is concerned with the design of *linear* multivariable dynamic systems [1-3]. One advantage of linear

---

The U.S. Government retains a nonexclusive, royalty-free license to the use of this paper.

systems analysis is that there are many analytical results, very strong rigorous statements can be made, and a wide choice of control system design methods are available for use in applications. Clearly there are many important practical control problems which may be analysed in this *linear* framework and result in successful control system designs.

Strictly speaking, most chemical processes are *nonlinear*, although these nonlinearities are often sufficiently weak that model approximations or local linearization followed by application of linear multivariable design procedures result in practical control system designs for this class of systems as well. Nevertheless, there are some types of chemical processes (usually chemical reactors) which have especially strong nonlinear behaviour and often do not respond well to standard linearized control systems analysis. The purpose of this paper is to point out, by examples, some of the difficulties which may arise in such strongly nonlinear systems and to stimulate thought and discussion over what new control system design procedures may be required to deal with these exotic process dynamics.

To illustrate the essential ideas, the continuous stirred tank reactor (CSTR) will be used as the principal example. Industrial reactors are often more complicated, but have the important features of these simpler nonlinear reaction processes.

AN EXAMPLE SYSTEM—THE CSTR

One of the simplest chemical reactors one can envision is a continuous-stirred-tank-reactor (CSTR) in which a single first order, exothermic irreversible reaction is taking place (A → B). We shall choose to study this simple system because it has all the qualitative features of similar industrial reactors, but this system is much easier to analyse. If the chemical reaction has heat of reaction sufficiently high, then it is well known that the system will exhibit extreme parametric sensitivity, ignition and extinction behaviour (i.e.; multiple steady states), and nonlinear oscillations for some operating conditions [4–8]. It is important to realize that these phenomena are not academic curiosities (cf recent reviews [6, 7] and references therein), but arise surprisingly frequently in chemical reactors of industrial importance.

For the CSTR system to be used as an example, the modelling equations take the form [8]

$$V \frac{dc_A}{dt'} = F(c_{Af} - c_A) - Vk_0 \exp\left[-\frac{E}{RT}\right] c_A \qquad (1)$$

$$V\rho C_p \frac{dT}{dt'} = \rho C_p F(T_f - T) + V(-\Delta H)k_0 \exp[-\frac{E}{RT}] c_A \quad (2)$$
$$-hA(T - T_c)$$

As indicated in Figure 1, V is the volume of the reactor, F the feed rate, c_{Af}, T_f the inlet reactant concentration and temperature respectively and c_A, T the values in the reactor. The quantity T_c is the jacket coolant temperature. The physical parameters ρC_p, h, A, E, ΔH are all assumed constants and the reaction rate is $r = k_0 \exp[-\frac{E}{RT}] c_A$. Here the rate constant preexponential factor, k_0, depends on the amount of catalyst in the reactor (either homogeneous or immobilized on a support).

Now let us consider the single input/single output problem of controlling the reactor temperature, T, in the face of disturbances in feed temperature, T_f, by adjusting the coolant temperature, T_c, in a boiling liquid jacket. It is assumed that the reactant concentration c_A will

$$\frac{dx_1}{dt} = -x_1 + Da(1-x_1)\exp\left\{\frac{x_2}{1+\frac{x_2}{\gamma}}\right\}$$

$$\frac{dx_2}{dt} = -x_2 + BDa(1-x_1)\exp\left\{\frac{x_2}{1+\frac{x_2}{\gamma}}\right\} - \beta(x_2 - x_{2c}) + d + \beta u$$

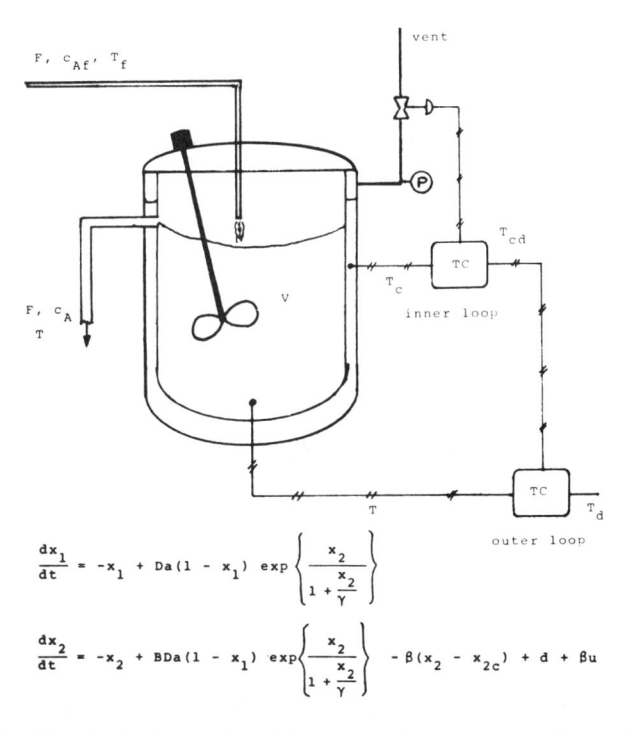

Figure 1. Simple Schematic of Reactor Temperature Control System.

ultimately settle down to the desired value if the reactor temperature is controlled properly. A schematic sketch of the process is shown in Figure 1 and the block diagram showing both inner and outer loop of the cascade control system is given in Figure 2. For our analysis, we shall assume the inner cascade loop is perfect and consider only the simplified block diagram shown in Figure 3. For a nonlinear system, the quantity g in the block diagram is a *nonlinear operator,* while g may be considered a *transfer function* for linearized analysis.

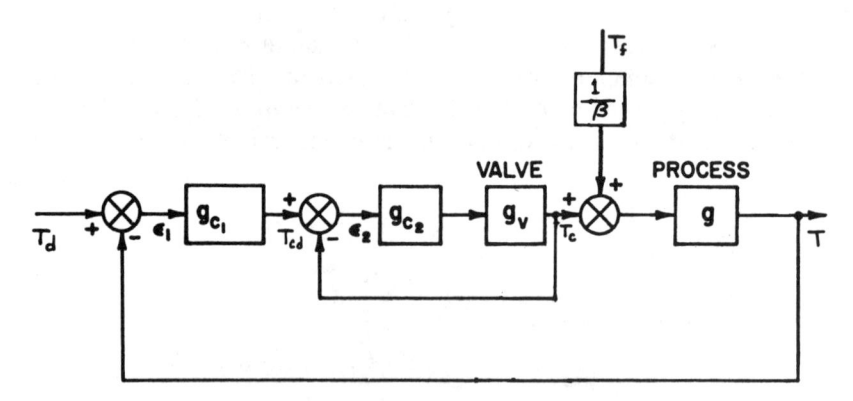

Figure 2. Cascade Controller Block Diagram.

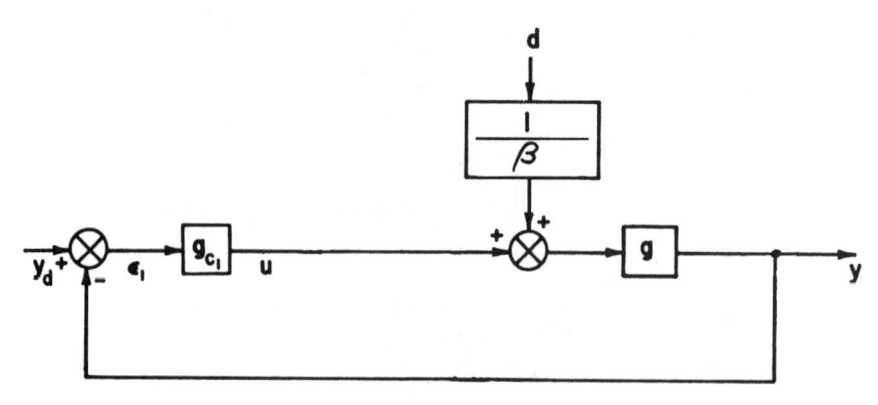

Figure 3. Simplified Block Diagram for Control System (perfect inner cascade loop).

To simplify the mathematics to follow, let us define the dimensionless variables

$$x_1 = \frac{c_{Af} - c_A}{c_{Af}}, \quad x_2 = \left[\frac{T - T_{f0}}{T_{f0}}\right]\left[\frac{E}{RT_{f0}}\right], \quad \tau = \frac{V}{F}, \quad \gamma = \frac{E}{RT_{f0}}$$

$$t = \frac{t'}{\tau} = \frac{t'V}{F}, \quad Da = \frac{k_0 \, e^{-\gamma}V}{F}, \quad B = \frac{(-\Delta H)c_{Af}}{\rho C_p T_{f0}}\left[\frac{E}{RT_{f0}}\right]$$

$$x_{2c} = \frac{T_{c0} - T_{f0}}{T_{f0}}\left[\frac{E}{RT_{f0}}\right], \quad \beta = \frac{hA}{F\rho C_p}, \quad d = \left[\frac{T_f - T_{f0}}{T_{f0}}\right]\left[\frac{E}{RT_{f0}}\right], \quad (3)$$

$$u = x_{2c}(T_c) - x_{2c}(T_{c0}) = \frac{T_c - T_{c0}}{T_{f0}}\left[\frac{E}{RT_{f0}}\right]$$

$$y = x_2 - x_{2s}, \quad \text{where } x_{2s} = \left[\frac{T_s - T_{f0}}{T_{f0}}\right]\left[\frac{E}{RT_{f0}}\right]$$

so that eqns. (1, 2) become

$$\frac{dx_1}{dt} = -x_1 + Da(1 - x_1)\exp\left[\frac{x_2}{1 + \frac{x_2}{\gamma}}\right] \quad (4)$$

$$\frac{dx_2}{dt} = -x_2 + BDa(1 - x_1)\exp\left[\frac{x_2}{1 + \frac{x_2}{\gamma}}\right] - \beta(x_2 - x_{2c}) + d + \beta u \quad (5)$$

Here the Damköhler number, Da, may vary as the reactor catalyst concentration changes. Note that x_1, x_2 represent dimensionless reactant conversion and temperature, while T_{f0}, T_{c0}, T_s are nominal design values for feed, coolant and reactor temperatures respectively. However, T_s also depends on Da for fixed T_{f0}, T_{c0}. The variables d, u, y are dimensionless deviation variables for the feed temperature disturbance (T_f), control (T_c), and output (T). The block diagram for the control system shown in Figures 1, 2 may be reduced to the simplified scheme given in Figure 3 (using y, u, d) if one assumes the inner cascade loop to be perfect.

OPEN LOOP BEHAVIOUR OF THE NONLINEAR REACTOR

It is useful to review the open loop nonlinear dynamic behaviour of the CSTR before discussing control system design. The dynamic behaviour of CSTR's has, of course, been well covered in the literature [4-8]; so only a brief summary is appropriate here.

Strong *parametric sensitivity* arises when the reactor conversion and temperature react drastically to small changes in a reactor parameter for some range of the parameter. An example of this for the CSTR is shown in Figure 4, where a small disturbance in feed temperature causes a large change in reactor temperature. The classical example of this is of course the case of a tubular reactor as reported by Bilous & Amundson [9] and shown in Figure 5. Note that a 2.5° change in feed temperature causes a ~60° change in reactor hot spot.

Parametric sensitivity may also arise in a dynamic sense; i.e. when a small change in a parameter causes a drastic change in the dynamic trajectory to steady state. These types of dynamics are possible even when the ultimate steady state behaviour is not significantly altered by the

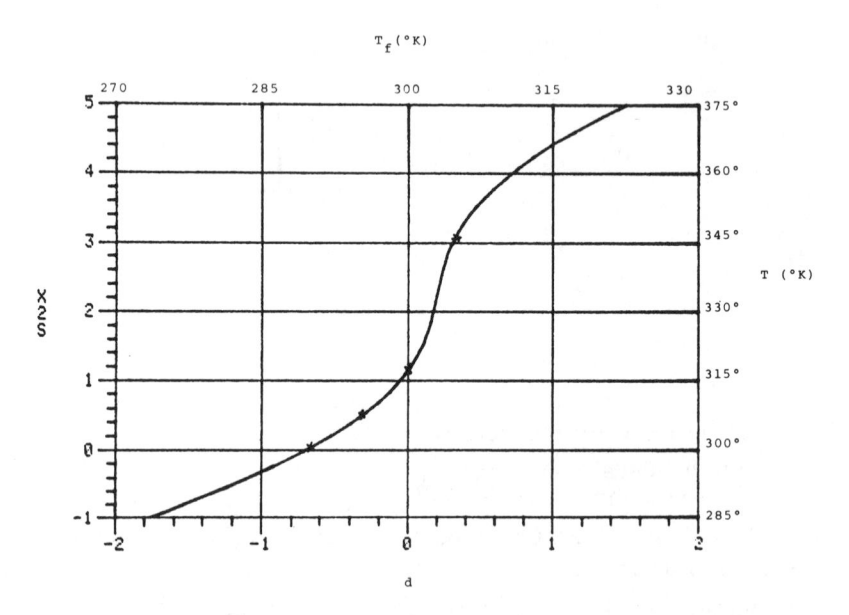

Figure 4. Steady State Response to Changes in Feed Temperature in Open loop (B = 7, β = 0.5, γ = 20, Da = 0.11), Evidence of Parametric Sensitivity.

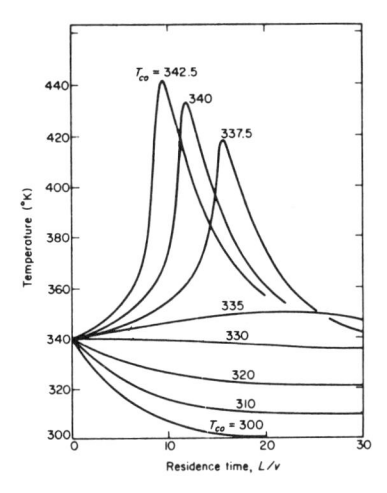

Figure 5. Parametric sensitivity of the tubular reactor steady state temperature profiles as a function of jacket coolant temperature, or batch reactor dynamic temperature as a function of jacket temperature.

parameter change. Because the dynamic batch reactor equations are the same as the steady state equations for a plug flow tubular reactor, Figure 5 also illustrates dynamic parametric sensitivity for a batch reactor.

Even more dramatic are changes in the open loop behaviour due to bifurcation of steady states or periodic solutions. One may determine the parameter values for which such bifurcation occurs by examining the eigenvalues of the linearized model. For the CSTR model (4, 5), the linearized equations have two eigenvalues, λ_1, λ_2.

Bifurcation occurs under the following conditions:

(i) Steady state branching occurs when either λ_1 or λ_2 passes through zero. This causes ignition/extinction behaviour due to multiple steady states.
(ii) Periodic branching occurs when λ_1, λ_2 become purely imaginary. This leads to sustained oscillations.

A detailed discussion of bifurcation analysis, ignition/extinction behaviour (multiplicity), and sustained oscillations in the CSTR has been given in [6–8]. The parameter dependence of the dynamic behaviour has been carefully mapped so that one may predict exotic dynamics in advance. For the CSTR, Figure 6 shows the types of multiplicity behaviour, and oscillatory solutions which are possible and the reactor

252   Chemical Process Control II

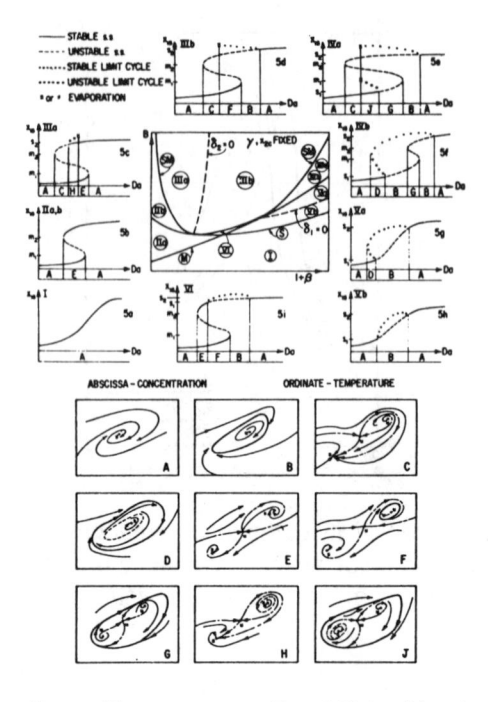

Figure 6. Bifurcation with respect to Damköhler Number (catalyst concentration) and resulting phase portraits.

parameter ranges over which such phenomena occur. For each value of Da (dimensionless catalyst concentration), a certain phase portrait in x_1, x_2 space is possible. Figure 6 shows 9 types which are known to exist and have been confirmed by numerical simulation. Other types have been postulated, but to date no dynamic simulations illustrating their existence have been published.

When the reactor mean residence time, $\tau = \frac{V}{F}$, is allowed to vary, the bifurcation behaviour is even more exotic [8]. This is illustrated in Figure 7. Obviously, other parameters could also be varied to produce similar bifurcation diagrams. In the following examples we shall fix all parameters except feed temperature, T_f, and demonstrate the effect of a disturbance in T_f on the reactor behaviour both with and without a control system in place.

Let us now use some specific calculations to illustrate the effect of feed temperature disturbances on the open loop reactor behaviour. The reported parameters are all evaluated at the nominal feed temperature (T_{f_0} = 300°K in these examples).

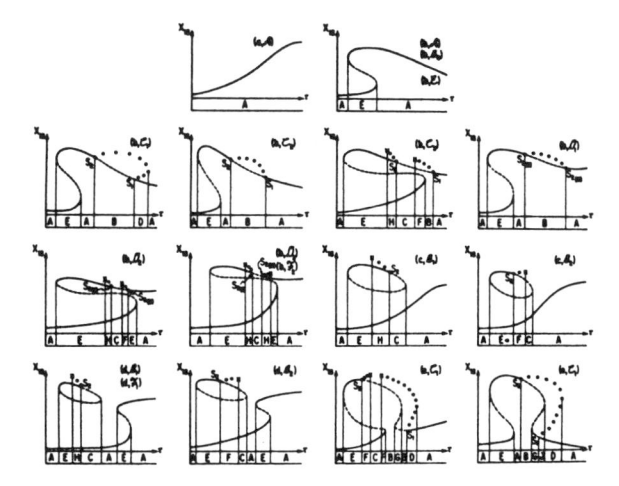

Figure 7. Bifurcation with respect to mean reactor residence time (feed flowrate)$^{-1}$.

(i) Case 1 (B = 7.0, β = 0.5, γ = 20, Da = 0.11):

This case, which falls in region I of Figure 6, has no multiple steady states or oscillations but does illustrate steady state parametric sensitivity quite well. Figure 4 shows the steady state response of reactor temperature to changes in feed temperature. Note that for Da = 0.11 there is strong parametric sensitivity for T_f about 300°K, while for T_f = 290°K there is little parametric sensitivity. This may be seen more dramatically in Figure 8, where the dynamic response to a 5°K increase in feed temperature at T_f = 290°K and at T_f = 300°K is plotted. In each case the reactor is started up from x_1 = 0, T = T_f, and then after it has lined out, a 5°K step change in feed temperature is made. Note how much more strongly the reactor responds to the disturbance at T_f = 300°K than at T_f = 290°K. However, there is little dynamic parametric sensitivity for this example.

(ii) Case 2 (B = 8.0, β = 0.3, γ = 20, Da = 0.072)

For this case in region II of Figure 6, there is ignition and extinction behaviour corresponding to multiple steady states. These effects are seen suddenly at certain values of the parameters as shown in Figures 9–11 for Da = 0.072. For example, note that when T_f = 295°K, a 5° increase in feed temperature to 300°K produces a small change (~10°K) in reactor temperature. By contrast, a 5°K change in feed temperature from T_f = 300°K to T_f = 305°K produces an ~70°K increase in reactor temperature

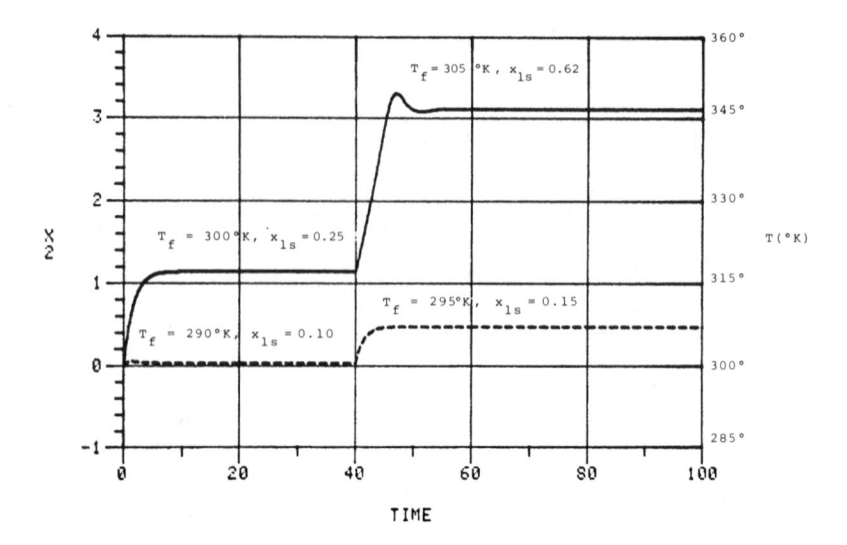

Figure 8. Open Loop Response to a 5°K Increase in Feed Temperature (B = 7, β = 0.5, γ = 20, Da = 0.11).

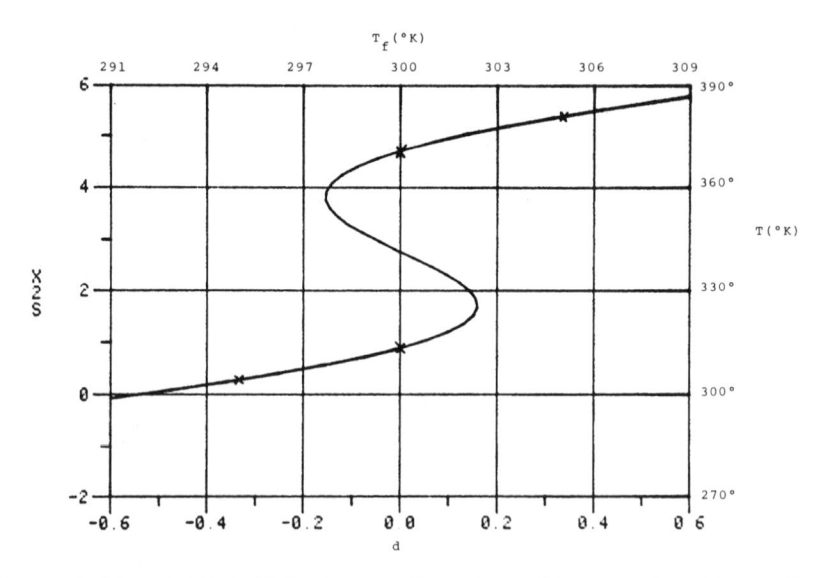

Figure 9. Steady State Behaviour in Open Loop (B = 8.0, β = 0.3, γ = 20, Da = 0.072).

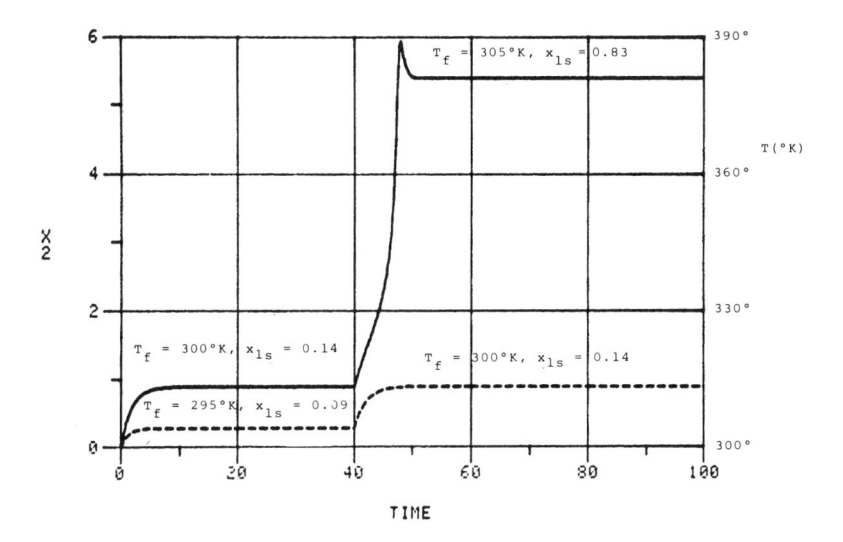

Figure 10. Open Loop Response to Step Increase of 5°K in Feed Temperature (B = 8, β = 0.3, γ = 20, Da = 0.072).

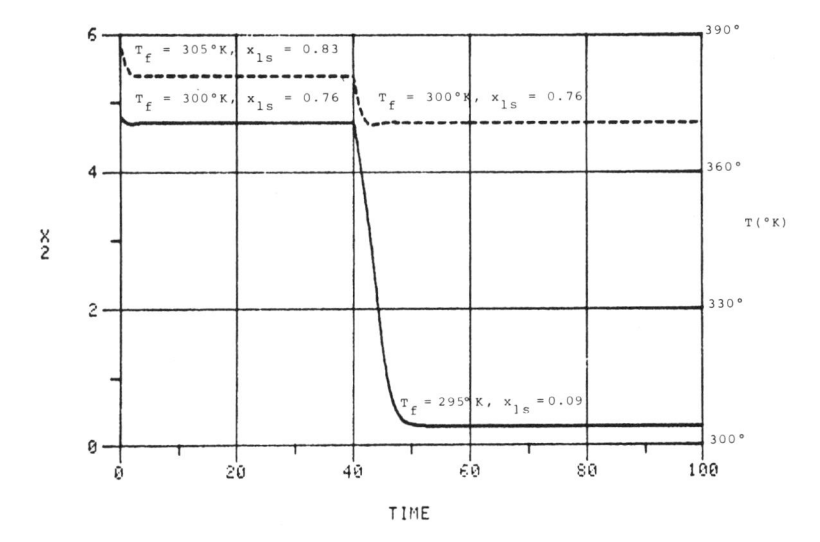

Figure 11. Open Loop Response to Step Decrease of 5°K in Feed Temperature (B = 8, β = 0.3, γ = 20, Da = 0.072).

due to ignition. The opposite effect occurs at extinction. Note that a 5°K decrease in feed temperature from $T_f = 305°K$ to $T_f = 300°K$ produces only about a 10°K drop in reactor temperature. However, another 5°K drop in feed temperature from $T_f = 300°K$ to $T_f = 295°K$ causes the reactor to extinguish itself and the reactor temperature drops ~70°K.

(iii) Case 3 (B = 11.0, β = 1.5, γ = 20, Da = 0.35)

This case which falls in region V of Figure 6 illustrates the situation when there are self generated reactor oscillations. Figure 12 illustrates the steady state situation for Da = 0.135 and shows that limit cycle oscillations exist for feed temperatures in the approximate range, 302°K < T_f < 307°K. Thus a step increase of 5°K in T_f at $T_f = 300$ causes the reactor to oscillate while a similar 5°K increase in T_f for $T_f = 295°K$ has little effect on the reactor temperature. A *step decrease* in feed temperature can also be interesting. At $T_f = 315°K$, a 5°K decrease in feed temperature causes only a small decrease in reactor temperature. However, another 5°K decrease in feed temperature to 305° produces oscillatory behaviour in the reactor. This dynamic behaviour is illustrated in Figures 13, 14.

From these simple examples, it is clear that relatively small parameter changes can produce dramatic changes in the steady state and dynamic

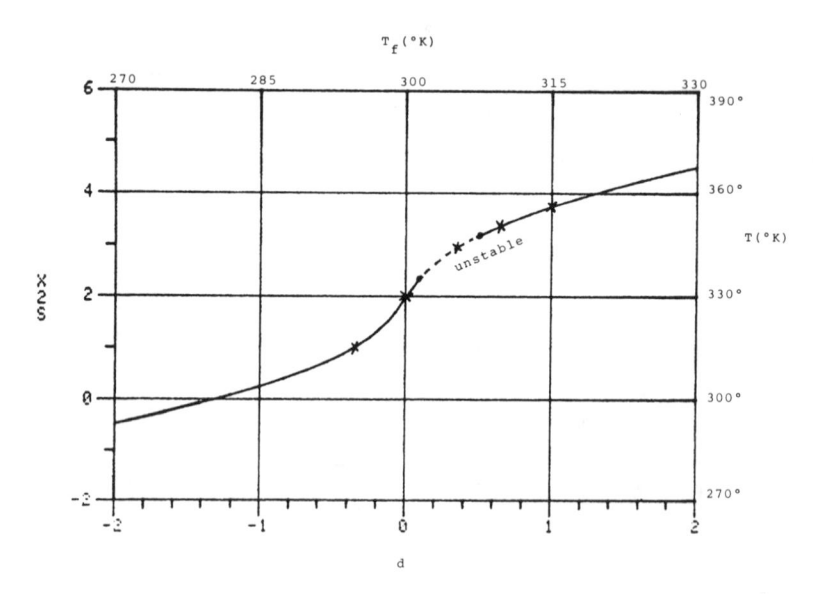

Figure 12. Steady State Behaviour in Open Loop (B = 11, β = 1.5, γ = 20, Da = 0.135).

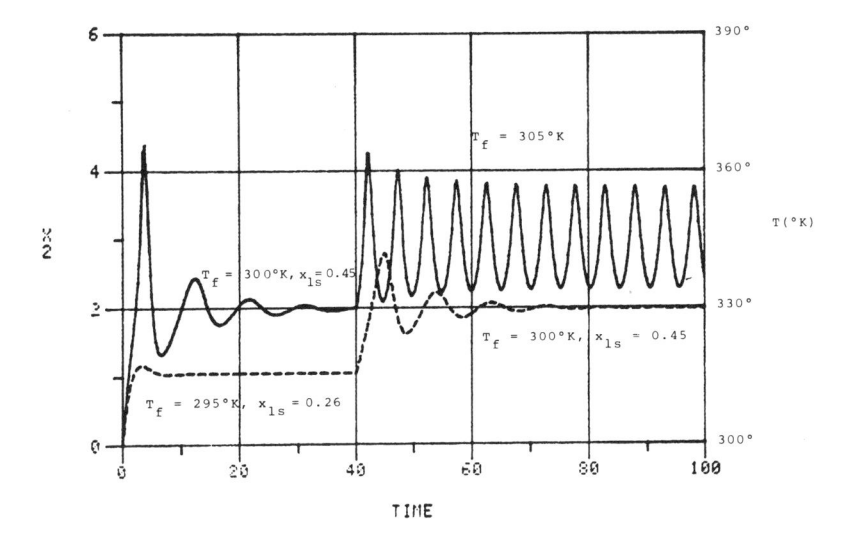

Figure 13. Open Loop Response to a 5°K Increase in Feed Temperature
(B = 11, β = 1.5, γ = 20, Da = 0.135).

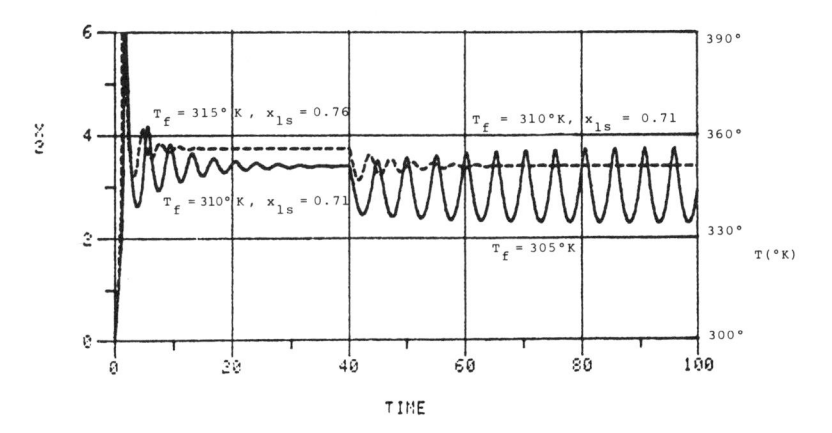

Figure 14. Open Loop Response to a 5°K Decrease in Feed Temperature
(B = 11, β = 1.5, γ = 20, Da = 0.135).

behaviour of the CSTR in the open loop. Such effects are sure to test the robustness of any control system one might apply to the system. In the next section we illustrate this point.

CLOSED LOOP BEHAVIOUR

It is interesting to observe how the nonlinear CSTR will behave when conventional control system design methods are used together with the linearized model equations. To see how the linearized system behaves, let us linearize (4, 5) about steady state values of conversion and temperature x_{1s}, x_{2s} corresponding to the nominal values of coolant and feed temperatures T_{c0}, T_{f0} and some choice of reactor parameters, Da, β. The resulting linearized system, in deviation variables.

$$z_1 = x_1 - x_{1s}$$
$$z_2 = x_2 - x_{2s} \tag{6}$$

is given by

$$\frac{d\mathbf{z}}{dt} = \mathbf{A}\mathbf{z} + \mathbf{B}u + \mathbf{\Gamma} d, \, \mathbf{z}(0) = \mathbf{0} \tag{7}$$
$$y = \mathbf{C}\mathbf{z}$$

where

$$\mathbf{z} = \begin{bmatrix} z_1 \\ z_2 \end{bmatrix}, \, \mathbf{B} = \begin{bmatrix} 0 \\ \beta \end{bmatrix}, \, \mathbf{\Gamma} = \begin{bmatrix} 0 \\ 1 \end{bmatrix}$$

$$\mathbf{A} = \begin{bmatrix} -\left(\dfrac{1}{1-x_{1s}}\right) & \dfrac{x_{1s}}{\left[1+\left(\dfrac{x_{2s}}{\gamma}\right)\right]^2} \\ -\dfrac{Bx_{1s}}{1-x_{1s}} & -(1+\beta) + \dfrac{Bx_{1s}}{\left[1+\left(\dfrac{x_{2s}}{\gamma}\right)\right]^2} \end{bmatrix}, \, \mathbf{C} = [0 \ 1] \tag{8}$$

The linearized model (7) may be converted to the Laplace domain for the case of temperature control to produce the transfer function relations

$$\bar{y}(s) = g(s) \, \bar{u}(s) + g_d(s)\bar{d}(s) \tag{9}$$

where

$$g(s) = \mathbf{C}(s\mathbf{I} - \mathbf{A})^{-1}\mathbf{B} \tag{10}$$

$$g_d(s) = \mathbf{C}(s\mathbf{I} - \mathbf{A})^{-1}\mathbf{\Gamma} \tag{11}$$

Here

$$(s\mathbf{I}-\mathbf{A})^{-1} = \frac{\begin{bmatrix} s + 1+\beta - \dfrac{Bx_{1s}}{\left[1 + \left(\dfrac{x_{2s}}{\gamma}\right)\right]^2} & \dfrac{x_{1s}}{\left[1 + \left(\dfrac{x_{2s}}{\gamma}\right)\right]^2} \\ -\dfrac{Bx_{1s}}{1 - x_{1s}} & s + \dfrac{1}{1 - x_{1s}} \end{bmatrix}}{D} \tag{12}$$

where

$$D = s^2 + \left[1+\beta - \frac{Bx_{1s}}{\left[1 + \left(\dfrac{x_{2s}}{\gamma}\right)\right]^2} + \frac{1}{1 - x_{1s}}\right]s$$

$$+ \left\{\left[1 + \beta - \frac{Bx_{1s}}{\left[1 + \dfrac{x_{2s}}{\gamma}\right]^2}\right]\left(\frac{1}{1 - x_{1s}}\right) + \frac{Bx_{1s}^2}{(1 - x_{1s})\left[1 + \dfrac{x_{2s}}{\gamma}\right]^2}\right\} \tag{13}$$

so that

$$g(s) = \frac{K\left(\dfrac{s}{z} + 1\right)}{\Upsilon^2 s^2 + 2\zeta\Upsilon s + 1} \tag{14}$$

$$g_d(s) = \frac{K_d\left(\dfrac{s}{z} + 1\right)}{\Upsilon^2 s^2 + 2\zeta\Upsilon s + 1} \tag{15}$$

where

Chemical Process Control II

$$K = \frac{\beta}{1 + \beta - \dfrac{B(x_{1s} - x_{1s}^2)}{\left[1 + \dfrac{x_{2s}}{\gamma}\right]^2}}, \quad K_d = \frac{K}{\beta},$$

$$z = \frac{1}{1-x_{1s}}, \quad \Upsilon^2 = \frac{1 - x_{1s}}{1 + \beta - \dfrac{B(x_{1s} - x_{1s}^2)}{\left[1 + \dfrac{x_{2s}}{\gamma}\right]^2}}$$

$$2\zeta\Upsilon = \frac{\left(1-\beta - \dfrac{Bx_{1s}}{\left[1 + \dfrac{x_{2s}}{\gamma}\right]^2} + \dfrac{1}{1-x_{1s}}\right)(1 - x_{1s})}{\left(1+\beta - \dfrac{B(x_{1s} - x_{1s}^2)}{\left[1 + \dfrac{x_{2s}}{\gamma}\right]^2}\right)} \tag{16}$$

$$= \Upsilon^2 \left(1+\beta - \dfrac{Bx_{1s}}{\left[1 + \dfrac{x_{2s}}{\gamma}\right]^2} + \dfrac{1}{1-x_{1s}}\right)$$

The values of x_{1s}, x_{2s} in these expressions must be calculated from the steady state solutions of the non-linear equations (4, 5).

A proportional controller was applied to the reactor as noted in Figures 1–3. The controller gain, k_c, was chosen in each case to provide good performance* at the nominal feed temperature $T_f = 300°K$. The manipulated variable is the coolant temperature, x_{2c} (i.e.,; T_c) which is constrained to lie between upper and lower bounds. In practical situations there is also a maximum heat load on the boiling jacket (limited by explosive boilup). However, to keep things simple here we shall assume the feedback controller is only constrained at rapid reaction rate by coolant temperature limits. With these conditions on the feedback controller, let us simulate the behaviour of the closed loop system under each of the conditions discussed in the last section.

---

* The linearized model, g(s), is stable for all controller gains if it is stable in the open loop. Thus Ziegler-Nichols or similar tuning criteria are not helpful.

(i) Case 1 (B = 7.0, β = 0.5, γ = 20, Da = 0.11, $-0.5 \le u \le 0.5$):

This case was used to illustrate parametric sensitivity in Figures 4 and 8. In Figure 15, the closed loop response to a 5°K step change in feed temperature is shown for the two situations presented in Figure 8. In each case the controller gain, k_c = 10, was chosen to reflect good control system design for T_{f0} = 300°K. Comparison of Figures 4 and 8 shows that the feedback controller is excellent for T_{f0} = 290°K, but it has more difficulty handling the 5°K feed temperature disturbance when T_{f0} = 300°K. The offset noted is not due to poor controller tuning and cannot be corrected by integral action; it is due to the cooling system operating at full capacity and is thus constrained control at u = 0.5. Nevertheless, operation with the feedback controller is a considerable improvement over open loop operation and one might be willing to accept the performance shown in Figure 15.

(ii) Case 2 (B = 8.0, β = 0.3, γ = 20, Da = 0.072, $-0.5 \le u \le 0.5$):

The case when there is the possibility of ignition and extinction due to multiple steady states is illustrated here. The open loop situation was shown in Figures 9–11. Figure 16 shows that a feedback controller designed at T_{f0} = 300° on the lower branch (k_c = 10) will also perform well

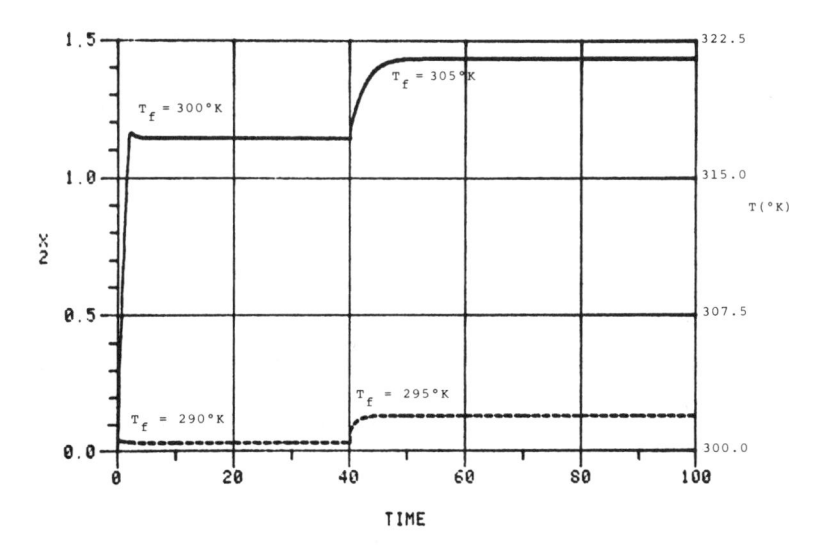

Figure 15. Closed Loop Response to a 5°K Increase in Feed Temperature; the Set Point of the Controller is the Temperature before Disturbance (B = 7, β = 0.5, γ = 20, Da = 0.11, k_c = 10, $-0.5 \le u \le 0.5$).

262   Chemical Process Control II

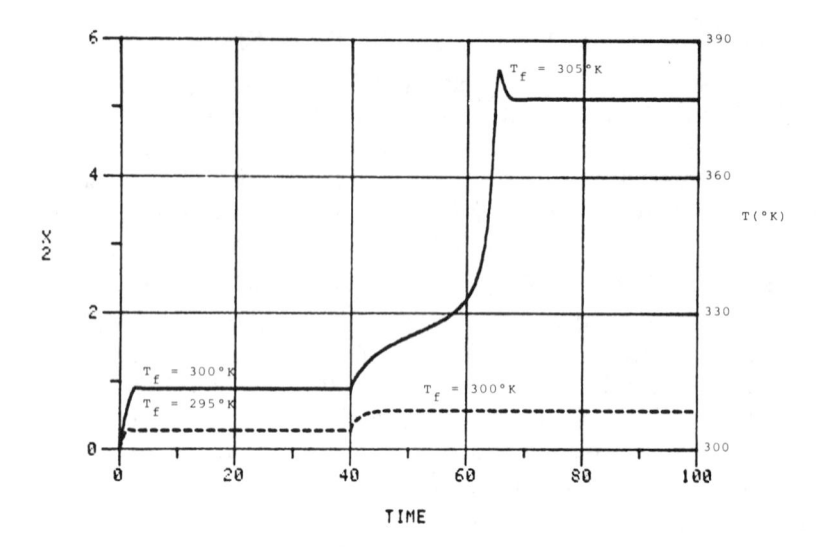

Figure 16. Closed Loop Response to a Step Increase of 5°K in Feed Temperature; the Set Point of the Controller is the Temperature before the Disturbance (B = 8, β = 0.3, γ = 20, Da = 0.072, k$_c$ = 10, −0.5 ≤ u ≤ 0.5).

for T_{f_0} = 295°K, showing only a slight amount of offset. However a step increase in feed temperature from 300° to 305° leads to a huge increase in reactor operating temperature even with a feedback controller. This is, of course, due to the limited ability of the jacket to remove the heat release after ignition of the reaction. When one operates on the upper branch (cf Figures 9, 11), the danger is extinction of the reaction. Figure 17 illustrates what may happen for the closed loop situation. For T_{f_0} = 305°K on the upper branch and a 5°K *decrease* in feed temperature to 300°K, the reactor operating temperature is relatively unaffected. However another 5°K decrease in feed temperature to 295°K causes a sharp drop in reactor temperature and reactant conversion. Here the control system is unable to hold the ignited state due to constraints on the coolant temperature. In this case the closed loop operation is not significantly different from the open loop situation shown in Figures 10, 11.

(iii) Case 3 (B = 11.0, β = 1.5, γ = 20, Da = 0.135, −0.1 ≤ u ≤ 0.1):

This case illustrates the influence of the feedback control system on the oscillating reactor. Recall that the open loop situation is illustrated in

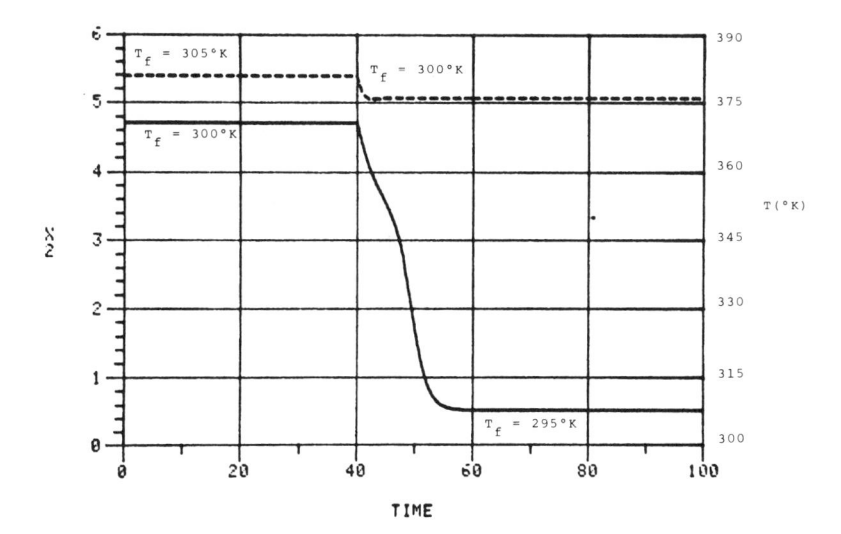

Figure 17. Closed Loop Response to a Step Decrease of 5°K in Feed Temperature; the Set Point of the Controller is the Temperature before the Disturbance (B = 8, β = 0.3, γ = 20, Da = 0.072, k_c = 10, $-0.5 \leq u \leq 0.5$).

Figures 12–14. The control system has a value of $k_c = 3.11$ which is a good choice of controller gain for $T_{f_0} = 300°K$. Figure 18 demonstrates the closed loop response when there is a 5°K step increase in feed temperature. For a change from $T_f = 295°$ to $T_f = 300°$ the closed loop system behaves much better than the open loop case shown in Figure 13. However, a step change from $T_f = 300°K$ to $T_f = 305°K$ still produces strong oscillations in temperature even with the feedback controller. Similar things can happen for a *decrease* in feed temperature. For example, when the feed temperature is *decreased* from $T_f = 315°K$ to $T_f = 310°K$, the control system holds the reactor temperature almost constant (cf Figure 19). However, a further decrease from $T_f = 310°K$ to $T_f = 305°K$ leads to oscillations in reactor temperature which the control system is unable to control. These oscillations are smaller in amplitude than in the open loop. They are also smaller than the ones shown in Figure 18, because of the different controller set points used. In both cases the feed temperature is 305°K; however in Figure 18, the control system is constrained at u = −0.1 trying to hold the 330°K set point. By contrast, in Figure 19, the control system is constrained at u = +0.1 trying to hold the 351° set point. When the controller is constrained in this way, it is even

264   Chemical Process Control II

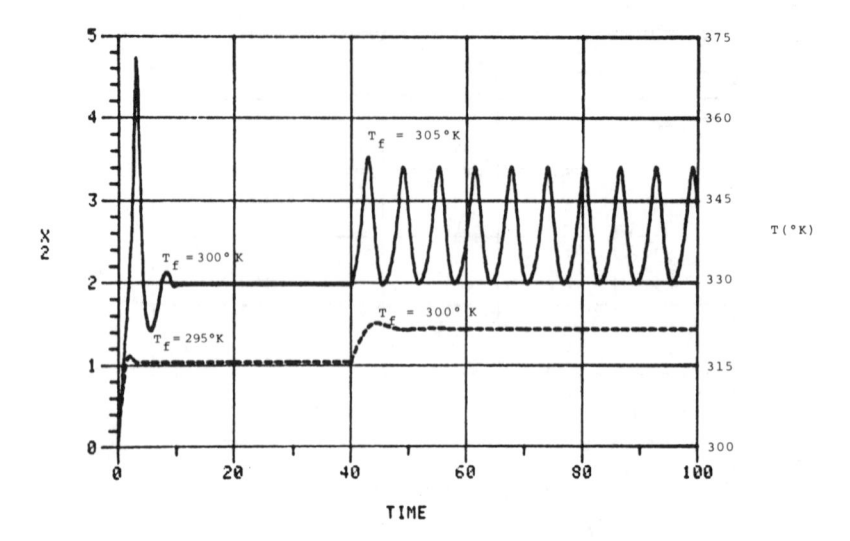

Figure 18. Closed Loop Response to a Step Increase of 5°K in Feed Temperature; the Set Point of the Controller is the Temperature before the Disturbance (B = 11, β = 1.5, γ = 20, Da = 0.135, k_c = 3.11, −0.1 ≤ u ≤ +0.1).

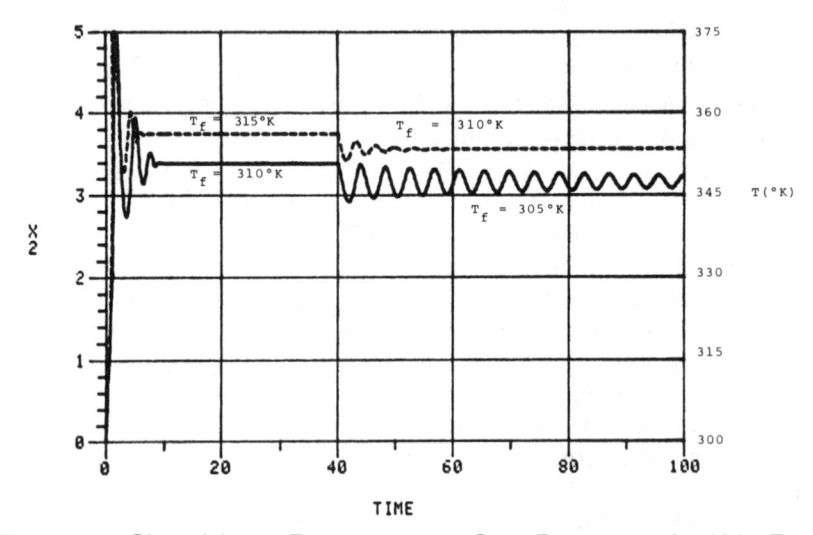

Figure 19. Closed Loop Response to a Step Decrease of 5°K in Feed Temperature; the Set Point of the Controller is the Temperature before the Disturbance (B = 11, β = 1.5, γ = 20, Da = 0.135, k_c = 3.11, −0.1 ≤ u ≤ +0.1).

sometimes possible to have the control system *cause* oscillations when the open loop system has no oscillations. This is quite a different phenomenon from that nrmally encountered when sufficiently high controller gains in linear systems cause control system instabilities. Here we have seen that the linearized model, Eqns. (9, 14, 15), cannot be made unstable by high controller gains; but the actual nonlinear system can be unstable for some system parameters and controller gains. If it were possible to increase controller power sufficiently, then the oscillations present when $T_f = 305°K$ could be damped out by the controller in two ways: (i) if the temperature set point *was not* an open loop unstable steady state (e.g.; $T_f = 300°K$ here), then the controller would keep the reactor from oscillating by simply choosing u to compensate for disturbances in T_f, (ii) if the set point *was* an open loop unstable steady state, then the controller would have to be always active to compensate for the natural oscillations of the system. In practice this means constant value motion and mechanical wear.

These simple examples were chosen to illustrate the problems of nonlinear control; however more complex practical reactor systems can have more severe difficulties than shown here. Nevertheless these case studies do show that limited controller power can lead to serious loss of control when the open loop system passes into a region of parametric sensitivity, ingnition/extinction, or autonomous oscillation. A secondary problem is the question of optimal controller tuning in the face of strong nonlinearities. This becomes even more serious in the case of multivariable control because the relative gains and the dynamic parameters of the loops may change drastically with sustained disturbances.

Methods for the synthesis of control systems for strongly nonlinear processes are available for single input/single output problems. In these cases, careful dynamic analysis may be used to define the troublesome parameter ranges (e.g. through the use of bifurcation theory as was done here). Then the control system can be carefully constructed with variable gains, overrides, and failsafe shutdown procedures to insure safe operation. Sometimes, even redesign to eliminate the most difficult parameter regions is possible. However, the nonlinear multivariable control system design problem is much more difficult and no proven control system design procedures seem to be available.

CONCLUDING REMARKS

Although illustrations of control system design problems due to parametric sensitivity, multiplicity, and nonlinear oscillations have been

266   Chemical Process Control II

presented for a simple CSTR, such problems exist for many reactors of industrial interest. Heterogeneous catalytic reactors such as packed or fluidized beds, bulk solution and emulsion polymerization reactors, gas-liquid reactors, etc. are just a few industrial reactor systems exhibiting these problems in the plant.

To illustrate that we have only scratched the surface in our understanding of nonlinear dynamic behaviour, a particularly interesting case of exotic dynamics was reported recently by Rossler and coworkers [10] for the reaction sequence

$$A \rightarrow B \rightarrow C$$

carried out in a CSTR with jacket cooling. Their simulations, one of which is shown in Figure 20, show cascading bifurcations of periodic solutions as the jacket heat transfer coefficient (or heat transfer area) is increased past the Hopf bifurcation point. These cascading periodic solutions degenerate into chaos at certain values of the heat transfer coefficient. The understanding and control of these types of dynamics certainly requires more study and research.

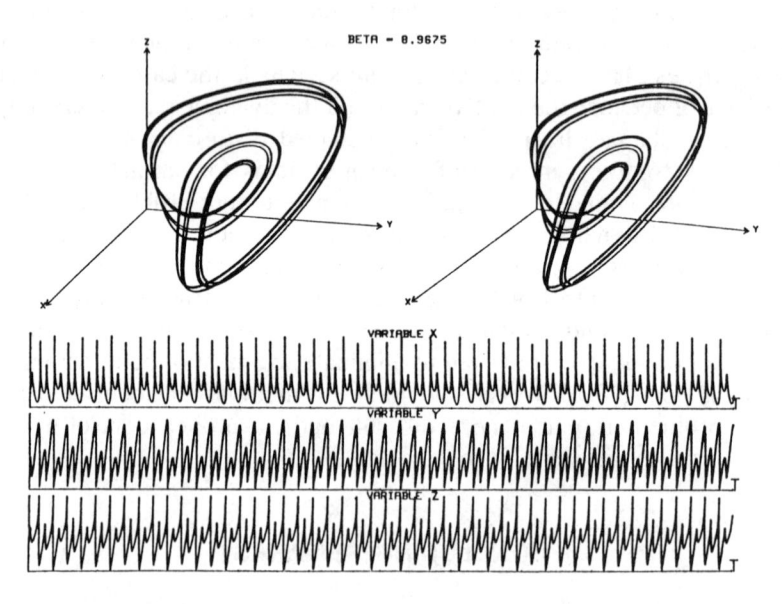

Figure 20. Complex Oscillations Found in the Open Loop Operation of a CSTR for the Reaction A → B → C. x = [A], y = [B], z = Temp. (Taken from Ref. 10).

From the practical point of view, one should realize that bifurcation analysis allows troublesome nonlinear dynamic behaviour to be foreseen and provides guidelines for redesign of the process to minimize the severity of the problem. However, in the final design, good control systems are needed to insure the safety and profitability of these reactors. Unfortunately, the tools available for arriving at such designs are limited, particularly for multivariable control, so that innovative research is sorely needed in this area.

ACKNOWLEDGMENTS

This work was partially supported by the Department of Energy and the National Science Foundation. The author is indebted to Mr. David C.-C. Pai, who carried out the calculations for the figures.

REFERENCES

[1] Proceedings IFAC Symposium on CAD of Control Systems, Zürich, Pergamon Press (1979).
[2] MacFarlane, A. G. J. edit., "Frequency Response Methods in Control Systems," IEEE Press (1979).
[3] Ray, W. H., "Advanced Process Control," McGraw-Hill (1981).
[4] Aris, R. and N. R. Amundson, *Chem. Eng. Sci.* 7:121, (1958).
[5] Perlmutter, D. D., "Stability of Chemical Reactors," Prentice-Hall (1972).
[6] Ray, W. H., chapter in "Applications of Bifurcation Theory," P. Rabinowitz edit., Acad. Press (1977).
[7] Stewart, W. E., W. H. Ray, C. C. Conley edit., "Dynamics and Modelling of Reactive Systems," Acad. Press (1980).
[8] Uppal, A., W. H. Ray, A. B. Poore, *Chem. Eng. Sci.* 29:967 (1974); also 31:205 (1976).
[9] Bilous, O. & N. R. Amundson, *AIChE J.* 2:117 (1956).
[10] Kahlert, C., O. E. Rössler, and A. Varma, Proceedings of Workshop on *Modelling of Chemically Reacting Systems*, Heidelberg, Sept. 1980.

A REVIEW OF SOME ADAPTIVE CONTROL SCHEMES FOR PROCESS CONTROL

Pierre R. Bélanger
Department of Electrical Engineering
McGill Uniersity
Montreal, Que.

ABSTRACT

The object of this paper is adaptive control and its application to the process industries. The basic ideas and theoretical concepts of current methods are discussed, with some emphasis on the self-tuning regulator. The two functions of control and identification are addressed separately, then together in an adaptive loop. The issues of convergence are outlined. There follows a section on some of the practical considerations to be addressed in the implementation. Finally, a few successful applications are discussed.

INTRODUCTION

The idea of adaptive control follows quite naturally from the possibility of real-time identification. If one can generate and update a system model, the logical next step is to use the model to update the controller "on the fly." Thus, adaptive control has been with us since the fifties. For the most part, it has been the province of experience and art—not to say tinkering. It is only in the last decade that sound theoretical results have been obtained to provide a firm foundation.

To describe adaptive control, imagine an experiment where input and output data are generated, and used to find (e.g. by maximum likelihood) the estimates of the parameters of a model with a given structure. Use the results of this offline experiment to design a control algorithm appropriate for the model, and implement this control algorithm on the plant. To go from this to adaptive control, we simply do this on line. New parameter estimates are generated at every sampling interval and used to "redesign" the controller, i.e. to update the controller parameters. (See Fig. 1)

It is often the case in process control that process dynamics are subject to change. For example, the time constant and delay of the basic weight dynamics of a paper machine change substantially from one grade to another. If a single, fixed controller is used, it must be very loosely tuned, and the performance will probably be mediocre for all dynamic conditions.

One way to solve this problem is to build a lookup table of control parameter settings appropriate to each operating condition, and to refer to this table to set the control parameter: in the aerospace industry, this is known as "gain scheduling." If a good dynamic model exists whose parameters depend on the operating conditions in an explicit and known fashion, the lookup table can be constructed off-line by designing

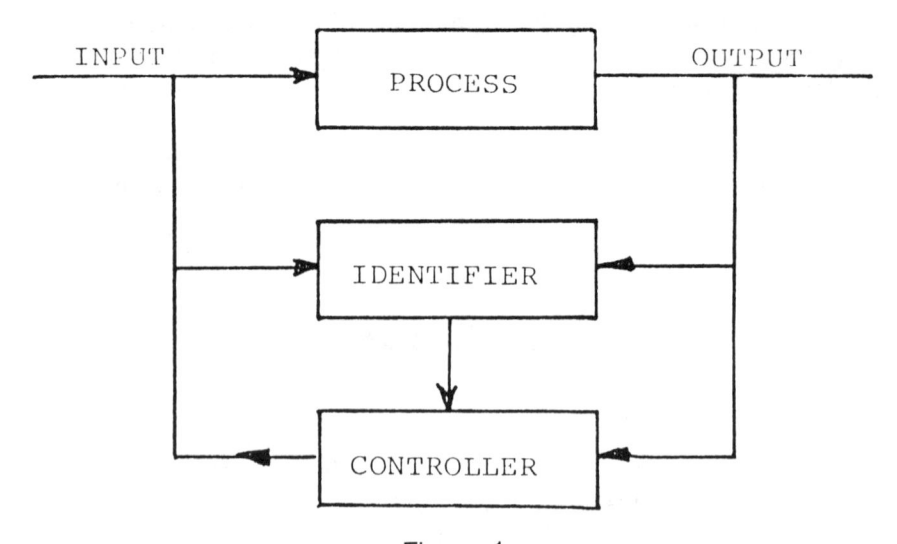

Figure 1.

controllers for many different operating conditions; high-performance aircraft control systems are designed this way, with the air data (dynamic pressure and altitude, both measurable) defining the operating condition.

In order for gain scheduling to work, the system dynamics must depend only on known, measurable factors, e.g. set points, ambient temperature. If it is to be practical, the work involved in the construction of the lookup table must not be too great. If adequate dynamic models are available, the table can be constructed off-line; if not, the table must be built up by experimentation, and the number of factors must be small enough to yield a manageable number of operating points.

There are many instances in process control where these conditions are not satisfied. Good dynamic models are often expensive and difficult to get; it is not always possible to identify all factors affecting process dynamics; finally, dynamics sometimes change with time. Such factors make adaptive control a strong candidate solution to the problem, if high performance standards are to be achieved.

MODELS AND CONTROL LAWS

Discrete-Time Models

The dynamic models under consideration are linear, discrete-time, stochastic models of the form

$$y(t) = \frac{B(q^{-1})}{A(q^{-1})} u(t-k) + \frac{C(q^{-1})}{A(q^{-1})} e(t) \tag{1}$$

where q^{-1} is the unit delay operator;
$e(t)$ is a zero-mean white sequence;
k is an integer, the time delay;
A,B,C are polynominals of order n in q^{-1}.

The polynominals A and C may be assumed monic (leading term of unity); it may also be assumed that $C(q^{-1})$ has all its zeros within the unit circle.

Pole-Zero Placement

Two basic types of control laws have been used: pole-zero placement and linear-quadratic. The pole-zero method ignores the disturbance dynamics, being based only on the dynamics of the deterministic portion.

The deterministic portion of (1) has a transfer function

$$G(z) = \frac{B(z^{-1})z^{-k}}{A(z^{-1})} \tag{2}$$

The linear regulator of Fig. (2) yields the transfer function [1]

$$\frac{y}{u_d} = \frac{z^{-k} T B}{AR + z^{-k} BS} \qquad (3)$$

The desired transfer function may not be arbitrarily chosen. The following rules must be respected:

(i) The pure delay must be at least equal to that of the plant;
(ii) The excess of poles over zeros must be at least as great as that of the plant;
(iii) The plant zeros outside or close to the unit circle must be retained (or else they would need to be cancelled by unstable or poorly-damped poles).

Given the desired transfer function

$$G_d = \frac{B^- Bm_1 z^{-k}}{Am} \qquad (4)$$

where B^- has as its roots the zeros of B that lie outside or near the unit circle, it is shown in [2] that the polynominals T, R and S that define the control law can be found as follows:

(i) Solve for R_1 and S

$$A R_1 + q^{-k} B^- S = A_m A_o \qquad (5)$$

Here, A_o is a stable polynominal containing closed-loop modes not excitable from u_d: it is arbitrary.

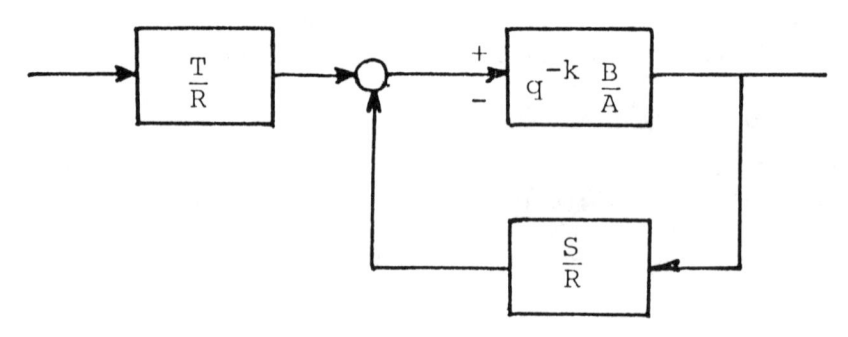

Figure 2.

(ii) If $B = B^+ B^-$,

$$R = R_1 B^+ \tag{6}$$

$$T = A_o Bm_1 \tag{7}$$

Linear-Quadratic Control

The linear-quadratic control law takes into account both the process dynamics and the noise spectrum. The regulator version is given here, but the algorithm can be modified to include a changing set point. It can be shown [23] that the control law that minimizes $J = E[y^2(t) + \rho u^2(t)]$ leads to the closed-loop characteristic polynominal

$$P(q^{-1}) = C(q^{-1})[\rho A^*(q^{-1})A(q^{-1}) + B^*(q^{-1})B(q^{-1})]_+ \tag{8}$$

where $A^*(q^{-1}) = q^{-n} A(q)$, and similarly for B, [ ]$_+$ = monic polynominal in q^{-1} having as its roots the roots of [ ] within the unit circle. In words, the closed-loop poles are i) the roots of C and ii) the stable roots of $\rho A^*A + B^*B$.

If $\rho = 0$, we have minimum-variance control. If the plant zeros are all inside the unit circle, then

$$[B^*B]_+ = B$$

and

$$P(q^{-1}) = B(q^{-1})C(q^{-1}) \tag{9}$$

Predictor Models

For the models and control schemes presented so far, the calculation of the control law is not simple. We shall now formulate models, called *predictor models*, where the control law parameter follows directly from the model parameters. It is shown in Appendix A that the system (1) may be rewritten as

$$C\, y(t+k) = G\, y(t) + B\, F\, u(t) + F\, C\, e(t+k) \tag{10}$$

with G and F satisfying the polynominal equation

$$C = A\, F + q^{-k} G \tag{11}$$

Here, F is a monic polynominal of order k−1 in q^{-1}, while G is of order n. If the control law

$$G\,y(t) + B\,F\,u(t) = 0 \qquad (12)$$

is used, them, from (10),

$$y(t+k) = F\,e(t+k) \qquad (13)$$

i.e. y(t+k) is a linear combination of the future values of the white sequence. It is not difficult to show that $E[y^2(t+k)]$ is minimized in that case.

If the model is given at the outset in the form of (10) rather than (1), the control law (12) is seen to come directly from the model, without any computation. It turns out that the closed-loop modes using (12) are the zeros of C and of B; if the plant has zeros outside the unit circle (i.e. is nonminimum phase), the closed-loop is unstable and this minimum-variance law cannot be used.

This idea may ge generalized. Define

$$\phi(t) = A_m y(t) - q^{-k} B_m u_d(t) + \rho\, q^{-k} u(t) \qquad (14)$$

where A_m and B_m are polynominals in q^{-1}, ρ is a positive scalar and $u_d(t)$ is an external input. Using (1),

$$\begin{aligned} A\,\phi(t) &= A_m[q^{-k}Bu + Ce] - q^{-k} A\, B_m\, u_d(t) + \rho\, A q^{-k}\, u \\ &= q^{-k}(A_m B + \rho A)u - q^{-k} A\, B_m\, u_d + A_m\, Ce \end{aligned} \qquad (15)$$

A transformation similar to the one used to get (10) yields

$$A\,C\phi(t+k) = G\phi(t) + (A_m B + \rho A)F\,u(t) \\ - A\,B_m\,F\,u_d(t) + F\,A_m\,C\,e \qquad (16)$$

where

$$A_m\,C = A\,F + q^{-k}\,G \qquad (17)$$

and G is now a polynominal of order 2n. Substituting (14) for $\phi(t)$ on the RHS of (16) and using (17),

$$\begin{aligned} C\phi(t+k) = &\,G\,y(t) - C\,B_m\,u_d(t) \\ &+ (B\,F + \rho C)\,u(t) + F\,C\,e(t) \end{aligned} \qquad (18)$$

The control law

$$G\, y(t) - C\, B_m\, u_d(t) + (BF + \rho C)\, u(t) = 0 \tag{19}$$

yields

$$\phi(t+k) = F\, e(t) \tag{20}$$

It is shown in [10] that this minimizes

$$J = E[\phi^2(t+k)] \tag{21}$$

If the objective is to make the system behave like the model $\frac{B_m}{A_m} q^{-k}$, then $\rho = 0$; the use of $A_m = 1$, $B_m = 0$ yields a minimum variance-plus-weighted control law.

The closed-loop poles for this control law are the roots of C plus the roots of $BA_m + \rho A$. If A is a stable polynominal, it is always possible to make $BA_m + \rho A$ stable by choosing a sufficiently large ρ, even if B has roots outside the unit circle.

If the form (18) is given, the control law follows directly from the parametric description.

IDENTIFICATION

Basic Algorithms

There are two types of adaptive control. The first is called *explicit* or *indirect*, the other *implicit* or *direct*. In explicit adaptive control, we work from a model such as (1) and identify A, B and C. The identification is followed by a calculation of the control law parameters, such as (5), (6), (7). The method is called "explicit" because the plant parameters are explicitely identified, and "indirect" because the control parameters are obtained via calculation.

In implicit adaptive control, we start from a model in the form of (10) or (18). The identification is performed directly on the control law parameters G and BF, and is therefore only implicitly addressing the plant identification problem. Since implicit adaptive control has been used almost exclusively in process control, we shall now concentrate on its properties.

For convenience, rewrite (10):

$$C\, y(t+k) = G\, y(t) + B\, F\, u(t) + F\, C\, e(t+k) \tag{10}$$

Now, recall some basic notions from least-squares estimation. Suppose that

$$x(t) = \underline{\theta}^T \underline{\phi}(t) + \epsilon(t) \tag{22}$$

with $\underline{\theta}$ a constant vector of parameters to be estimated from knowledge of $x(t)$ and $\underline{\phi}(t)$. Suppose that a prior estimate $\hat{\underline{\theta}}(0)$ is given, together with a prior error covariance matrix $P(0)$ i.e.

$$P(0) = E[(\underline{\theta} - \hat{\underline{\theta}}(0))(\underline{\theta} - \hat{\underline{\theta}}(0))^T]$$

It is well known that the following recursive algorithm yields a least-squares estimate of $\underline{\theta}$:

$$\hat{\underline{\theta}}(t) = \hat{\underline{\theta}}(t-1) + \underline{K}(t)(x(t) - \underline{\phi}^T(t)\hat{\underline{\theta}}(t-1)) \tag{23}$$

$$\underline{K}(t) = P(t)\underline{\phi}(t)(1 + \underline{\phi}^T(t)P(t)\underline{\phi}(t))^{-1} \tag{24}$$

$$P(t) = P(t-1) - \underline{K}(t-1)(1 + \underline{\phi}^T(t-1)(P(t-1)\underline{\phi}(t-1))\underline{K}^T(t-1) \tag{25}$$

This algorithm is known to yield an unbiased estimate of θ under the conditions that: 1) $\epsilon(t)$ be uncorrelated with the components of $\underline{\phi}(t)$ and

2) $\dfrac{1}{N}\sum\limits_{t=1}^{N}\underline{\phi}(t)\underline{\phi}^T(t)$ tends to a nonsingular matrix.

A somewhat simpler algorithm is given by the stochastic approximation method, with $\underline{K}(t)$ given by [3]:

$$\underline{K}(t) = \frac{a}{r(t)}\underline{\phi}(t) \tag{26}$$

$$r(t) = r(t-1) + \underline{\phi}^T(t)\underline{\phi}(t) \tag{27}$$

where a is a constant.

Combined Identification and Control

Rewrite (10) as

$$y(t+k) = G\,y(t) + B\,F\,u(t) + (1-C)\,y(t+k) + F\,C\,e(t+k) \tag{28}$$

Suppose that $C = 1$. Then,

$$y(t+k) = G\,y(t) + Q\,u(t) + F\,e(t+k) \tag{29}$$

where $Q(q^{-1}) = B(q^{-1}) F(q^{-1})$ for ease of notation.
We write

$$\underline{\phi}^T(t) = [y(t)\ y(t-1)\ ..\ y(t-n+1)\ u(t)\ u(t-1)\ ..\ u(t-k-n+1)] \quad (30)$$

$$\theta = [g_0 g_1\ ..\ g_{n-1}\ q_0 q_1\ ..\ q_{k+n-1}] \quad (31)$$

Then, (29) may be writen as

$$y(t+k) = \underline{\phi}^T(t)\ \underline{\phi} + F\ e(t+k) \quad (32)$$

Since the noise term $F\ e(t+k)$ contains only $e(t+1)$, $e(t+2)$, ... $e(t+k)$, it is uncorrelated with $\underline{\phi}(t)$; therefore, the least-squares method applied to (30) will yield unbiased estimates that converge to θ, provided $\dfrac{1}{N} \sum_{t=1}^{N} \underline{\phi}(t)\ \underline{\phi}^T(t)$ tends to a positive definite matrix.

This latter condition is easily satisfied in an open-loop context, when $u(t)$ is chosen a priori: essentially, it is sufficient to make $u(t)$ satisfy the so-called "persistent excitation" condition [4]

$$\begin{bmatrix} R_u(0) & R_u(1) & \cdots & R_u(n-1) \\ R_u(1) & R_u(0) & & \\ & & \ddots & \\ R_u(n-1) & & & R_u(0) \end{bmatrix} \text{ is positive definite;} \quad (33)$$

where $R_u(\ell) = \lim_{N \to \infty} \dfrac{1}{N} \sum_{t=1}^{N} u(k)\ u(k+\ell)$ for zero-mean inputs. Roughly speaking, the input spectrum must contain enough frequencies.

The situation is not so clear cut in the adaptive context. First, $u(t)$ may not be assumed bounded, since we do not know whether or not the adaptive system is stable (to prove this is one of our objectives). If $u(t)$ blows up, so does $\dfrac{1}{N} \sum \underline{\phi}(t)\ \underline{\phi}^T(t)$. Second, we do not know whether the $u(t)$ generated by adaptive control satisfies (33).

Local stability addresses the following question: given an initial $\hat{\theta}(0)$ close to θ, will the algorithm converge to θ? Ljung [5] has developed methods to deal with local stability. In the present case, the answer is positive, at least if the minimum-variance control law is used and the system is minimum-phase. This means that $B(q^{-1})$ has all its roots inside

278   Chemical Process Control II

the unit circle, and the control is generated by using (12) with G and Q replaced by their estimates, i.e.

$$\hat{G}(t)\, y(t) + \hat{Q}(t)\, u(t) = 0$$

or

$$u(t) = -\frac{1}{\hat{q}_0}\, [\hat{g}_0 y(t) + \ldots + \hat{g}_{n-1} y(t-n+1) + \hat{q}_1 u(t-1) \\ + \ldots + \hat{q}_{n+k-1} u(t-n-k+1)] \tag{34}$$

For global stability, we require convergence from any initial estimate $\hat{\theta}(0)$. Results for deterministic systems are given in [6,7], while stochastic systems are studied in [3]. For the case at hand, the answer is positive also.

If $C \neq 1$, use (10) to write

$$y(t+k) = \frac{1}{C}\, [G\, y(t) + Q\, u(t)] + F\, e(t+k)$$
$$= \frac{1}{C}\, \underline{\phi}^T(t)\, \underline{\theta} + F\, e(t+k) \tag{35}$$

Because $F\, e(t+k)$ is a linear combination of $e(t+1), e(t+2) \ldots e(t+k)$, all uncorrelated with $\phi(t)$, we have

$$\hat{y}(t+k) = \frac{1}{C}\, \underline{\phi}^T(t)\, \underline{\theta}$$

and

$$C\, y(t+k) = C\, \hat{y}(t+k) + F\, C\, e(t+k) \tag{36}$$

Using this (28) yields

$$y(t+k) = \underline{\phi}^T(t)\, \underline{\theta} + (1-C)\, \hat{y}(t+k) + F\, e(t+k) \tag{37}$$

Now, suppose we use the minimum-variance control law, i.e.

$$\hat{G}(t)\, y(t) + \hat{Q}(t)\, u(t) = \underline{\phi}^T(t)\, \underline{\hat{\theta}}(t) = 0 \tag{38}$$

Then,

$$\hat{y}(t+k) = \frac{1}{C}\, [\underline{\phi}^T(t)\, \underline{\theta} - \underline{\phi}^T(t)\, \underline{\hat{\theta}}(t)] = \frac{1}{C}\, \underline{\theta}^T(t)[\underline{\theta} - \underline{\hat{\theta}}(t)] \tag{39}$$

If $\hat{\underline{\theta}}$ is close to $\underline{\theta}$, the term $(1-C)$ ŷ(t+k) in (37) is small, and the identification problem is nearly the one studied already. (Note that it is the use of the minimum-variance law that makes this happen).

For this problem, $\hat{\underline{\theta}}(t) \to \underline{\theta}$ locally if $C(\lambda_i) > 0$, where λ_i are the zeros of $B(z^{-1})$. Global stability can also be proved, under suitable similar positivity conditions on C[5].

Similar conclusions on stability are feasible if (18) is used as a starting point.

SOME PRACTICAL CONSIDERATIONS

Model Structure

The first order of business is the choice of the model delay k and order n. The delay k is often known (e.g. transport delay), at least if the sampling period is not too small; in any case, it should never be underestimated.

The order n may be overestimated. It has been suggested that the polynominal $C(q^{-1})$ in (1) may be approximated by $1/H(q^{-1})$ (for instance by long division) Using this in (1) gives

$$A(q^{-1}) H(q^{-1}) y(t) = B(q^{-1}) H(q^{-1}) u(t) + e(t) \qquad (40)$$

In effect, an increase in model order makes the situation closer to the case C=1, i.e. uncorrelated residuals. The price to be paid is an increase in the computational burden. This may also not work if the noise variance is small, because $H(q^{-1})$ would be estimated with relatively large uncertainty and could be non-minimum phase.

In most process-control applications. The models have been of order 2 to 6.

Control Objectives

In most process control applications, the objective is *regulation*, i.e. holding a measurement close to the desired value (set point). This usually requires that the measurement converge to the set point in the absence of perturbations. This can be achieved by using as input

$$\Delta u(t) = u(t) - u(t-1) \qquad (41)$$

Thus, the model identified as Δu(t) as input and y(t) as output, and the control law yields Δu(t). This is a "velocity" algorithm; integration (reset) is introduced, and steadystate convergence is assured if stability is. This approach was taken in [8].

A second method, used in [9, 10], consists of using adaptive control on a plant already in closed loop.

If the fixed controller contains integration, proper steady-state behaviour is assured. This scheme also offers the possibility of having a fixed regulator as backup, should the adaptive controller fail.

Finally, notice that the linear dynamical equation (1) really stands for

$$A(q^{-1}) [y(t)-y^*] = B(q^{-1}) [u(t-k)-u^*] + C(q^{-1}) e(t) \quad (42)$$

where y^* is the set point value and u^* is that value of $u(t)$ which, when applied to the process in the absence of disturbances, will yield a steady state output y^*. Since u^* is contant,

$$-B(q^{-1})u^* = -(b_0 + b_1 + \ldots + b_n)u^* = U \quad (43)$$

The constant U is unknown, but may be identified as an extra parameter and is easily included in the control calculations: this method was suggested in [11].

The next task is to select the control scheme. Classical lore (e.g. Ziegler—Nichols) dictates that tuning parameters be chosen to shape the response to set point changes, rather than to minimize the effect of disturbances. This classical strategy does ensure stability, and is probably best if frequent set point charges are expected or if disturbances come as pulse-like "upsets" rather than as stationary random noise. Pole-zero placement may be seen as an extension of classical methods and may be used in those cases.

On the other hand, there are many situations where disturbances are best characterized by stationary random processes. Because it is tuned to the disturbance spectrum, the minimum-variance control law would outperform pole-zero placement in such cases.

Almost all applications so far have used the direct method. Some caution must be exercised if non-minimum phase behavior is expected: as was pointed out in Section 3, direct methods lead to instability unless special precautions are taken.

The use of a weighting ρ on the control effort, as in (21), can stabilize, provided that open-loop system poles are stable. This condition would unfortunately preclude the use of an integrator in cascade with the plant to handle the steady state (or, equivalently, using Δu as an input). It has also been suggested that non-minimum phase effects be modeled by additional delays. As a last resort, an indirect method can always be used.

Algorithm Modifications

The minimum-variance control law (34) is

$$u(t) = -\frac{1}{\hat{q}_0}[\hat{g}_0 y(t) + \ldots + \hat{g}_{n-1} y(t-n+1) + \hat{q}_1 u(t-1)$$
$$+ \ldots + \hat{q}_{n+k+1} u(t-n-k+1)] \quad (34)$$

It is possible for the estimate $\hat{q}_0$ to be small at some time, implying large control values. To counter this, notice that u(t) is chosen to force $\underline{\phi}^T(t) \underline{\theta} = 0$. If $\hat{\underline{\theta}}(t) \to \underline{\theta}$, then from (32) y(t+k) → F e(t+k); however, this will also be true if $\hat{\underline{\theta}}(t) \to k \underline{\theta}$, because k $\underline{\phi}^T(t) \underline{\theta}$ would also go to zero. Therefore, the estimate has a degree of freedom.

This degree of freedom may be used to fix one element of $\underline{\theta}$: the natural choice is to set $\hat{q}_0 = \beta_0$, a constant. It has been shown [12] that the algorithm will still converge if i) β_0 is of the same sign as q_0 and (ii) is sufficiently large in magnitude. It can be shown that q_0 is the first non-zero sample of the system step response, a quantity for which a rough guess will usually be available.

Another common modification to the basic algorithm is the addition of a discount, or "forgetting" factor to the identification algorithm. As defined, the gain $\underline{K}(t)$ in (23) goes to zero, and the adaptation stops. This causes no problem if the objective is the occasional use of the algorithm to tune a controller for a process where parameter variations are slow.

In some cases, however, it is desirable to track parameter changes (through a grade change, for example). This requires that the update gain $\underline{K}(t)$ be prevented from tending to zero. In the least-squares algorithm, this is done by multiplying the RHS of (25) by $1/\lambda$, where $0 < \lambda < 1$. This amounts to minimizing $\sum_{i=0}^{t} \lambda^{t-i} \epsilon^2(i)$; i.e., data from the more remote past is weighted more lightly than more recent data. The factor λ has been called the "discount factor" or the "forgetting factor." Values between .9 and 1 have been used, although a value of .70 was used in [9] to track rapid variations during a grade change. In general, a low λ speeds up the response of the algorithm, but causes the estimates to wander in the steady state.

Startup

The simplest startup procedure is to make all initial estimates zero and to use $P(0) = \alpha I$. This was done in [8], with $\alpha = 10$. A large α speeds up the algorithm, since it implies that one's parameter estimates are poor and in dire need of updating.

Another startup method consists of using initial estimates leading to an initial control law identical to one already in use. For example, the minimum-variance control for a second order system with one delay is

$$u(t) = -\frac{\alpha_1}{\beta_0} y(t) - \frac{\alpha_2}{\beta_0} y(t-1) + \frac{\beta_1}{\beta_0} u(t-1) \frac{\beta_2}{\beta_0} u(t-2) \qquad (43)$$

Suppose a PI controller was used, with

$$u(t) = -k_1 y(t) - k_2 y(t-1) + u(t-1) \qquad (44)$$

Identity of (44) and (45) would be ensured if

$$\hat{\alpha}_1(0) = k_1 \beta_0, \ \hat{\alpha}_2(0) = k_2 \beta_0, \ \hat{\beta}_1(0) = \beta_0, \ \hat{\beta}_2(0) = 0$$

This startup method was used in [13].

Finally, the parameter estimates and the covariance matrix may be initialized by a trial period of straight identification: this was the course taken in [8]. If this takes place under fixed linear feedback control, care must be taken to ensure the identifiability of the parameters. A criterion for identifiability under feedback conditions is given in [14].

Numerical Implementation

The implementation on a machine with a high-level language is relatively straightforward. One application reported a program length of 35 FORTRAN statements with an execution time of 20 m sec on a PDP-15 for least-squares identification with 7 parameters a microprocessor implementation is discussed in [6], and some results on fixed-point implementation are given in [16].

Care must be taken to ensure the positive definiteness of P(t), especially if $\lambda=1$: failure to do this will drive the algorithm into instability. The best way to do this is to use UD or square-root factorization methods [17].

APPLICATIONS

We now give a quick overview of a few applications of adaptive control to the process industries.

One of the earlier applications was to the control of moisture on a paper machine. Moisture is controlled by steam flow, with thick stock flow used as a feedforward variable [13]. Six parameters were used, and good control was obtained in about 12 samples. In the steady state, the control was qualitatively comparable to that resulting from a well-turned PI controller. This result seems modest, until one is reminded that the tuning from grade to grade is automatic: PI controllers are often loosely turned to avoid re-tuning at each grade.

A more recent application in the paper industry is the control of a chip refiner [18]. Wood chips are fed near the axis between two counter-rotating grooved plates; the pulp is created by mechanical action and discharged at the periphery. The plate gap is manipulated to control the motor power. Due to wear, the gain between these variables drops by a ratio of 20 to 1 over about 600 hours of operation. Further, the pulp pad sometimes collapses, leading to a negative incremental gain: i.e., a gap decrease leads to decreased power. A self-tuning regulator was successfully used both to compensate for large gain variations and to take appropriate action when pad collapse occurred. Only one parameter was tuned, but a variable discount factor was used, as proposed in [19].

An application in mineral processing was presented in [8], where the self-tuning regulator was used to control an ore crusher. A ball mill is followed by a separator, and the coarse particles are recirculated into the mill. The problem is that the grinding action decreases if the feed is increased too much: more material is recirculated to increase the feed further, and so on until the mill "plugs up." The strategy is to keep the mill power at a given level, in the face of changes in grindability. A seven-parameter algorithm was used, and satisfactory control was achieved in 20 samples. Appreciable economic gains were realized from the ability to hold a power set point closer to the stability limit, i.e. 200 kw rather than 170 kw.

In [9], the problem was to control a T_iO_2 kiln, where the natural anatase form is cnverted to rutile. The control variable is fuel flow, and the output is rutile composition, sampled on a half-hour basis. A two-parameter self-tuning regulator was used to control a plant already controlled by a fixed regulator. The fixed regulator, by itself, lifted the in-spec percentage (percent of the time when the composition is within the desired range) from 68% to 75%; the adaptive controller has lifted it further, to above 90% [20].

Finally, the top composition of a binary distillation column was controlled with a self-tuning regulator [21]. Due to nonlinear behavior, the best tuning for a PI controller is different for increases and decreases in feed: a 4-parameter adaptive controller proved adapt at handling both. The same column was controlled by the model-reference scheme, [22] with similar results. In addition, adaptive feedforward control was used: this proved successful at minimizing perturbations due to feed variations.

CONCLUSIONS

Although the theoretical advances of the past few years have been spectacular, it is too soon to assert that adaptive control is now an "off-the-shelf" algorithm. Successful applications, as far as we know, have all received a fairly heavy academic input, usually involving researchers in the actual field trials.

Although much has been learned about the practical aspects of adaptive control, there are still some important questions to be answered. The question of non-minimimum phase systems has only received partial answers: there is no really completely satisfactory way of dealing with such systems. Another theoretical problem concerns the use of a low-order model to control a high-order plant; this is aways the case in practice, of course, the adaptive control has been successful despite the lack of theory. We must point out, however, that most applications have involved a fairly large number of parameters, i.e. 7 or 8 rather than 2 or 3. It would be desirable to have a theory that can cope with a small number of adapted parameters.

Finally, the adaptive control of multivariable systems has barely been scratched [24, 25]. The basic limitation is really the same as with scalar systems: the theory can deal only with plants that have causal inverses. In the scalar case, this eliminates non-minimum-phase systems. In the multivariable case, this implies limitations on the time delays affecting the inputs.

There is every reason to believe that the remaining questions about adaptive control will soon be answered. The theoretical advances, together with the practical lore coming out of further pilot applications, should make adaptive control a fairly common tool within a few years.

REFERENCES

[1] Aström, K. J. and Björn Wittenmark, IEE Proc., Vol. 127, Part D, No. 3, pp. 120–130 (1980).
[2] Aström, K. J., "Direct methods for nonminimum phase systems" Preprints of the 19th IEEE Conf. on Decision and Control, Albuquerque, Dec. 1980.
[3] Goodwin, G. C., P. J. Ramadge and P. E. Caines, "Descrete time stochastic adaptive control" Preprints of the 18th IEEE Conf. on Decision and Control, Fort Lauderdale, FL, Dec. 1979.
[4] Aström, K. J. and Peter Eykhoff, Automatica, Vol. 7, pp. 123–162 (1971).
[5] Ljung, Lennart, IEEE Trans. on Automatic Control, Vol. AC-22, pp. 551–575 (1977).
[6] Morse, A. S., IEEE Trans on Automatic Control, Vol. AC-25, pp. 433–439 (1980).
[7] Narendra, K. S., Y. H. Lin and L. S. Valavani, IEEE Trans. on Automatic Control, Vol. AC-25, pp. 440–448 (1980).

[8] Borrisson, Ulf and Rolf Syding, Automatica, Vol. 12, pp. 1–8 (1976).
[9] Dumont, G. A. and P. R. Bélanger, IEEE Trans. on Automatic Control, Vol. AC-23, pp. 532–537 (1978).
[10] Wittenmark, Björn, "Self-tuning regulators" Report TFRT-3054, Dept. of Automatic Control, Lund Inst. of Technology (1973).
[11] Clarke, D. W. and P. J. Gawthrop, Proc. IEEE, Vol. 122, pp. 924–934 (1975).
[12] Aström, K. J. and Björn Wittenmark, Automatica Vol. 9, pp. 195–199 (1973).
[13] Cegrell, Torsten and Torbjörn Hedqvist, Automatica, Vol. 11, pp. 53–60 (1975).
[14] Söderström, Torsten, Lennart Ljung and Ivar Gustavsson, IEEE Trans. on Automatic Control, Vol. AC-21, pp. 837–840 (1976).
[15] Clarke, D. W., S. N. Cope and P. J. Gawthrop, "Feasibility study of the application of microprocessors to self-tuning controllers" Report No. 1137/75, Dept. of Engineering Science, Oxford University.
[16] Bélanger, P. R. and Youssef Gonheim, "Fixed-point arithmetic microprocessor implementation of self-tuning regulators" Preprints, 1980 Joint Automatic Control Conf., San Francisco, Aug. 1980.
[17] Bierman, G. J., *Factorization methods for discrete sequential estimation*, Academic Press, New York, 1977.
[18] Dumont, G. A., "Self-tuning control of a chip refiner motor load" Pulp and Paper Research Inst. of Canada Report PPR 288, July 1980.
[19] Fortescue, T. R., L. S. Kershenbaum and F. G. Ydstie, "Implementation of self-tuning regulators with variable forgetting factors," submitted to Automatica.
[20] Dumont, G. A. and P. R. Bélanger, "Successful application of advanced control theory to a chemical process," Preprints, 1980 Automatic Control Conference, San Francisco, Aug. 1980.
[21] Sastry, V. A., D. E. Seborg and R. K. Wood, Automatica, Vol. 13, pp. 417–424 (1977).
[22] Martin-Sanchez, J. M., "A general solution to the adaptive control of linear time-variant processes with time delays" Dept. of Chemical Engineering, University of Alberta.
[23] Kailath, Thomas, *Linear Systems*, Prentice Hall, 1980.
[24] Borisson, Ulf, Automatica, Vol. 15, pp. 209–216 (1979).
[25] Goodwin, G. C., P. J. Ramadge and P. E. Caines, "Discrete-time multivariable adaptive control," Preprints of 18th IEEE Conf. on Decision and Control, Fort Lauderdale, Florida, Dec. 1979.

APPENDIX A

Write

$$y(t+k) = \frac{B}{A} u(t) + \frac{C}{A} e(t+k) \qquad (A.1)$$

Now, $\frac{C}{A} e(t+k) = q^k \frac{C}{A} e(t)$

286   Chemical Process Control II

$$q^k \frac{C}{A} = q^k \frac{q^n C}{q^n A} = q^k \frac{C^*}{A^*} \qquad (A.2)$$

Since $q^k C^*$ is a polynominal of order n+k and A^* is a polynominal or order n, by long division:

$$q^k \frac{C^*}{A^*} = F^* + \frac{G^*}{A^*} \qquad (A.3)$$

where $F^*(q) = q^k + f_1 q^{k-1} + \ldots + f_{k-1} q$ and G^* is a polynominal of order n. Now,

$$q^k \frac{C^*}{A^*} = q^k \frac{C}{A} = q^k F + \frac{q^n G}{q^n A}$$

or

$$\frac{C}{A} = F + q^{-k}\frac{G}{A} \qquad (A.4)$$

This may be written

$$C = AF + q^{-k} G \qquad (A.5)$$

Using (A.4) in (A.1),

$$y(t+k) = \frac{B}{A}u(t) + q^k F\, e(t) + \frac{G}{A}e(t) = \frac{B}{A}u(t) + F\, e(t+k) + \frac{G}{A} e(t) \quad (A.6)$$

Note that $F\, e(t+k)$ contains only future samples of the noise. Shifting (A.1) backwards by k steps,

$$e(t) = \frac{A\, y(t) - B\, q^{-k}\, u(t)}{C} \qquad (A.7)$$

This gives present and past noise values. Insert in (A.6) to get

$$y(t+k) = \frac{B}{A}u(t) - F\, e(t+k) - \frac{G}{A} \frac{A\, y(t) - B\, q^{-k} u(t)}{C}$$

or

$$A\, C\, y(t+k) = A\, G\, y(t) + B[C - q^{-k}\, G]\, u(t) + A\, F\, C\, e(t+k) \quad (A.8)$$

Use (A.5) to get $C - q^{-k} G = A\, F$, cancel A; there results Equation (10).

MODEL ALGORITHMIC CONTROL (MAC); REVIEW AND RECENT DEVELOPMENTS*

Raman K. Mehra, Ramine Rouhani and John Eterno
Scientific Systems, Inc.
54 Rindge Avenue Extension.
Cambridge, MA 02140

J. Richalet and A. Rault
Adersa/Gerbios, France

ABSTRACT

The purpose of this paper is to present a review of Model Algorithmic Control (MAC) approach along with recent developments and applications. After a short historical overview, the main features of MAC are discussed. Next, MAC implementation and algorithms are outlined. This is followed by a discussion of applications and several recent developments. A fairly complete bibliography is included for further information.

HISTORICAL OVERVIEW

Model Algorithmic Control was first conceived by Adersas/Gerbios in the late 60's. (Richalet and Gimoret 1968, Richalet and Lecamus 1968;

---

*The work reported in this paper was supported by the Office of Naval Research under contract N00014-C-76-1024.

Richalet, Lecamus and Hummel 1970; Richalet, Rault, Pouliquen 1971). The method was subsequently developed and applied successfully to several industrial processes involving multivariable plant dynamics. With the growth of digital technology in the 70's and particularly the availability of powerful microprocessors with large memories, the application of MAC to industrial processes became more practical and economical. For instance, in the past five years, MAC has been successfully used for Aircraft Flight Control (Rault et al. 1975), Superheater Control (Testud 1977), On-Line Control of Steam Generator (Lecrique et al. 1978), European Transonic Wind Tunnel Control (Mereau and Littman, 1978), F-100 Jet Engine Multivariable Control (Mehra et al. 1977), Adaptive Flight Control (Mehra et al. 1978) and Electric Plant Control (Mehra et al. (1980)). Recently attention has also been focused on theoretical properties of MAC to gain a better understanding for design purposes. These theoretical results are reported in Richalet et al. (1978), Mehra et al. (1979), Praly (1979), Mehra et al. (1980), Mehra and Rouhani (1980) and Reid et al. (1980).

BASIC PRINCIPLES OF MODEL ALGORITHMIC CONTROL (MAC)

The MAC strategy relies on the following four features: (i) internal model of the system; (ii) hierarchical decomposition; (iii) reference trajectory and output constraints; and (iv) control trajectory computation.

Internal Model of the System

The multivariable system to be controlled is represented by a mathematical model in time-domain of the input-output type (see Figure 1). For linear systems, the model is of the impulse response type, a representation which has certain distinct advantages over the state space representation or the transfer function representation for multivariable control. For nonlinear systems, both the state space and the input-output representations have certain advantages and disadvantages. In applications, one may use either or both depending upon the nature of nonlinearities and the complexity of the resulting controller. The purpose of the internal model is to have a flexible representation of the controlled system stored in the computer memory, which can be updated as the system changes and which can be used at any instant to predict the future behavior of the system under different control inputs. The internal model of the system is used to compute optimal inputs, to detect process changes, sensor malfunctions and severe faults. The inputs and current output of the internal model are updated according to the actual observed values of

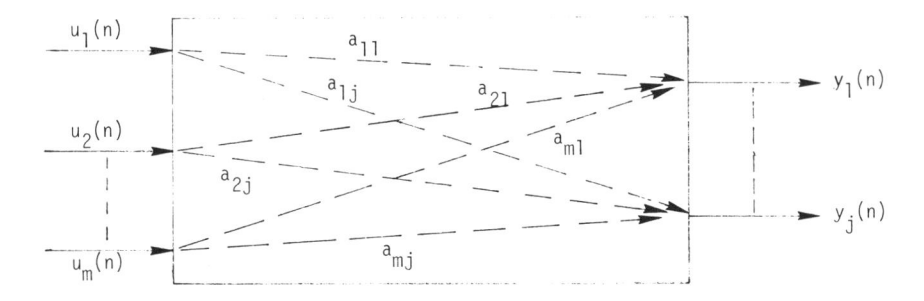

Figure 1. Impulse Response Representation of a Linear System.

these variables, but any large difference between the computed and the actual values gives important clues as to the malfunctioning of sensors and actuators.

The internal model of the system is generally obtained via off-line identification, using either a physical model structure when one is easily available or an input-output representation model such as the impulse response model, which may change with the operating point. Some form of on-line parameter identification may also be done in those cases where large random variations of system parameters are expected. It has been found from experience that the robustness of MAC is sufficient to take care of small parameter changes.

Hierarchical Decomposition

In an industrial process such as a Boiler-Steam Generator System, several levels of control can be specified based on a decomposition in time and space. At the lowest level, called level zero, fast regulation of flow rates and feed rates is required. This is typically a single-loop control problem and can be solved satisfactorily using a PID controller. The control action is very local and fast so that analog controllers are quite suitable for this application. Level 1 control typically involves the use of level 0 output variables as the control variables for regulation and set point control of temperatures, pressures and levels. The control action at this level is more spread out spatially and is multivariable in nature. Linear models can be used satisfactorily at this level as long as very large set point changes are not required. The MAC design has been used very satisfactorily for level 1 control of processes. Notice that one may use a number of level 1 controllers, one for each group of controlled variables, as long as the interactions between the groups are slow and can be handled at a higher level. For example, in a boiler system combustion dynamics is much faster than the steam generation dynamics so that at level 1, combustion process and the steam generation process may be controlled separately. Similarly if the interaction at level 1 between the water level control problem and the steam temperature control problem is small (e.g., the effect of changes in feedwater flow on temperatures may be small), one may design separate controllers for the two processes. One-way effects may still be taken into account by using feedforward action.

The control at level 2 requires coordination between level 1 controllers with the objective of producing large changes in the plant output. For example, runback to houseload from full load is a level 2 control problem since large changes in MW demand are required and a coordination between all the control variables is necessary. The behavior of the system is highly nonlinear at level 2 and the payoff from improved control is direct and very significant. Typically, the economic payoff is much more obvious at higher control levels, such as the prevention of a shut-down can be easily converted into dollars whereas the economic payoff from better control of temperatures and flows is not immediately obvious. However, since all the control levels are connected, poor performance at lower levels leads to poor performance at higher levels. In other words, the performance at level 2 cannot be fully optimized unless the performance of controllers at levels 1 and 0 is optimized first. The situation is very similar to that in a battleground where a good commander is ineffective without good soldiers and vice versa.

It may be noticed that the use of hierarchical levels of control simplifies the problem of developing internal models, by reducing the number of inputs and outputs at each level and for each controller. Furthermore, the controllers at higher nonlinear levels run at slower sampling rates so that more complex control schemes can be used. This balance between the complexity of the controller and that of the controlled process is a very desirable feature.

Reference Trajectory and Output Constraints

The desired response of the closed-loop system is specified in the form of a reference trajectory and constraints which are updated on-line using the actual output of the sytem. It is possible to handle state-dependent constraints in this fashion and to eliminate the steady state offset error. It should be noticed that the specification of reference trajectories and constraints is much easier and natural than the specification of a scalar performance index. Typically, the requirements on controlled outputs are stated as "no overshoot," "fast response," "within maximum and minimum limits," etc. These requirements are difficult to express in a scalar quadratic performance index, but they are easily converted into desired trajectories and constraints which can also be explained to the operators without any difficulty. In a hierarchical control system, reference trajectories for lower levels may be specified by the higher levels, through an optimization process since the controlled outputs of lower levels are the control variables for the higher levels. The reference trajectories at the highest level would generally be obtained via a combination of on-line and off-line optimization.

The concept of using reference trajectories is more general than model-following. Firstly, it may not be possible to represent a reference trajectory by a simple model, and secondly, under sensor or actuator fault conditions, one may have to relax the system requirements to control within a band. Certain control problems such as water level control involving inexact measurement of the water level are formulated more naturally as band-control problems rather than model-following or scalar performance index problems.

Control Trajectory Computation

Controls are computed, in general, for a number of future time points using an iterative optimization technique which minimizes the distance between the desired reference trajectory and output trajectory predicted by the internal model, while keeping all the output, state and control constraints satisfied. The complexity of the control algorithm is directly dependent on the structure of the internal model, the number of inputs and outputs and the constraints. For linear systems, the impulse response representation results in a simple projection-type quadratic programming solution which can be implemented quite fast in micro/mini-computers of the present generation. The actual dimensionality of the state does not increase the complexity of the algorithm as it would in a state vector representation.

SPECIAL FEATURES OF MAC AND IDCOM (IDENTIFICATION AND COMMAND) FOR LINEAR MULTIVARIABLE CONTROL

The MAC design approach has been found to be flexible and robust. It is well suited for the evolving microprocessor technology providing high speed memory and fast computation times for basic calculations such as convolutions. The universality of the impulse response representation leads to a unified design approach for systems of all orders. Furthermore, the parameter-linearity of this representation leads to a duality between identification and control.

The MAC is implemented by a program called IDCOM. The special features of IDCOM are: (i) no model order reduction is required since an impulse response representation is used. (ii) Input magnitude and rate constraints are handled directly and exactly. (iii) The control law is time-varying and the closed loop response is robust to parameter changes. (iv) Gain scheduling is replaced by on-line updating of the internal model using operating data and parameter estimation techniques, thereby reducing reliance on theoretical models of the system. (v) Same algorithm is used for impulse response identification and for control law computation, thereby simplifying the hardware requirements. (vi) Different sampling rates can be used for controlling different outputs. (vii) The control laws can be modified on-line in case of sensor failures or degraded system performance.

IDCOM IMPLEMENTATION

A block diagram of IDCOM is shown in Figure 2. As indicated in the figure, when a new measurement is made, it is fed to two blocks of IDCOM and used to compute a reference trajectory and a "zero input" prediction of the future outputs over a short horizon (for optimization). The reference trajectory is usually a first-order exponential drawn from the current (measured) output to a given set point. The designer supplies a time constant (τ_i) for this exponential for each output i. The "zero-input" prediction uses the past inputs, measurements, and internal impulse-response model to predict the future outputs *in the absence* of future control inputs.

These two trajectories (reference and zero-input prediction) are differenced to obtain an error trajectory to be minimized by the future controls. The control calculation block then performs this minimization in one of several ways. Once an input sequence has been computed, the first input is applied to the plant, and the cycle starts again after the next measurement.

BASIC CONTROL COMPUTATION

The first control computation is essentially a simple inversion routine (in the absence of constraints) which finds inputs to zero the error between the reference and zero-input trajectories. To make the problem tractable, the error is only considered at a few points in the future, and the

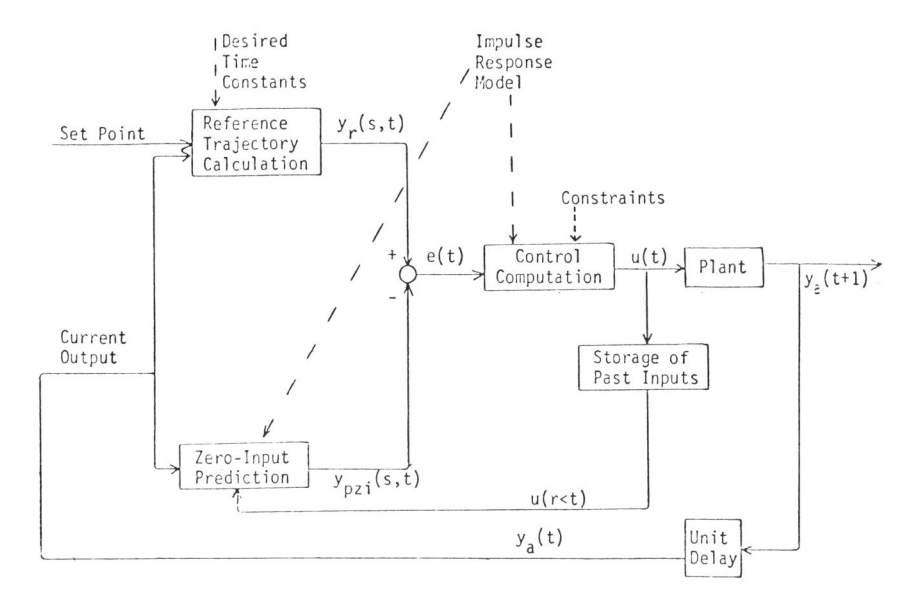

Figure 2. IDCOM Components.

future controls are required to be constant over intervales ("blocks") between the output-matching points. For example, three input values u_1, u_2 and u_3 may be computed to define the next five future controls as:

$$u(1,t) = u_1$$
$$u(2,t) = u_2$$
$$u(3,t) = u_2$$
$$u(4,t) = u_3$$
$$u(5,t) = u_3$$

where $u(s,t)$ is the computed control at time t to be applied s steps in the future. This is shown pictorially in Figure 3.

For this case, the outputs to be controlled (the error to be reduced) occur at the *endpoints* of each block, i.e., 1, 3 and 5 steps in the future. The advantage of this blocking is that it permits reasonable long optimization horizons with low-dimensional required calculations. In this example, only three numbers are computed for five steps ahead. Since the controls will generally be recalculated one step ahead, this technique sacrifices little.

The optimization routine used in the basic algorithm tries to invert the system to find inputs which result in perfect output matching at the

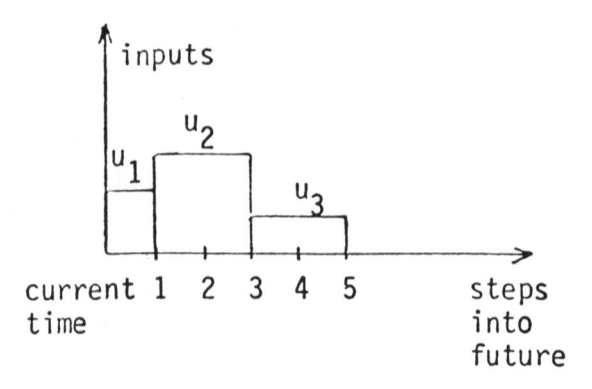

Figure 3. Input Blocking.

chosen points. If this is possible, then the first control causes the first output to be correct independent of the future inputs and outputs. If constraints are encountered, or if the computation time is shortened, however, the computed solution for one step ahead differs from the perfect control in a way that results in good performance at the future selected points.

GRADIENT ALGORITHM

In order to improve the control calculation for general systems, a gradient-projection algorithm has recently been developed. This algorithm retains the input blocking discussed above (to reduce the dimension of the problem) but does not block the outputs. Instead, the new algorithm minimizes the sum of the squares of the output errors at *each* step in the future up to the end of the last input block. Thus, for the example above, the new algorithm would compute three control members to minimize five future errors.

The capability also exists in this algorithm for adding input and output weighting matrices if desired. In general, we use only output weighting unless control cost is meaningful. Output weighting alone is very convenient, permitting easy tuning of the controller to achieve desired performance for each output. Finally, we note that, for only one block with endpoint one step ahead, the gradient algorithm and inversion routine are equivalent (without control weights).

COMPARISON OF IDCOM WITH STATE SPACE CONTROL TECHNIQUES

IDCOM, like most of the State Space control techniques is a multivariable time-domain technique. It is different, however, in several important ways. These differences are important since they account for the success of IDCOM in industrial process control applications where the lack of good models has prevented the use of other modern control techniques. The important differences are:

(i) *Robustness.* State Space techniques generally use full state feedback based on a parametric state vector model. IDCOM, on the other hand, uses output feedback based on an impulse response model. It can be argued that the effect of modeling errors on IDCOM will be less compared to that on state vector techniques.

(ii) *Implementation.* Modern control techniques based on state vector models generally require the solution of matrix Riccati equations. Since these equations are computationally time-consuming to solve for systems of high order, the practical implementation of these techniques requires model reduction and off-line solution of the Riccati equation. In practice, only the steady state gains from the Riccati equation are implemented and these gains are scheduled as a function of the operating point (temperature, pressure, etc.). This implementation does not allow for on-line changes in system model and performance criteria. The IDCOM approach, on the other hand, is much more flexible and adaptive, since the model, criteria, and sampling rates can be adjusted on-line. This flexibility comes from the use of the impulse response representation, which has the additional advantages of conceptual simplicity, ease of identification and elimination of the model order reduction problem.

(iii) *Criterion Function and Constraints.* Most of the State Space techniques require specification of a scalar criterion function. In practice, this is very difficult or impossible to do. The specification of the information used in IDCOM is easier and more natural. Basically, the specification of desired responses and constraints is simpler than the specification of a scalar performance index. It is also possible to set up a priority list on the control of outputs in IDCOM and even to make them conditional on the occurrence of some future events. For example, the failure of a sensor can be detected and the control strategy can be shifted accordingly. The problems of sensor blackout or computer overload can also he handled since controls are computed and stored for several future time points and these can be used till new information or computational capability becomes available. The exact satisfaction of control and state constraints is absolutely essential in many applications. Such constraints are handled much more easily in IDCOM, thanks to the impulse response representation of the system.

(iv) *Duality of Identification and Control.* From a practical standpoint, the duality of identification and control for an impulse response model is much more useful than the duality between estimation and control for state vector models. The former can be used directly to increase the robustness of the controller by on-line identification whenever the residuals between the model output and the system output are consistently outside some prespecified confidence bands.

Some of the other modern control approaches which are closer to MAC are Dynamic Matrix Control (Cutler et al. (1980) Model Reference Adaptive (Landau (1974)) and Self Tuning Regulators (Astrom (1980)). They are, however, not exactly similar since the specification of models (reference, prediction, control) and algorithmic computations are done differently. Model Reference Adaptive (MRA) techniques try to reduce the error between the system output and the model output by using error feedback. No computation of predictive controls is done and no reference trajectories are specified. The extensions of these techniques to nonlinear system and to hierarchical levels of control are much more difficult both from a conceptual and an algorithmic viewpoint. Furthermore, by on-line adaptation of the internal model in MRA and STR the ability to detect failures is lost. The Dynamic Matrix Control approach is conceptually similar to MAC, but lacks proper use of reference trajectories and internal models which are especially important for monminimum phase systems. Similar problems exist with the use of other Inverse Control type of methods (Meyer (1980)).

APPLICATIONS

Several applications of MAC have been reported in the literature over the last four years. More importantly, MAC has been operational on over 10 industrial processes in France for the last 6 years. Brief summaries of a few applications are presented here.

MULTIVARIABLE CONTROL OF A STEAM GENERATOR USING MAC (LECRIQUE ET AL. (1979))

A description of the process is given in Fig. 4. It represents a 250 MW natural circulation boiler of a Martigues-Ponteau oil power plant in the south of France. It was decided to control steam pressure P_v, superheated steam temperature T_s and the reheated steam temperature T_r, using as

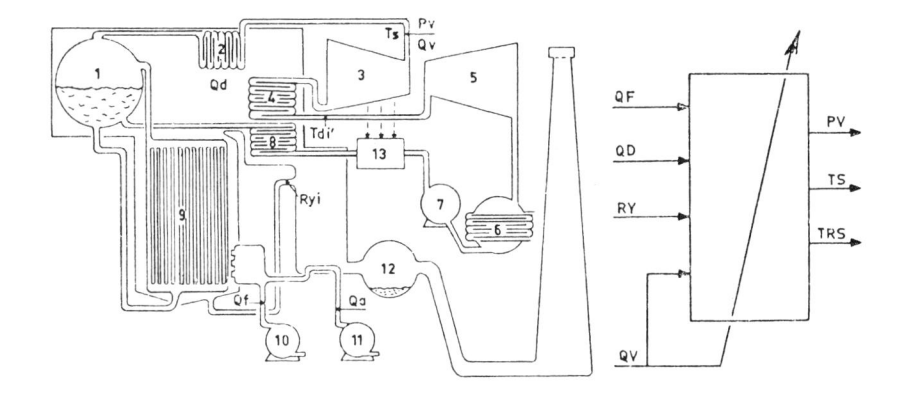

Figure 4. Power Plant Description; Input and Output Signals.

control the fuel-oil flow, Q_f, the gas recirculation flow, R_y, and the spray water flow, Q_d, under conditions of changing load, Q_v. Twelve impulse responses between the four inputs and three outputs were identified from plant test data using square-wave type inputs and the identification algorithm described in Richalet et al. (1978). Eight of the impulse responses for temperature variables are shown in Fig. 5a and 5b. The structure of the controller is shown in Fig. 6 where the anolog PI controller is retained as backup. Since pressure dynamics is faster, the pressure control loop is sampled ever 20 sec. at all operating loads. The temperature dynamics is a function of load, becoming faster with increasing load and the gains varying nonlinearly as a function of load. The variation in time constants is handled by changing the sampling rate as a function of load in such a way that the impulse response function in terms of samples stored in the computer does not change. The gains of the impulse responses are changed according to a predetermined function of load. In this fashion, the nonlinear temperature dynamics is easily taken into account. Actuator nonlinearities such as those on R_y are also handled easily by computing the output of the actuator and then inferring the input from the actual nonlinear characteristic. The actual control computation takes about 3 sec., which is a small fraction of the fastest sampling interval of 20 sec.

The IDCOM controller has been operational on the Martigues-Ponteau plant since April 1976 and the results are very encouraging. Some of the results and comparisons with an adaptive analog controller are shown in Figs. 7, 8 and 9. Table 1 shows further results, indicating a reduction in maximum temperature deviations by a factor larger than 1.6 in case 5 of

298   Chemical Process Control II

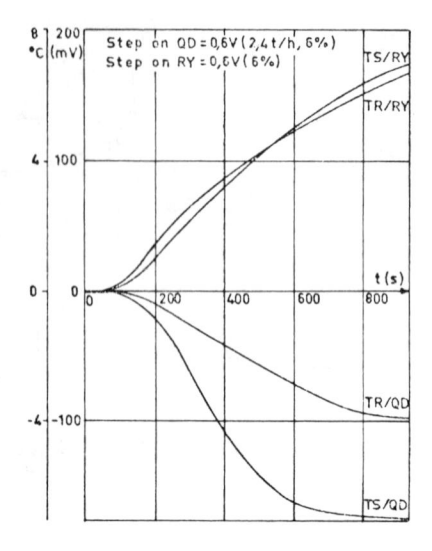

Figure 5a. Identification—Step Responses: TS/QD, TS/RY, TR/QD, TR/RY.

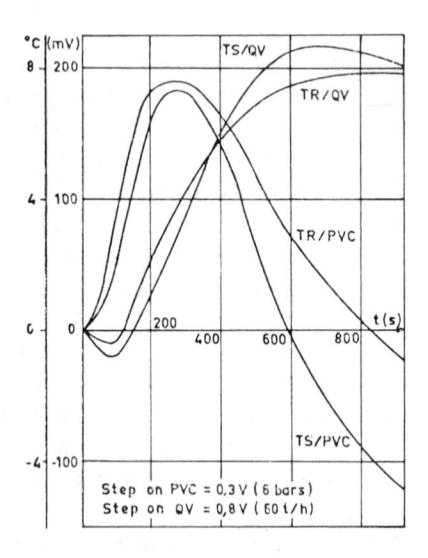

Figure 5b. Identification—Step Responses: TS/PVC, TS/QV, TR/PVC, TR/QV.

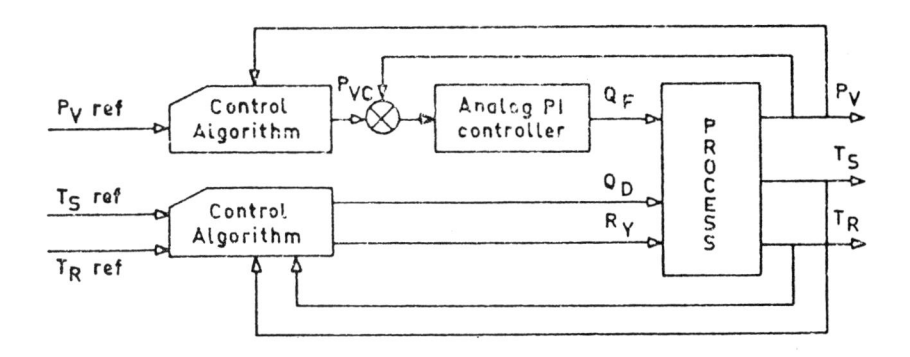

Figure 6. Control Implementation.

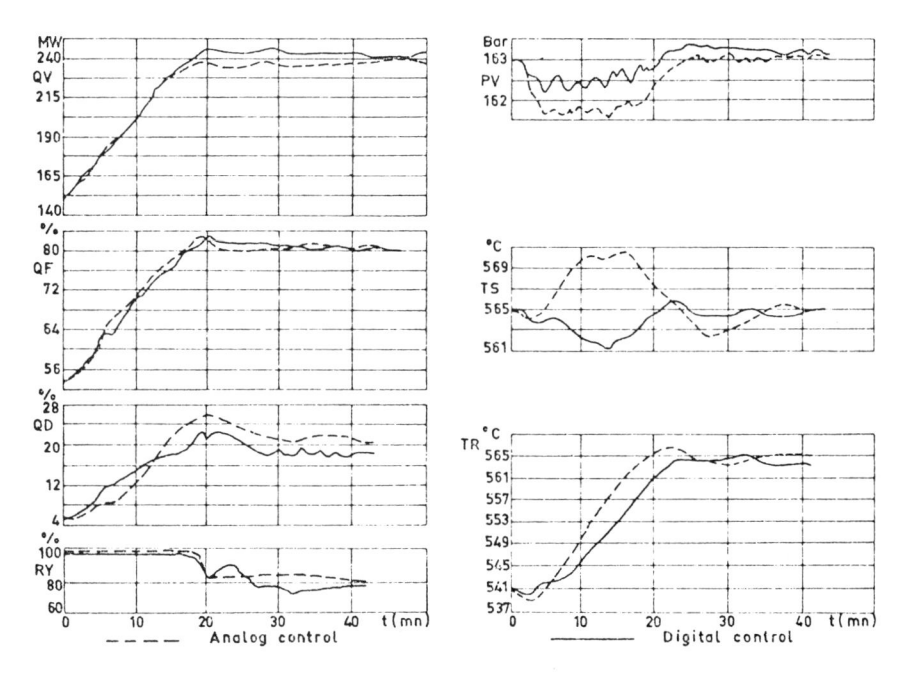

Figure 7. Load Rise from 140 to 240 MW at a Rate of 5 MW/min.

300   Chemical Process Control II

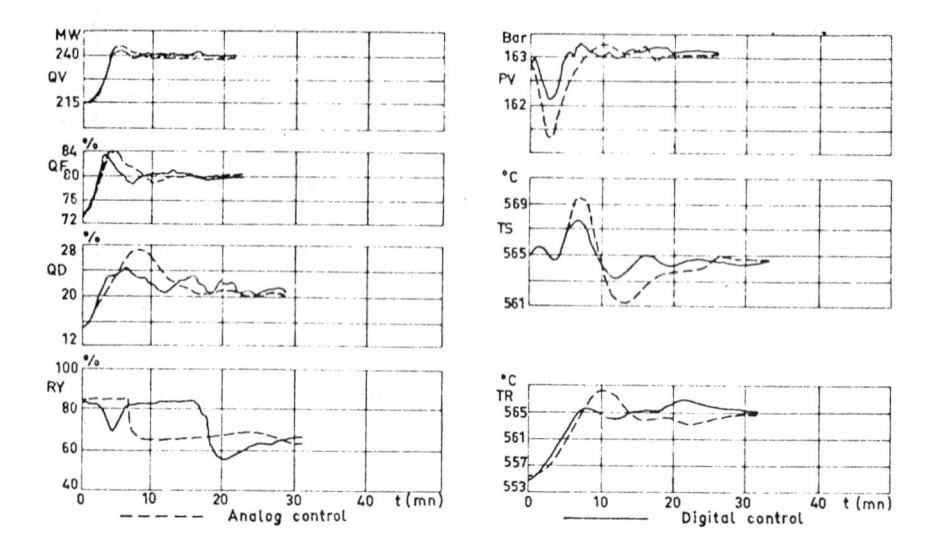

Figure 8. 25 MW Load Rise at a Rate of 10 MW/min).

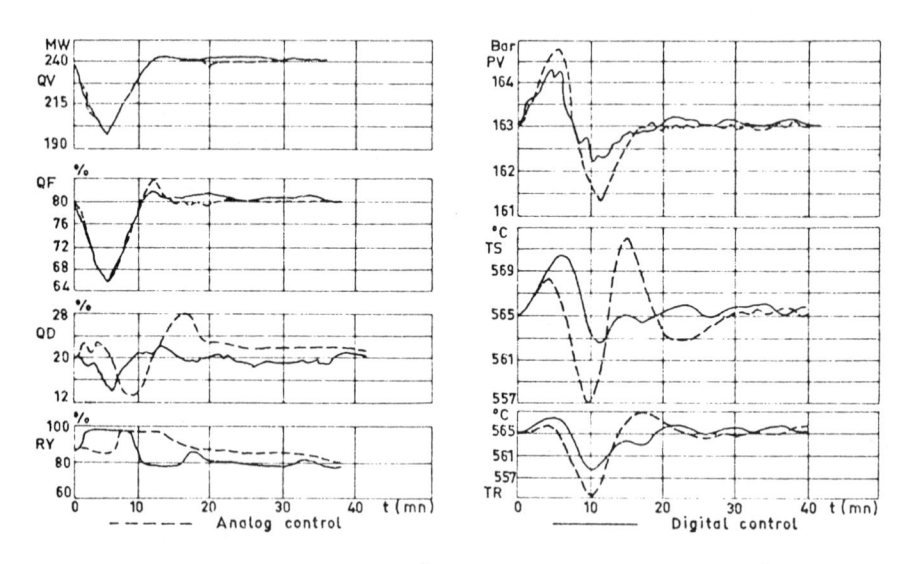

Figure 9. Triangular Load Variation (45 MW at a Rate of 7.5 MW/min.)

Table 1. Results

Test		Steam Pressure (bar) ANA.	Steam Pressure (bar) IDCOM	Superheated Temperature (°C) ANA.	Superheated Temperature (°C) IDCOM	Reheated Temperature (°C) ANA.	Reheated Temperature (°C) IDCOM
1st Test	d_1	+1,6	+1,1	+ 1,6	+3,2	+ 0,5	+ 2,0
	d_2	—	—	− 6,8	−2,0	−25,3	+23,0
Decrease of load:	d_3	1,6	1,1	+ 2,0	—	—	—
40% in 19 min	Δ			8,4	5,2	25,8	25,5
2nd Test	d_1	−1,5	−0,8	− 1,6	−4,0	—	—
	d_2	—	—	+ 5,6	+0,8	+ 1,0	—
Increase of load:	d_3	1,5	0,8	− 3,2	—	—	—
40% in 19 min	Δ			7,2	4,8	1,0	—
3rd Test	d_1	+2,0	+2,0	+ 4,0	+4,0	+ 2,0	+ 2,0
	d_2	—	—	− 5,2	−4,4	−12,0	− 9,0
Decrease of load:	d_3	—	—	+ 2,4	—	—	—
10% in 30s	Δ	2,0	2,0	9,2	8,4	14,0	11,0
4th Test	d_1	−1,5	−1,1	—	—	—	—
	d_2	—	—	+ 5,6	+2,8	+ 5,0	+ 0,5
Increase of load:	d_3	1,5	1,1	− 3,6	−2,4	− 0,5	− 2,0
10% in 150 s	Δ			9,2	5,2	5,5	3,5
5th Test	d_1	+1,8	+1,3	+ 4,0	+6,0	+ 2,4	− 2,8
	d_2	−1,5	−0,7	− 7,6	−2,8	−10,0	− 6,8
Triangular load vari-	d_3	—	—	+ 7,6	—	+ 4,8	—
ation: 16% in 11 min	Δ	3,3	2,0	15,2	8,8	14,8	9,6

d_1, d_2, d_3 = first, second and third peak amplitudes.
Δ = peak to peak amplitude.

V-shaped load variations. In the third test where load is decreased at 20%/min., IDCOM eliminates third peaks and reduces maximum temperature deviations.

UTILITY BOILER CONTROL (TESTUD (1979))

The plant is shown in Fig. 10. It is a steam generator, connected to a large crude oil distillation column, and uses as a primary heat source the fumes of the vaporizing furnace. The goal is to control the steam pressure PV and the water level NIV so the PV can be reduced as close as possible to 4 bars, resulting in better extraction of energy. The various plant inputs and outputs are shown in Fig. 10. It is required to control the level NIV within a band, while trying to maintain pressure PV constant at a value as

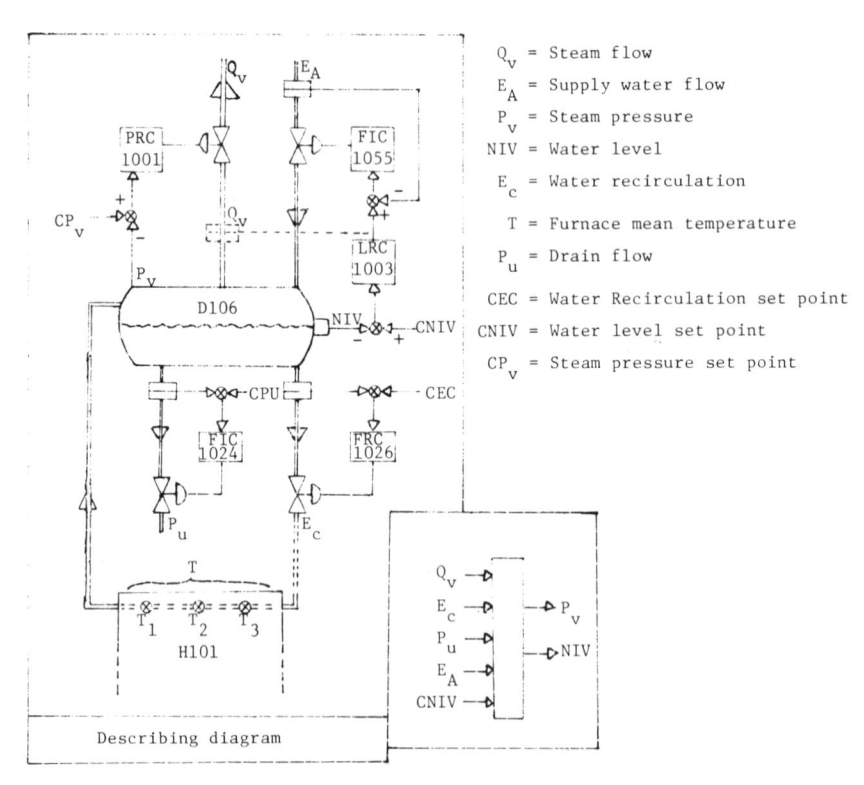

Q_v = Steam flow
E_A = Supply water flow
P_v = Steam pressure
NIV = Water level
E_c = Water recirculation
T = Furnace mean temperature
P_u = Drain flow
CEC = Water Recirculation set point
CNIV = Water level set point
CP_v = Steam pressure set point

Figure 10. Steam Generator Description.

close as possible to 4 bars. The identifed model is shown in Fig. 11. Using this as an internal model for IDCOM, it was possible to reduce the pressure PV from 7 bars to 5 bars as a result of a tenfold reduction in the variance of regulation error.

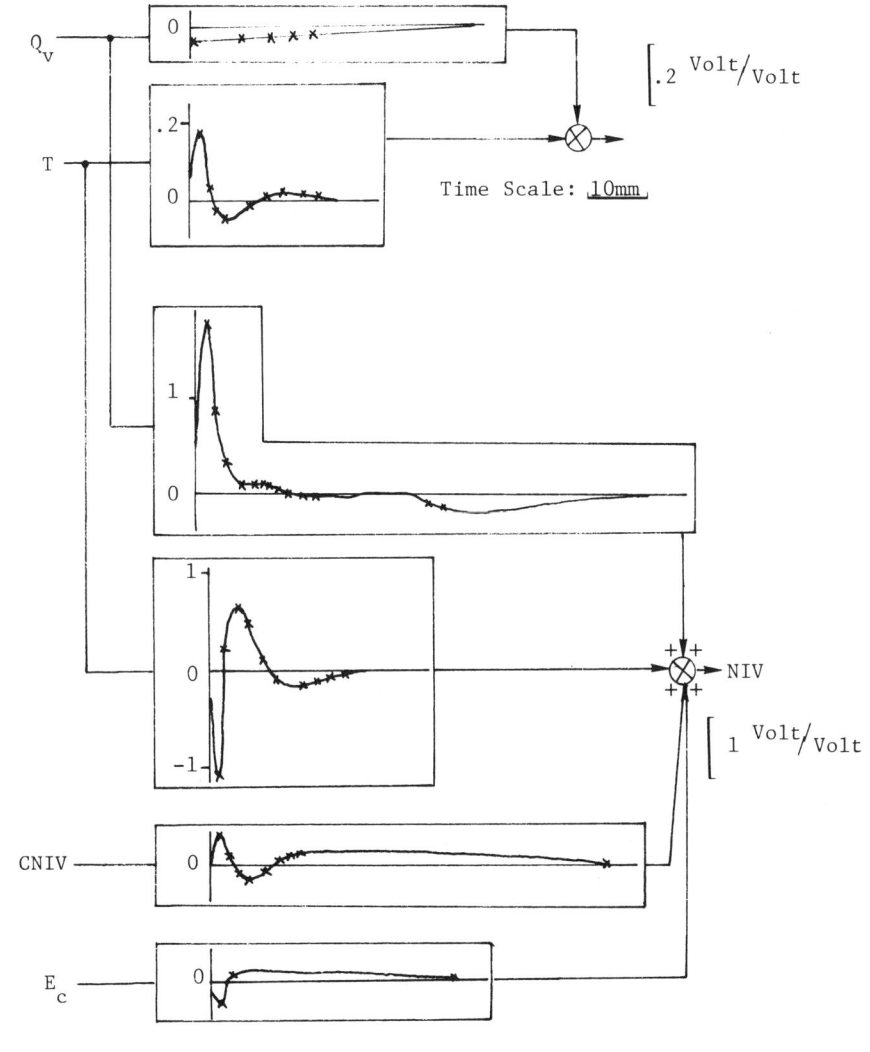

Figure 11. Impulse Responses of Model.

DISTILLATION COLUMN CONTROL
(RICHALET ET AL. (1978), ENGRAND (1980))

This application involves two levels of control. At level 1, IDCOM controls two temperatures and pan level using flow rates as the control variables. At level 2, the temperatures are set to control the distillate cloud point under changes in crude quality and other plant variables. Closed loop identification and gradient type optimization algorithms are used to move the plant towards an optimum operating point.

Other on going applications of MAC consists of the Control of a Deminer Boat (Daclin (1980)), Mirage Control (Mehra et al. (1978)), F-100 Jet Engine Control (Mehra et al. (1978)), Adaptive Missile Control (Mehra et al. (1980)), Glass Furnace Control (Bordier et al. (1979)), Cryogenic Wind Tunnel Control (Mereau et al. (1978)) and Green house temperature—humidity control.

RECENT DEVELOPMENTS

In this section we describe some of the recent developments of MAC. These consist of (i) theoretical properties (ii) Hierarchical MAC and STP (Singular Perturbation Theory) (iii) Associative Control (iv) Alternative Prediction Schemes and (v) Sample Rate Selection.

THEORETICAL PROPERTIES (MEHRA ET AL. (1978, 1979, 1980))

The application of Model Algorithmic Control to industrial processes involves consideration of three models. The Actual Model—denoted by ho—is supposed to represent the exact impulse response of the plant to be controlled. The Internal Model—denoted by h—is used to predict the behavior of the plant ho. The Internal Mode Impulse response h is usually the identified version of ho. In general h differs from ho because of the identification and modeling error. Finally, there is a third model called Reference Model, usually of first order, which represents the desired ideal path of the output from its current value to a given set point C. The parameters of this last model are chosen by the operators.

To avoid the bias on the output, one has to use a closed loop prediction. The single-input single-output versions of the above are:

Plant:

$$y_o(t+1) = h_o\, u(t) \qquad (6.1)$$

where ho is an nxl vector of truncated impulse response parameters $u(t)$ is nxl vector of past inputs

Prediction Model: (closed loop)
$$y(t+1) = h\ u(t) + (y_0(t)-y(t)) \quad (6.2)$$

Reference Model:
$$y_r(t+1) = \alpha y_0(t) + (1-\alpha)C \quad (6.3)$$

$0 < \alpha < 1$
C: desired set point
α: A parameter related inversely to the trajectory time constant (small α implies fast approach to set point C).

Stability Results:

For the closed loop system to be stable, it is necessary and sufficient that the polynomial:
$$z^{-1}(H_0(z) - H(z)) + H(z) - \alpha z^{-1} H_0(z) \quad (6.9)$$

has all its roots within the unit circle, where $H_0(z)$ ($H(z)$) is the z-transform of $ho(h)$. It has been shown, that: (i) If the identification is nearly perfect ($ho \simeq h$), then the stability of the plant ho implies the stability of the closed loop controller. Besides the output of the plant is of first order with a pole at α. (ii) If the identification is not perfect, then the designer has to choose both α (reference model) and $H(z)$ (the internal Model) in a way that (6.4) has all its roots in the unit circle. In most cases, if the plant is stable, and minimum phase, then relatively simple and crude choices of h stabilize the closed loop system. Of course the closer h is to ho, the shorter is the convergence time of the output to its desired set point C.

ROBUSTNESS

One of the most useful properties of MAC is its robustness, which appears at least partially to result from the use of impulse response model for plant representation. It can be shown that when the mismatch between the plant ho and the Internal Model h is a pure gain, as long as the gain factor q is less than $\frac{1}{1-\alpha}$, the system is stable and the output converges to its desired value C. In particular, a slow reference model (α close to 1) enhance the robustness of the systems. Further experiences on the control of a F-100 Jet Engine (Mehra et al. (1978)) has confirmed the robust character of the control system in the presence of phase shift.

STOCHASTIC ENVIRONMENT

Properties of MAC have been studies in presence of both white and colored additive output noise. In particular it is shown that the variance on the output of the controlled system decreases as the reference model becomes slower (α increases). For colored noise the results are more complex depending on the correlation properties of the noise. For instance with a highly positively correlated noise, it is necessary to speed up the Reference Model (decrease α) to decrease the variance on the output. In general it is often advantageous to perform an output filtering, particularly when the output noise level is large. A proper choice of an output filter decreases the output variance while allowing the selection of a faster reference trajectory.

NON-MINIMUM PHASE PLANT

When the plant ho is non minimum phase, difficulties arise in the sense that the convergence of the output to its desired value require inputs of infinite magnitude. Within the framework of MAC, it is relatively simple to come up with a solution which consists of selecting an optimum Internal Model h such that the Euclidean distance of the output to the reference trajectory is minimized while the input energy remains finite. This requires the off-line solution of a Ricatti equation which determines h. (Mehra et al. (1980)). The important concept here is that the internal model used for control calculations is different from the internal model used for prediction calculations.

HIERARCHICAL MAC AND SPT

In a recent application of Singular Perturbation Theory (Mehra et al. (1979)), it was shown that the optimization of systems with multiple time scales leads naturally to a hierarchical control structure, in which the reference trajectories for the lower (faster) levels are specified by higher (slower) levels. This eliminates the arbitrariness regarding the specification of reference trajectories and constraints. A combination of MAC and SPT appears to provide a powerful tool for overall optimization of plant performance through proper coordination and control of different levels.

ASSOCIATIVE CONTROL

For control of complex plants where on-line optimization is not feasible or too expensive, one may rely on the cheap availability of computer memory to store precomputed solutions. The basic idea is to associate the present situation of the plant with a previously observed situation and retrieve previously stored control values or laws. The technique is particularly attractive for those problems in which analytic plant models are not available. The basic principles of MAC are still applicable, but the control computations are done differently. These techniques are currently under development at Scientific Systems, Inc. and Adersa/Gerbios, France.

ALTERNATIVE PREDICTION SCHEMES (MEHRA ET AL. (1980))

For linear systems, the predicted response can be written as the sum of Zero Input Response (y_{pzi}) and Zero State Response (y_{pzs}). The former is based on the assumption that the inputs in the future are zero, whereas the latter represents the response purely due to future inputs. Current implementations of MAC use the impulse response model for computations of both y_{pzi} and y_{pzs}. However, if a state model of the system is available y_{pzi} can be obtained from a State Estimator (Kalman Type) or a Luenberger Observer. This is particularly attractive for continuous time and/or nonlinear systems. Even if a state representation is not available, an ARMA (Autoregressive Moving Average) model may be more suitable for prediction due to its lower order. The main advantage of the Impulse Response representation is in the control computation. Thus it is possible to use two different models viz. a state vector model for prediction purposes and an impulse response model for control calculations. The two models need not necessarily have the same input-output properties since the objective of the prediction model is to produce accurate predictions whereas the objective of the control model is to produce satisfactory, stabilizing and robust control values. In particular, for nonminimum phase systems, the two models are different since the true plant model does not have a stable inverse.

SAMPLE RATE SELECTION

For continuous-time systems with discrete observations, the selection of sampling rate is important. A very high sampling rate is expensive and wasteful of system resources whereas a very slow sampling rate would lead to large excursions of the controlled variables over the sampling interval. A procedure for sample rate selection in MAC has recently been given by (Mehra et al. (1980)). In this procedure, the reference trajectories and the predicted plant outputs are compared on a continuous-time basis, even though the observations and control actions are performed discretely in time. The optimal sampling rate is determined by solving two Ricatti equations, one representing estimation cost and the other representing control cost for a given sampling scheme (Kwaakernak and Sivan (1972)).

CONCLUSION

Model Algorithmic Control (MAC) design is highly flexible and versatile for plant wide process control applications. It makes full use of the current microprocessor capabilities. Field experience confirms its robustness properties. The ability of MAC for handling constraints exactly is not only important from an operational point-of-view, but is also valuable for plant design, actuator selection and the determination of constraints that limit overall plant performance. The major difference between MAC and other classical as well as modern control techniques lies in on-line optimization which permits on-line changes by operators of desired closed loop plant behaviour under changing plant conditions and under emergency conditions.

REFERENCES

Adersa/Gerbios: IDCOM—Conduite algorithmique des processers industriels de fabrication. Proceedings Conference on the Algorithmic Control of Industrial Systems, Paris 1976.
Astrom, K. J., "Self-Tuning Regulators—Design Principles and Applications," in Applications of Adaptive Control, K. S. Narendra and R. V. Monopoli eds., New York: Academic Press, pp. 1–68, 1980.
Bordier, M., J. L. Testud and A. Rault, "Commande D' Un Systeme A Grand Retard Pur Application A La Commande D' Une Ligne D'Etriage De Tube De Verre" Adersa/Gerbios Tech. Report, Oct. 1979.
Cutler, C. R. and B. L. Ramaker, "Dynamic Matrix Control—A Computer Control Algorithm," Preprints, Joint Automatic Control Conf, San Francisco, Calif., (1980)
Daclin, E., "Application of Multivariable Predictive Control Techniques to a Deminer Boat" August, Proceed. JACC, 1980.
Engrand, J. C., "Applications of Multivariable Control in a Refinery and Implementation on a Dedicated Minicomputer," August, Proceed. JACC, 1980.
Kwaakernak, H. and R. Sivan, *Linear Optimal Control Systems*, Wiley, (1972).

Landau, I. D., "A Survey of Model Reference Adaptive Techniques (Theory and Applications)," *Automatica*, Vol. 10, pp. 353–379, 1974.

Le Bourgeois, F., "IDCOM Application and Experiences on a PVC Production Plant," August, Proceed. JACC, 1980.

Lecamis, F., J. Richalet, "Identification des systemes discrets linearies monovariables por minimisation d' une distance de structure," Electronics Letters, Vol. 4 No. 24 (1968).

Lecrique, M., A. Rault, M. Tessier, J. L. Testud, "Multivariable Regulation of a Thermal Power Plant Steam Generator," IFAC World Congress, Helsinki 1978.

Mehra, R. K., P. Mereau and D. Guillaume, "Flight Control Application of Model Algorithmic Control with IDCOM (Identification and Command)," Proc. IEEE CDC, pp. 977–982, (1978).

Mehra, R. K., R. Rouhani, "Theoretical Considerations on Model Algorithmic Control for Non-minimum Phase Systems," Proceed, JACC (1980).

Mehra, R. K., R. Rouhani, L. Praly, "New Theoretical Developments in Multivariable Predictive Algorithmic Control," San Francisco, Proceed. JACC (1980).

Mehra, R. K., et al., "Model Algorithmic Control Using IDCOM for the F100 Jet Engine Multivariable Control Design Problem," in *Alternatives for Linear Multivariable Control* (ed, Sain et al.), Chicago: NEC, Inc. (1978).

Mehra, et al., "Application of Model Algorithmic Control To Fossil-Fueled Power Plants," Scientific Systems Interim Report for Contract DE-AC01-78ET29328, (1980).

Mehra, et al., "Basic Research In Digital Stochastic Model Algorithmic Control," Scientific Systems Interim Report AFWAL-TR-80-3125, (1980).

Mehra, et al., "Study of the Application of Singular Perturbation Theory," Scientific Systems NASA Contract Report 3167, (1979).

Mereau, P., J. P. Littnaun, "European Transsonic Wind Tunnel, Dynamics Simplified Model," Adersa/Gerbios, 1978.

Meyer, G., "Use of System Inverses in the Design of Flight Control Systems for Aircraft of Nonlinear Characteristics," Preprints JACC, San Francisco, (1980).

Praly, L., "General Study of Single Input Single Output Linear Time Variant Control Laws: Application to an Adapted Model Algorithmic Control (AMAC)," Adersa/Gerbios report, (1979).

Rault, A., J. Richalet and P. LeRoux, "Commands Auto-Adaptive d' un Avion," Adersa/Gerbios, 75/195, (1975).

Reid, J. G., D. E. Ghaffin and J. T. Silverthorn, "Output Predictive Algorithmic Control: Precision Tracking with Application to Terrain Following," JACC, (1980).

Reid, J. G., R. K. Mehra and E. Krikwood, "Robustness Properties of Output Predictive Dead-Beat Control: SISO Case," CDC, Ft. Lauderdale, Fl., (1979).

Richalet, J., "General Principles of Scenario Predictive Control Techniques," August Proceed. JACC, (1980).

Richalet, J., B. Gimonet, "Identification des systemes discrets linearines monovariables par minimisation d' une distance de structure," Electronics Letters, Vol. 4, No. 24 (1968).

Richalet, J., F. Lecamus, P. Hummel, "New Trends in Identification, Minimization of a Structural Distance," Weak Topology, 2nd IFAC Symposium on Identification, Prague (1970).

Richalet, J., A. Rault, R. Pouliquen, "Identification des processus par la methode du modele," Gordon and Breach (1971).

Testud, J. L., "Commande Numerique Multivariable du Ballon de Recuperation de Vapeur," Adersa/Gerbios (1979).

APPLICATION OF MODELLING, ESTIMATION AND CONTROL TO CHEMICAL PROCESSES

E. D. Gilles

ABSTRACT

This paper shall attempt to provide a brief survey of some current research being conducted at the Institute of Systemdynamics and Control at the University of Stuttgart.

The first part will treat the problem of state estimation in a polymerization reactor. Following that we shall discuss a method of model reduction which can be applied to a class of chemically reacting systems with distributed parameters such as fixed-bed reactors. Finally we will study the dynamics and control of an extractive distillation column. It will be shown that model reduction and state estimation allowed us to develop a control system which has been successfully implemented in a large-scale plant. Further experimental studies using a laboratory-type multi-component distillation column will also be outlined in this report.

ESTIMATION OF STATES IN A POLYMERIZATION REACTOR

A CSTR in which a polymerization reaction is performed can be considered an interesting example of the application of state estimation techniques to chemical processes. The product quality of the produced polymer depends largely upon the operating conditions in the reactor. This quality is mainly determined by the chemistry of the monomer and the chain length distribution of the polymer. In more complicated polymerization processes other measurements concerning tacticity, chain branching, sequence distribution, particle size distribution etc. are needed in order to judge the produced product. All of these measurements of polymer product quality can only be achieved with much experimental effort. Often, it is only possible to determine some means; for example those of the chain length distribution. These measurements are not usually continuous in time and show a considerable time lag between the beginning of sampling and the end of the analysis. Because of this, analytical measurements can hardly be used in an on-line supervision or control of the polymerization process.

However, there are some other measuring variables, such as the temperature and the refractive index which are easy to obtain on-line from the process. Together with a process model these measurements can be used in order to estimate the states describing the product quality of the polymer. This will be demonstrated in the course of this presentation.

In the last few years, an increasing number of papers has been published on the modelling of polymerization reactors [1–10]. Models of non-isothermal CSTR's in which a homogeneous free radical polymerization occurs generally have the structure depicted in Fig. 1.1 [11, 12].

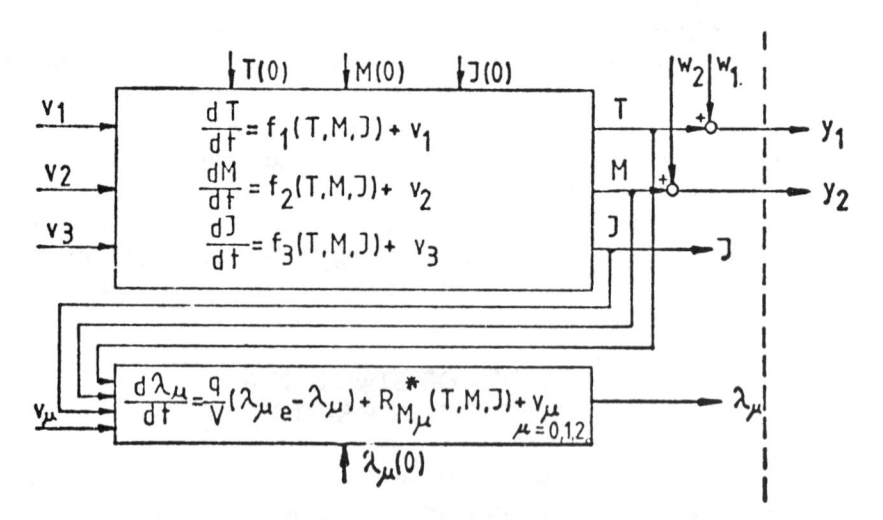

Figure 1.1. Structure of a CSTR-model for polymerization.

The entire system is divided into two subsystems. The states of the first subsystem are the concentrations of the various species such as monomer M and initiator J. and the temperature T. These quantities are obtained from mass and energy balances which can be given in form of non-linear ordinary differential equations. The states λ_μ of the second subsystem are the moments of the chain length distribution. This infinite set of equations is linear and stable. The coupling between the two subsystems is directed from the first to the second. This means that conversion and temperature changes effect the distribution, but the distribution does not retroact upon these states. Using as measuring signals the temperature y_1 and the refractive index y_2 means that the measurements depend on the state of the first subsystem only. Hence, the states of the second subsystem are not observable by means of these measurements. As the second subsystem is stable, the influence of its unknown initial conditions decays exponentially with time. Consequently, one can simulate the evolution of the chain length distribution using the estimated states of the first subsystem. The structure of the state estimation algorithm is shown in Fig. 1.2.

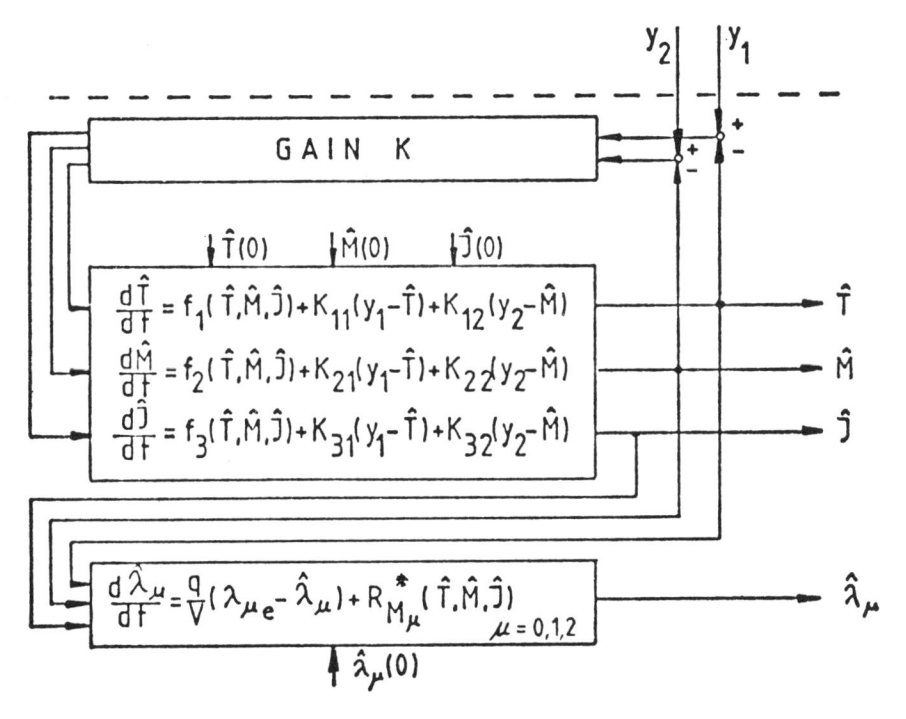

Figure 1.2. Structure of the state estimation.

314   Chemical Process Control II

The gain matrix K of the estimator can be obtained in different ways. Both, non-linear observers and extended Kalman-Bucy-filters [11, 12] have been used for the design. Fig. 1.3 shows the chain length distribution estimated by a non-linear observer. In this estimation only temperature measurements from the process have been taken into account. Fig. 1.4 shows the comparison of Kalman-filter-estimates where the temperature, or both the temperature and the refractive index have been measured.

Fig. 1.4 gives the responses of the states and the estimates resulting from a stepwise increase in the cooling temperature. The quantities $\bar{m}_n$, $\bar{m}_w$ and u denote the number average chain length, the weight average chain length and the polydispersity of the produced polymer. The estimates of these quantities have been improved by the refractive index measurement.

As we have seen, the chain length distribution does not retroact upon the measuring variables $y_{1/2}$. Therefore, the rate of convergence of the estimated distribution toward the distribution of the polymer within the reactor is determined by the parameters of the reactor. It, thus, cannot be influenced by any choice of the gain matrix K.

Modelling errors in the second subsystem lead to a divergence of the CLD-estimation. Therefore a sampling of the CLD at certain time intervals will lend more support to the estimation. The chain length distribution is usually measured with a gelpermeation chromatograph

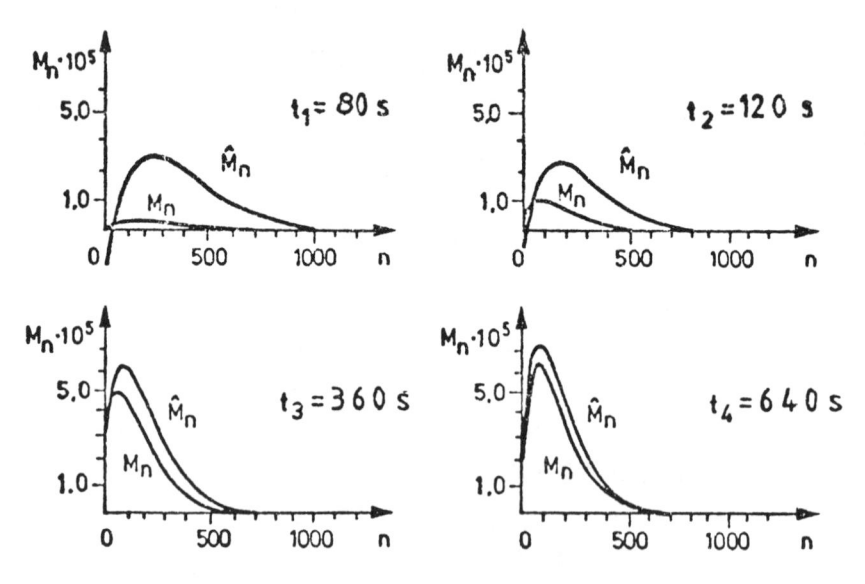

Figure 1.3. Estimated chain length distribution.

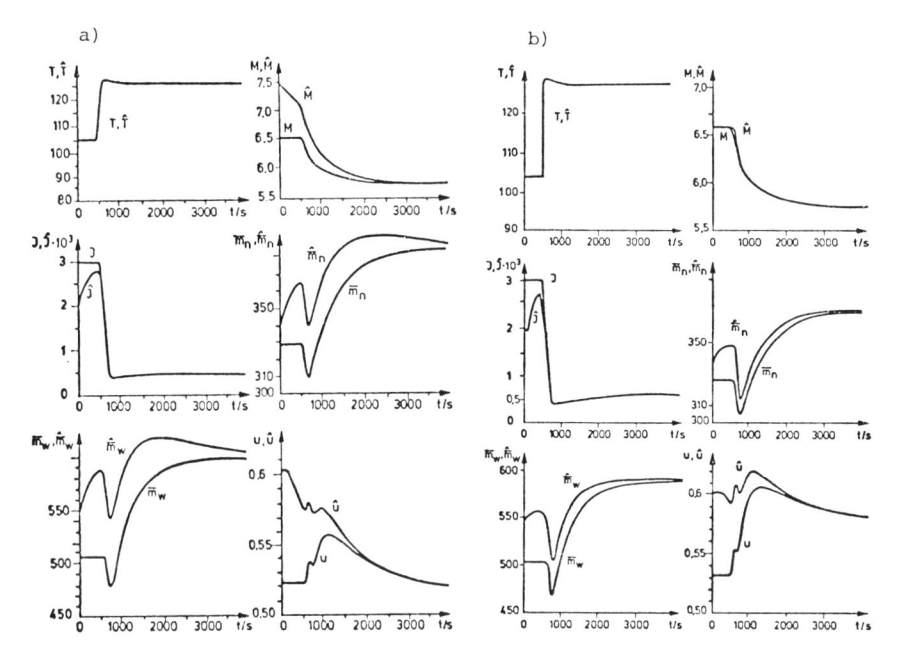

Figure 1.4. Kalman-filter-estimates; a) temperature measurement only; b) measurement of temperature and refractive index.

(GPC). The GPC analysis of a sample will take at least 10 minutes. Thus, the measuring results are obtained with a time delay τ_A. Fig. 1.5 gives the structure of an estimator taking into account the GPC-measurements.

At time $t_1, t_2, \ldots$ samples are taken from the product. These samples are fed into the GPC. The analysis of each sample takes the time τ_A. After that time the filter up-dates the estimation of $\bar{m}_w$. It is self-evident that this technique greatly improves the estimation result, as it is shown in Fig. 1.6.

MODEL REDUCTION OF FIXED-BED REACTORS

A large number of papers published in recent years deal with problems of the mathematical modelling of fixed-bed reactors [13–16].

Figure 2.1 shows the fixed-bed reactor. The gaseous components react on the active surface of the catalyst, where the heat is released in the presence of an exothermic reaction. A rather simple model which enables us to describe the reactor behavior in the case of a single reaction

$$A \rightarrow \text{products}$$

316   Chemical Process Control II

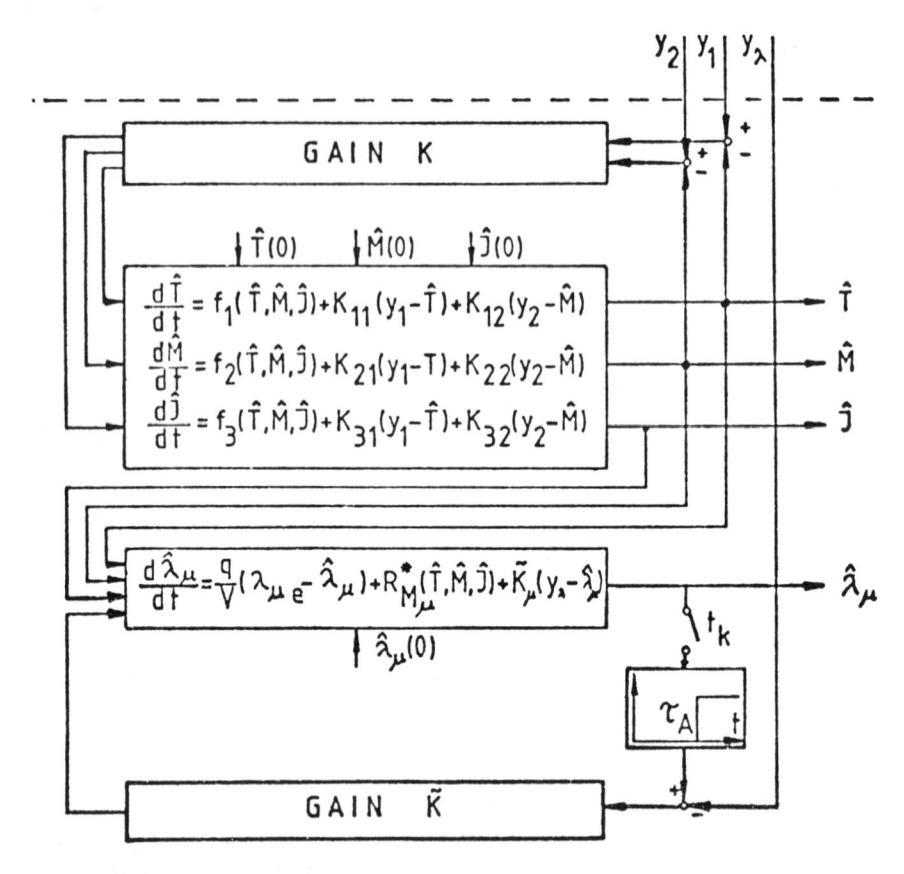

Figure 1.5. Structure of the estimator with the consideration of the GPC-measurement.

comprises a mass- and an energy balance equation. These balance equations read in dimensionless form as follows:

$$\frac{\partial y}{\partial z} = -\text{Da } r(x,y); \quad y(0,t) = y_o \tag{2.1}$$

$$\frac{\partial x}{\partial t} + \frac{\partial x}{\partial z} - \epsilon \frac{\partial^2 x}{\partial z^2} + B(x-x_k) = \text{Da } r(x,y) \tag{2.2}$$

$$x(0,t) = x_o; \quad \left(\frac{\partial x}{\partial z}\right)_{z=1} = 0.$$

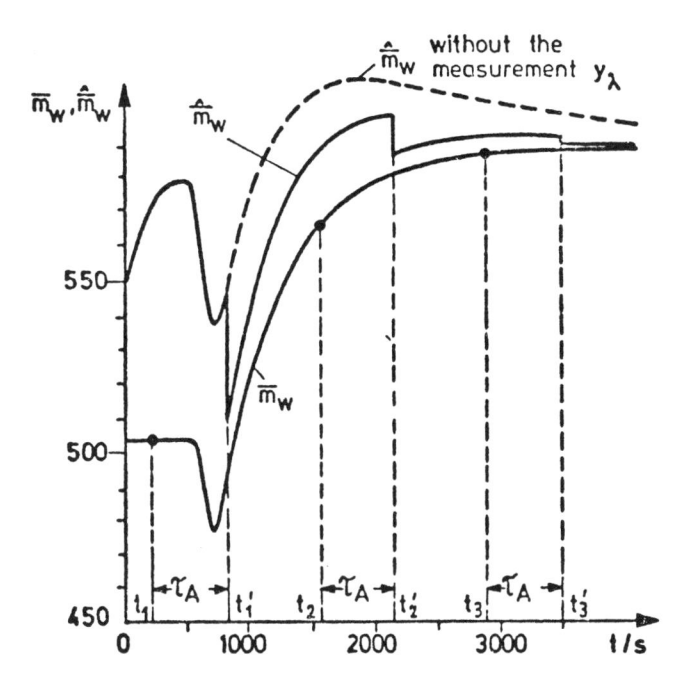

Figure 1.6. Estimation of $\overline{m}_w$ with and without GPC-measurements.

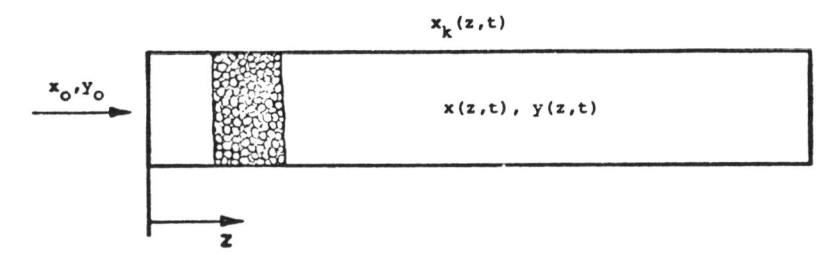

Figure 2.1. Fixed-bed reactor.

The state variables of the reactor are—both dimensionless—the concentration y of the component A and the temperature x. It must be emphasized that the heat capacity of the catalyst is about a thousand times larger than that of the fluid. This difference causes a temperature disturbance to propagate within the reactor at a velocity which is about a thousand times slower than that of a concentration disturbance. It allows us to use only the quasistationary form of the mass balance equation in describing the reactor.

In dealing with the control of the fixed-bed reactor a reduced model is extremely useful on the basis of which the controller can be designed. This also includes the design of observers and filters for state estimation.

In order to achieve such a model reduction, we are going to take advantage of one important property of the reactor. This property is called form stability and means that the principal shape of the reaction rate- and temperature profile remain nearly unchanged during a dynamic process [17, 18].

For the sake of simplicity it is assumed that the reaction rate obeys a first order law, so that

$$r = k_o \, e^{-\gamma/x} \cdot y. \quad (2.3)$$

The reactor behavior as given by equ. (2.1) and (2.2) is interpreted now as the result of the superposition of two subsystems S_1 and S_2. In order to achieve this splitting, we postulate that with

$$x = x_1 + x_2; \quad y = y_1 + y_2$$

the following relations are valid:

$$S_1: \begin{array}{l} \dfrac{\partial x_1}{\partial t} + \dfrac{\partial x_1}{\partial z} - \epsilon \dfrac{\partial^2 x_1}{\partial z^2} + B(x_1 - x_k) = 0 \\[6pt] x_1(0) = x_o; \left(\dfrac{\partial x_1}{\partial z}\right)_{z=1} = 0 \\[6pt] y_1(0) = y_o. \end{array} \quad (2.4)$$

$$S_2: \begin{array}{l} \dfrac{\partial x_2}{\partial t} + \dfrac{\partial x_2}{\partial z} - \epsilon \dfrac{\partial^2 x_2}{\partial z^2} + Bx_2 = Da \, r(x,y) \\[6pt] x_2(0) = 0; \left(\dfrac{\partial x_2}{\partial z}\right)_{z=1} = 0. \end{array} \quad (2.5)$$

$$\dfrac{\partial y_2}{\partial z} = -Da \, r(x,y); \quad y_2(0) = 0. \quad (2.6)$$

The advantage of this splitting must be seen in the fact that subsystem S_1 comprises all input variables affecting the reactor behavior either through the boundary conditions (x_o, y_o) or through the source function (x_k). The subsystem is linear and hence can be solved analytically. The influence of the heat conduction may usually be disregarded in this part of the system, so that S_1 can be considered a pure flow system from which the solution is easily obtained.

The subsystem S_2 represents the autonomous part of the reactor. This part of the system is primarily determined by the internal sources through the reaction and hence exhibits a strongly nonlinear dependency upon the state variables of the reactor. Fig. 2.2 illustrates the splitting of the system. There is only a unidirectional influence of S_1 upon S_2, and this takes place via the reaction rate. The subsystem S_1 can be interpreted as a basis upon which the reaction zone of subsystem S_2 moves up- or downstream. S_1 serves as a connection between the input variables of the reactor and the moving reaction zone.

Our further considerations can be confined to the subsystem S_2. In order to achieve a model reduction for S_2, we use the "form stability" property which characterizes the behavior of the moving zone. According to Fig. 2.3, the space coordinate of S_2 is split into three domains:

(1) the ignition domain: $z \in [0, z_Z]$
(2) the reaction domain: $z \in [z_Z, z_L]$
(3) the extinction domain: $z \in [z_L, 1]$.

In order to explain principally the proceeding of model reduction, the following considerations may be confined to the ignition domain.

With $z' = z - z_Z$ a coordinate system is introduced which moves with the ignition point z_Z. The motion of the ignition point is chosen in such a way as to verify a quasistationary state behavior of the ignition domain in comparison to the processes within the reaction domain. The local profile of the reaction rate within the ignition domain is approximated by

$$\bar{r}^{(1)}(z',t) = r_Z(t) \, e^{\beta_1(t)z'}. \tag{2.7}$$

Due to the form stability, we may expect $\bar{r}^{(1)}$ to be a valid approximation in both, the steady and the unsteady state. Thus, we obtain for the subsystem S_2 within the ignition domain:

$$\frac{\partial x_2^{(1)}}{\partial t} + (1-w_Z) \frac{\partial x_2^{(1)}}{\partial z'} - \epsilon \frac{\partial^2 x_2^{(1)}}{\partial z'^2} + B \, x_2^{(1)} = Da \, \bar{r}^{(1)}$$

$$x_2^{(1)}(-z_Z) = x_{2o} = 0; \quad \left(\frac{\partial x_2^{(1)}}{\partial z'}\right)_{z'=0} = q_{2Z}^{(2)}. \tag{2.8}$$

320   Chemical Process Control II

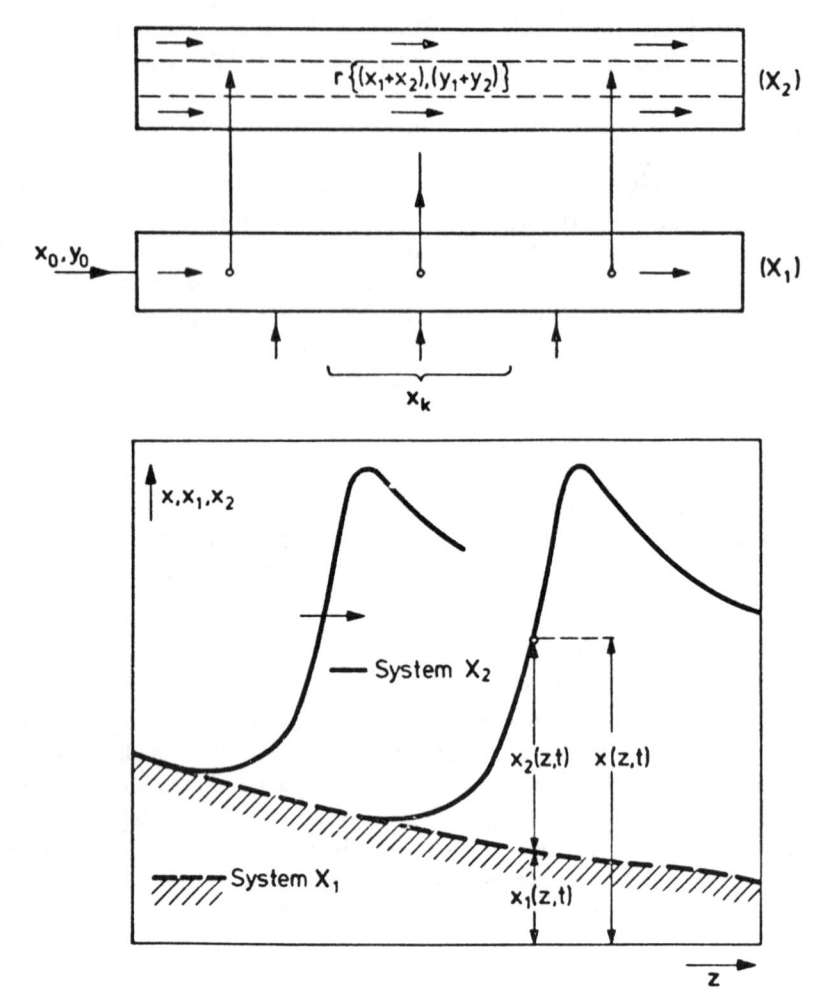

Figure 2.2. Splitting of the reactor into two subsystems $S_{1/2}$.

$$\frac{\partial y_2^{(1)}}{\partial z'} = -Da\, \bar{r}^{(1)}; \quad y_2^{(1)}(-z_Z) = y_{20} = 0. \tag{2.9}$$

Fig. 2.3 shows the corresponding block diagram. The quantity $x_{2Z}^{(1)}$ represents the contribution of S_2 to the ignition temperature x_Z. The input $q_{2Z}^{(2)}$ to the ignition domain takes into account the retroaction of the

E. D. Gilles 321

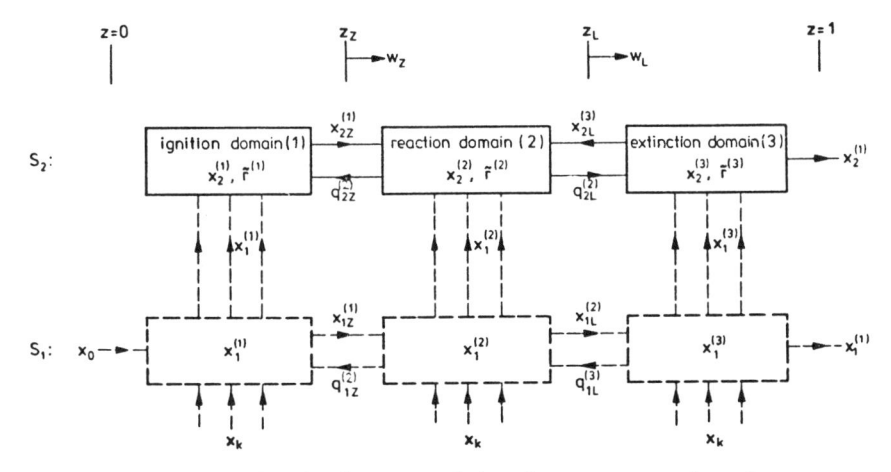

Figure 2.3. Block diagram of the three reactor domains.

reaction domain. This quantity can be considered the part of the temperature gradient which is generated by S_2.

It is advisable to split the variable $x_2^{(1)}$ into a particular integral $\xi^{(1)}$ comprising the influence of the reaction and into a homogeneous part to satisfy the boundary conditions. Thus,

$$x_2^{(1)} = \xi^{(1)} + \varphi^{(1)}. \qquad (2.10)$$

The particular integral is assumed to obey the following differential equation:

$$\frac{\partial \xi^{(1)}}{\partial t} + (1-w_Z) \frac{\partial \xi^{(1)}}{\partial z'} - \epsilon \frac{\partial^2 \xi^{(1)}}{\partial z'^2} + B\, \xi^{(1)} = \text{Da}\, \bar{r}^{(1)}. \qquad (2.11)$$

With

$$\lim_{z' \to -\infty} \xi^{(1)} (z',t) = 0$$

the particular integral is chosen to be

$$\xi^{(1)} (z',t) = \xi_Z^{(1)} (t)\, e^{\beta_1(t) z'} \qquad (2.12)$$

322   Chemical Process Control II

In the quasistationary state $\xi_Z^{(1)}$ can be given as follows:

$$\xi_Z^{(1)} = \frac{Da \; r_Z}{(1-w_Z) \, \beta_1 - \epsilon\beta_1^2 + B}. \tag{2.13}$$

The reaction rate r_Z as a function of the ignition temperature reads:

$$r_Z = \frac{k_0 \, e^{-\gamma/xZ} \, y_0}{1 + \frac{Da}{\beta_1} k_0 \cdot e^{-\gamma/xZ}(1-e^{-\beta_1 z_Z})}. \tag{2.14}$$

In order to determine the time dependent parameter β_1 of the particular integral, we postulate that

$$\left(\frac{\partial \bar{r}^{(1)}}{\partial z'}\right)_{z'=0} = \left(\frac{\partial r\,(x,y)}{\partial z'}\right)_{z'=0}. \tag{2.15}$$

This yields

$$\beta_1 = \frac{\gamma}{x_Z^2} q_Z - Da \, k_0 \cdot e^{-\gamma/xZ}. \tag{2.16}$$

With the particular choice of $\xi^{(1)}$ the following conditions must hold for the homogeneous part of $x_2^{(1)}$ in order to satisfy equ. (2.8):

$$\frac{\partial \varphi^{(1)}}{\partial t} + (1-w_Z) \frac{\partial \varphi^{(1)}}{\partial z'} - \epsilon \frac{\partial^2 \varphi^{(1)}}{\partial z'^2} + B \, \varphi^{(1)} = 0 \tag{2.17}$$

$$\varphi^{(1)}(-z_Z) = -\xi_Z^{(1)} e^{-\beta_1 z_Z}; \quad \left(\frac{\partial \varphi^{(1)}}{\partial z'}\right)_{z'=0} = q_{2Z}^{(2)} - \beta_1 \xi_Z^{(1)}$$

The quasistationary solution of this equation reads:

$$\varphi^{(1)}(z',t) = \varphi_{1Z}^{(1)}(t) \, e^{-\lambda_{11} z'} + \varphi_{2Z}^{(1)}(t) \cdot e^{\lambda_{12} z'}. \tag{2.18}$$

The Eigenvalues can be given as follows:

$$\lambda_{1\frac{1}{2}} = \frac{1-w_Z}{2\epsilon} \left[\sqrt{1 + \frac{4B\epsilon}{(1-w_Z)^2}} \mp 1 \right]. \tag{2.19}$$

In general $\lambda_{12} \gg \lambda_{11}$. The first term on the right hand side of equ. (2.18) serves to satisfy the boundary condition of $x_2^{(1)}$ at the front of the reactor. Hence, we obtain:

$$\varphi_{1Z}^{(1)} = -\xi_Z^{(1)} \cdot e^{-(\beta_1 + \lambda_{11})z_Z}. \tag{2.20}$$

The second term of this equation allows to connect the two spatial domains at $z' = 0$ in an appropriate manner. This means that

$$\varphi_{2Z}^{(1)} = \frac{q_{2Z}^{(2)} + \lambda_{11}\varphi_{1Z}^{(1)} - \beta_1 \xi_Z^{(1)}}{\lambda_{12}} \tag{2.21}$$

comprises the retroaction of the reaction domain upon the ignition domain.

From these relations a global energy balance can be formulated for the ignition domain. It reads in its quasistationary form

$$\epsilon \left\{ \lambda_{12} x_{2Z}^{(1)} - q_{2Z}^{(2)} - (\lambda_{11} + \lambda_{12}) \varphi_{1Z}^{(1)} \right\} = Da \frac{r_Z}{\lambda_{11} + \beta_1}. \tag{2.22}$$

The right-hand side of this balance equation can be interpreted as the heat production curve, while the terms on the left side define the heat removal line.

In order to satisfy the moving velocity w_Z, we proceed from the assumption that the ignition domain is always found to be at its stability- or its ignition boundary. This means that the heat removal line of this domain is assumed to be the tangent to the heat production curve as shown in Fig. 2.4. This condition reads:

$$\epsilon \lambda_{12} = Da \frac{\partial}{\partial x_{2Z}^{(1)}} \left[\frac{r_Z}{\lambda_{11} + \beta_1} \right]. \tag{2.23}$$

In applying this stability criteria both the quantities $q_{2Z}^{(2)}$ and $\varphi_{1Z}^{(1)}$ in equ. (2.2) are assumed to be input variables.

The stability condition (2.23) and the energy balance (2.22) allow us to determine both w_Z and $x_{2Z}^{(1)}$, provided that the temperature gradient $q_{2Z}^{(2)}$ is known. One can obtain the quantity from a corresponding analysis of the reaction domain which, however, is not considered to be in a quasistationary state.

The considerations concerning the extinction domain are very similar to those of the ignition domain. Again, we make use of a stability

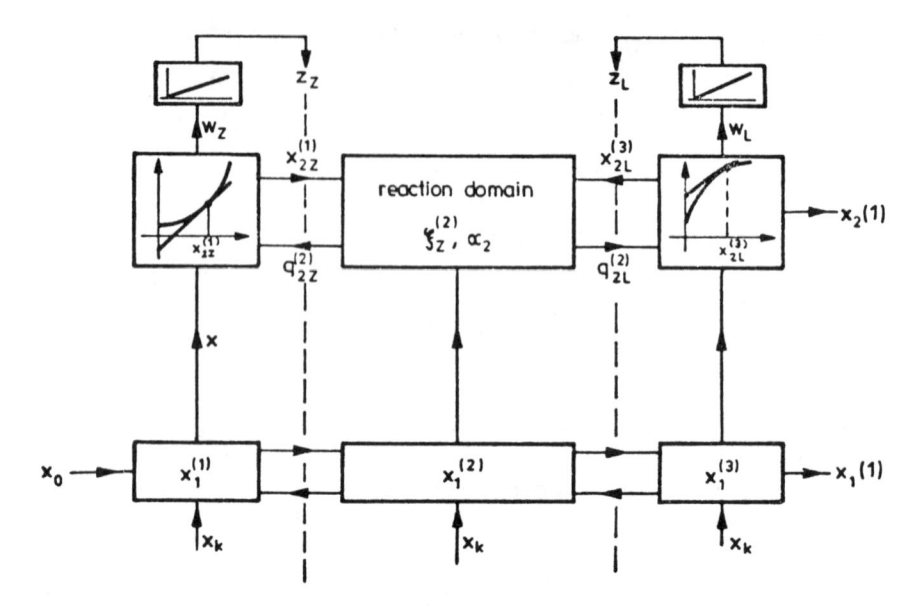

Figure 2.4. Block diagram of the reduced reactor model.

criterium—which in this case is an extinction condition—in order to determine the moving velocity w_L of the extinction point z_L.

Figure 2.4 shows the complete block diagram of the fixed-bed reactor based upon the reduced model. It can be seen that the reaction domain is limited by two moving stability boundaries located at z_Z and z_L. The coordinates of these two points can be considered two important state variables of the reduced model. Two further state variables are used to describe the reaction domain. Thus the reduced model of the subsystem S_2 comprises a total of four state variables only.

The steady state and dynamic behavior of the fixed-bed reactor were simulated by means of the reduced model. Fig. 2.5 shows some steady-state profiles of both the reaction rate approximation $\bar{r}$ and the result r for this quantity obtained from the calculated temperature and concentration profile. For various values of x_o and x_k which are assumed to be equal the profiles of $\bar{r}$ and r are always in good accordance. This means that $\bar{r}$ can be considered an appropriate approximation of the real reaction rate profile.

In Fig. 2.6 a comparison is made between the reduced model and the detailed model for a stepwise increase of the inlet and the cooling temperature. The reduced model needs a little more time to reach the new steady state. It can be seen from these results that the form stability of the

E. D. Gilles 325

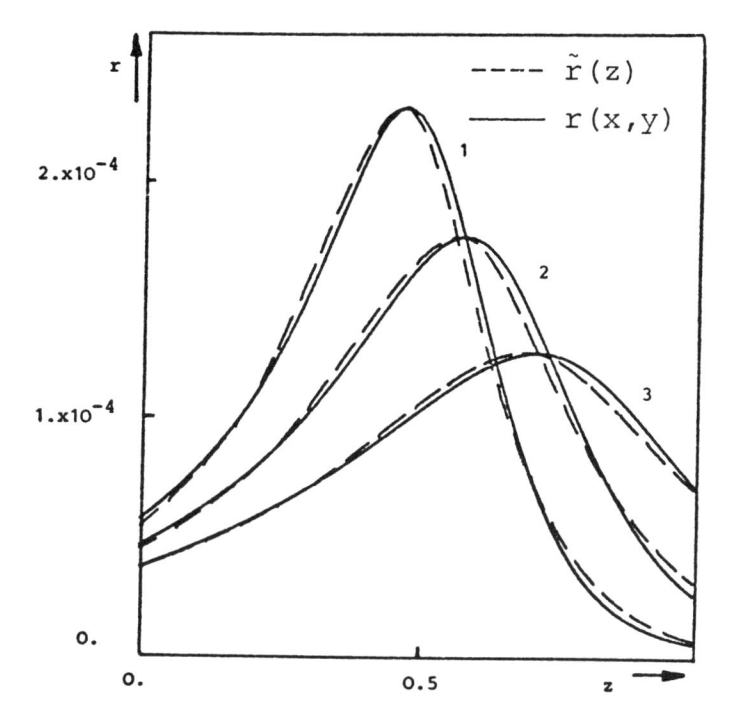

Figure 2.5. Steady-state profiles of $\tilde{r}$ and r.

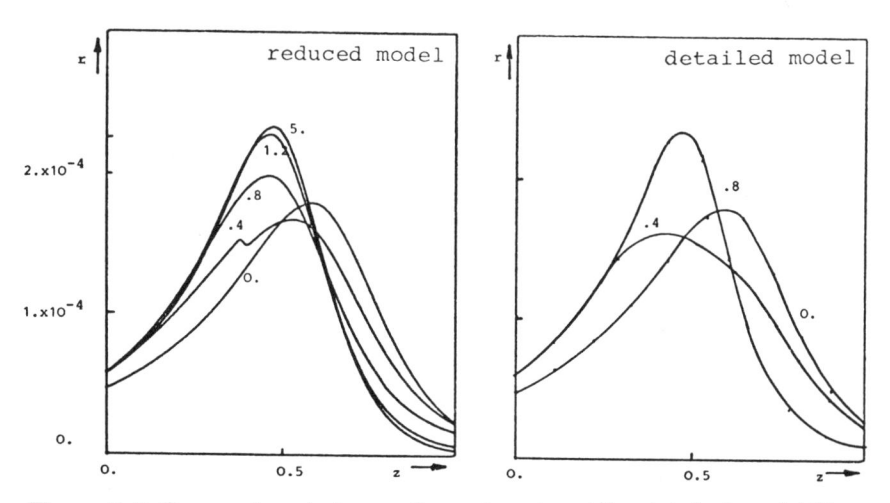

Figure 2.6. Comparison between the reduced and the detailed model Step increase of inlet and cooling temperature.

reaction rate profile indeed is a characteristic feature of the reactor behavior.

The reduced model allows us to describe the reactor behavior at a sufficient accuracy in order to use such a model for the solution of control and filtering problems. A further advantage must be seen in the fact that the state variables of the reduced model are chosen in such a way as to show a weakly non-linear dependency of the input variables. This property supports an optimistic outlook toward linearization and further reduction of this model.

MODELLING AND CONTROL OF DISTILLATION COLUMNS

There are several reasons for which distillation columns are a worthwhile topic of research in the field of modelling and control of chemical systems: They are the most-used separation processes in chemical plants, but due to the complex inherent interaction of countercurrent two-phase flow, thermodynamics and heat- and mass-transfer, the dynamic behavior is not yet fully understood. Despite the great progress in control theory, it should be noted that most of the industrial plants are equipped with empirically-developed control systems. Many years of practical experience have shown the good operating characteristics of most of these units, but occasionally serious problems do arise. These problems could be surmounted and further improvements would be possible, if some more up-to-date concepts such as observers, filters and optimal control could be used at a reasonable cost. The implementation of such control systems, which are based on a state variable representation of the plant equations, requires low-order models for reasons of practicality. Hence the problem of model reduction must be solved.

The investigations done at our Institute were initiated by serious control problems which arose in a large-scale extractive distillation unit of HOECHST AG (Fig. 3.1a). In the column, the binary azeotrop isopropanol/water is separated by using the extractant glycol. A considerably simplified dynamic model for this column was deduced and employed in the design of a new control system. After the implementation, field tests and continuous operation during three years proved the high reliability of the control.

While simulating the steady state and the dynamic behavior of the large-scale column, it was recognized that a binary mixture of water and glycol, containing some 98 mole percent water, can be obtained by using a vapour side stream. This side stream must be located at the stage of the maximum concentration of the intermediate boiling component water

E. D. Gilles   327

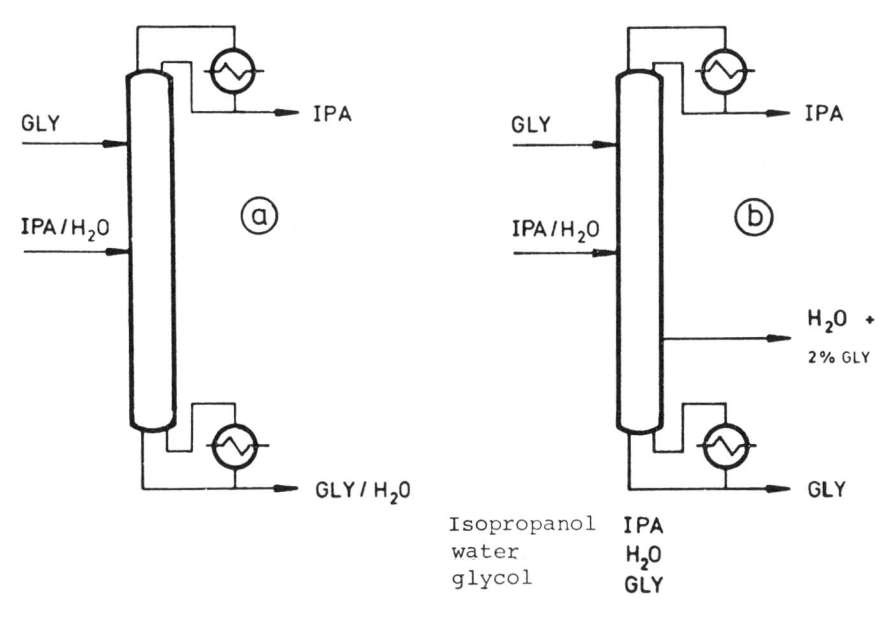

Figure 3.1. Extractive distillation column; a) large scale; b) with vapour side stream.

(Fig. 3.1b). We chose this case to study how the simplified model could be modified and used for the controller design for columns with a side stream.

To support the theoretical investigations and to test models and control systems, a laboratory distillation column was constructed at the Institute.

Mathematical simulation model

Digital simulation based on an appropriate mathematical model is the best tool for understanding the dynamic behavior of a column. We use a set of equations which models every stage as an heterogeneous two-phase system, consisting of the two totally mixed homogeneous phases vapour and liquid (Fig. 3.2) [19, 20].

Constant molal holdup n' and quasi-steady-state behavior of the vapour phase are presupposed. The mass balance equations of the k-th stage are therefore

$$n' \frac{dx_{ik}}{dt} = L_{k-1} x_{ik-1} - L_k x_{ik} - j_{ik} + F_k x_{Fik}; i=1, \ldots, I-1 \quad (3.1)$$

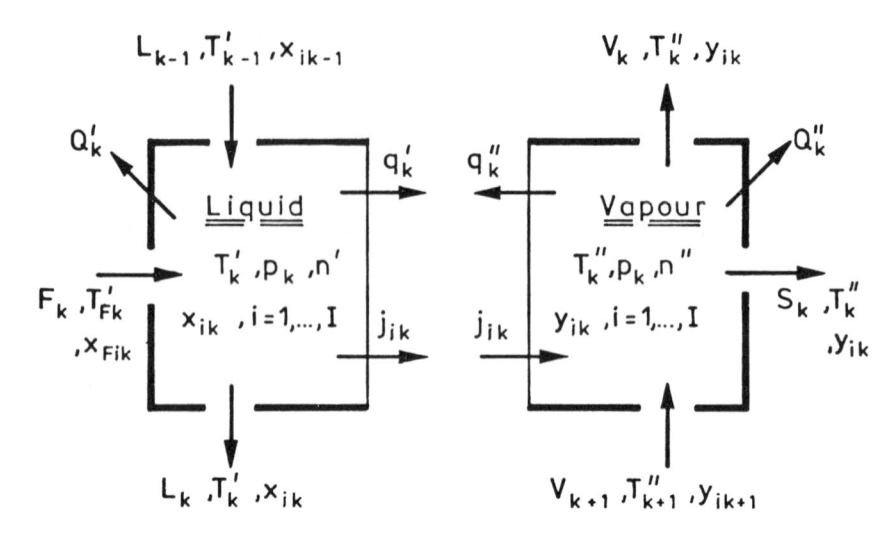

Figure 3.2. Model system of stage k.

$$0 = L_{k-1} - L_k - \sum_{i=1}^{I} j_{ik} + F_k \qquad (3.2)$$

$$0 = V_{k+1} y_{ik+1} - (V_k + S_k) y_{ik} + j_{ik}; \; i=1, \ldots, I-1 \qquad (3.3)$$

$$0 = V_{k+1} - V_k - S_k + \sum_{i=1}^{I} j_{ik}. \qquad (3.4)$$

Further simplifications are introduced to compute the mass transfer rates j_{ik} and the influence of energy on them. The heat transfer is assumed to be perfect and all mass transfer resistances are concentrated on the vapour phase side of the phase boundary. With these assumptions, the stage temperature $T_k' = T_k'' = T_k$ may be computed from the liquid compositions. This so-called boiling-point procedure also yields the equilibrium composition of the vapour y_{ik}^*, $i=1, \ldots, I$. The difference between y_{ik}^* and the bulk composition y_{ik} is considered to be the driving force of the mass transfer. As energy and mass transfer are connected by the energy balance equation of the phase boundary, there is an additional influence due to the energy impressed from the surroundings. Writing r_{ik} for the heat of vapourization, one derives:

$$J_{ik} = \frac{r_{IK}}{r_{ik}} \alpha_I (y^*_{ik} - y_{ik}) + y^*_{ik} \frac{\Delta E_k}{r_{ik}} ; i=1,\ldots,I \qquad (3.5)$$

$$\Delta E_k = L_{k-1} c'_{pk-1} \Delta T_{k-1} - V_{k+1} c''_{pk+1} \Delta T_k + F_k c_{pFk} \Delta T_{Fk} - Q_k \quad (3.6)$$

Here, the ΔT's denote the temperature differences $\Delta T_j = T_j - T_{j+1}$. The loss of heat to the surroundings is summarized by $Q_k = Q'_k + Q''_k$.

The pressure drop Δp_k of a stage is mainly affected by the vapour flow V_k. Taking only this flow affect into consideration, the pressure of the k-th stage is approximately

$$p_k \simeq p_o + \xi \sum_{m=1}^{k} V_m^2 \qquad (3.7)$$

where p_o denotes the top pressure.

In the condenser, the vapour is condensed totally, whereas the dynamic behavior of the reboiler is described by two lag elements. One of them is caused by the heat transfer through the heat exchanger. The other lag element represents the energy balance equation of the reboiler.

With this model, we were able to match measured date at a sufficient accuracy by adjusting the parameters holdup n', the vapour flow resistance ξ and the mass transfer coefficient α_I.

Model reduction and control of the large-scale column

Below the feed F_A, in the steady state temperature profile of the large-scale column (Fig. 3.3), two regions of saturation, separated by a steep increase of temperature, a so-called temperature front, can be recognized. The temperature front is caused by the interphasial exchange of isopropanol and water.

If a step disturbance reduces the feed rate F_A and—by means of a ratio control—also F_E and the reflux R, then the region of exchange and the temperature front are forced to move upward the column (Fig. 3.4). The state variables of all other regions are only slightly altered.

Order reduction means that one has to replace the large number of state variables—i.e., the various liquid mole fractions—obtained by the balance equations, through the introduction of a smaller set of suitably chosen state variables. As the movement of the region of mass exchange isopropanol/water turned out to be the essential dynamic quality of the large scale column, an appropriate choice is apparent: The locus z_1 of the region of high mass transfer has to be chosen as a state variable. A dynamic model that describes the movement of the region of mass

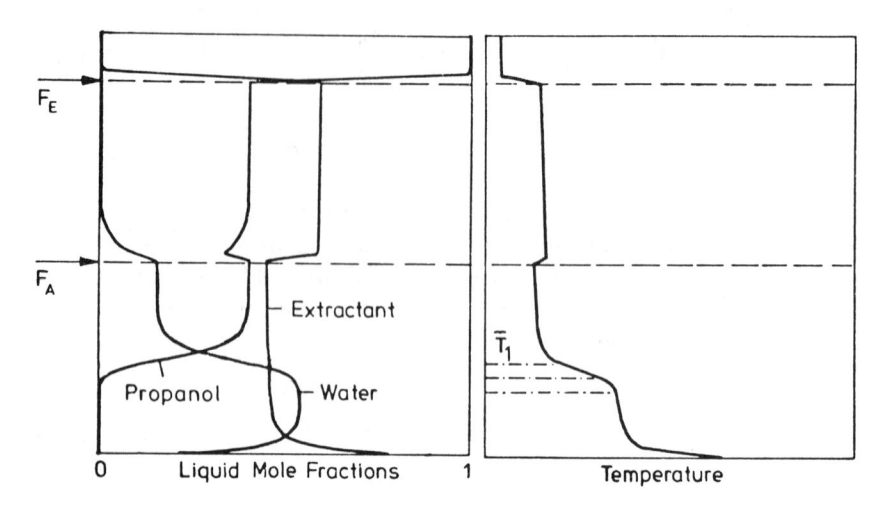

Figure 3.3. Steady state temperature and mole fraction profiles of the large scale extractive column.

transfer can be deduced from the overall balance equations of the lower part of the column. One obtains a simple integrating element (Fig. 3.5).

To complete the simplified dynamic model, we take into account the two lag elements representing the reboiler and a non-linear measuring element which is determined by the shape of the temperature front.

This simplified dynamic model was used to design a state observer and an optimal control. An INTEL 8080 micro-processor was programmed and implemented in the large-scale plant. Fig. 3.6 shows the dynamic response at the closed loop, if a step disturbance of the feed rates occurs (same amplitude as in Fig. 3.4). Due to observer and optimal control, the maximum of deviation of the region of mass transfer is about half a plate and the product specifications are within the prescribed limits.

The new control configuration has been in continuous operation for nearly three years. It has proven itself to be highly reliable and efficient [19, 20]. The former difficulties have been surmounted.

Model reduction and control of the column with side stream

From the column with a vapour side stream, the pure extractant can be obtained as the bottom product by properly adjusting the flow rates of the heating steam and the side stream (Fig. 3.7).

In the stages below the side stream, the increasing concentration of the extractant causes the temperatures to rise and a second temperature front

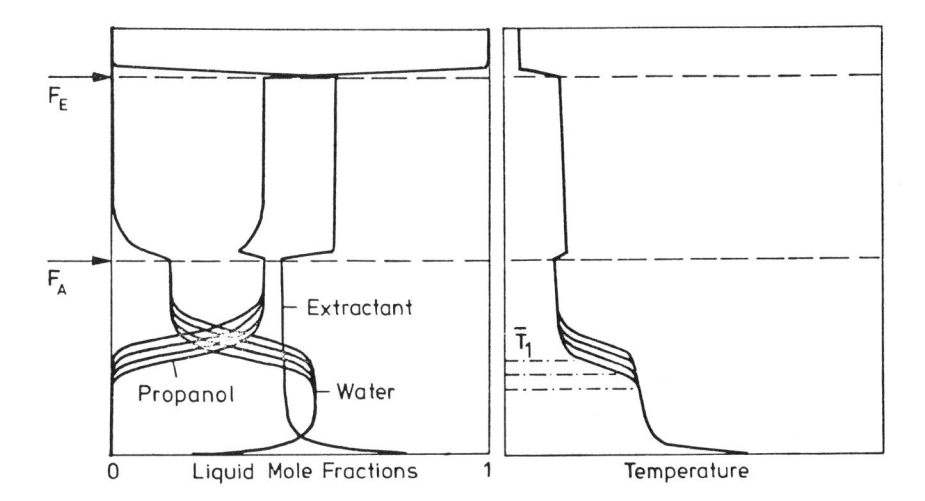

Figure 3.4. Large scale column: step decrease of feed rates.

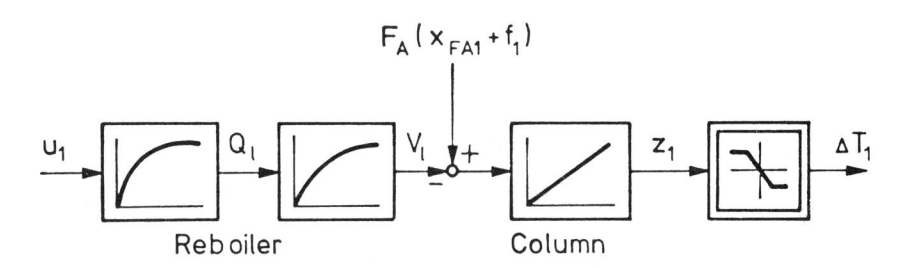

Figure 3.5. Reduced model of the large scale distillation column.

occurs. This front is due to the interphasial exchange of the extractant and the water.

A stepwise reduction of the feed rate now forces both regions of high mass transfer and both temperature fronts to move upward in the column (Fig. 3.8).

As in the former case, the moving regions of high mass transfer turned out to be the essential dynamic quality of this column and hence the loci of the regions are suitable state variables of a reduced-order model. Simulation results show, that an observer and an optimal control designed by use of the reduced model ensure the product specifications, even if large disturbances occur (Fig. 3.9) [21].

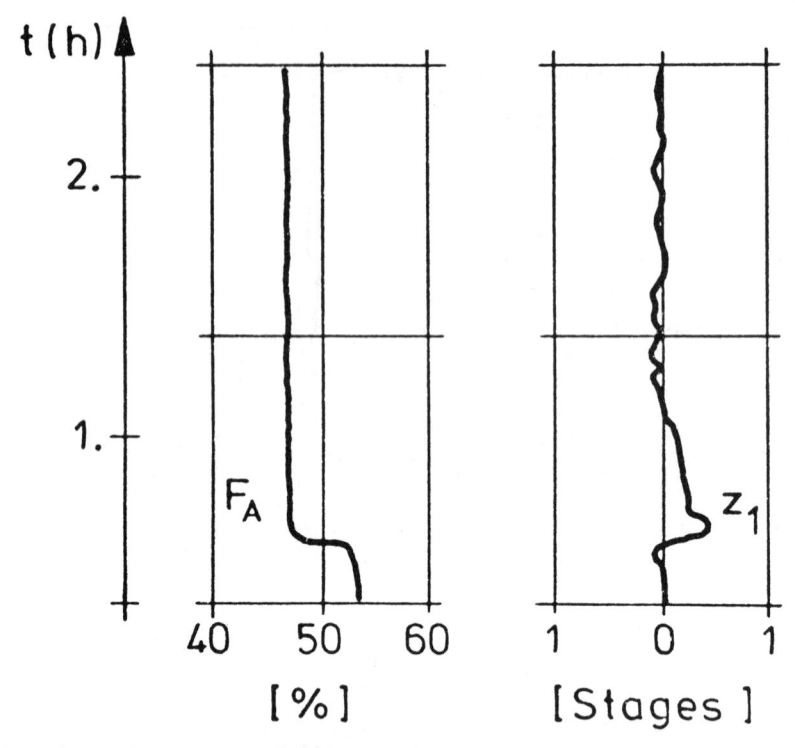

Figure 3.6. Control system based on the simplified model: 10% step disturbance of the feed F_A.

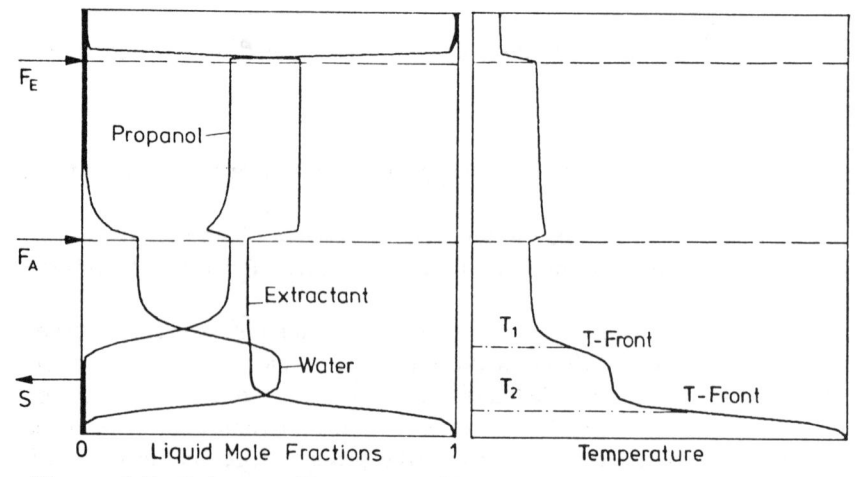

Figure 3.7. Column with vapour side stream: steady state profiles.

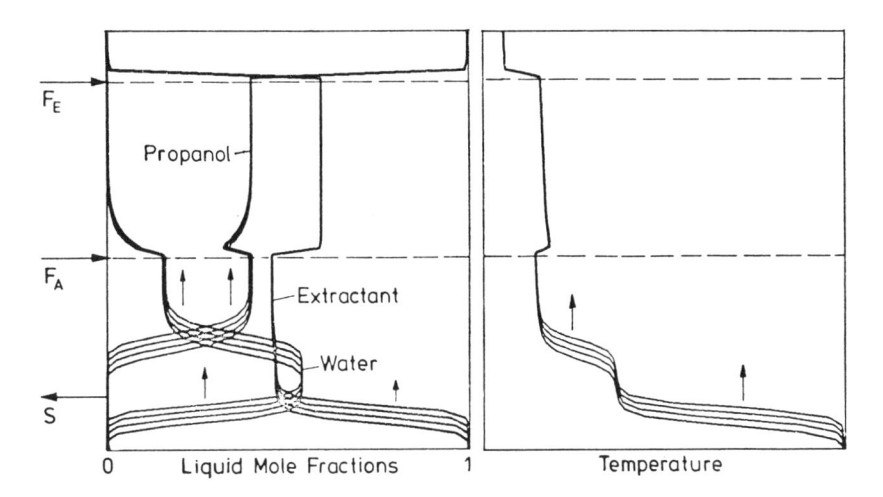

Figure 3.8. Column with vapour side stream: 10% decrease of feed rates.

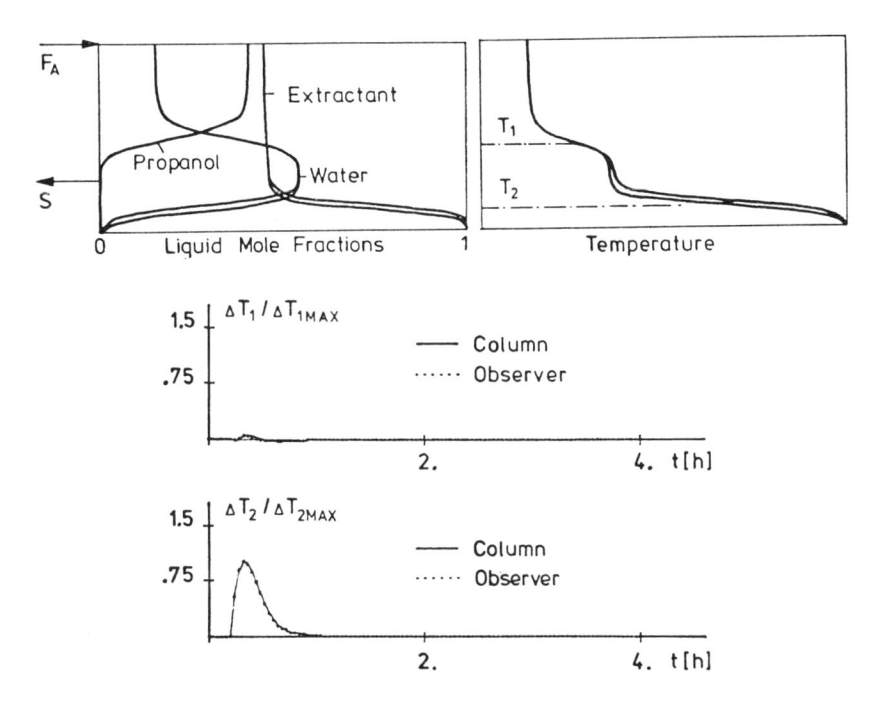

Figure 3.9. Column with side stream, observer and optimal control based on the reduced model: 10% step decrease of feed rate F_A.

Laboratory distillation column

The flow sheet of the laboratory distillation column is shown in Fig. 3.10. The important technical data can be found in the table below.

The column was designed for a multi-purpose use in research and education of students of Control Theory and Chemical Engineering. Two component systems are separated and remixed to allow for a continuous operation. By switching between two feed streams, step disturbances of feed concentration and feed temperature may be achieved. To study the behavior of an extractive distillation column, the additional feed is used for the extractant. A liquid or vapour side stream is installed to conduct the experimental investigation on separations with a side product.

In order to provide the full information about the temperature profile of the column, every tray is equipped with a thermocouple and can be optionally equipped with a device to take liquid or vapour samples.

For routine supervision and secondary control loops, a Z 80 microprocessor is implemented (Fig. 3.11). This microprocessor also drives a scanner to switch between the thermocouples and serves a flow sheet display. A terminal is used to enter values of setting variables and to display current values of measurement variables.

All data stored in the microprocessor may be transferred to a PDP 11/34 process computer and vice versa by means of a serial data line. The PDP is used for storing the measured data and for implementing control systems with high-level FORTRAN.

Current research

Current topics of interest in the field of distillation columns include the generalization of the model reduction procedure and the experimental investigation of column dynamics and control of columns with side products.

REFERENCES

[1] Gilles, E. D., and U. Knöpp (1967), Die Dynamik von Rührkesselreaktoren bei Polymerisationsreaktionen, Teil 1 und 2. Regelungstechnik 15:199–203, 262–269.
[2] Duerksen, J. S., A. E. Hamielec, and J. W. Hodgins (1967). Polymer Reactors and Molecular Weight Distribution-I Free Radical Polymerization in a Continuous Stirred Tank Reactor. AIChE J., 13:1081–1085.
[3] Gerrens, H. (1976). Polymerization Reactors and Polyreactions. A Review. Chemical Reaction Engineering. *4th ISCRE,* Heidelberg 1977 Survey Papers.

E. D. Gilles   335

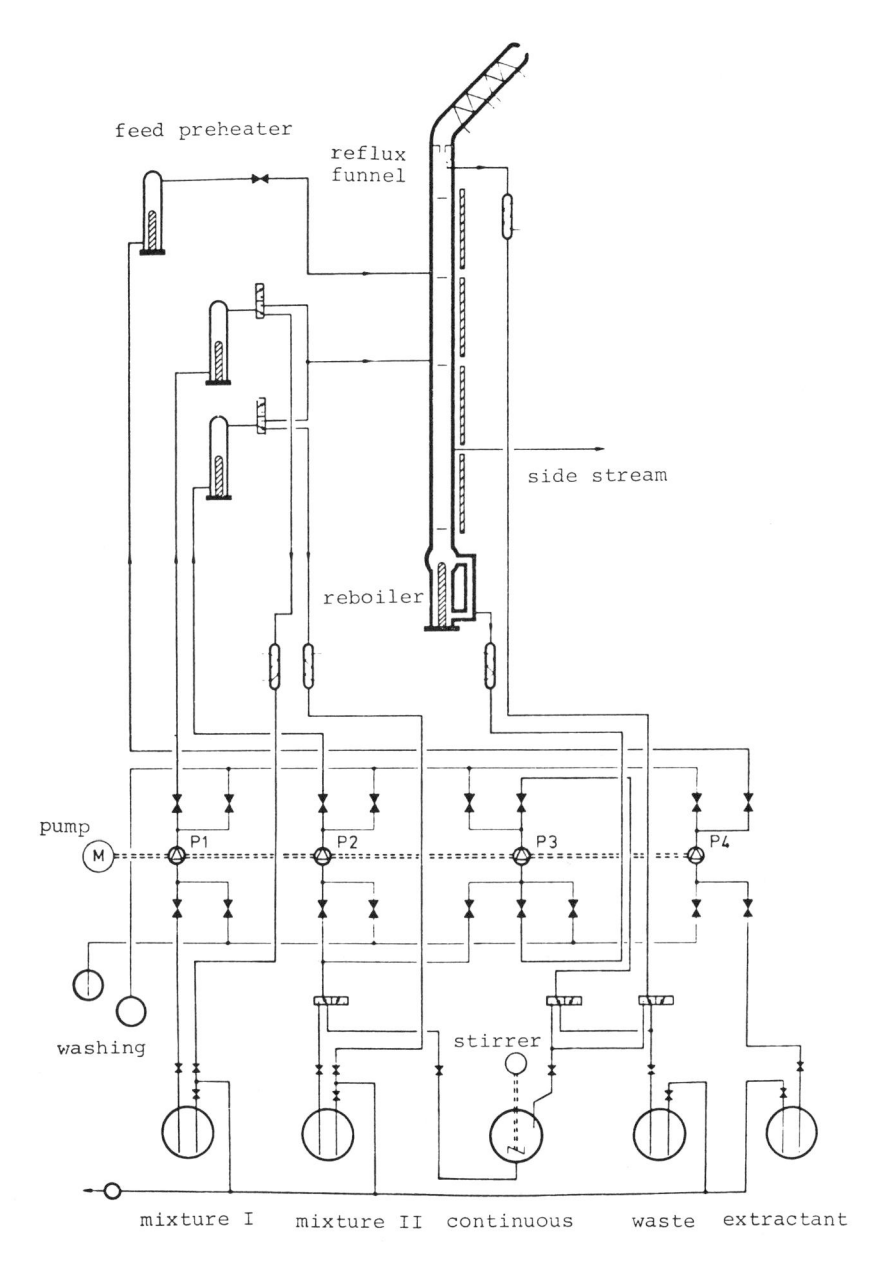

Figure 3.10. Flow sheet of the laboratory distillation column.

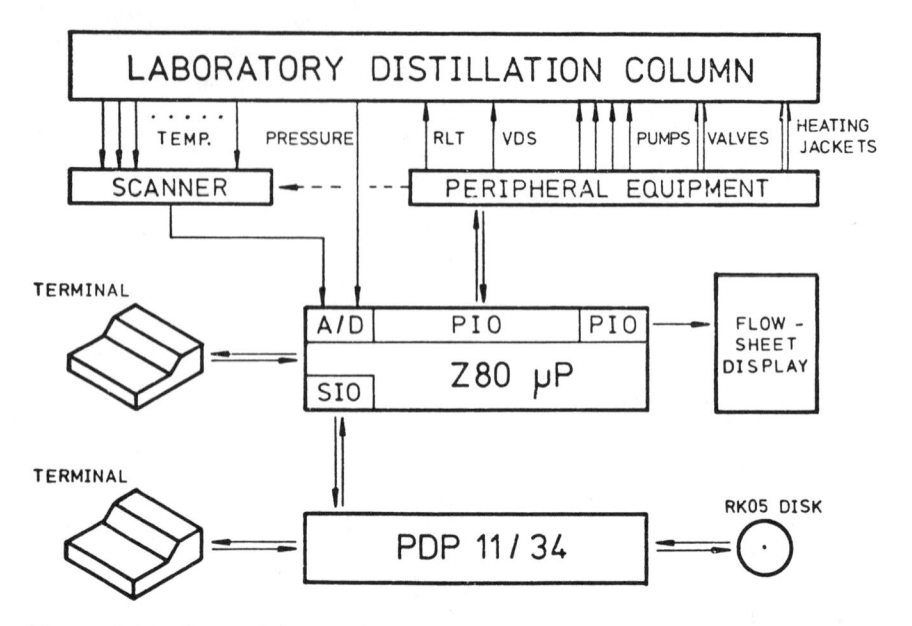

Figure 3.11. Supervision and control system of the laboratory distillation column.

Table 1. Technical data of the laboratory distillation column.

column material	glass
trays	40 bubble cap trays one cap each, efficiency $> 70\%$
column diameter	10 cm without insulation
column height	6.8 m including reboiler and condenser
reboiler	electrically heated O, . . . ,6 kW level control by overflow pipe
condenser	total condensed, water cooled, reflux ratio adjustable by switching funnel
feed preheaters	electrically heated O, . . . ,1 kW
feed pumps	4 adjustable piston pumps
heating jacket	10 sectors independently adjustable

[4] Hui, A. W., and A. E. Hamielec (1972). Thermal Polymerization of Styrene at High Conversions and Temperatures—An Experimental Study, J. Appl. Polymer Sci., 16:749–769.
[5] Jainsinghani, R., and W. H. Ray (1977). On the Dynamic Behaviour of a Class of Homogeneous Continuous Stirred Tank Polymerization Reactors. Chem. Eng. Sci., 32:811–825.
[6] Kenat, T. A., R. J. Kermode, and S. L. Rosen (1967). Dynamics of a Continuous Stirred Tank Polymerization Reactor. I & EC Process Des. Developm., 6:363–370.
[7] Liu, S. L., and N. R. Amundson (1961). Polymerization Reactor Stability. Berichte Bunsenges. (Z. Elektrochemie), 65:276–282.
[8] Ray, W. H. (1972). On the Mathematical Modelling of Polymerization Reactors. J. Macromol. Sci.-Revs. Macromol. Chem. C 8(1):1–56.
[9] Wittmer, P., T. Ankel, H. Gerrens, and H. Romeis (1965). Zum dynamischen Verhalten von Polymerisationsreaktoren. Chem. Ing. Techn., 37:392–399.
[10] Zeeman, R. J., and N. R. Amundson (1965), Continuous Polymerization-Models— Polymerization in Continuous Stirred Tank Reactors. Chem. Eng. Sci., 20:331–361.
[11] Schuler, H. (1978). An Observer for the Chain Length Distribution in Polymerization Reactors. 6th International Congress of Chemical Equipment Design and Automation (CHISA'78), Praha, August 21–25, 1978, Paper K 4.3.
[12] Schuler, H. (1980). Estimation of States in a Polymerization Reactor. IFAC prp 4 Ghent, Belgium.
[13] Schmitz, R. A. Multiplicity, Stability and Sensitivity of States in Chemically Reacting Systems. Adv. Chem. Ser. 148, (1975).
[14] Ray, W. H. (1976). Bifurcation Phenomena in Chemically Reacting Systems. Proceedings Advanced Seminar on Applications of Bifurcation Theory. Math. Res. Center, Univ. of Wisconsin.
[15] Aris, R. (1975). The Mathematical Theory of Diffusion and Reaction in Permeable Catalysis. Vol. I and II, Clarendon Press, Oxford.
[16] Gilles, E. D. (1976). Reactor Models. Proceedings 4th Intern. Symp. on Chemical Reaction Eng., DECHEMA.
[17] Gilles, E. D. and Ruppel, W. Model Reduction of Chemically Reacting Systems in "Dynamics and Modelling of Reactive Systems." Academic Press 1980.
[18] Gilles, E. D. and Epple, U. Non-linear wave propagation and model reduction of the fixed-bed reactor. Workshop on Modelling of Chemical Reaction Systems, Heidelberg 1980 (to be published).
[19] Silberberger, F. Modellbildung, Simulation und Regelung einer Extraktiv-Destillationskolonne. Dissertation, Universität Stuttgart, 1979.
[20] Gilles, E. D., Retzbach, B. and Silberberger, F. Modelling, Simulation and Control of an Extractive Distillation Column. ACS Symposium Series, No. 124, 1980.
[21] Gilles, E. D. and Retzbach, B. Reduced Models and Control of Distillation Columns with Sharp Temperature Profiles. Preprints of the 19th IEEE Conference of Decision and Control, Albuquerque, 1980.

"WHAT NEW CONTROL STRATEGIES ARE NEEDED FOR INDUSTRIAL CONTROL?"

Round Table Discussion

Moderator: W. Harmon Ray—University of Wisconsin—Madison, Wis.

Panelists:
  Raymond H. Ash—Proctor & Gamble Co.—Cincinatti, Ohio
  Karl J. Astrom—Lund Inst. of Technology—Lund, Sweden
  Page Buckley—Dupont Co.—Wilmington, Delaware
  Magne Fjeld—Accuray Corp.—Columbus, Ohio
  John J. Haydel—Shell Development Co.—Houston, Texas
  Raman K. Mehra—Scientific Systems, Inc.—Cambridge, Mass.
  Roger Sargent—Imperial College—London, England
  Dale Seborg—University of California—Santa Barbara, Calif.

OPENING STATEMENTS BY THE PANEL:

Dale Seborg:
  Much of the process control research to date has been concerned with taking developments from electrical engineering and the aerospace industry and attempting to apply them to process control problems. Although it is clearly desirable to evaluate new developments in other fields, it is also necessary for process control researchers to look at their own problems, identify unique attributes of these problems, and then develop creative new solutions. Examples of the types of problems that should be considered include constraints in process control, and adaptive control problems where one has some knowledge of the process such as time-

varying gains or dynamics. We also need more emphasis on control strategies for large systems, for instance, plant-wide control systems.

Roger Sargent:

Let me start with an example. In our laboratories we have a pilot-plant crystallizer which exhibits a nonlinear limit cycle. Simulation of the plant showed that this could be exploited to double the throughput compared with steady-state operation, but we were frustrated in demonstrating this experimentally by a more mundane problem. Unless we maintained a liquid level in the evaporator's cyclone separator, the downpipe dried out and was blocked solid with crystals in a few moments. Measuring the level of a swirling slurry of crystals in saturated solution is not easy. We solved the problem by using a self-tuning regulator, estimating the down-flow from easier-to-measure flow-rates and a simple mass and energy balance model. Only after this could we get on to studying the optimal throughput problem.

In most plants we are faced with problems at two levels like this: the simple problems of regulating pressures and liquid-levels to enable the plant to be run at all, and the larger problem of adjusting operating conditions to meet production targets. There is even a third level, to do with monitoring possible hazards and automatic fault detection, and we certainly need new strategies to deal with this area for we know very little about it.

The PID controller is the classical answer to regulatory problems, and I am sure we shall go on using PID controllers, but they really are an anachronism with present-day computing capability. We are most of the way to a better solution with the Astrom self-tuning regulator, but it needs to be more robust. Progress on robustness is coming about through the use of adaptive filters and the "receding-horizon" controller, and use of the same ideas with state-space models gives a simple self-tuning controller for interacting multivariable problems. Problems remain on dealing with non-minimum phase systems without complicated calculations and on robust self-starting of the controllers.

A word on process models is necessary here. Too often in the past "advanced" control has been rejected because of the time and effort required to build and validate accurate models. But self-tuning regulators do not need such models—and the chemical engineer's process insight can be used to produce very simple nonlinear models which will outperform the standard linear models, which of course are always available as a default option.

At the level of optimal process operation there is a greater economic incentive to use more elaborate models, and today we are a long way from the indiscriminate use of steady-state LP models, which gained their reputation in the early days from oil refinery planning. With better modelling techniques and better computational algorithms we can do orders of magnitude better than only a few years ago, but the work required is still substantial enough to repay more effort in this area. We also have to learn the proper balance between off-line and on-line computation, and more comprehensive and easier-to-use computer packages which integrate process design with dynamic simulation for operability studies are going to be an important tool for the process designers.

Karl Astrom:

I would like to bring up some problems relating to the use of advanced control algorithms which deserve attention. The development of modern control theory has so far largely been concentrated on development of *pure* control algorithms. Very little attention has been given to the operational aspects of the algorithms.

To illustrate what I mean let us consider a simple PID algorithm. You all know what the pure version of the algorithm looks like. You are also well aware that it is necessary to consider a whole range of auxiliary issues like mode switching, reset windup, saturation, limitation, gap, selectors, operator interfaces, etc. After the simple three term controller was conceived it took a considerable time before all these problems were fully understood and solved. It is also well-known that a proper consideration of these operational issues is at least as important for the performance of the controller as the pure algorithm and its tuning.

There is clearly a need to consider the analogous problems for advanced control algorithms. This does not seem too hard to do. I have seen good solutions in specific cases. As far as I know very little has, however, been published in this area. The problem of windup is, for example, clearly related to the problem of resetting the state of the regulator. Concepts and algorithms for doing this are available. The details should however be worked out and published.

Turning to the adaptive algorithms there are many more operational issues to be considered than for simple PID controllers. To start with there are many different ways of using adaptive techniques. Adaptation can be used as a tuning device which is called in on demand. It can be used to build up gain schedules and it can be used in truly adaptive control to continuously update the parameters of a regulator. The operator interface can also be done in many different ways. While a PID controller only has manual and automatic modes a self-tuner can in addition have fixed gain, estimation but no tuning, and tuning modes. A crucial problem is if there should be any adjustments on an adaptive controller. Although it has been a longstanding dream of control engineers to have *knobless* black box controllers it is my personal opinion that such a solution is possible only in very special applications. In particular it must at least be possible to tell the regulator what we expect it to do. This leads to adaptive regulators with *performance related knobs*. For process control it is for example possible to design adaptive regulators which have one major dial labeled *closed loop response time*. Such performance related knobs are meaningful and easy to adjust. The concept also offers many interesting possibilities for design of new systems.

In summary I think that there are many operational issues related to the use of advanced control algorithms which deserve attention. They may in fact be crucial for the acceptance of advanced control.

Ray Ash:

Process design practice puts in huge tanks to eliminate problems. Chemical engineers could save huge amounts of capital if they were better able to handle feedforward control and integrated processes. Many of these intermediate storage tanks could be eliminated.

80–90% of control loops present no problem. These are tuned in the field by operating personnel without the use of any optimization techniques. It may be that we could do a lot better. We do worry, for example, about temperature control on exothermic reactors.

The tools one needs to develop for process applications include feedforward control, deadtime compensation, double and triple cascade loops, etc. These are now easy to implement at no extra cost using the computer systems currently available. However, practitioners are frightened by this power. They do not know how to tune these more complex loops.

I have a challenge to academics for the self-tuning regulator. We require a simple-minded approach to self-tuning. All that is needed is first or second order deadtime compensation systems and tuning of parameters. It is probably desirable to have a button that says "tune." This is a hardware and design packaging problem but a solution could provide better control than the simple PID. Why not have a simplified self-tuning regulator which shows constants? The algorithms are very similar to standard PID controllers.

John Haydel:

The "gap" seems to be a popular discussion item at this conference. Is this the gap between theory and practice? Is it between academe and industry? Is it between control and computer engineers? There seem to be several gaps which are impeding progress in process control. Computer process control requires a holistic approach. Professionals working in this field are very much part of a system, which requires an overview beyond these gaps.

The basic control problem in the chemical and petroleum industries is multivariable with constraints on the inputs and outputs. Instability must be accounted for but it is not the dominant theme, as implied by research in academe. These systems are normally damped, perhaps over-designed. Deadtime is always present.

We have a need to implement huge numbers of control loops using very robust and easy to use control algorithms. This would be facilitated with better identification tools, e.g., a microcomputer which could plug into a process and provide process models directly.

Page Buckley:

There are three areas for improvement in control: multivariable control, identification techniques, and adaptive control. When one looks at the last two, one can't calculate process characteristics in advance. One also has parametric forcing as for example with pH control problems with variable buffering.

Magne Fjeld:

Before we can consider the real control problems, one has to examine the function of the controllers and the personnel involved. Are we examining the problem from the point of view of a vendor, user, high-volume product, or one-shot project. One needs to examine the characteristics of the process, the quality control requirements, etc. before one can define what is really needed. Another important question involves what kind of personnel will be doing the installation and maintenance of the control system? Are the users operators or research engineers?

A major control problem in large integrated plants involves graceful degradation. In this area, one could be much more elegant than the approach used by current systems. We also need good utility software for control system design.

Raman Mehra:

In asking the question, what are the needs for chemical process control, one often gets different answers from different groups. For example, in a nuclear plant control problem with which I am involved, the question arose is there a need for advanced control strategies? The study produced two responses: users said yes, vendors said no. The vendors only agreed that there were some control opportunities.

There is a tremendous need for emergency control. This is not usually thought of as feedback control but it has similar characteristics. There is no doubt that one would encounter major non-linear conditions.

There are major problems in control structure including input/output pairing, hierarchial levels and structure, and information reliability.

GENERAL DISCUSSION

The above comments summarized the panel's presentation. Following these, there was discussion from the audience. A few of the major comments are summarized below:

Jim Douglas (University of Massachusetts):

We have a major shortage of trained people. NSF funding in control would probably increase if industry helped to publicize the shortage and if industry would publish problem areas where additional research is needed.

Page Buckley:

DuPont trains their own people. It takes 2–3 years in the field before a control engineer can make major contributions.

Ray Ash:

There will be unprecedented opportunities for process control engineers in the 80s. By contrast, in the 70s there was reluctance by companies to use new control systems because they did not trust the hardware. Chemical engineers in process control are obtaining the highest salary offers presently. There is a very large demand. We have a need to put current theory to work and the 1980s should provide that opportunity.

John Rijnsdorp (Technical University of Twente, Netherlands):

A central problem in chemical process operation is product quality control. Usually the variables, which really are important, cannot continuously be measured. As a consequence, the automatic control system deals with easily measurable but less important variables.

A solution is inferential control: design the automatic control system for reducing fluctuations in product qualities. However, operators do not like this because the variables shown in the control room show large fluctuations.

Another problem is the integrity of sophisticated control algorithms. Automatic control should be maintained when measuring devices become unavailable, control valves become saturated, or when the operator takes over control functions.

Distillation Column Control

CONTROL OF HEAT-INTEGRATED DISTILLATION COLUMNS

Page S. Buckley
Engineering Department
E. I. du Pont de Nemours & Co.
Wilmington, Delaware 19898

ABSTRACT

Heat integrated columns, those with energy recovery features, are more difficult to operate than conventional columns and present challenges to the control engineer. Experiences on several recent projects will be reviewed, and suggestions offered for future installations.

INTRODUCTION

Distillation is a very energy intensive unit operation. It is not surprising, therefore, that in the present era of rapidly increasing energy costs there is an increasing emphasis on both energy conservation and energy recovery. The main thrust of this paper is on the latter, but it is pointless to recover energy unless we also try to conserve it. Consequently, we will discuss conservation at least briefly.

ENERGY CONSERVATION

For distillation, this means designing and operating a column so that it makes the specified separation with the least amount of energy per pound of feed. To accomplish this we have a number of techniques:

a. Automatic control of composition of product streams. Operators commonly over reflux conventional columns with a single top product and a single bottom product. Extra heat is used to ensure meeting or exceeding specified product purities.

   Geyer and Kline [1] give as an example, a 70-tray column separating a mixture with a relative volatility of 1.4 and with specifications of 98% low boilers overhead and 99.6% high boilers in the base. If the operator adds enough boilup and reflux to increase overhead purity to 99% and base purity to 99.7%, an increase of 8% in energy consumption results.

b. Feed provided at the proper feed tray. It can be shown that this results in a lower energy requirement per pound of feed than would feeding on any other tray. As feed composition or enthalpy deviates from design values, the optimum feed tray location changes also.

c. Column operation at minimum pressure [2]. Lower pressure usually means higher relative volatility which means that the necessary separations can be accomplished with lower boilup/feed and reflux/feed ratios. Condenser capacity may be limited, however, and the column may flood at lower boilup rates than it would when operating at higher pressure.

d. Use of lowest pressure steam available [1]. In many plants, excess low pressure steam is available which would otherwise be vented to the atmosphere. Where reboiler ΔT might be too small if the steam were throttled, one may use a partially-flooded reboiler and throttle condensate. Since low pressure steam is seldom available at constant pressure or steam quality, pressure and temperature compensation of flow measurements is highly desirable if steam is throttled instead of condensate.

e. In many plants steam traps require considerable maintenance and have significant leakage. The use of steam condensate receivers instead of traps reduces maintenance and steam losses.

f. For vacuum columns, there is some opinion [1] that mechanical vacuum pumps offer energy savings over steam jets.

g. Dry distillation—for columns using live steam it is sometimes economical to switch to a steam-heated reboiler.

h. Insulation—older columns, designed before the energy crunch, can often benefit by new, increased insulation.

HEAT RECOVERY

Energy recovery in a distillation column means, practically speaking, recovering or reusing heat contained in the column product streams, whether they be liquid or vapor. A number of schemes have appeared in the literature. The two chief ones involve (a) "multiple effect" distillation, analogous to multiple effect evaporation, and (b) vapor recompression. But regardless of the scheme, there are five design factors which must be considered:

a. Reserve capacities which may be required:

   extra heating capacity
   extra cooling capacity
   extra distillation capacity

These are important for startups and shutdowns, changes in production rate, changes in feed composition, and changes in product specifications. Generally speaking, however, auxiliary* reboilers and condensers should be avoided if at all possible. Their use increases investment and increases instrumentation and control complexity.

Some users have found that turning auxiliary reboilers and condensers on and off is enough of a problem that they prefer not to do it. Instead, they always maintain at least a small load on these heat exchangers. This obviously wastes energy.

b. Priorities—if recovered energy is to be distributed to several loads, what is the order of priorities?
c. Interactions—elaborate heat recovery schemes are often highly interactive; how is this to be dealt with?
d. Overall heat balance—how to maintain it?
e. Inerts (low boiler) balance—with elaborate heat recovery schemes, this is sometimes a problem. Too much inerts or low boilers will blanket process-process heat exchangers; too little will result in product losses through the vents.

In view of the above, it is probably apparent that control of columns with heat recovery schemes is more difficult than control of conventional columns.

Let us now look at three types of multiple-effect distillation.

---

*"Auxiliary" condensers and reboilers are those installed in parallel with "normal" condensers and reboilers for startup or peak load purposes.

Multiple Loads Supplied by Single Source

Sometimes, as shown by Figure 1, a column which is a very large energy user, becomes the energy source for a number of loads, each of which acts as a condenser. Two methods have been used to allocate the energy to be recovered: (a) throttling the vapor heating medium to each condensor, or (b) operating the condenser partially flooded by throttling the condensate. Some priority scheme must be established for startups and any other occasion when vapor supply is temporarily short.

One method of handling the priority problem is to use overrides and to split-range the various valves involved as shown by Figure 2. The scheme shown illustrates the use of pneumatic devices, but the concept may readily be implemented with some digital or some analog electronic controls. For the six loads in Figure 2, we employ six gain 6 relays. For load 1 which had the highest priority, the gain 6 relay is calibrated to have an output span of 3–15 psig for an input span of 3–5 psig. Load 2 has the next highest priority so its gain 6 relay is calibrated for 5–7 psig input, 3–15 psig output. This continues until the load with lowest priority, load 6, has a gain 6 relay calibrated for 13–15 psig input, 3–15 psig output. At the design stage of such a system, considerable care must be exerted to get a suitable value of controller gain and to get proper valve sizing.

Process-process heat exchangers are commonly designed for very small temperature differences. If vapor throttling is used, it should be recog-

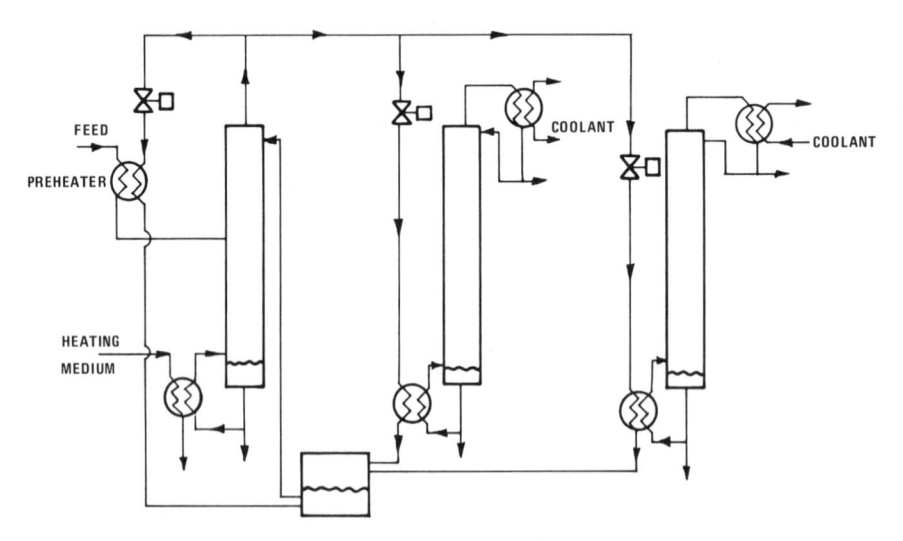

Figure 1. Heat Recovery Scheme—Single Source, Multiple Loads.

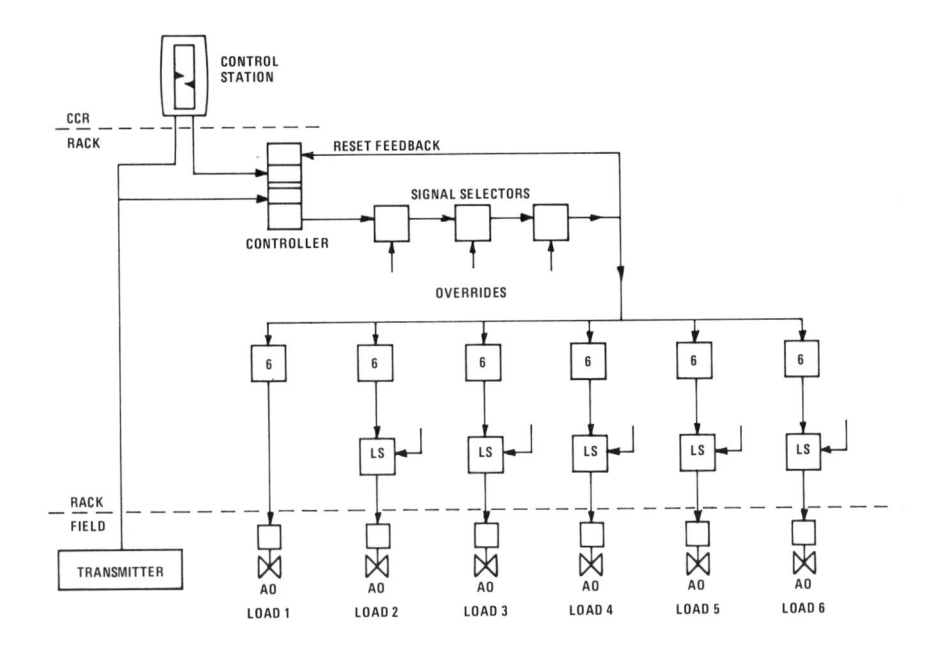

Figure 2. Scheme for Establishing Heat Load Priorities.

nized that the vapor supply valves will tend to have small pressure drops. Accordingly, it is advisable to have vapor flow control to each load, with a setpoint from a primary controller.

In the absence of vapor flow control, interactions may be severe and very close control of supply and load pressures may be required.

Single Source, Single Load

When there is only one source and one load (See Figure 3), contol may be both simpler and more flexible. The column which is the source does not need to be operated at a constant pressure—in the scheme shown, it will find its own pressure. For the illustrative example, the overhead composition of the supply column is controlled via reflux; the base composition of the load column is controlled by boilup in the supply column.

The scheme of Figure 3 has an interesting dynamics problem. The controls must be so designed that changes in vapor flow from the supply column must reach the condenser reboiler at about the same time as feed flow changes from the supply column. If there is a serious discrepancy,

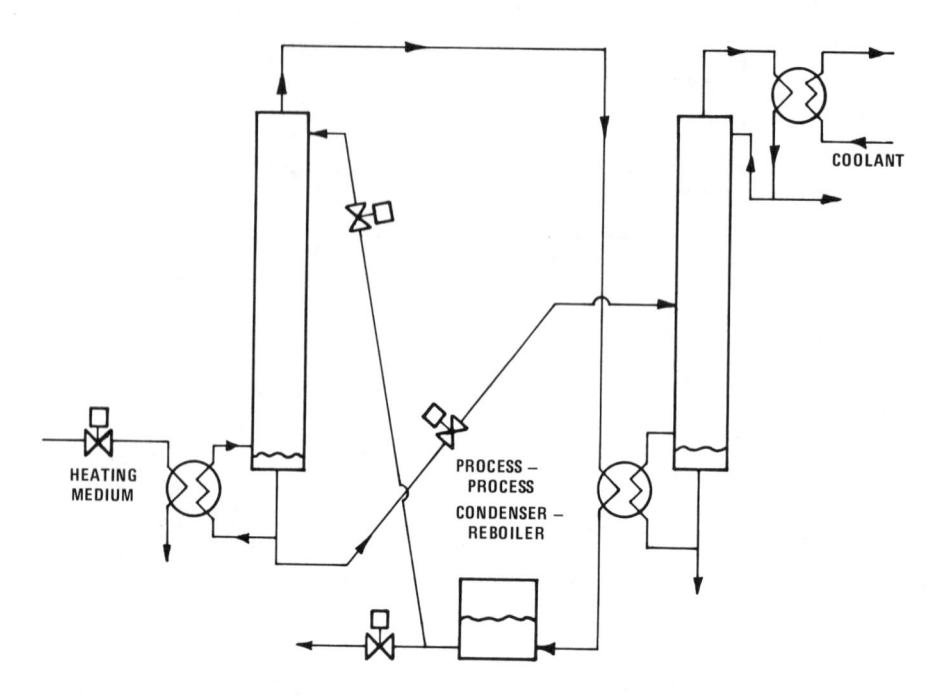

Figure 3. Heat Recovery—Single Source, Single Load—Scheme 1.

particularly if the second column bottom product flow is small, base level in the second column may experience serious upsets.

Another problem associated with this scheme is the selection and sizing of the feed valve to the first column. This column will run at a low pressure at low feed rates and a higher pressure at high feed rates. Assuming that the feed comes from a centrifugal pump, one can see that valve pressure drop will be very high at low flow, and low at high flow. The variation in valve pressure drop with flow will be much greater than that normally encountered in a pumped system.

In another version, a following column in the train supplies heat to a preceding column as shown by Figure 4. In this particular case, the first column gets only part of its heat from the second column; the remainder comes from an auxiliary reboiler. Interactions between the two columns may be severe. Again, for the cases studied, we have found it advantageous to let pressure find its own level in the second column, i.e., the one supplying heat.

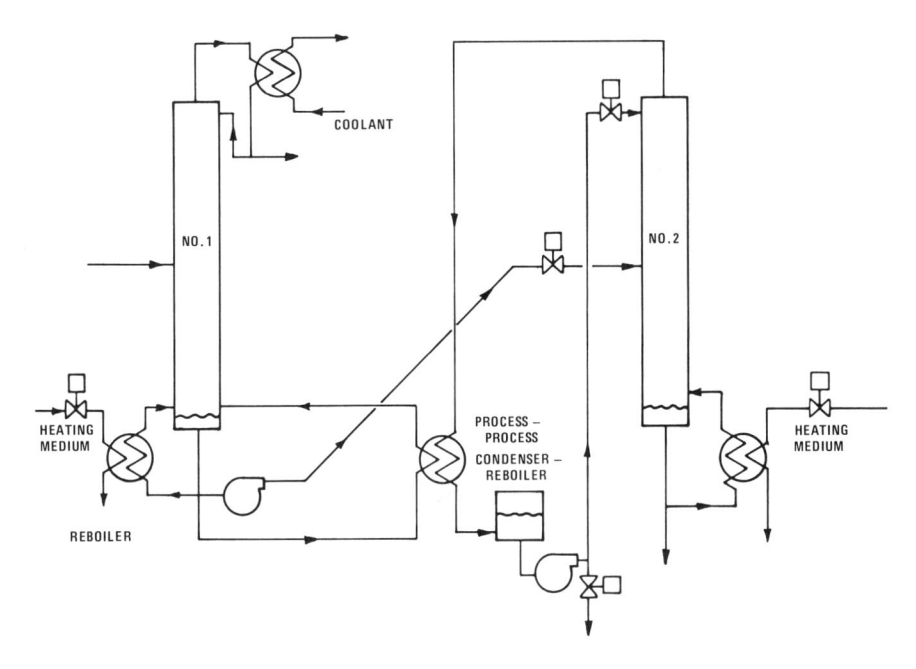

Figure 4. Heat Recovery—Single Source, Single Load—Scheme 2.

An interesting practical problem here is how to adjust the auxiliary reboiler on the first column. After examining some complex heat balance schemes, we decided the simplest approach was to use column ΔP. Vapor flow to the first column from the condenser-reboiler will not be constant, but the ΔP control will provide a rapid method of ensuring constant boilup. The ΔP control in turn may have its setpoint adjusted by a composition controller for the lower section of the first column.

Split Feed Columns

A third arrangement which has its advocates [3] involves splitting feed between two columns which make the same separation (Figure 5). The supply column, however, runs at a higher pressure than the load column. The feed split is controlled to maintain a heat balance.

Combined Sensible and Latent Heat Recovery

In addition to the recovery of the latent heat of vapor streams, in many cases it is practical to recover part of the sensible heat in the column

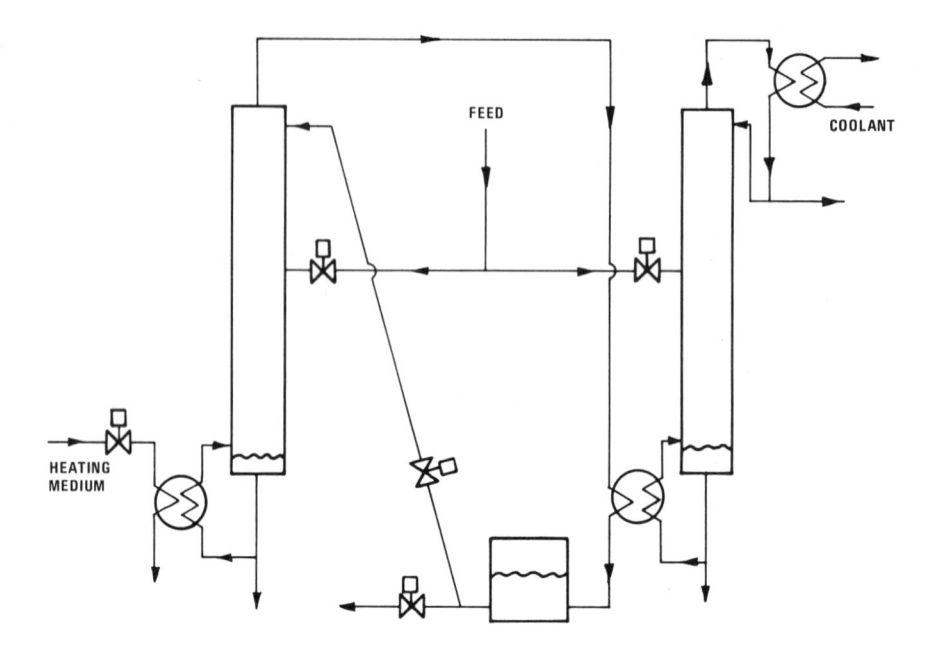

Figure 5. Heat Recovery—Split Feed, Single Load.

bottom product by exchange with column feed and with steam condensate. Such schemes have been used by the chemical and petroleum industries for years. Since feed flow is typically set by level controllers or flow ratio controllers, feed flow rate will not be constant. The feed enthalpy or temperature is therefore apt to be variable. This may make column composition control difficult unless one employs either feedforward compensation or a feed trim heater with control for constant temperature or enthalpy.

Energy Recovery by Vapor Recompression

In the past, vapor recompression has often been considered for distillation of materials boiling at low temperatures. The incentive was to be able to use water-cooled condensers, thus avoiding the expense of refrigeration. Today the interest is centered on getting the column vapor compressed to the point where its temperature is high enough to permit using the vapor as a heat source for the reboiler [4, 5]. An auxiliary, steam-heated reboiler may be necessary for startup, see Figure 6. A

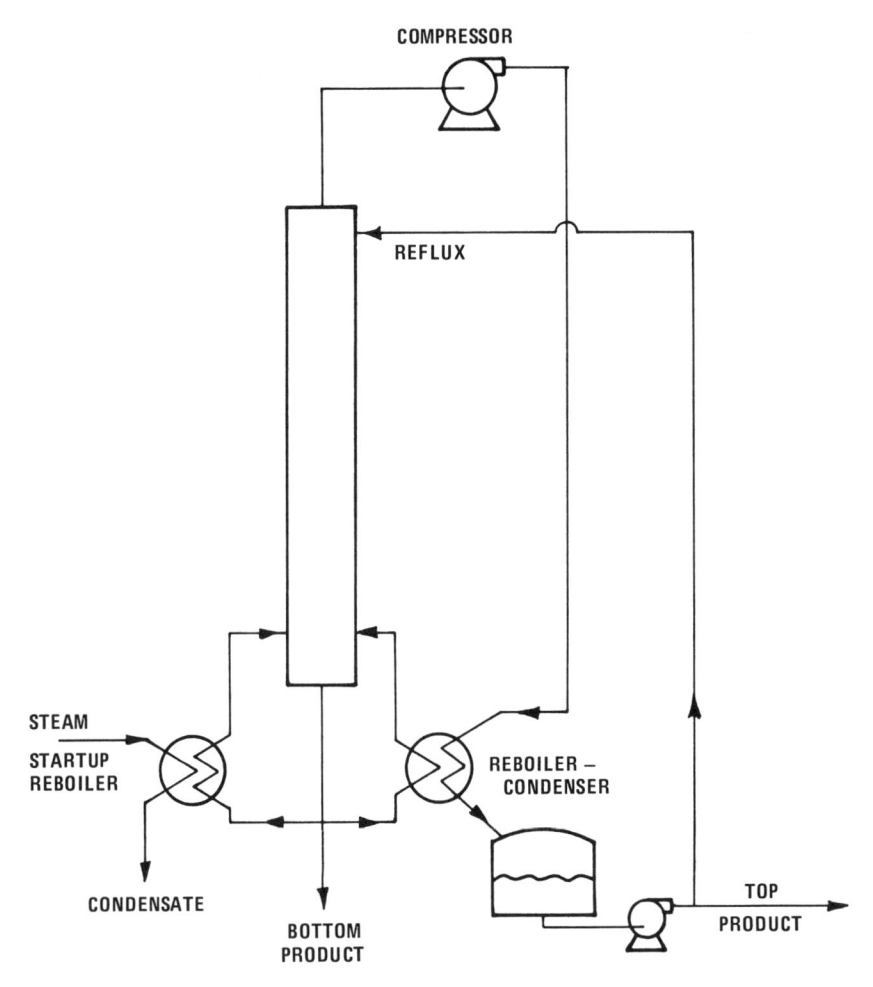

Figure 6. Heat Recovery Via Vapor Recompression.

review of compression equipment and methods of estimating operating costs has recently been presented by Beesley and Rhinesmith [6].

SOME CONTROL CONSIDERATIONS

A basic philosophy of process control has previously been presented in several places [7, 8]. We have used it extensively for distillation control,

whether for individual columns, for columns in distillation trains, or for heat-integrated columns. It has three main facets:
1. Material balance control,
2. Protective controls (overrides and interlocks),
3. Product quality control.

As applied to distillation columns, this philosophy implies control of the composition of each product stream and extensive use of feedforward compensation, usually steam/feed, etc., ratio controls. The implementation, however, often requires dealing with the interactions between composition controls at the two ends of the column, and also composition control of any side draw streams. If we can find suitable methods for dealing with the distressing (from a control engineer's standpoint) nonlinearities and nonidealities of many columns, it may be possible to use advantageously some of the newer technology for multivariable control. As mentioned earlier, dual composition control helps minimize steam consumption. Professor Waller will present at this meeting a paper on dual composition control for binary distillation columns; consequently, we will not pursue it further here.

Regardless of the type of composition measurement, a well known practical problem is that of getting adequate resolution and specificity. The latter is particularly important when compositions cannot be measured at the column ends; inferential measurements must be made at interior trays.

Less well known is the problem of getting adequate resolution of manipulated variables. When a column is so designed that product specifications for impurities are less than, say, 100 ppm, adequate resolution of both measurement and manipulation may be difficult to achieve.

Let us now turn our attention to some practical equipment considerations involving flooded condensers and reboilers.

FLOODED CONDENSERS

The petroleum industry has apparently used flooded condensers for some years, but their use in the chemical industry is probably more recent.

A basic, horizontal condenser with coolant in the tubes is shown on Figure 7. The curved, dotted line is to suggest condensate liquid level. Some of the problems we have encountered include the following:
1. If liquid level gets too high there is apt to be severe hammering in the shell. Wild [4] has suggested that a cure is to introduce a little inerts

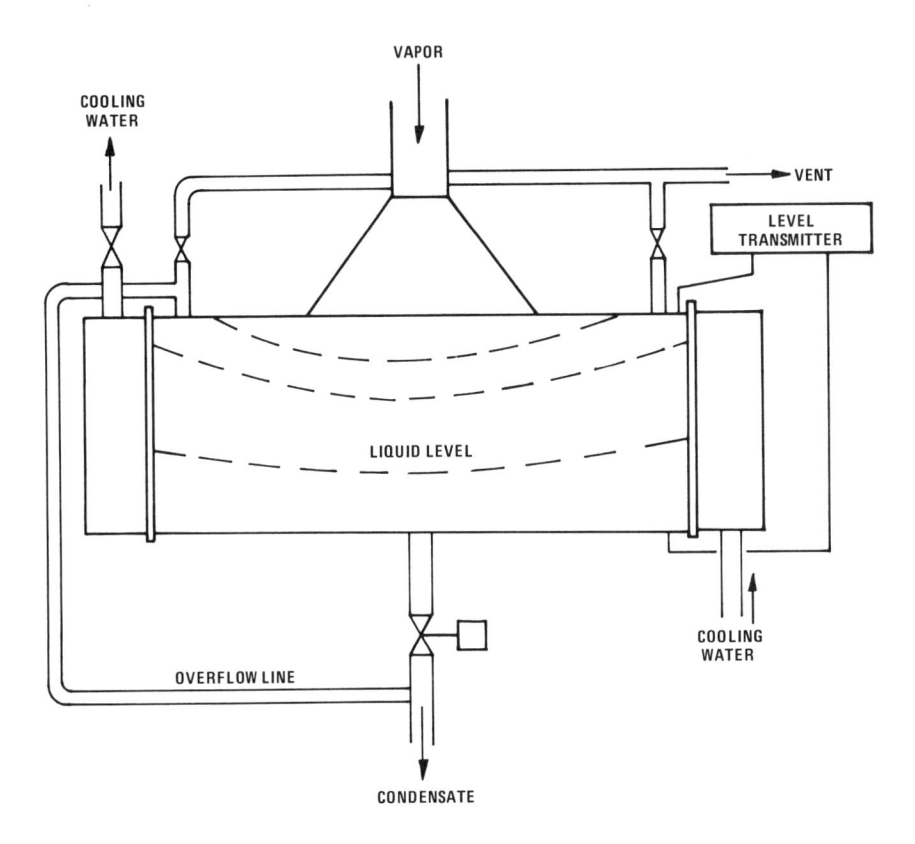

Figure 7. Horizontal Condenser, Vapor in Shell.

into the inlet vapor. Since it is necessary to maintain an inerts balance on the column anyway, one might throttle the vent to control liquid level. This will require a level measurement, and may or may not interfere with column pressure control (if used).
2. For flooded condensers used as auxiliaries which are normally out of service there is the problem of how to turn them off. Some auxiliary water-cooled condensers like those of Figure 7 were originally installed without the valves shown in the vent line. When condensate and cooling water were shut off, large amounts of liquid were entrained in the vent system. Incoming vapor depressed liquid level in the center of the condenser thereby flooding the vent nozzles. A related problem was the occasional flooding of condensate up into the vapor line. Calculations showed that schedule 5

pipe might not take the load, so an overflow from the top of the shell to a point below the condensate valve was installed.

3. Flooded condensers are much more sluggish than non-flooded condensers. A mathematical analysis has shown that response characteristics are roughly those of a first order lag. Derivative compensation has therefore been shown to be helpful.

FLOODED REBOILERS (Figure 8)

Flooded vertical thermosyphon reboilers are being used increasingly in the chemical and petroleum industries. With such reboilers, steam is supplied to the shell at a pressure which is hopefully constant but is often variable. Control is via steam condensate throttling which raises and

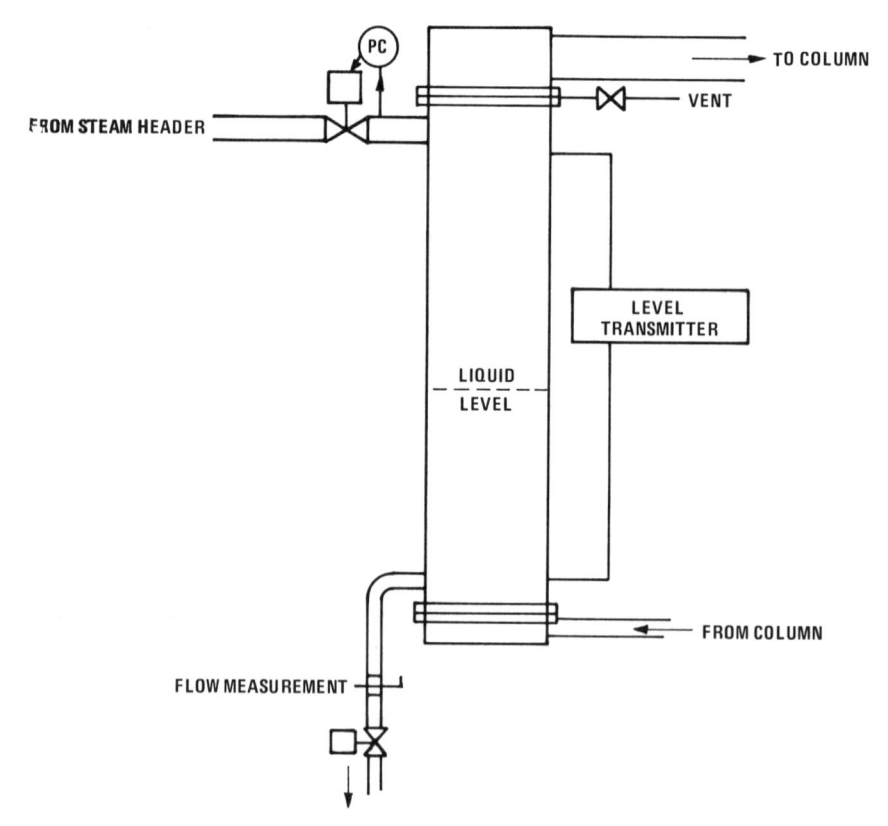

Figure 8. Flooded Reboiler.

lowers condensate level in the shell to expose more or less heat transfer area. Advantages include:

1. Reduced building steel cost—column and reboiler may often be located closer to the ground, particularly if the alternative is a condensate pot with a pump.
2. Lower head condensate pump—suction head is increased compared to that available for steam throttling, and a cheaper pump and drive may be used. Pumping costs are also reduced.
3. Condensate control valves are smaller for a given flow than those used for steam. They are also less likely to be noisy.
4. Simpler and cheaper flow measurement.
5. Easier to make effective use of low pressure steam.

Against these are two disadvantages: (a) one must be careful not to let column liquid base level get too low when reboiler shell liquid level is high, and (b) the flooded reboiler may have a significant lag.

We have derived a mathematical model somewhat similar to that of the flooded condenser which indicates that flooded reboiler dynamics are approximately those of a first-order lag where the time constant is typically 2–6 minutes. Computer simulations indicate that derivative compensation is fairly effective.

Again we have found it useful to put a level transmitter on the reboiler shell.

REFERENCES

[1] Geyer, G. R. and Kline, P. E., CEP May, 1976, pp 49–51.
[2] Shinskey, F. G., "Distillation Control", McGraw-Hill, 1977.
[3] Tyreus, B. D. and Luyben, W. L., CEP September, 1976, pp 59–66.
[4] Rush, F. E., CEP July, 1980, pp 44–59.
[5] Shaner, R. L., CEP May, 1978, pp 47–52.
[6] Beesley, A. H. and Rhinesmith, R. D., CEP August, 1980, pp 37–41.
[7] Buckley, P. S., "Techniques of Process Control", book 1964, John Wiley, NYC.
[8] Buckley, P. S., INTECH, September, 1978, pp 15–122; INTECH, October, 1978, pp 49–53.
[9] Wild, N. H., Chem. Eng., April 21, 1969, p 132.

EXPLICIT VERSUS IMPLICIT DECOUPLING IN DISTILLATION CONTROL

Carroll J. Ryskamp
The Foxboro Company
Foxboro, MA

ABSTRACT

Decoupling can be explicit, with added control functions, or implicit through the judicious use of process and control system knowledge. Explicit decoupling requires accurate quantative loop interaction data, and trained people at the plant site to keep the system updated and in service. Implicit decoupling is used when the interaction is not exactly known, is difficult to determine, or changes with frequently imposed shifts in operating conditions.

A number of methods are used to achieve some decoupling implicitly. Variables can be paired to create loops with minimal interaction. One loop may be fast and the other made slow. Simple nonlinear dynamic elements may be used. Measured and/or manipulated variables can be altered.

When explicit decoupling is used, it should be reduced to a simple subset of computational elements which are added to create stability at an acceptable speed of response.

SITUATION

Decoupling is a term which describes methods used to reduce interaction between control loops. It can be explicit, with added computational elements, or implicit through the judicious use of process and control system knowledge.

The rising cost of energy has focused attention to distillation column operation and control strategies. Higher purity products require higher reflux ratios with corresponding increased energy consumption. Conversely, making less pure products lowers energy requirements. In order to keep products as impure as specifications will allow, to minimize energy cost, on-line analyzers are used in closed loop control. Dual composition control is the closed loop control of both the top and bottom product compositions. These two composition controllers (AC's) will always interact. Any change in the mass balance (Figure 1A or 1B) or the heat input per unit of feed (Figure 2A or 2B) affects both product compositions. The operating data is given on the figures. The column has 50 trays. Tray 0 represents the reboiler and tray 51 the overhead receiver. The curves represent the composition of the liquid leaving the trays for each case. The probability axis is used to better indicate the small changes near the ends of the curves, and it also tends to make straight lines if there is no distortion at the feed tray.

Almost any process variable which is adjusted to force one composition to set point will cause a shift in the other product composition. The degree of interaction, which is unique for each column, is a function of process design, control strategy, product specifications, and operating conditions. Temperature control may be substituted for one or both of the AC's as an inferred composition control. The interaction will be different but still exists. Adequate decoupling is required to keep the loops stable and in automatic control.

Also, efforts to save energy have encouraged the use of heat integrated columns. This practice introduces interaction between the columns, increasing the need for some method of decoupling.

PROBLEM

Complete explicit decoupling requires accurate quantative data for the dynamic and steady state interaction between process variables. If tuning constants are not adequately matched to the process response, the explicit decoupling system can cause more upset by its action than would have occurred without it. Sometimes both AC's must be in automatic, if either one is, because the tuning constants are adjusted for this condition.

C. J. Ryskamp 363

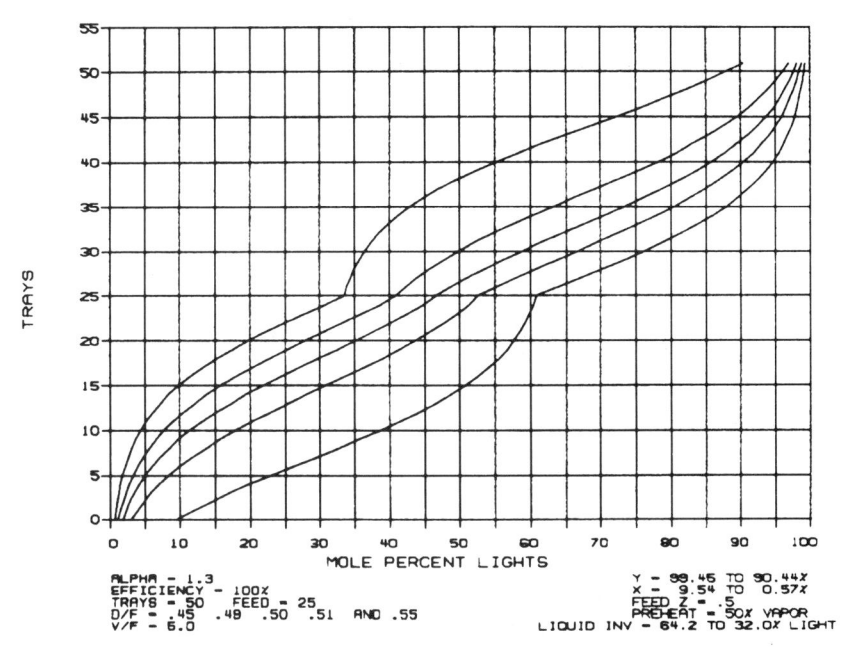

Figure 1A. Liquid Composition Profiles Variable Mass Balance.

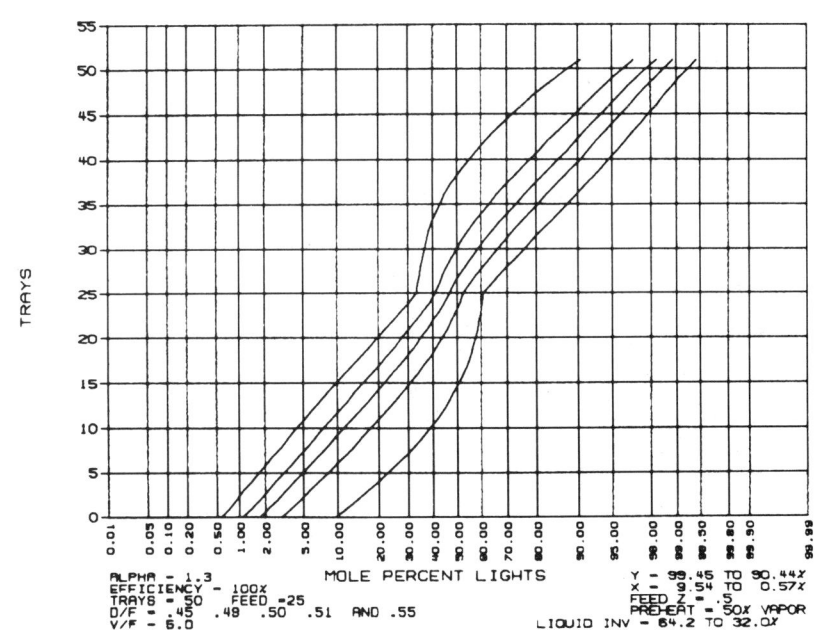

Figure 1B. Liquid Profiles on Probability Axis Variable Mass Balance.

364   Chemical Process Control II

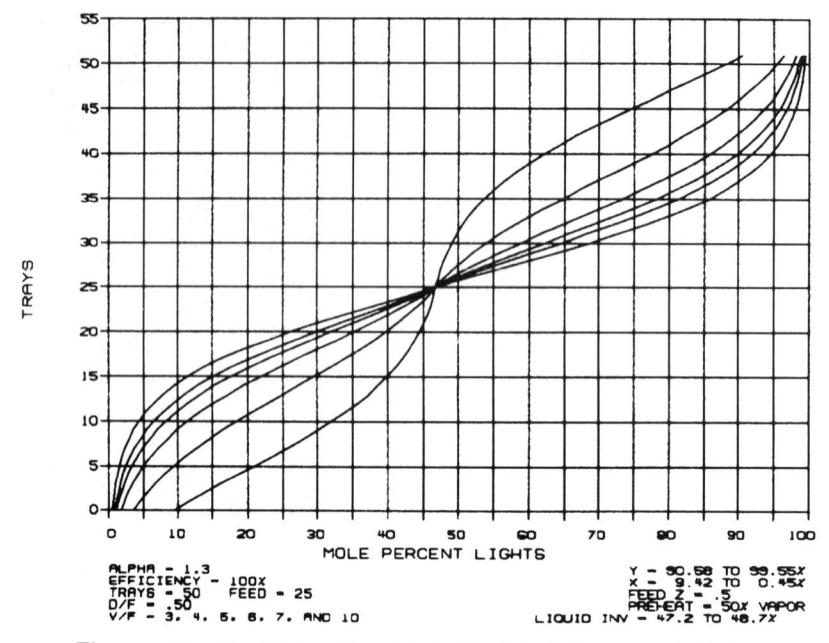

Figure 2A. Liquid Profiles Variable Heat Input and Reflux.

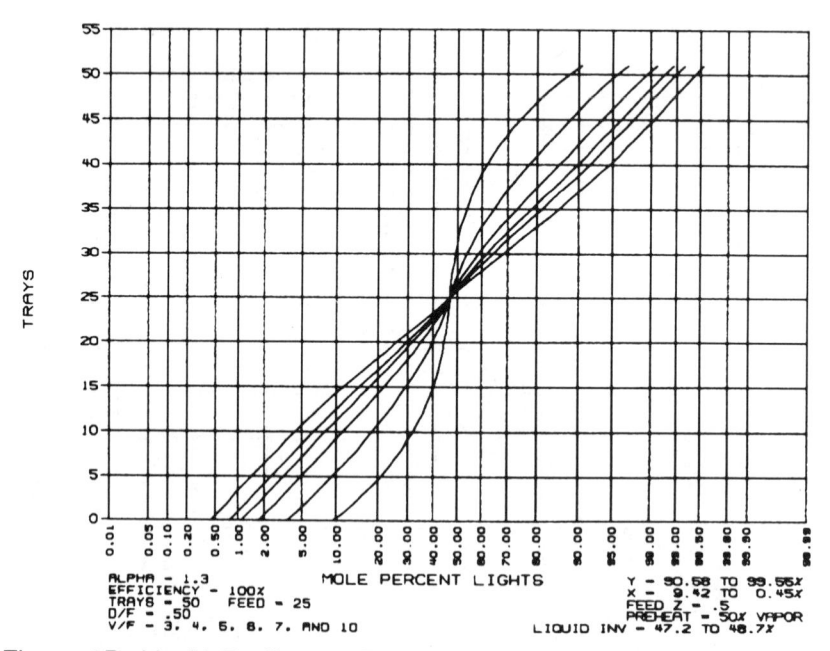

Figure 2B. Liquid Profiles on Probability Axis Variable Heat Input and Reflux.

Complete explicit decoupling also requires people available at the plant site with the knowledge and time to keep it in service, and update it for changes in operating conditions. Bumpless transfer from manual to automatic may be an added problem, calling for more operator training.

RESOLUTION

Complex explicit decoupling should generally be avoided because of the lack of current accurate process data and of the availability of trained people at the plant site. It should only be used where it can be kept relatively simple, and calibrated using only transmitter ranges with elementary heat or mass balance equations. If it is based upon the results from a distillation model or plant test data, probably it will be taken out of service and not updated when operating conditions change.

Implicit decoupling, on the other hand, can be studied in detail during the control strategy design phase. Once an adequate solution is found and implemented, it can remain in service over broad changes in operating conditions without further analysis and only minor adjustment, if any. Generally, implicit decoupling adds few computational elements and little or no additional complexity for the operator. One can often define in simple terms the functions being used, and easily visualize the entities represented by the signals. The resulting strategies are easier to understand and keep in service.

INFORMATION

The rest of this paper describes methods to achieve implicit decoupling by:
- Relative gains
- Dynamic decoupling
- Nonlinear dynamic element
- Change of measured variable
- Change of manipulated variable

Also some examples are given of simple explicit decoupling.

Relative Gains

The relative-gain concept [1] has been applied to dual composition control of distillation columns [2]. The measurement and manipulated variables are paired to minimize the steady state interaction. Actual process data are seldom used in these calculations because time and cost of upsetting the process are prohibitive. Generally, a steady state model of the process is used. The method is well documented and will not be described further here. However, the more significant phenomena is the relative gains at the frequency which the loops interact! This is very difficult to calculate and may be much different from the steady state relative gains.

Dynamic Decoupling

If one loop responds very fast and the other very slow, interaction is not a problem because the fast loop can compensate for any changes introduced by the slow loop. This method can be called dynamic decoupling. It has been used for years by instrument men when tuning loops which interact. The less important loop is made slower (wide proportional band and long reset), if both loops initially have about the same response. The slow loop cannot compensate for actions of the fast loop but the system is stable.

Dual composition control can also benefit from dynamic decoupling, especially if the residence time of the light key component is much different from the heavy key. The residence time on the trays is different for each component. If the inventory (moles) of a component in the liquid on the trays is divided by the flow rate (moles/hour) of that component in the feed, one calculates the average residence time for that component, ($T_i = V_i/F_i$). For a binary system, if top product is high purity (99.9% lights) and bottom product is low purity (10.0% lights), the heavy component inventory will be small, for example 28% of total liquid inventory on the trays (Figure 3A or 3B). The upper curve is the liquid profile, and the lower curve is the vapor profile. If the top product is low purity (90% lights) and the bottom product is high purity (0.1% lights), then the light component inventory will be small, for example 25% (Figure 4A or 4B). When products are near the same purity, the inventory will be split about 50/50 between light and heavy components. If the inventory of a component on the trays is small and its concentration in the feed is large, the residence time for that component will be relatively short. Conversely, if the inventory of a component is large and its concentration in the feed is small, its residence time will be relatively long. On multicomponent columns, the difference in residence times can be very significant, because the key components accumulate on the trays and their concentrations build up much higher than in the feed or either of the product streams.

If sample handling and analysis times are kept short, the loop controlling the key component with the shorter residence time can be made the faster loop, especially if vapor flow through the trays is manipulated. The

C. J. Ryskamp 367

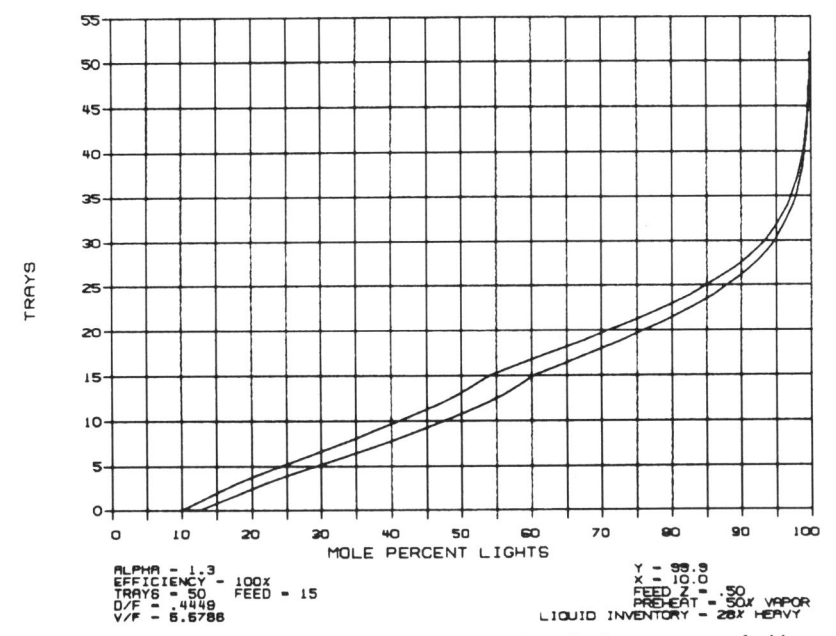

Figure 3A. Liquid and Vapor Profiles Small Inventory of Heavy Component.

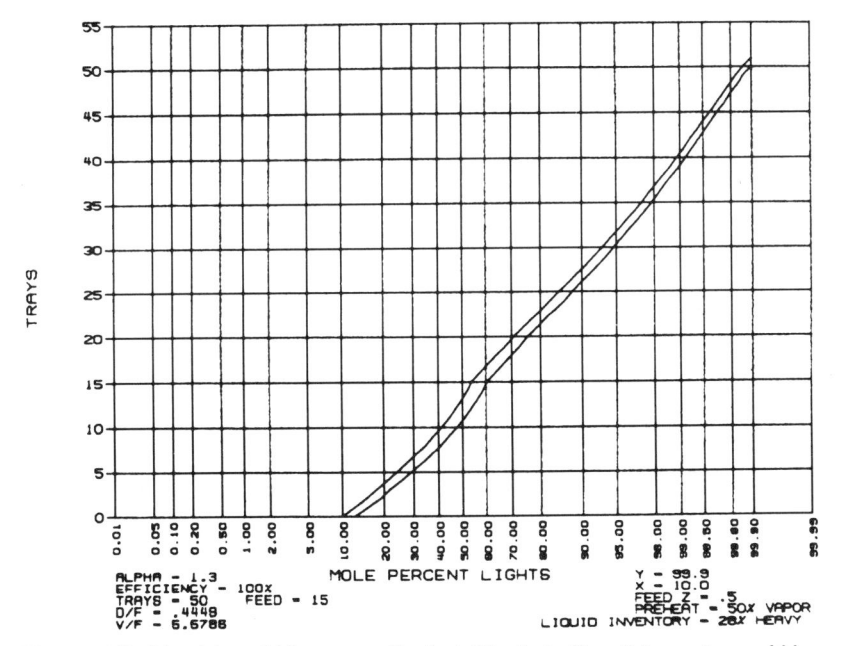

Figure 3B. Liquid and Vapor on Probability Axis Small Inventory of Heavy Component.

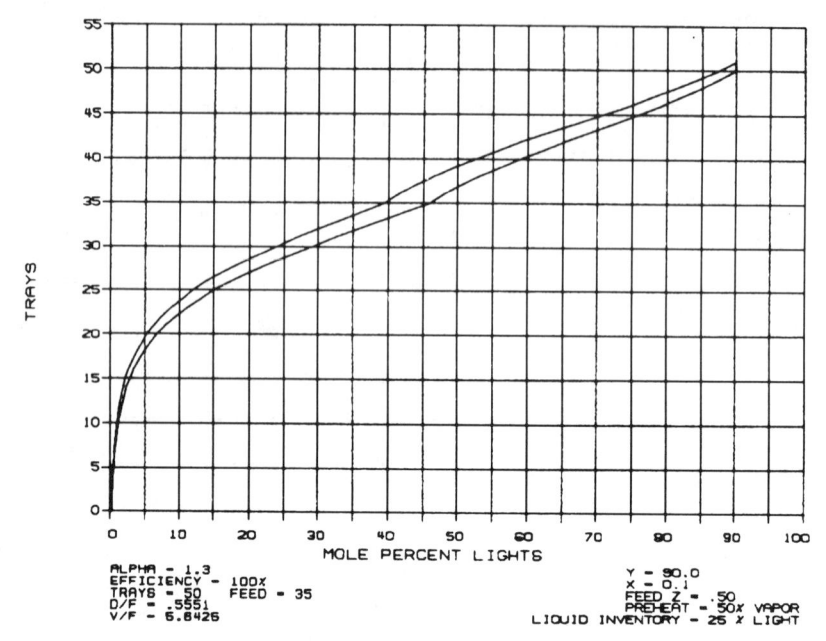

Figure 4A. Liquid and Vapor Profiles Small Inventory of Light Component.

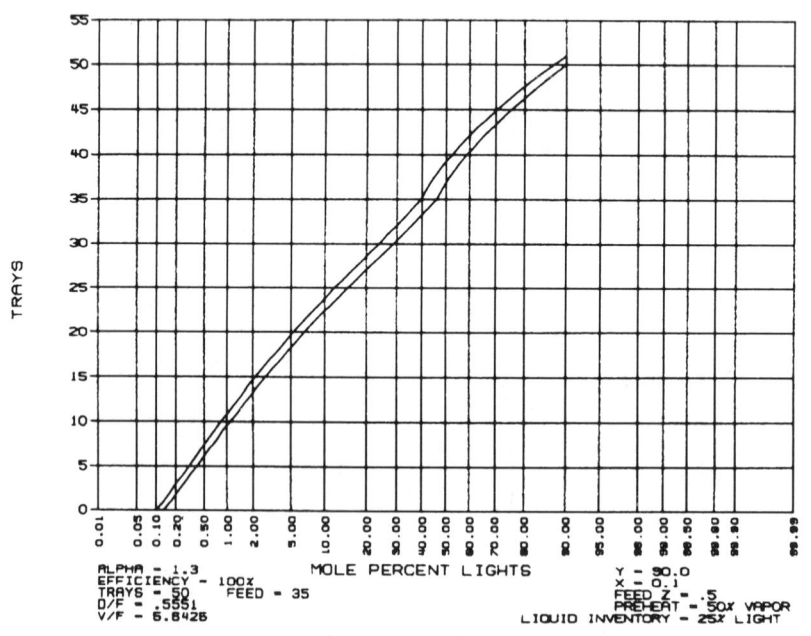

Figure 4B. Liquid and Vapor on Probability Axis Small Inventory of Light Component.

loop controlling the key component with the longer residence time manipulates reflux, or better reflux ratio, and is anticipated to be slow to respond. For example, if the heavy key has the shorter residence time, the bottom AC (measurement light key) should set heat input, with pressure setting heat removal in the condenser. Then the top AC (measurement heavy key) can set reflux ratio. The bottom AC will be fast and the top AC slow. If the light component has a much shorter residence time, then the top AC (measurement heavy key) should set vapor flow. If it sets vapor flow off the top of the column, then reboiler heat input must be on pressure control. The bottom AC (measurement light key) would then set reflux ratio. The top AC will be fast. The bottom AC would be expected to be slow, and the transport lag through the trays (normally 10 seconds per tray) will only make it a little slower. Overall, the fast loop maintains the position of the profile, and the slow loop adjusts its shape or separation on the column.

Packed columns have a much different dynamic response. The liquid holdup can be relatively small. Another consideration is the sensible heat of the packing material, especially on columns with a high temperature difference from top to bottom. If the composition profile is to be moved up or down, so must the temperature profile be moved. Therefore, some packing in the center must be heated or cooled respectively. The extra vapor generation in the reboiler or extra reflux required to heat or cool the packing can exhibit a distributed integrating lag that is greater than the time that would be required to change only liquid or vapor compositions. The packing can act as a filter which attenuates interaction between the top and bottom AC's. Each column will have unique characteristics.

Nonlinear Dynamic Element

A nonlinear dynamic element [3] can be built from a lag and a signal selector which is connected to the input and output of the lag (Figure 5). This provides a fast up versus slow down or a fast down versus slow up response depending upon whether the selector chosen is high or low respectively. With this device, the top AC can increase reflux ratio quickly but reduce it slowly, and the bottom AC can increase heat input quickly but reduce it slowly. Either composition will then be forced more pure quickly, but allowed to become less pure slowly. The nonlinear dynamic element can be installed directly in the AC output, which has been done many times in the past few years. Performance may be better installed in the reset signal path so as to avoid the phase shift associated with fast reset, except when it is needed to force product purity up to specification. This approach, however, may be too difficult for a typical instrument man to keep properly tuned.

370   Chemical Process Control II

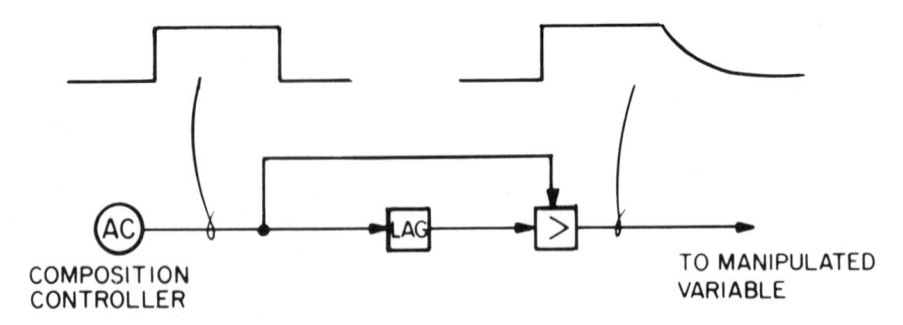

Figure 5. Non-linear Dynamic Element.

Change Measured Variable

A tray temperature above the feed and one below the feed may be used as the two measurements which infer compositions. If the lower one sets heat input and the upper one sets reflux, the interaction will be severe. However, if the sum of the temperatures is used to set heat input and their difference is used to set reflux, the interaction is greatly reduced. The sum of the temperatures indicates the position of the composition profile (or mass balance), and the temperature difference shows the separation. A similar approach might be used with top and bottom composition measurements, if they are converted to a logarithmic or probability scale. This is especially applicable to columns which have a symmetrical profile with little distortion at the feed tray. Both ends will change about the same with a shift in mass balance or separation.

A double differential temperature measurement has been applied to a multicomponent column [4]. Here, the temperature difference between two trays below the feed is subtracted from the temperature difference between two trays above the feed. The double differential temperature is then used as the measurement for a composition controller. This measurement corresponds to profile position, and the method transforms a predominant dead time into a response closer to an exponential lag. The application described used a constant heat input. However, the method has been used successfully in dual composition contol, without explicit decoupling.

The logarithm of the composition of both products can also be used as a measurement, which will make the loop response less nonlinear. This change will also affect the gain of the interaction.

Change Manipulated Variable

Another method to reduce interaction is to select a different manipulated variable or a combination of variables. Combined reflux and distillate control [3] is an example (Figure 6). The pressure controller (PC) is explicitly decoupled from the top composition controller (AC) with one summing element. The PC output represents the total liquid flow leaving the condenser, which equals vapor flow off the column in the steady state. If the AC changes distillate flow (D), the summing element changes reflux

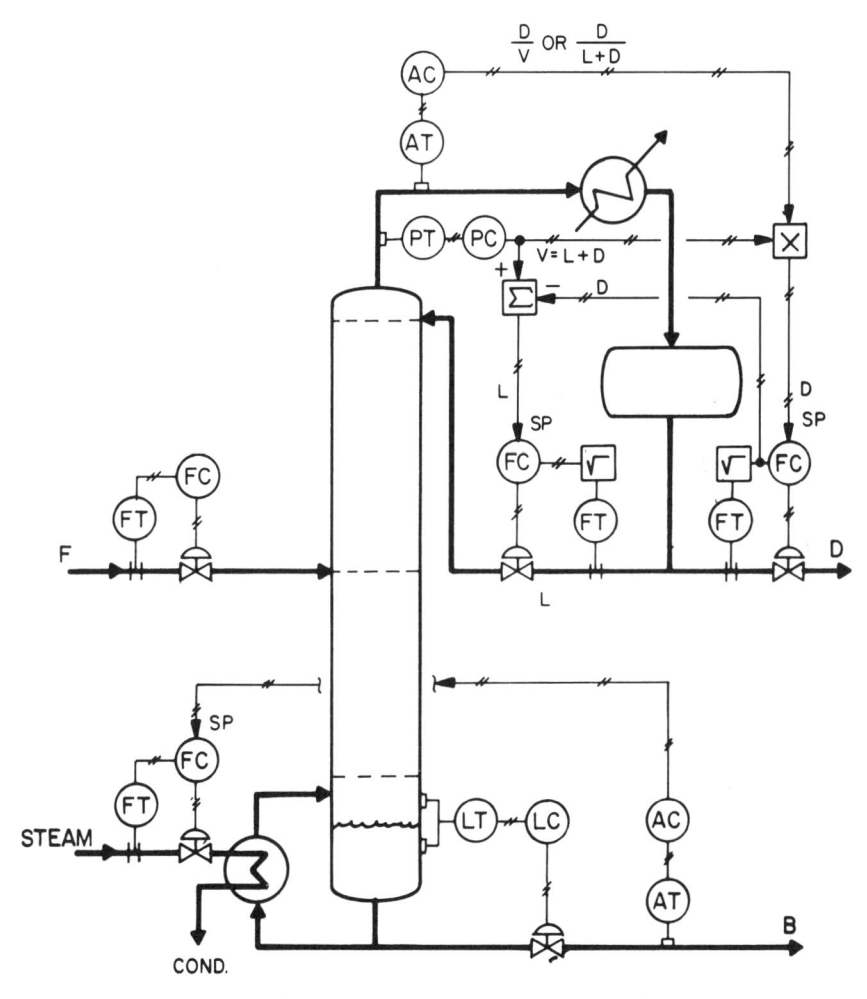

Figure 6. Combined Reflux & Distillate Control.

(L) an equal amount in the opposite direction, so that the total is constant and pressure is not affected.

Partial decoupling of the top AC and the bottom AC is achieved implicitly with the multiplying element. It holds reflux ratio constant if the top AC output is constant. An increase of heat input from the bottom AC does not make top product as impure as would occur with reflux constant (conventional control) nor as overpure as would occur with distillate flow constant (material balance control). Since top product is affected less with reflux ratio constant, upon changes in heat input from the bottom AC, decoupling is implied, and exists without signal connections between the top AC and bottom AC.

The computational elements in this system are easily scaled using only transmitter ranges. No distillation calculations or plant tests are required. Also, the scaling will be valid for all operating conditions and calibration does not have to be changed unless a flow transmitter range is changed. Also, no dynamic compensation is required.

Examples of Simple Explicit Decoupling

When interaction occurs through the utility supply, cascade loops can be used to provide decoupling. For example, a tray temperature controller should set steam flow to the reboiler, rather than the steam valve directly, to avoid upset from steam header pressure changes. On reboiler systems with hot oil heat, a tray temperature controller can set reboiler heat transfer rate and the heat transfer rate controller set the hot oil flow. This inhibits interaction between heat users due to changes in hot oil pressure or temperature.

Often heat is removed from a crude column with a pump-around stream which transfers the heat to cold crude and to a reboiler on another column, such as a stabilizer. This can cause interaction between the crude column and the stabilizer. However, explicit decoupling is achieved by adjusting the heat transferred to crude to compensate for changes in stabilizer heat input, and keeping heat removal from the crude column constant [5].

The operator on a crude column is well aware of the interaction between the side draws. Typically, the top product and three side product compositions are controlled. The reflux and three side draw flow rates are manipulated. A 4 × 4 decoupling matrix is needed if complete explicit decoupling is used. If the pump-around streams for heat removal are included, the matrix is even larger. This scheme cannot be kept in service unless an engineer is assigned to the unit, and he must be competent working with control systems.

A much simpler strategy [6] has been implemented using an elementary mass balance approach, which is nearly identical to the normal operator actions. It decouples in a downward direction to the next side draw only. Instead of 16 elements, only 3 are required (Figure 7). Dynamic compensation is not shown.

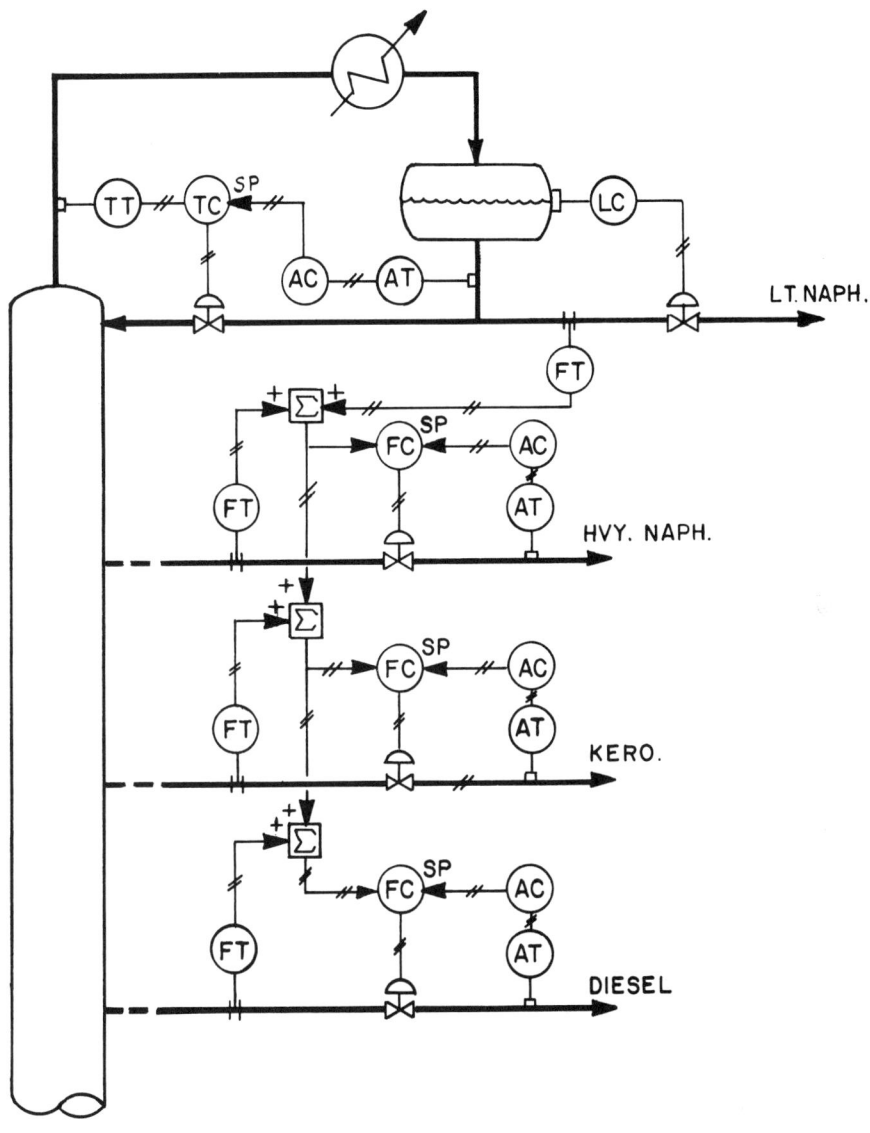

Figure 7. Simplified Crude Column Side Draw Decoupling.

374 Chemical Process Control II

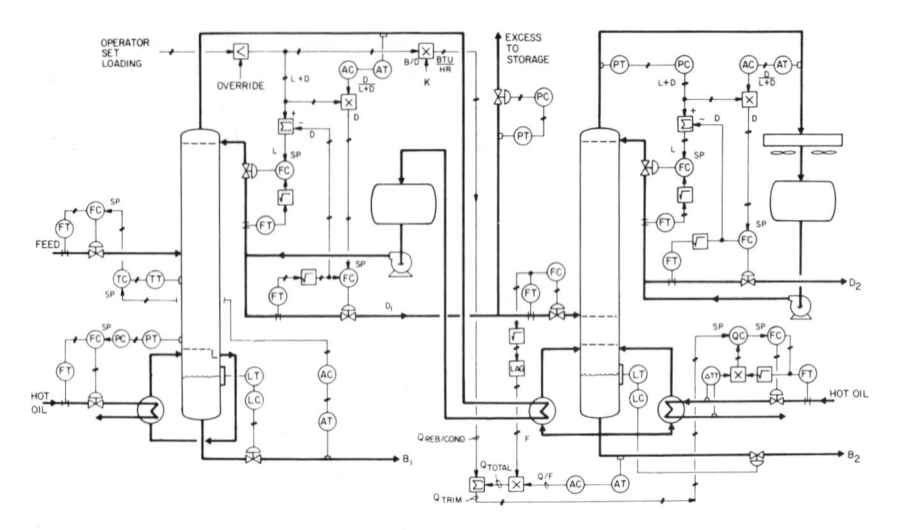

Figure 8. Dual Composition Control on Thermally Coupled Columns.

Another example of success with simple decoupling instead of a complete explicit method is the controls applied to a thermally coupled debutanizer and butane splitter [3]. The overhead vapor from the debutanizer provides most of the heat input to the butane splitter, through a common reboiler-condenser. The original control strategy did not stay in service because there was too much interaction between the columns and between the control loops. Essentially, feedforward material balance control had been used, with the pressure control on the first column setting the heat transfer rate in the common reboiler-condenser. The interaction was so severe, it was impossible to keep compositions on control. The key to stabilizing the operation of these columns was to hold the heat transfer rate in the common reboiler-condenser constant (Figure 8). This is the operator set loading of the columns.

The common reboiler-condenser is the major path which links the two columns together. Holding a constant heat transfer rate stops interaction between them and allowed each column to be controlled almost independently. A very complex explicit decoupling system, involving at least four measurements and four manipulated variables, could have been designed. If properly implemented and tuned, it may have worked. However, this could not have been easily understood or updated by plant people and would not have stayed in service. The system described [3] was implemented using combined reflux and distillate control, and tuned so that both top and bottom compositions on both columns could stay on closed loop control. The bottom loop was faster than the top on both columns.

It is recognized that the crude column side draw decoupling does not compensate for changes of separation in the column, nor does it adjust for changes in crude composition. Feedback from the composition controls is required for these functions. Feedforward was used only on crude flow rate. Also, combined reflux and distillate control does not compensate for changes in subcooling of reflux, feed preheat or feed composition. Feed flow rate was used in a feedforward loop to set heat input only. Again, feedback from the composition controllers does compensate for the other upsets. Adding more sophistication is feasible, and has been done, but does not stay in operation unless people at the plant site want it, understand it, and devote the time required to keep it updated.

REFERENCES

[1] Bristol, E. H., "On a New Measure of Interaction for Multivariable Process Control," *IEEE Transactions on Automatic Control*, January 1966.
[2] Shinskey, F. G., *Distillation Control for Productivity and Energy Conservation*, McGraw-Hill, New York, pp. 289–321 (1977).
[3] Ryskamp, C. J., *Hydrocarbon Procession*, 60(6):51–59 (1980).
[4] Boyd, D. M., *Chemical Engineering Progress*, 71(6):55–60 (1975).
[5] Ryskamp, C. J., Wade, H. L., Britton, R. B., "Computer Control of Crude and Vacuum Units for Energy Conservation," API 41st Midyear Meeting, Los Angeles (1976).
[6] Wade, H. L., Ryskamp, C. J., Britton, R. B., "Crude Tower Control Techniques for Better Productivity," ISA International Conference, Houston (1976).

ON THE DIFFERENCE BETWEEN DISTILLATION COLUMN CONTROL PROBLEMS

Thomas J. McAvoy
Department of Chemical and Nuclear Engineering
University of Maryland
College Park, MD 20742

ABSTRACT

Recent work has shown that there is a significant difference between distillation columns in regard to what control techniques work well and what techniques do not work well. For dual composition control of high purity towers, decoupling cannot be achieved and material balance control is superior to conventional control. For low purity towers the reverse conclusions hold. These results are significant in terms of shaping university research in the distillation control area. Most academic studies to date have focused on low purity towers and therefore their conclusions may not hold for high purity towers. Such towers are common and are more industrially important in that they are large energy consumers.

INTRODUCTION

Distillation is the most common of all the separation techniques used in the process industries. In addition, distillation tends to be energy inefficient since most of the energy which is put into a column at the reboiler is rejected to the environment at the condenser. Both of these facts result in distillation columns being large energy consumers. It has been estimated [1] that distillation columns typically consume 40% of the energy used in a chemical plant. An energy audit on all of the important towers is the United States [2] indicated that 3% of the total energy used in the U.S. in 1977 went into distillation. The increasing cost of energy has forced companies to look more closely at energy conservation in distillation systems.

Improved process control is one of the most significant ways of saving energy in distillation columns. The use of computer control has produced dramatic energy savings in some cases [3]. Minimum pressure control [4] also results in significant energy savings for systems where relative volatilities are low and reflux ratios are high. Lastly, Luyben [5] has shown that dual composition control, results in minimum energy usage in a tower. With dual composition control both top and bottom product compositions are controlled simultaneously.

This paper centers around dual composition control. In the past industry avoided dual composition control because of the possibility of interaction between the two loops. A typical industrial approach was to control one product purity and over reflux the tower to achieve product recovery. This process wastes energy. However, until recently there was no economic incentive to improve upon it. The use of dual composition control in industry has increased sharply in the last few years.

Recent work [6, 7] has shown that there is a significant difference between towers in terms of the ability to achieve dual composition control through the use of various control strategies. Techniques which work well for lower purity towers are poor when applied to high purity towers. Such high purity towers are common and they are more industrially important since they are large energy consumers. Many of the academic studies to date have been carried out on low purity towers. Thus, it can be questioned whether or not the results achieved in these academic studies can be extrapolated to high purity towers.

In this paper, the recent work [6, 7] on the difference between achieving dual composition control in varous towers is briefly reviewed. A simple explanation of the reason for the differences between towers is then presented. This explanation is based on the steady state gain characteristics of the distillation composition loops. Lastly, the implications of these results in terms of shaping future academic research in the area are discussed. Guidelines as to what problems should be studied from an energy conservation perspective are presented.

PREVIOUS WORK ON DUAL COMPOSITION CONTROL

The work which has been carried out to date on dual composition control has paralleled that which has been tried industrially. Two approaches to achieving dual composition control have been studied. The first approach involves selecting the best pairing of manipulated and controlled variables [6]. An example of the first approach is shown in Figure 1 where reflux and steam are used to control top and bottom product compositions. The second approach involves decoupling the two composition loops [7]. When simplified decouplers [8] are added to Figure 1, the control system shown in Figure 2 results. The results on decoupling will be discussed first since these were the first to be obtained.

A. Feasibility of Decoupling Dual Composition Control Loops

Decoupling is one of the control strategies that has been proposed to avoid the loop interaction with dual composition control. Steady state results [9, 10, 11], based on the relative gain array (RGA) [12], however,

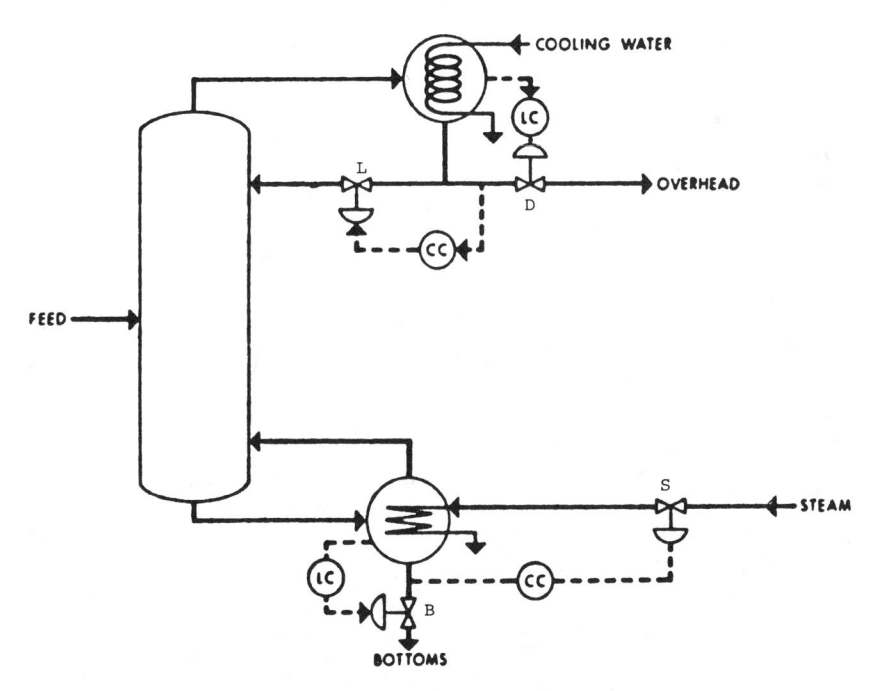

Figure 1. Dual Composition Control Scheme.

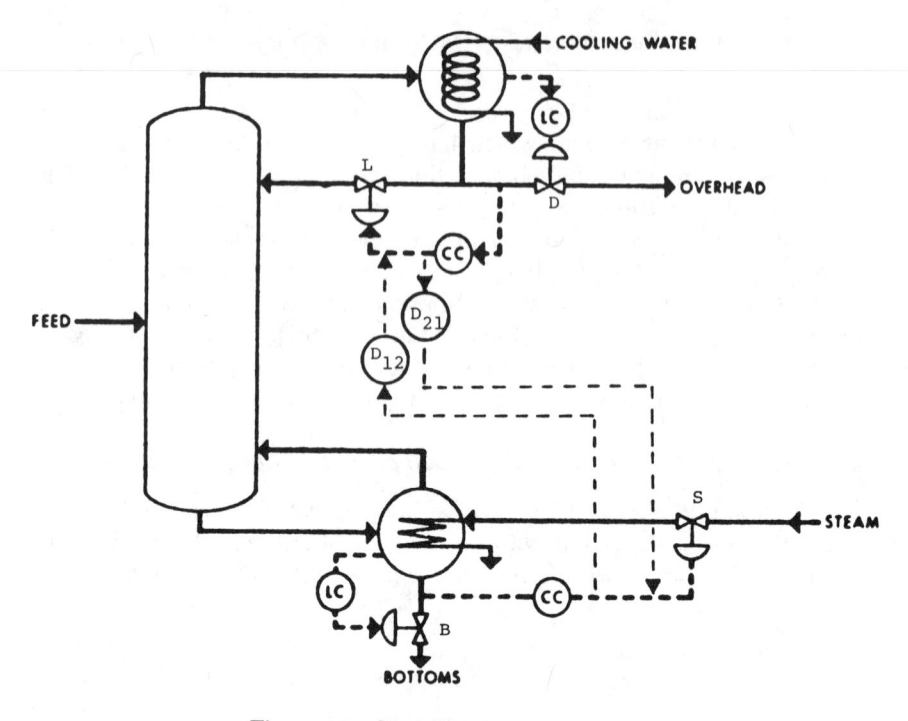

Figure 2. Simplified Decoupling.

indicated that decoupling would not be feasible in columns where the RGA is large. To test the validity of these steady state predictions, dynamic column simulations were set up. The details of the dynamic simulations can be found in [7]. Three columns (A, B, C) having a low (A), moderate (B) and large (C) RGA were examined. The essential features of these three towers are shown in Table 1. Transfer function models were developed for the towers by step testing the simulations. These transfer functions were used to synthesize the decouplers. To test whether decoupling could be achieved, the bottom composition set point was step changed. If the top and bottom composition loops were decoupled then the top composition should remain constant for such forcing. The results obtained for the low purity column A and the high purity column C are shown in Figures 3 and 4. The fact that x_D remains essentially constant for column A indicates that decoupling has been achieved in this case. For column C, however, x_D shows a significant deviation from steady state. In addition, the decoupled response was found to be inferior to the response achieved by simply eliminating the decouplers, which is shown in Figure 5.

Table 1. Parameters for Distillation Columns Studied.
($q = 1$, $x_f = .5$)

Column	Components	Approx. α	Product Split	# of Trays	Reflux Ratio	RGA* Reflux and Boilup Manipulated	RGA* Distillate and Boilup Manipulated
A	Benzene-Toluene	2.4	.07–.93	13	1.30	2.30	.611
B	Benzene-Toluene	2.4	.02–.98	17	1.71	5.04	.634
C	Methanol-Ethanol	1.65	.01–.9896	27	4.51	18.6	.600

*RGA values are approximate. They were determined from step tests.

The results shown in Figures 3 to 5 are particularly interesting because they show that a control technique which works well for one tower does not work well for another. If one only studied low purity towers then he might be tempted to draw the conclusion that decoupling works well in dual composition control systems. The results in Figures 4 and 5 show that this conclusion is not valid.

B. Comparison of Control Schemes with Different Variable Pairings

The simulations of towers A, B, C were also used to study various methods of pairing manipulated and controlled variables [6]. In addition to the conventional scheme shown in Figure 1, results for distillate flow controlling x_D and steam controlling x_B will be discussed here. Such a scheme is called a material balance scheme. A hybrid computer simulation of the transfer functions for towers A, B and C was used to simultaneously tune the two feedback controllers shown in Figure 1. These controllers were then applied to the non-linear digital simulations for each tower. Feed flow changes were used to test each control system. Results for columns A and C are given in Figures 6 and 7. These results show that for column A conventional control is superior to material balance control. For column C, the reverse is true. Thus, again a technique which works well for one column does not work well for

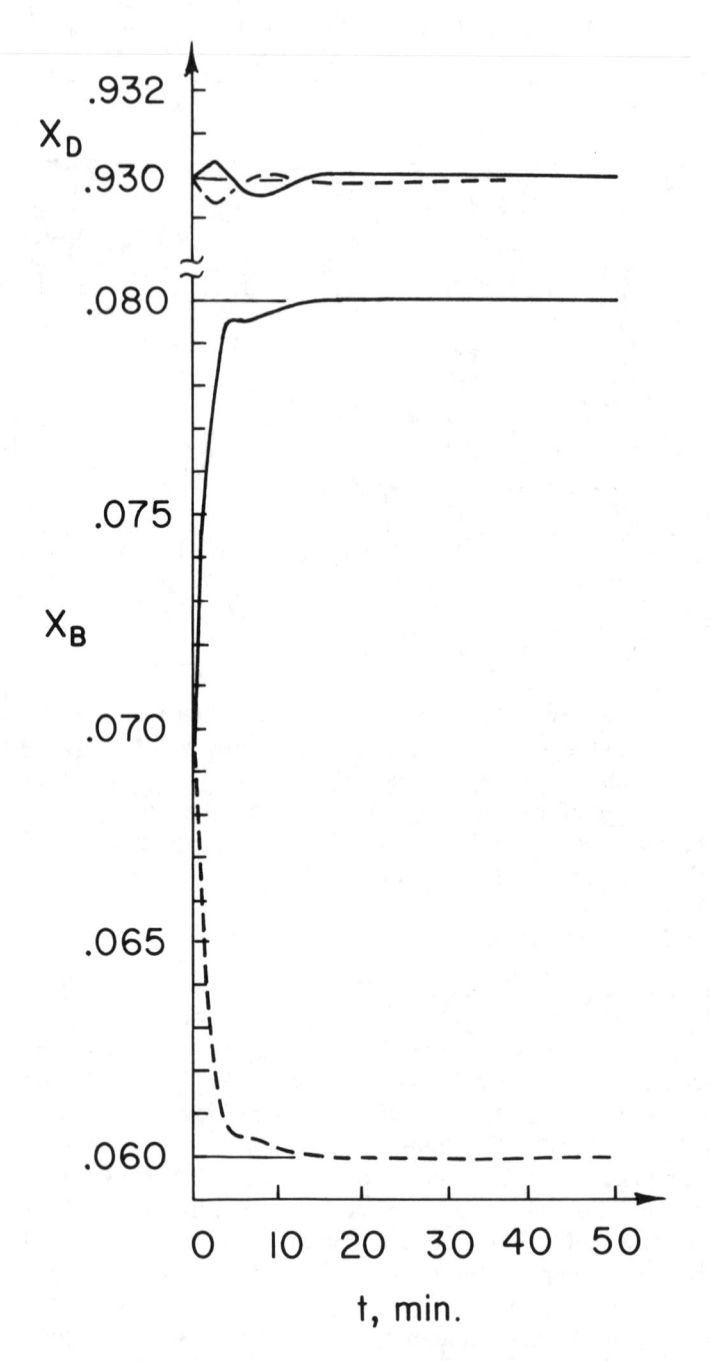

Figure 3. Column A Simplified Decoupling (taken from [7]).

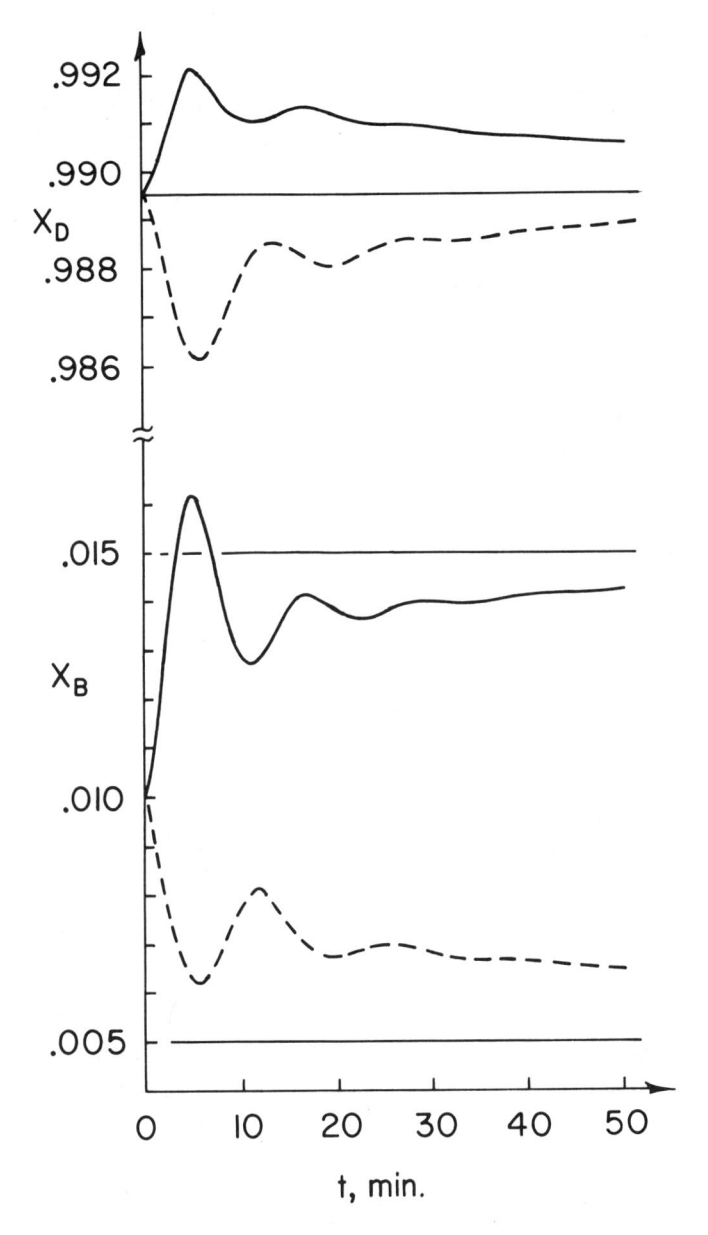

Figure 4. Column C Simplified Decoupling (taken from [7]).

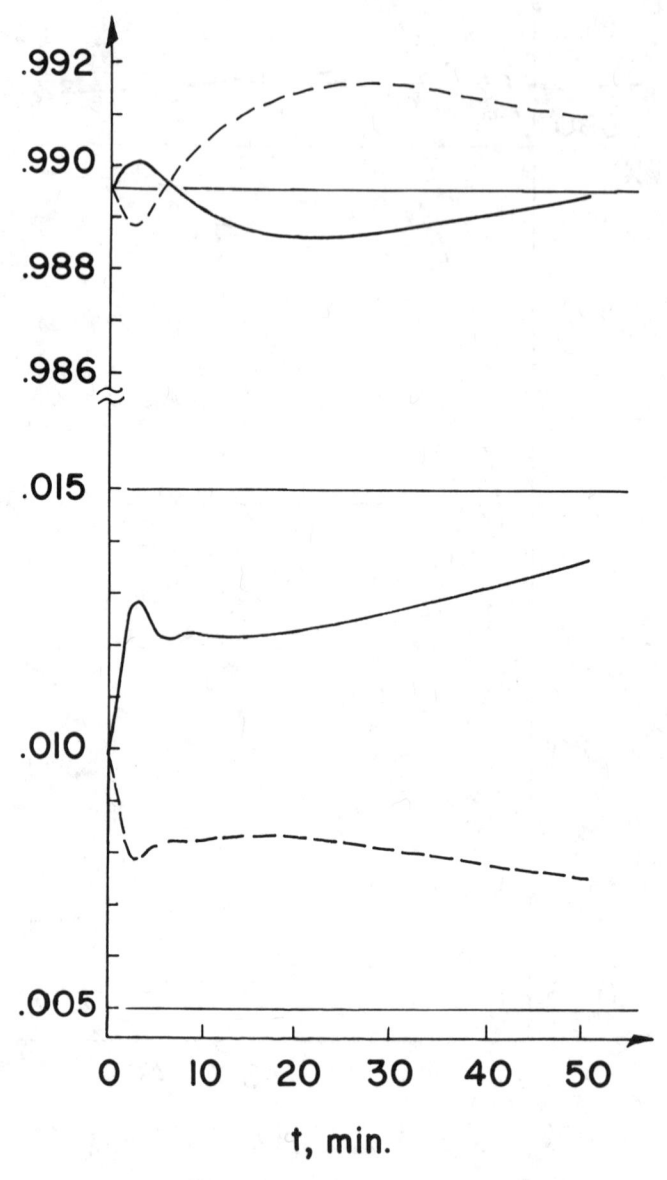

Figure 5. Column C No Decoupling (taken from [7]).

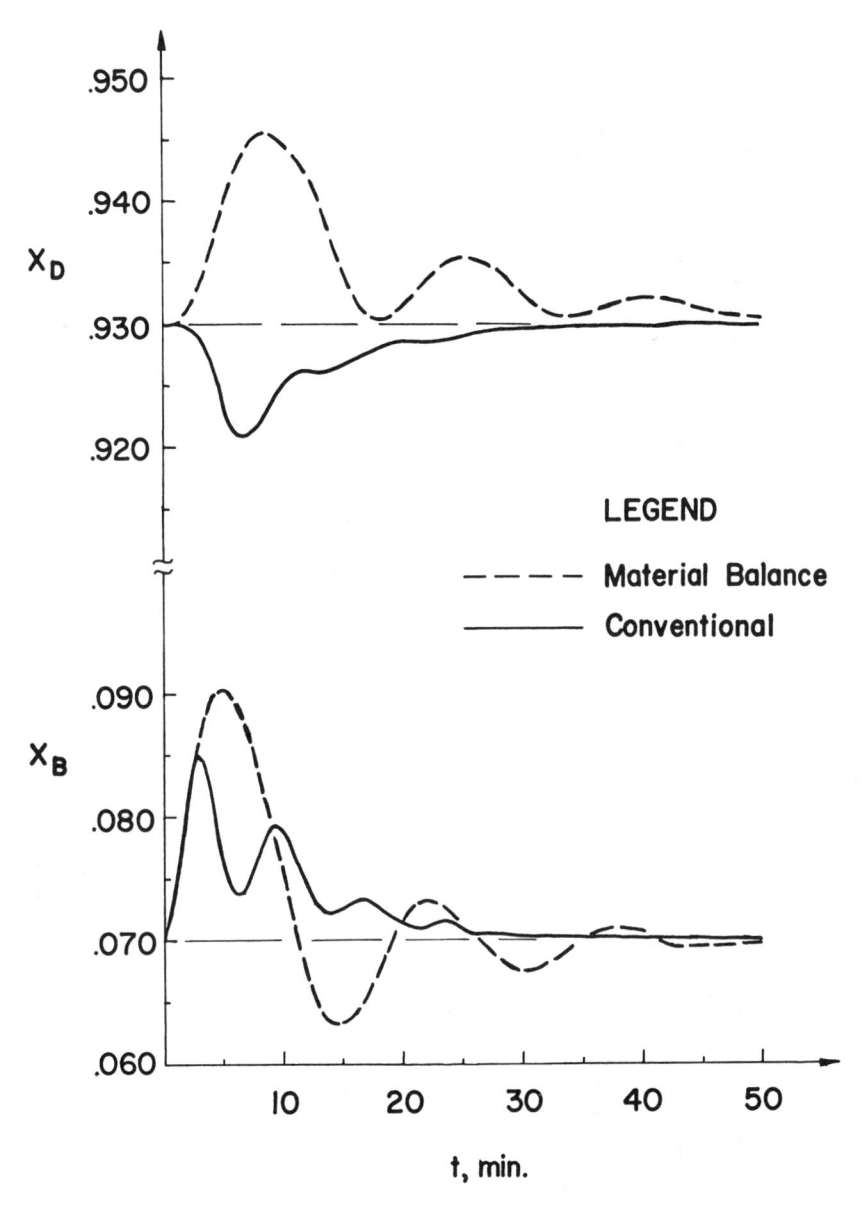

Figure 6. Column A 10% Increase In Feed Flow (taken from [6]).

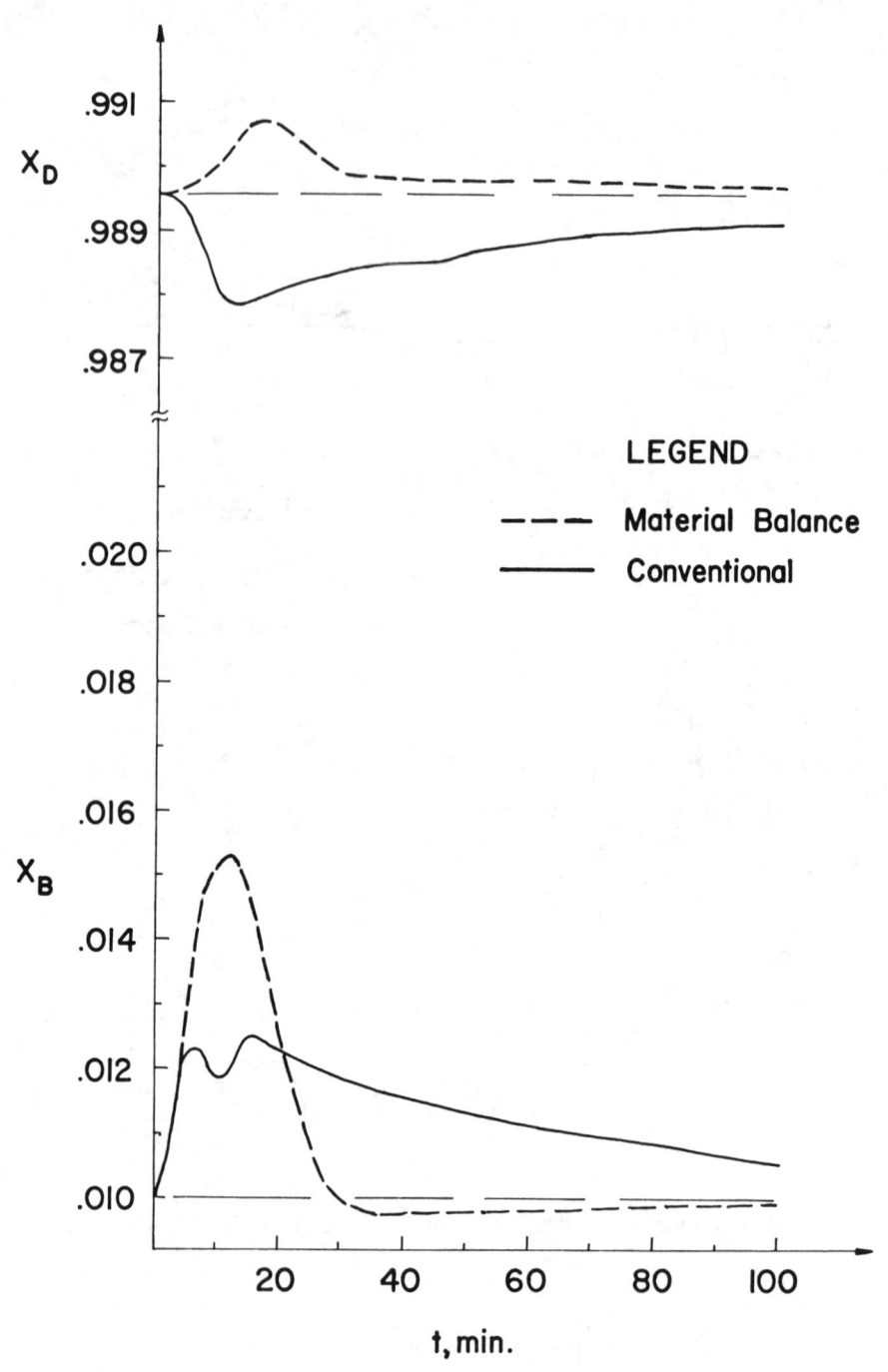

Figure 7. Column C 10% Increase In Feed Flow (taken from [6]).

another. It would be incorrect to extrapolate the results obtained for column A to column C.

Explanation of the Difference in the Ability to Control Various Towers

There are several reasons for the effectiveness of a control technique to change from one tower to another. First, the nature of the interactions can change substantially from tower to tower. The RGA [12] provides a measure of this interaction. For towers A to C the RGA values are shown in Table 1. As can be seen when reflux and boilup are manipulated the RGA increases sharply from column A to column C. This sharp increase indicates that the control system will become increasingly sluggish in its response as one proceeds from column A to C. By contrast, when distillate and boilup are manipulated the RGA remains relatively constant as one proceeds from column A to C. Jafarey, McAvoy and Douglas [13] have presented analytical expressions for the RGA for the two control strategies illustrated in Table 1. In addition, McAvoy [14] has published numerical RGA results for two additional variable pairings. These results can be used to estimate potential control difficulties with a particular variable pairing. The analytical expressions are especially easy to use in this manner since they show the trends in the RGA as product purity, relative volatility, reflux ratio and feed conditions change.

A second reason for the change in effectiveness of a control strategy is the variable gain which occurs in the composition control loops. This variable gain is the major reason for the failure of the linear decoupling approach to work on column C. In order to illustrate this non-linear gain effect a simple steady state approach will be taken.

Consider a binary column whose material balance is given as

$$x_F \, F = x_D \, D + x_B \, B \qquad (1)$$

A simple, approximate equation describing a column operation [15] is

$$N = \frac{\ln\left(\dfrac{x_D}{1-x_D} \dfrac{1-x_B}{x_B}\right)}{\ln(\alpha \sqrt{Rx_F/(Rx_F+1)})} \qquad (2)$$

Equation 2 is based on the assumptions of constant molal overflow and constant relative volatility. It will be assumed that reflux and boilup are manipulated and that high purity towers are considered. By differentiating eqs. 1 and 2, the gain of the distillate loop is given approximately as

388   Chemical Process Control II

$$\left.\frac{\partial x_D}{\partial (L/F)}\right|_{V/F} \simeq \frac{\left(\frac{F}{F-D}\right)\left(\frac{1-x_D}{x_B}\right)}{1 + \left(\frac{D}{F-D}\right)\left(\frac{1-x_D}{x_B}\right)} \qquad (3)$$

The derivation of eq. 3 is given in the Appendix. A similar expression can be derived for $\partial x_B/\partial V/F|_{L/F}$. Figure 8 shows a plot of eq. 3 for the case where $D = B$. As can be seen the distillate loop gain varies with $(1-x_D)/x_B$. Thus, changing tower conditions results in changes in the distillate and bottom composition loop gains.

Equation 3 and Figure 8 can be taken to apply to both columns A and C. However, the gain changes for column A are much smaller than those for column C. The reason for this difference is that $\left(\frac{1-x_D}{x_B}\right)$ changes much less for column A than for column C. A simple calculation will illustrate this point. If the distillate loop gain increases to 1.5, Figure 8 shows that

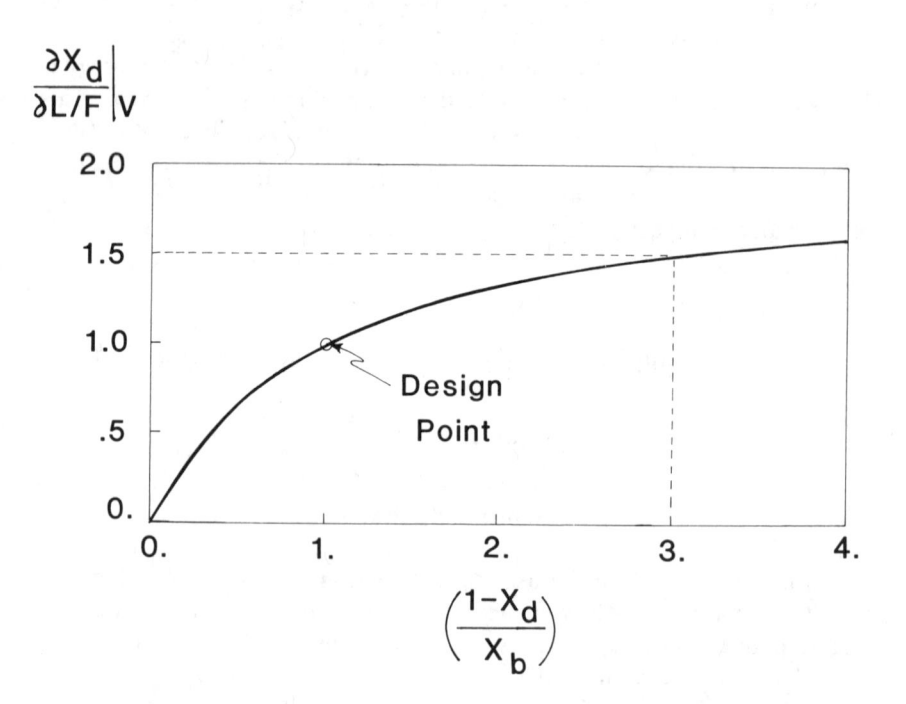

Figure 8.

$(1-x_D)/x_B$ has to have a value of 3.0. Further if x_B is taken as its steady state value (.07 for column A and .01 for column C), the resulting value of x_D can be calculated. For column A x_D has to be .79 to increase the gain to 1.5 while for column C x_D has to be .970. When these compositions are compared to the steady state values for x_D (.93 for column A and .9896 for column C) it can be seen that much smaller composition changes in column C result in significant gain changes than in column A. For column A the feedback controllers can keep compositions within 1 to 2% of steady state and for this range loop gain changes are small. For very high purity towers the gain changes become even more severe with small changes in composition. For a .999–.001 tower, changing x_D to .997 while keeping x_B constant will increase the gain to 1.5.

In addition to the interaction effects and the variable loop gain effects discussed here, asymmetric dynamic responses have also been reported for towers [16-18]. The tower treated in [16] is a very high purity, high reflux ratio tower, $R = 4 \times R_{min}$, with an $\alpha = 5.0$. Dynamic simulations showed that for changes in R and x_F the top plate sometimes responded faster than the bottom plate and sometimes slower than the bottom plate. Thus, the time constant for a particular response depended on both the direction of the forcing and the initial condition for the tower. In general, the slowest responses were obtained when adjacent plates had compositions that were very close to one another and nearly pure. References [17, 18] discuss additional towers in which asymmetric dynamic behavior has been observed.

Implications for Further Research on Distillation Control

Table 2 shows the key features of other studies involving dual composition control. For the most part these studies focused on low purity towers with a small number of trays. An exception is the work of Luyben [8] where a moderate purity tower is treated. For Wood and Berry's [22] and Brosilow and Tong's [26] towers the distillate purity is low while the bottoms purity is moderate to high. Brosilow and Tong's study is based on an industrial tower separating 5 components. The remaining studies are academic in nature and the experimental work all involves methanol and water. Recently Fuentes and Luyben [18] presented an excellent paper on controlling high purity towers. Although they discussed dual composition control, the results they actually presented were all for single composition control. Thus, very little work has been carried out on high purity towers.

These statements should not be taken as a criticism of past research on the dual composition control problem. Rather the goal of the work was to test various control strategies. Existing unit operation laboratory towers

Table 2. Other Studies Involving Dual Composition Control.

Reference No.	$x_D - x_B$	x_F	α	# of Trays*	Type of Study
[8]	.95 −.05	.5	2	20	Simulation
	.98 −.02	.5	2	20	Simulation
[19]	.793−.187	.57	—	3	Simulation
[20]	.871−.129**	.57	2.23	5	Simulation
[21]	.95 −.047	.272	3.79	24	Experimental
[22]	.808−.282	.50	1.5	10	Simulation
[23]	.931−.00282	.328	3.79	8	Experimental
[24, 25]	.949−.0430	.535	3.79	8	Both
[26]	.093−.021***	.351	3.25	16	Simulation

*Excluding condenser and reboiler.
**Estimated from information in reference.
***Based on an industrial 5 component tower, x_D = heavy key composition in D and x_B = light key composition in B.

were used and simulations which did not require excessive computer time were employed. For further work in the area it is proposed that more emphasis should be placed on the particular problem treated so that industrially important towers can be studied.

Towers which are large energy consumers are prime candidates for the application of advanced controls. In a recent survey paper [27] Tolliver and Waggoner contrasted towers in the chemical and petroleum industries. In addition energy consumption estimates for all of the important towers in the U.S. can be found in a recent Department of Energy publication [2]. Both the chemical and petroleum industries have towers which are large energy consumers. In the petroleum industry product purities are often moderate and large energy consumption is the result of large thoughputs. In the chemical industry product purities are high to very high which results in high energy consumption. In addition latent heats of vaporization are larger for chemical products than for petroleum products [2].

Given this picture then further work on distillation control should be directed toward moderate and high purity towers. Of the two classes the high purity towers will present the most difficulty for academic workers. Thus towers will exhibit the highest degree of non-linearity in their behavior. In addition they often involve highly non-ideal components and a significant number of trays. Given these facts and the safety considerations which are necessary in a university it probably will be impossible to study high purity towers via experimentation. One means of carrying out experimental work would be through a joint industry/university cooperative program on existing commercial or pilot plant towers.

If simulation is used then computer times may become restrictive for many towers because of the number of trays and product purities involved. This problem will be more severe for components with low α's. Dynamic models for towers are stiff [28] and this further complicates the simulation approach. Digital techniques for calculating tower dynamics are improving [29-31], and efficient computer codes are available from authors [29, 31]. Such improvements may allow researchers to focus on more industrially significant problems without excessive computer usage.

CONCLUSIONS

This paper has focused on the problem which researchers have in shaping a research program in the distillation control area. Some towers, namely those with low product purities, are easy to study either through experimentation or simulation. However, it has been shown that the results for these towers cannot be extrapolated to higher purity and more industrially significant towers. In an academic environment the primary means of studying these high purity towers probably will be via digital simulation. Improvements in digital techniques for integrating stiff equations will help facilitate such studies.

ACKNOWLEDGMENTS

This work was supported by the National Science Foundation under grant ENG-76-17382.

NOMENCLATURE

α = Relative volatility
B = Bottoms flowrate
D = Distillate flowrate
F = Feed flowrate
L = Reflux flowrate
N = Number of ideal plates
R = Reflux ratio (L/D)
V = Vapor boilup
x_B = Bottoms mole fraction
x_D = Distillate mole fraction
x_F = Feed mole fraction

REFERENCES

[1] Shinskey, F. G., "Process Control Systems," 2nd Edition, p. 277, McGraw-Hill, New York (1979).
[2] Radian Corp., "Energy Conservation: A Route to Improved Distillation Profitability," Technology Transfer Manual (DOE/CS/4431-T2), published by Dept. of Energy (May 1980).
[3] Williams, J. S., "Installation of a Computer Process Controller," presented at 57th Annual Gas Processors Assoc. Convention, New Orleans (March 1978).
[4] Shinskey, F. G., "Distillation Control for Productivity and Energy Conservation," Chaps. 1 and 2, McGraw-Hill, New York (1977).
[5] Luyben, W. L., I&EC Fund., 14:321–325, (1975).
[6] McAvoy, T. J. and Weischedel, K., "A Dynamic Comparison of Material Balance Versus Conventional Control of Distillation Towers," presented at VIII IFAC Congress, Kyoto, Japan (Aug. 1981).
[7] Weischedel, K. and McAvoy, T. J., "Feasibility of Decoupling in Conventionally Controlled Distillation Columns," I&EC Fund, 19:379, (1980).
[8] Luyben, W. L., AIChE J., 16:198, (1970).
[9] Jafarey, A. and McAvoy, T. J., I&EC Process Design Develop., 17:485, (1978).
[10] McAvoy, T. J., I&EC Fund., 18:269, (1979).
[11] Jafarey, A., McAvoy, T. J., and Douglas, J. M., I&EC Process Design Develop., 19:114, (1980).
[12] Bristol, E. H., IEEE Trans. Auto Control, AC-11:133, (1966).
[13] Jafarey, A., McAvoy, T. J. and Douglas, J. M., I&EC Fund., 18:181, (1979).
[14] McAvoy, T. J., ISA Trans., 16(4):83, (1977).
[15] Jafarey, A., Douglas, J. M., and McAvoy, T. J., I&EC Process Design Develop., 18:197, (1979).
[16] Mizuno, H., Watanabe, Y., Nishimura, Y. and Matsubara, M., CES, 27:129, (1972).
[17] Nazmul, K. and Stainthorp, F. P., "Asymmetric Dynamics and Feedforward Control of a Fractionating Column," presented at JACC Meeting, Boulder, Colorado (June 1979).
[18] Fuentes, C., and Luyben, W. L., "Control of High Purity Distillation Columns," presented at 73rd Annual A.I.Ch.E. Meeting, Chicago, Ill. (Nov. 1980).
[19] Rafal, M. D. and Stevens, W. F., AIChE J., 14:85, (1968).
[20] Hu, Y. C., and Ramirez, W. F., AIChE J., 18:479, (1972).
[21] Luyben, W. L. and Vinante, C. D., Kem. Teollisuus, 29:499, (1972).
[22] Toijala, K. and Fagervik, K., Kem. Teollisuus, 29:1, (1972).
[23] Wood, R. K. and Berry, M. W., CES, 28:1707, (1973).
[24] Oakley, D. O. and Edgar, T. F., "Optimal Feedback Control of Binary Distillation Column," presented at JACC Meeting (1976).
[25] Schwanke, C. D., Edgar, T. F., and Hougen, J. O., ISA Trans., 16(4):69, (1977).
[26] Brosilow, C. and Tong, M., AIChE J., 24:492, (1978).
[27] Tolliver, T. L. and Waggoner, R. C., "Distillation Column Control; A Review and Perspective from the CPI," presented at short course on distillation dynamics and control, Lehigh University (May 1980).
[28] Tyreus, B. D., Luyben, W. L. and Schiesser, W. E., I&EC Process Design Develop., 14:427, (1975).
[29] Gallun, S. E. and Holland, C. D., "Extractive Distillation Column Described at Unsteady State Using Gear's Procedure," presented at 73rd Annual AIChE Meeting, Chicago, Ill. (Nov. 1980).

[30] Prokopakis, G. J. and Seider, N. D., "Dynamic Simulation of Distillation Towers," presented at 73rd Annual AIChE Meeting, Chicago, Ill. (Nov. 1980).
[31] Ballard, D. and Brosilow, C., "Dynamic Simulation of Multicomponent Distillation Columns," presented at 71st Annual Meeting of AIChE, Miami (Nov. 1978).

APPENDIX

DERIVATION OF AN APPROXIMATE EXPRESSION FOR STEADY STATE GAIN — $\partial x_D / \partial L |_V$

In deriving an expression for $\left.\frac{\partial x_D}{\partial L}\right|_V$, a high purity tower is assumed ($x_D \to 1.0$, $x_B \to 0$.). For such a tower the numerator in eq. 2 can be written as

$$\frac{x_D}{1-x_D} \cdot \frac{1-x_B}{x_B} \simeq \frac{1}{1-x_D} \cdot \frac{1}{x_B} \tag{A-1}$$

If eq. A-1 is substituted into eq. 2, if D is replaced with V−L and if equations 1 and 2 are differentiated with respect to L holding V constant, one gets

$$0 = \left.\frac{D \partial x_D}{\partial L}\right|_V - x_D + \left.\frac{(F-D)\partial x_B}{\partial L}\right|_V + x_B \tag{A-2}$$

$$\frac{(N/2)(R+1)}{(Rx_F+1)(R)D} = \frac{1}{(1-x_D)} \left.\frac{\partial x_D}{\partial L}\right|_V - \frac{1}{x_B} \left.\frac{\partial x_B}{\partial L}\right|_V \tag{A-3}$$

Eq. A-2 can be used to eliminate $\partial x_D / \partial V |_L$ from eq. A-3 to give

$$\left.\frac{\partial x_D}{\partial x_L}\right|_V = \frac{1-x_D}{x_B} \cdot \frac{\dfrac{x_D - x_B}{F - D} + \dfrac{N(R+1)x_B}{2RD(Rx_F+1)}}{1 + \dfrac{(1-x_D)D}{x_B(F-D)}} \tag{A-4}$$

For high purity towers the first term in the numerator of eq. A-4 is much more significant than the second term. Further, $x_D - x_B$ in the first term can be approximated as 1.0 to give

Chemical Process Control II

$$\left.\frac{\partial x_D}{\partial L}\right|_V \approx \frac{\frac{(1-x_D)}{x_B (F-D)}}{1 + \frac{(1-x_D)D}{x_B (F-D)}} \qquad (A\text{-}5)$$

Multiplying eq. A-5 by F gives eq. 3.

UNIVERSITY RESEARCH ON DUAL COMPOSITION CONTROL OF DISTILLATION: A REVIEW

Kurt V. Waller
Process Control Laboratory
Department of Chemical Engineering
Åbo Akademi, 20500 ÅBO FINLAND

ABSTRACT

The popular research object of dual composition control of distillation is reviewed. Various feedback strategies are treated in order of increasing complexity. Some aspects on feedforward control are discussed, and a survey is given of experimental facilities for university research on computer control of distillation. The review ends with a list of unsolved problems and suggestions for future research.

INTRODUCTION

For a long time distillation control has been the most popular subject in chemical process control research. Dual-composition control in distillation has received more attention than any other interacting chemical process, but deservedly so, because of its industrial importance [1].

The distillation process has features such as high dimensionality, nonlinearity, and significant dead time. Usually only few measurements are available. Multicomponent distillation and columns with side streams or multiple feeds have barely been touched upon. Future systems will be designed to be more integrated and will be operated closer to constraints. These facts taken together will maintain distillation a stimulating challenge for chemical process control research for a long time to come.

This review of the research on dual composition control of distillation is complementary to Edgar's and Schwanke's review of the application of modern control theory to distillation columns [80].

CONTROL BY SINGLE-INPUT SINGLE-OUTPUT (SISO) CONTROLLERS

The traditional industrial control strategy for multivariable control problems is to use a multiloop control scheme consisting of several PI or PID controllers connected in SISO-loops [2]. This is the approach with which more complicated control strategies should be compared.

The standard scheme for dual composition control of distillation consists of two feedback loops. The top composition x_D (or an indirect measure of it, like a temperature in the top section of the column) is connected to the reflux flow L or to the distillate flow D, (L←x_D) or (D←x_D). Analogously the bottoms composition is controlled by manipulation of boilup V or bottoms flow B in response to the bottoms composition x_B: (V←x_B) or (B←x_B). Of the four possible combinations, only three are useful. The scheme (L←x_D) (V←x_B) is usually referred to as the "conventional" or "standard" approach, whereas (D←x_D) (V←x_B) and (L←x_D) (B←x_B) are called (Shinskey's) "material balance control" schemes [3, 4, 5].

The two control loops are coupled through the distillation column: the control actions taken by the loop at one end propagate through the column so as to enter the loop at the other end as disturbances and are returned through the column to the first loop. An early paper illustrating the detrimental effects of the interaction is the one by Rosenbrook [6], where also the high sensitivity that such systems may have to process parameter changes is illustrated.

The strength of the interaction and the control quality obtainable by SISO-loops is dependent on the choice of manipulated variables. For conventional as well as for material-balance controlled columns the interaction in the steady-state sense is investigated by McAvoy [7] using Bristol's relative-gain-array (RGA) concept [8]. In a later simulation study where also dynamics is included, McAvoy and Weischedel [5] conclude that it depends on the system at hand whether conventional or material balance control is to be preferred. Shinskey [9] summarizes his experience on this issue: D-V manipulation is more effective with high reflux-ratio columns, whereas L-V manipulation seems to be better with those requiring a low reflux ratio.

Several workers have illustrated that neglecting process dynamics in the RGA analysis may lead to wrong pairing, and a number of extensions of the RGA concept to make it include also dynamic effects have been suggested, a favorite illustrative example being the distillation process [10, 11, 12, 2].

Between the two extremes conventional control and material balance control, there are a number of combinations, hybrids. The potential of using various ratios between the flows as manipulators has long been recognized. An early example is given by Rijnsdorp [13], who suggests the ratio of reflux and top vapor flow as manipulator for the top loop: (L/V←x_D). Stainthorp and Jackson [14] experimentally studied a scheme in which the top loop manipulator was L/D. For a number of columns McAvoy [7] studied the steady-state interaction also for Rijnsdorp's suggestion, both in combination with (V←x_B) and—as a hybrid between Rijnsdorp's and Shinskey's approaches—with (B←x_B). McAvoy found the smallest amount of steady-state interaction for the last scheme, (L/V←x_D) (B←x_B). He also extended the idea in search of such functional combinations of manipulated variables as would make the loops noninteracting, the final result was degeneracy [15].

A large number of ratio control schemes have actually been suggested in the literature, and a survey is given in the book by Rademaker et al. [16]. The book also discusses various ratio-control schemes, among them also two-ratio schemes, in which both the manipulators are flow ratios.

New interest in ratio control has recently been shown. Ryscamp [17] suggests the scheme (D/V←x_D) (V←x_B). Shinskey [9] analyses the scheme and finds it to leave a relative gain typical in the range of 1.5–3.5. Through simulation McAvoy and Weischedel [5] found it superior both to the conventional scheme and to the material balance control scheme.

These investigations in which are sought SISO-systems which show only mild interaction will have bearing not only on SISO-design. They will probably be of major interest also to those working on holistic multivariable approaches. One recent illustration: Dahlqvist [18] compared a multivariable self-tuning regulator utilizing L and V as manipulators with two SISO-selftuning regulators using L/V and V as manipulators. Thus although the latter scheme did not contain any terms accounting for the multivariable interaction, Dahlqvist reports that it showed performance comparable to or even better than the one obtained by the general multivariable selftuning regulator. This was reported both regarding simulations and the experiments on an 11-tray pilot scale column.

It thus seems most likely that these ratio control schemes will be shown interest also in the future, both as such and combined with more complicated schemes like various decoupling schemes.

Another problem in distillation control that is likely to meet with an increased activity is the nonlinearity of the process. It has been reported many times that high-purity columns can be very nonlinear, recently by Fuentes and Luyben [19].

The whole field of *variable transformation to facilitate control* is a most stimulating challenge for research [20]. One example was discussed above: choice of various flow ratios as manipulators for *reduction of the interaction* between loops. Another example is the *reduction of the nonlinearity* of distillation systems by a choice of new suitable variables [16]. A third purpose, discussed by Waller and Mäkilä [21] for chemical reactor analysis and control, is the one for *reduction of the dimensionality* of the system.

DECOUPLING

Since the interaction between the two composition loops as a rule is detrimental, it is intuitively appealing to insert some kind of interaction compensators, much like feedforward controllers, to cancel out the effects of each manipulative variable on the composition at the opposite end of the column [22]. This is the approach labeled decoupling in the distillation control literature.

Distillation decoupling has been shown great interest. 10 years ago Luyben [22], giving credit to Buckley, initiated a series of papers on the subject [23] to [32].

In these studies the standard manipulators L and V were used. Two decoupling elements were used to eliminate the interaction. In recent studies these two choices have been investigated.

As discussed above, the interaction is strongly dependent on the choice of manipulated variables. Since the decouplers are designed on the basis of process models, perfect decoupling will never be achieved in practice due to model mismatching. Then it is natural that the system sensitivity to decoupler errors will be different for different choices of manipulators [33]. Therefore the choice of manipulators is of importance also in the decoupling approach, and future work can be expected in decoupling in connection with various ratio-control schemes.

The other item of increased recent interest concerns the number of decouplers to use. The first-used approach to have two decouplers between the two feedback loops has been called *two-way decoupling* (also complete or bilateral decoupling). Simplifying the scheme by putting one of the decouplers to zero results in *one-way* (or partial, unilateral) *decoupling*. The rationale for one-way decoupling has been formulated by Shinskey [34] so that one decoupler is enough to break the "third loop" in the system, i.e. the loop through the process connecting the two primary feedback loops.

One-way decoupling was experimentally tried by Stainthorp and Jackson [14] but no significant improvement in control was obtained compared to a SISO-scheme with L/D as manipulator in the top loop.

Oakley and Edgar [35] found from a linear-quadratic (LQ) control design that the resulting optimal scheme with good accuracy could be approximated by a scheme having the same structure as a one-way decoupled scheme, where changes in reboiler heat duty are fed to the top loop not only through the column but also through the decoupler. The same conclusion is obtained in a study by Tung and Edgar [11] on the extension of the RGA-concept to include dynamics.

Dahlqvist [18] found when applying a multi-input multi-output (MIMO) self-tuner to a pilot plant column that the controller could be reduced to the structure of one-way decoupling without significant loss in control quality.

One-way decoupling in distillation is recommended by Shinskey [34, 4] and McAvoy and coworkers [15, 36].

Recently some simulation comparisons between one-way and two-way decoupling have been presented. Gagnepain and Seborg [2] found that one-way decoupling was capable of achieving tight control of one controlled variable but sometimes produced poor control of the other one. Two-way decoupling did not necessarily provide better control than conventional multiloop schemes. Further research was concluded to be necessary.

For one simulated column, Waller et al. [37] compared six feedback strategies, among them the three studied by Gagnepain and Seborg. Waller et al. found that the best control quality does not necessarily correspond to the best decoupling and concluded that decoupling may be used to improve the control quality of a system, not because of what it is meant to do, but because of the additional degrees of freedom it introduces into the control system through the additional number of design parameters.

However, making a control system more complex might also increase its sensitivity. This was illustrated in the sensitivity analysis made by Toijala and Fagervik [24]: "There is quite a risk of decreasing the controllability properties of the process through the introduction of the decoupling elements" [29]. Naturally the risk increases with increasing nonlinearity, as recently illustrated by Weischedel and McAvoy [38].

For further study, one-way nonlinear or adaptive decoupling of non-minimum-phase columns has been suggested as the area which may have the greatest practical potential concerning decoupling in distillation [1].

Multivariable Frequency Domain Methods

Some of the multivariable frequency methods developed in Great Britain show some similarities—such as diagonal dominance—with the decoupling approach.

In dual composition control of distillation, Edgar and coworkers [31] tried the Characteristic Loci method but without much success. Bilec and Wood [39] investigated the Direct Nyquist Array method, but only minor improvement compared to SISO-control was obtained.

On the other hand, Tyréus [40] reports a successful application of the Inverse Nyquist Array method to a 30-tray, 2-meter diameter industrial column. Considerable dynamic improvements were reported to be achieved compared to a multiloop SISO system. Interactive computer programs are usually used in these frequency response methods, but Tyréus was able to perform the design without such programs.

HOLISTIC NONADAPTIVE MULTIVARIABLE CONTROL METHODS

It is quite natural that a theoretically optimal control system should have the property that all measurements are connected to all manipulators. In practice, however, some of the connections might as well be zero. Consideration of such aspects as model mismatching and system sensitivity may well be expected to lead to the result that some of the connections in practice *ought to be* zero.

This theoretically optimal structure where all outputs are connected to all inputs is the same as the one in two-way decoupling utilizing two sensors. The decoupling approach, however, has the questionable feature that it *a priori* tries to eliminate interaction, without any check of how beneficial the interaction may be for the system

The same structure as in two-way decoupling is also obtained when the (weighted) sum and the (weighted) difference between the signals from the two sensors (if two) are used for feedback. An early example is provided by Davison [41], other investigators have later discussed the same approach.

Linear-quadratic (LQ) Approaches

Neglecting control costs in the performance index in deterministic LQ problems results for continuous systems in bang-bang control. This approach has been studied by Brosilow and coworkers for one-product control [42, 43] and extended to two-product control by Toijala and Gustafsson [44], but found to be unsuitable for disturbance rejection in practice due to excessive switching of the controls [44, 45, 35].

It is more suitable to work in discrete time. Then the control is of dead-beat type. McGinnis and Wood [46] tried a simple deadbeat controller experimentally but satisfactory control was not obtained, a fact attributed to the noise in the bottoms composition measurements.

Rafal and Stevens [47] simulated a 3-plate column, optimizing a one-step ahead criterion. Hu and Ramirez [48] simulated a nonlinear 5-tray column and found optimal (LQ) controllers to give control superior to that of well-tuned SISO PI- controllers.

Hu and Ramirez [48] utilized a performance index in which the control costs were weighted through a weight put on u, the time derivative of the control signal. This was a very popular performance index formulation in the early 70s and well suited to cope with constant (step) disturbances (see Waller and Gustafsson [49] for a discussion of the formulation and other possible formulations for process control needs).

A performance index including u was also used by Waller et al. [50] in a simulation study of model mismatching in LQ distillation control. The controller was computed on a model reduced from the original model for the 10-plate column, and applied (through simulation) to the original high-order model. No significant reduction in control quality was obtained even if the model used in the controller computation was reduced to 4th order.

Extension to include also state estimation (Kalman-filtering) was treated by Hammarström and Waller [51] where special emphasis was put on how to treat the integrals in the PI-controllers utilizing state estimation.

A controller containing double integration of the outputs was also illustrated in [50]. This controller is interesting in process control, since it takes the integral of the output to zero, i.e. gives the correct *average* value for the outputs [75]. If the product from the process is stored and blended, the average value may be much more interesting than the momentary value.

Double integration was also illustrated by Chan and Talbot [52], who extended the study of Hu and Ramirez [48] to cope with set-point changes. Abedi [53] extended the analysis to include state estimation as well through Kalman-filters and Luenberger observers.

An experimental comparison of LQ-control with two SISO-loops was made by Pike and Thomas [54] on a 12-plate pilot column.

Oakley and Edgar [35] studied dual-composition control for set-point changes both through simulation and experimentally on an 8-tray column. The multivariable LQ-controller was found to perform better than tuned SISO controllers. Oakley and Edgar verified experimentally the simulation results of Waller and coworkers [50] which showed that neglecting the feedback coefficients for intermediate plates did not significantly reduce the control quality of the multivariable controller.

Dahlqvist [55] experimentally compared multivariable PI-control designed by LQ methods with SISO-controllers on an 11-tray pilot column. Although the multivariable controllers were derived from purely theoretical models they were reported in general to give better control results than the SISO-controllers tuned to the real process.

More extensive systematic comparisons between various control strategies for dual composition control have been reported by Edgar et al. [31, 56], Waller et al. [37], and Taiwo [32]. In all these studies the manipulators are the standard ones: reflux and reboil rate.

Edgar et al. *simulated set-point changes for one column* (8 plates). The control methods studied were: (1) SISO (2) Decoupling (3) Characteristic Loci (4) Continuous optimal (LQ) (5) Discrete optimal (LQ). The discrete optimal yielded the best simulated performance, but its sensitivity to model mismatching remained unknown. Also, for experimental use the problem of state estimation would have to be solved. Edgar et al. conclude that the decoupling approach offers nearly as good a control as optimal control offers, but that it involves much less design time.

Waller and coworkers [37] *simulated disturbance rejection for one column* (10 plates). The controllers studied were SISO, two-way decoupling, one-way decoupling as well as three approaches originating in LQ-control. In two of the three LQ approaches it was assumed that the inner structure of the controller was the same as the one in two-way decoupling with frequency independent decouplers, i.e. the two outputs were connected to the two inputs through PI-elements only. The parameters of these structure-constrained regulators were calculated from different process models: the first one from a high-order model, the second from reduced state-space models. In the last approach studied, the constraints on the inner structure of the controllers were removed. The resulting controllers then contain observers, which in the study were taken to be minimal-order Luenberger observers. This last approach resulted in a simulated control quality considerably better than that of the other 5 approaches. The issue of model mismatching was given special emphasis in the study.

Taiwo [32] *simulated disturbance rejection for 4 column models*, namely those of Luyben and Vinante [25], Wood and Berry [28], Edgar et al. [31], and a model of a packed bed pilot plant column. Three feedback control strategies were compared: (1) SISO-control by two PI-controllers (2) Two-way decoupling (3) Multivariable PI-control, consisting of 4 PI-controllers connected between the two outputs and the two inputs. The controllers were tuned by a procedure called "the method of inequalities," a procedure for trade-off between specifications such as settling time and overshoot.

Taiwo's results were in the line with those of Waller et al. [37], that is, that the increased degrees of freedom obtained by increasing the number of parameters can be utilized to improve the control quality.

Sensor Location and Optimal Process Model Complexity

The location of sensors may be a very important problem. A frequent approach in the control literature is to locate the sensors by some kind of observability considerations.

Waller et al. [50] illustrated through simulation the drastic effect the location of a third sensor (two fixed in the product lines) may have on distillation control quality. Also when state estimation (Kalman-filter) was used [51], the location of the third sensor strongly affected control quality in LQ-control. In the latter study, no correlation was found between control quality and an observability index suggested by Balchen et al. [57].

It thus seems that observability considerations are not enough for sensor location. This is also the opinion of Mellefont and Sargent [58] who suggest an implicit enumeration algorithm for the selection of measurements to be used in optimal feedback control of a linear stochastic system. Use of the algorithm on a distillation system gave results which were "considerably different from what would be expected for pure estimation" [58].

The problem is also treated in the so called Geometric Approach for control system design. An illustration is given by Takamatsu et al. [59] in a simulation study of a 9-plate column. Postulating proportional feedback the approach is used to find out which states are to be fed back for complete disturbance rejection. For the studied system it is found that feedback of only two plate compositions can completely eliminate the effect of disturbances in feed composition on both outputs (for all times). For changes in feed flow rate there exists no feedback that completely eliminates the effect of the disturbances on both outputs for the column studied.

The optimal size and complexity of the process model for controller design is directly related to the location and number of sensors. Dahlqvist [18] investigated, both experimentally and through simulation on an 11-plate pilot column, LQ-controllers based on various model sizes. The unmeasured states were obtained through state estimation. Using 2 sensors, a 2nd order model gave better results than a 6th order model. With 3 sensors, a 6th order model gave better control than a 13th order model, and also better than a 4th order model.

These results qualitatively agree with results obtained by Waller and Hammarström [60]. A heuristic explanation of the fact that higher-order models may degrade control quality is that more states have to be estimated in higher order models, and they will be more erroneous the more they are.

Optimal model complexity in controller design is a central problem in making way for multivariable control theory (like LQ) to successful industrial applications.

ADAPTIVE FEEDBACK CONTROL

There have been a number of discussions and investigations during the last years, both experimental and through simulations, of adaptive control of distillation, especially of self-tuning regulators for minimum variance control [61–65, 18].

Most of the mentioned papers treated the dual composition control problem, but the potential of the adaptive methods was not put to a severe test, since the process conditions were fairly constant.

It seems that the self-tuning regulator has considerable potential for dual-composition control of distillation. Results of Dahlqvist [18] indicate that the control quality can be significantly affected by the choice of manipulators. It would now be interesting to test the controller on industrial scale columns, where both changing process conditions and realistic disturbances would put the approach to a real test.

FEEDFORWARD CONTROL

Feedforward control has for a long time been considered to yield much control improvement for distillation, even when very crude and simple feedforward controllers are used. It is mainly the often easily measurable feed flow changes that are compensated by changed product flows as well as by changed internal flows. It is straightforward to include this kind of feedforward in the previously discussed feedback controllers, as for example illustrated by Dahlqvist for the self-tuning regulator [18].

An illustration of the application of linear feedforward controllers, based on column transfer functions for both products, is given by Wardle and Wood [66] in an industrial scale column control study (column diameter 1.2 meter, 30 plates, separating benzene and methyl-ethyl-ketone).

Jafarey and McAvoy [67] recently studied steady-state feedforward control to compensate also disturbances in feed composition. The approach is an interesting alternative to feedback for dual composition control, but presupposes that feed composition can be measured. The key to the strategy is an equation which describes how the internal flows, e.g. V/F, should be varied with changes in x_F to keep the separation x_D-x_B constant. Then fixing x_D (or x_B) at its set point and changing V/F with x_F according to the relation in question also fixes the other product composition. Thus the two feedback loops normally used for dual composition control are replaced by one feedback and one feedforward loop. It seems that there is still quite a potential in this kind of studies of *structural alternatives* for distillation control.

EXPERIMENTAL FACILITIES FOR UNIVERSITY RESEARCH IN DISTILLATION CONTROL

The distillation process is one of the most popular processes for university pilot plant research in process control. This is due to several reasons. One is the economic importance of this dominating separation process. Its importance is steadily increasing with increasing energy costs. Another reason is the complexity of the process. So far mainly binary mixtures have been studied, the whole world of multicomponent distillation is a virgin field for process control research. The process is also well suited for use in teaching of multivariable process control and estimation techniques.

Table 1 is not a complete list but provides a rough picture of the experimental situation in the university world. Some of the systems listed are not in use at present, and a number of new systems not listed are at the planning and building stage.

Table 1. Survey of experimental university facilities for computer control of distillation. (Systems in parenthesis are expected to be running in the near future.)

Country	University	Distillation Column No. of plates	Plate type	Column diameter	Mixture Distilled	Reference
Canada	U. of Alberta	8	bubble-cap	23 cm	methanol/water	[28]
	U. of Ottawa	10	bubble-cap	23 cm	methanol/water	[52]
England	U. of Newcastle upon Tyne	16	sieve	20 cm	2-methoxyethanol/water	[62] [64]
	U.M.I.S.T.	29	bubble-cap	30 cm	methanol/isopropanol	[14]
	U.M.I.S.T.	8.5 m high	packed bed	23 cm	methanol/isopropanol	[32]
Finland	Helsinki U. of Technology	12	sieve bubble-cap turbogrid	25 cm	ethanol/water	[73]
Japan	Kyoto U.	10	sieve	12 cm	ethanol/water	[76]
Norway	Technical U. of Norway	11	sieve	13 cm	ethanol/water	[74]
Sweden	Royal Institute of Technology	11	sieve	40 cm	ethanol/water	[18]
U.S.A.	Lehigh U.	24	bubble-cap	20 cm	methanol/water	[25]
	U. of California, Santa Barbara	12	sieve	15 cm	(1-butanol/2-butanol/tert.-butanol)	[78]
	U. of Colorado	5	sieve	13 cm	methanol/water	[48] [79]
	U. of Florida	12	bubble-cap		methanol/tert.-butanol	[54]
	U. of Texas at Austin	8	sieve	15 cm	methanol/water	[35]l
	U. of Wisconsin	20	bubble-cap	30 cm	ethanol/water (ethanol/methanol/water)	[77]

UNSOLVED PROBLEMS AND SUGGESTIONS FOR FUTURE RESEARCH

There is a wealth of unsolved problems in distillation control. The list below is largely based on opinions expressed in the fall of 1980 by a number of university researchers and industrial designers.

Fundamental Problems in Distillation Dynamics and Control

There are a number of unsolved problems concerned with the understanding and prediction of the internal behavior in distillation towers. Such are

—plate efficiencies from a dynamic point of view. The problem is treated in [68].
—nonideal mixtures that exhibit large temperature gradients. Then the thermal dynamics may become dominating.
—effects of pressure transients, especially in low-pressure distillation.
—system properties close to constraints or when exceeding constraints, such as for partially flooded columns.

Also the difficulties of getting fast and reliable composition measurements can be considered an important fundamental problem for distillation control.

Model Mismatching

Model mismatching is one of the central issues in process control. The failure of advanced multivariable control systems is often due to the mismatch between physical system and system model used for controller design. The optimal complexity of process models is one important item, already mentioned, another is the question of process nonlinearities, repeatedly reported for high-purity columns.

One way of solving the model mismatching question is through adaptive control approaches, certainly one issue worth more investigation.

Control System Structures

This problem contains several subproblems, already discussed above, such as choice of manipulators, and sensor number and location. The manipulator issue concerns using various flow ratios as well as the possibility of using feedforward/feedback systems instead of multiloop feedback systems.

Multicomponent Distillation

In multicomponent distillation dynamics and control, only some exploratory studies have been published. Here non-key components can show most irregular behavior, such as inverse response (right-half-plane zeroes) and extremely low or high gains [69]. Also, the non-key components can have a significant effect on the behavior of the key components. There are several groups now heading in this direction.

Integration of Columns

While there are many unsolved problems for single columns, industrial distillation often involves trains of columns. With increased emphasis on energy utilization, column integration will become more important. Increased integration will cause new control problems, for example those introduced by the positive feedback caused by heat integrated systems.

Operator Acceptance of Advanced Control Schemes

This important item can be facilitated by simulation packages which, using e.g. color graphics, can illustrate for the operators how composition profiles in distillation columns move up or down in the columns depending on various control actions taken.

Reference Process and Disturbance Library

Summing up the Asilomar Conference on Chemical Process Control, Foss and Denn [70] listed some recommendations to the U.S. National Science Foundation. The first of these concerned the need for a set of simulation models that could be used as standard examples by research workers to test control strategies.

Such a standard library is slowly evolving. In distillation there are a number of pilot plant models used quite extensively: Toijala and Fagervik [24], Luyben and Vinante [25], Wood and Berry [28], Edgar et al. [31]. Taiwo [32] recently added a packed bed column model to the list.

Although there are some differences between the models mentioned, they are in many respects quite similar. They are, for example, all linear. Thus, there is a need for a larger variety of models with respect to generic characteristics, such as models with right-half-plane zeroes and sharp changes in gain with change in nominal conditions (strongly nonlinear) [71]. McAvoy [72] has taken up the problem by studying a spectrum of columns with respect to nonlinearity.

The variety of generic characteristics of distillation columns is probably the reason why several investigators don't see much of a problem in dual composition control in distillation, while other consider it a most relevant problem.

In a process model library, the most valuable models would be those for industrial columns. Also, models for typical industrial disturbances are of equal importance, since in many cases one cannot a priori state which model is more important in an application, that of the process or that of the disturbance [20].

Thus, a library of industrially significant generically different and classified distillation column models as well as disturbance models would be an extremely valuable tool for the future university research in distillation control.

Future Experiments

From the above follows that experimental projects should be discriminating in their choice of mixture and type of experiments to insure coverage of the generic characteristics considered important [71].

ACKNOWLEDGMENT

Most valuable information and comments on the subject treated have been provided by a number of persons, among them O. A. Asbjornsen, C. B. Brosilow, P. S. Buckley, S. A. Dahlqvist, T. F. Edgar, A. S. Foss, W. L. Luyben, T.J. McAvoy, H. V. Nordén, J. E. Rijnsdorp, P. Scholander, F. G. Shinskey, O. Taiwo, B. D. Tyréus and R. K. Wood. This kind help is gratefully acknowledged.

Travel support to the Engineering Foundation conference on Chemical Process Control II was provided by the Engineering Foundation and the Neste Foundation. This support is gratefully acknowledged.

REFERENCES

[1] Shinskey, F. G., Personal communication, 1980.
[2] Gagnepain, J. P. and D. E. Seborg, "An analysis of process interactions with applications to multiloop control system design," presented at the 72nd Annual Meeting, AIChE, San Francisco (1979).
[3] Shinskey, F. G., *Oil Gas J.*, 67 (July 14), 76 (1969).
[4] Shinskey, F. G., *Distillation control*, McGraw-Hill, New York (1977).
[5] McAvoy, T. J. and K. Weischedel, "A dynamic comparison of material balance versus conventional control of distillation columns," paper submitted to VIII IFAC world congress, Kyoto (1981).

[6] Rosenbrock, H. H., *Trans. Instn Chem. Engrs*, 40:35 (1962).
[7] McAvoy, T. J., *ISA Trans.* 16(4):83 (1977).
[8] Bristol, E. H., *IEEE Trans. Auto. Contr.* 11:133 (1966).
[9] Shinskey, F. G., "Predicting distillation column response using relative gains," The Foxboro Co. (1980).
[10] Witcher, M. F. and T. J. McAvoy, *ISA Trans.*, 16(3):35 (1977).
[11] Tung, L. S. And T. F. Edgar, "Analysis of control-output interaction in dynamic systems," presented at the 71th Annual Meeting, AIChE, Miami (1978).
[12] Kominek, K. W. and C. L. Smith, "Analysis of system interaction," presented at the 86th Nat'l. Meeting, AIChE, Houston (1979).
[13] Rijnsdorp, J. E., *Automatica*, 1:15 (1965).
[14] Stainthorp, F. P. and C. B. Jackson, "Control of a fractionating column subject to product rate demand changes," Proc. III IFAC Symp. on Multiv. Techn. Syst., Paper S 39 (1974).
[15] Jafarey, A. and T. J. McAvoy, *Ind. Eng. Chem. Proc. Des. Dev.*, 17:485 (1978).
[16] Rademaker, O., J. E. Rijnsdorp, and A. Maarleveld, *Dynamics and control of continuous distillation units*, Elsevier, Amsterdam-Oxford-New York (1975).
[17] Ryscamp, C. J., *Hydrocarbon processing*, 59(6):51 (1980).
[18] Dahlqvist, S. A., "Control of a distillation column—application of sampled data control to a large pilot plant," Tech.Dr thesis, Royal Institute of Technology, Stockholm, Sweden (1980).
[19] Fuentes, C. and W. L. Luyben, "Control of high-purity distillation columns," presented at the 73rd Annual Meeting, AIChE, Chicago (1980).
[20] Waller, K. V., *Paper and Timber*, 62:128 (1980).
[21] Waller, K. V. and P. M. Mäkilä, *Ind. Eng. Chem. Proc. Des. Dev.*, 20:1 (1981).
[22] Luyben, W. L., *AIChE J.*, 16:198 (1970).
[23] Niederlinski, A., *Ibid.*, 17:1261 (1971).
[24] Toijala (Waller), K. V. and K. Fagervik, *Kem. Teollisuus*, 29:5 (1972).
[25] Luyben, W. L. and C. D. Vinante, *Ibid.*, 29:499 (1972).
[26] Changlay, Y. and T. J. Ward, *AIChE J.*, 18:225 (1972).
[27] Shunta, J. P. and W. L. Luyben, *Chem. Eng. Sci.*, 27:1325 (1972).
[28] Wood, R. K. and M. W. Berry, *Ibid.*, 28:1707 (1973).
[29] Waller, K. V., *AIChE J.*, 20:592 (1974).
[30] Schwanke, C. O., T. F., Edgar, and J. O. Hougen, "Problems in the application of multivariable control theory to distillation columns," presented at the 69th Annual Meeting, AIChE, Chicago (1976).
[31] Schwanke, C. O., T. F. Edgar, and J. O. Hougen, *ISA Trans.*, 16(4):69 (1977).
[32] Taiwo, O., *Chem. Eng. Sci.*, 35:847 (1980).
[33] McAvoy, T. J., *Ind. Eng. Chem. Fund.*, 18:269 (1979).
[34] Shinskey, F. G., "The stability of interacting control loops with and without decoupling," Proc. IV IFAC Symp. on Multiv. Techn. Syst. p. 21 (1977).
[35] Oakley, D. R. and T. F. Edgar, "Optimal feedback control of a binary distillation column," Proc. JACC p. 551 (1976).
[36] Jafarey, A., T. J. McAvoy, and J. M. Douglas, *Ind. Eng. Chem. Proc. Des. Dev.*, 19:114 (1980).
[37] Waller, K. V., L. G. Hammarström, and K. C. Fagervik, "A comparison of six control approaches for two-product control of distillation," Report 80-2, Process Control Lab., Abo Akademi, Abo, Finland (1980).
[38] Weischedel, K. and T. J. McAvoy, *Ind. Eng. Chem. Fund.* 19:379 (1980).
[39] Bilec, R. and R. K. Wood, "Multivariable frequency domain controller design for a binary distillation column," presented at the 86th Nat'l. Meeting, AIChE, Houston (1979).

[40] Tyréus, B. D., *Ind. Eng. Chem. Proc. Des. Dev.*, 18:177 (1979).
[41] Davison, E. J., *Trans. Instn Chem. Engrs.*, 45:T 229 (1967).
[42] Brosilow, C. B. and K. R. Handley, *AIChE J.*, 14:467 (1968).
[43] Merluzzi, P. and C. B. Brosilow, *Ibid.*, 18:739 (1972).
[44] Toijala (Waller), K. V. and S. Gustafsson, "Simulation of bang-bang control of continuous distillation," Report 71-5, Process Control Lab., Abo Akademi, Abo, Finland (1971).
[45] Toijala (Waller), K. V., *Acta Acad. Aboensis Math. Phys.* 32(2), (1972).
[46] McGinnis, R. G. and R. K. Wood, *Can. J. Chem. Eng.*, 52:806 (1974).
[47] Rafal, M. D. and W. F. Stevens, *AIChE J.*, 14:85 (1968).
[48] Hu, Y. C. and W. F. Ramirez, *Ibid.*, 18:479 (1972).
[49] Waller, K. V. and S. E. Gustafsson, *Chem. Eng. Sci.*, 30:1265 (1975).
[50] Waller, K. V., S. E. Gustafsson, and L. G. Hammarström, "Incomplete state-feedback integral control of distillation columns," presented at KemTek 3, Copenhagen, Denmark (1974)
[51] Hammarström, L. G. and K. V. Waller, "State estimation in multivariable PI-control," Report 74-10, Process Control Lab., Abo Akademi, Abo, Finland (1974).
[52] Chan, A. L. and F. D. Talbot, *Can. J. Chem. Eng.*, 53:91 (1975).
[53] Abedi, H., "Optimal control and identification of distillation tower," Ph.D. thesis, Univ. of Colorado at Boulder (1976).
[54] Pike, D. H. and M. E. Thomas, *Ind. Eng. Chem. Proc. Des. Dev.*, 13:97 (1974).
[55] Dahlqvist, S. A., "Control of the top and the bottom product compositions in a pilot distillation column," presented at the third international symposium on distillation, Instn Chem. Engrs., London (1979).
[56] Schwanke, C. O., "Experimental multivariable control of a pilot plant scale distillation column," Ph.D. thesis, Univ. of Texas at Austin (1976).
[57] Balchen, J. G., M. Fjeld, and O. A. Solheim, *Multivariable control systems* (in Norwegian), Tapir press, Trondheim, Norway (1970).
[58] Mellefont, D. J. and R. W. H. Sargent, *Ind. Eng. Chem. Proc. Des. Dev.*, 17:549 (1978).
[59] Takamatsu, T., I. Hashimoto, and Y. Nakai, *Automatica*, 15:387 (1979).
[60] Waller, K. V. and L. G. Hammarström, Unpublished results, (1980).
[61] Sastry, V. A., D. E. Seborg, and R. K. Wood, *Automatica*, 13:417 (1977).
[62] Badr, O., D. Rey, and P. Ladet, "Adaptive dual control for a multivariable distillation process," Proc. 5th IFAC/IFIP Symp. on Digital computer applications to process control, p. 141 (1977).
[63] Morris, A. J., T. P. Fenton, Y. Nazer, "Application of self-tuning regulators to the control of chemical processes," *Ibid.*, p. 447 (1977).
[64] Morris, A. J. and Y. Nazer, "Distillation column control using self-tuning techniques," Proc. EFChE congress on Contribution of computers to the development of chemical engineering and industrial chemistry, p. 50 (1978).
[65] Anbumani, K., L. M. Patnaik, and I. G. Sarma, "Multivariable self-tuning control of a distillation column," Proc. JACC, FA5-D (1980).
[66] Wardle, A. P. and R. M. Wood, "Problems of application of theoretical feedforward control models to industrial-scale fractionating plant," I. Chem. E. Symp. Series No. 32, Instn Chem. Engrs., London (1969).
[67] Jafarey, A. and T. J. McAvoy, *ISA Trans.*, 19 (1980) (in press).
[68] Asbjornsen, O. A., *Chem. Eng. Sci.*, 28:2223 (1973).
[69] Toijala (Waller), K. V. and S. Gustafsson, *Kem. Teollisuus*, 29:173 (1972).
[70] Foss, A. A. and M. M. Denn, *AIChE Symp. Ser.*, 72(159):230 (1976).
[71] Foss, A. S., Personal communication, 1980.

[72] McAvoy, T. J., "On the difference between distillation column control problems," to be presented at this conference (1981).
[73] Salminen, K. K. and A. Halmu, *Kemia-Kemi*, 4:465 (1977).
[74] Loe, I., "Distillation column modeling, dynamics, and control," Tech. Dr. thesis, Technical University of Norway, Trondheim (1976),
[75] Scholander, P., "Computer control of a pilot plant stripping column," Proc. 4th IFAC/IFIP Int. Conf. on Digital computer applications to process control, Zürich (1974).
[76] Hashimoto, I., Personal communication, 1981.
[77] Ray, W. H., Personal communication, 1981.
[78] Seborg, D. E. Personal communication, 1981.
[79] Hu, Y. C., Ph.D. thesis, Univ. of Colorado at Boulder, 1970.
[80] Edgar, T. F. and C. O. Schwanke, "A review of the application of modern control theory to distillation columns," Proc. JACC, p. 1370 (1977).

Control in Energy Management and Production

CONTROL ANALYSIS OF GASIFICATION/ COMBINED CYCLE (GCC) POWER PLANTS

George H. Quentin
Advanced Power Systems Division
Electric Power Research Institute
Palo Alto, California

ABSTRACT

A series of control analyses were initiated by EPRI using computer simulation to verify that integrated Gasification/Combined Cycle (GCC) power plants can maneuver effectively to meet load-following requirements, as well as providing the baseload economic benefits shown in earlier EPRI studies. Experimental pilot plant studies were conducted simultaneously to determine the inherent transient response capabilities of key process components. Such analyses are rather unique in that dynamic performance and controllability of future process technology have been evaluated well before commercial development is completed.

INTRODUCTION

Gasification/Combined Cycle (GCC) Power Plants represent a promising new technology for converting fossil energy in the form of coal to electric power. They offer many advantages over current power plants which burn coal directly to generate steam for a steam turbine-generator.

GCC plants will convert coal to gas, and selectively remove sulfur compounds before burning the clean fuel gas in a gas combustion turbine at temperatures of 2000°F or higher. Exhaust gases from the gas turbine will produce steam in a heat recovery steam generator (HRSG) to power a steam turbine. Additional steam may be produced in a gasifier/waste heat boiler, closely coupling the fuel gas process with the combined cycle power plant.

As a result, GCC power plants can achieve high thermal efficiencies (or low heat rate), making them economically competitive with other options. Furthermore, they can readily meet environmental requirements without serious penalty to plant efficiency, and have potential for even lower emissions levels if required. Land and water requirements will also be greatly reduced, helping to ease site restrictions. Additionally, such plants are expected to respond readily to changes in electric power demand, making them suitable for daily load-following service.

The development of this attractive new type of electric generating unit has been strongly supported by the utility industry. However, others have also recognized the commercial potential of the gasification process for generating synthesis gas for producing other fuels such as methanol and a variety of chemicals, including ammonia. This points out that GCC plants will, in reality, closely integrate a chemical process which converts coal to clean fuel gas, with a power generating combined cycle process consisting of gas turbine and steam turbine units. These plants will, therefore, introduce a distinctly new type of power plant operating environment, i.e., an integrated plant, which in all likelihood will be centrally operated and controlled using current control technology, consisting of distributed microprocessor-based control systems. Furthermore, such plants will have been designed with the prospect for relatively fast rates of response for possible load following service.

PROCESS DYNAMICS AND CONTROL

Existing oil or gas-fired combined cycle units are generally fast responding units which can satisfy daily variations in electrical demand. However, the transient performance of other interacting plant components including coal gasification reactors, acid gas absorption columns, and gaseous oxygen plants remained to be determined. It was necessary to show that power plants based on coal-derived fuel gas can operate effectively under such everchanging conditions, and when closely-coupled with very responsive CCP units. To satisfy this important goal,

EPRI initiated a program of transient response studies involving both pilot plant experiments and detailed computer simulation. As shown in Figure 1, such studies have formed the basis for comprehensive GCC plant simulation and control analyses.

Recent transient tests on the BGC/Lurgi slagging, moving-bed gasifier at the British Gas Corporation pilot plant at Westfield (Scotland), and similar tests on the Texaco entrained-flow gasifier at their Montebello (California) pilot plant have both demonstrated excellent load-following characteristics.

GCC PLANT PROJECTS

The plant control studies have provided timely results serving as baseline information for design and operation of future GCC plants. Most prominent of such ventures is the impending GCC demonstration plant which will come on-line in early 1984, generating 100 megawatts at the Southern California Edison Cool Water Station, near Barstow, California.

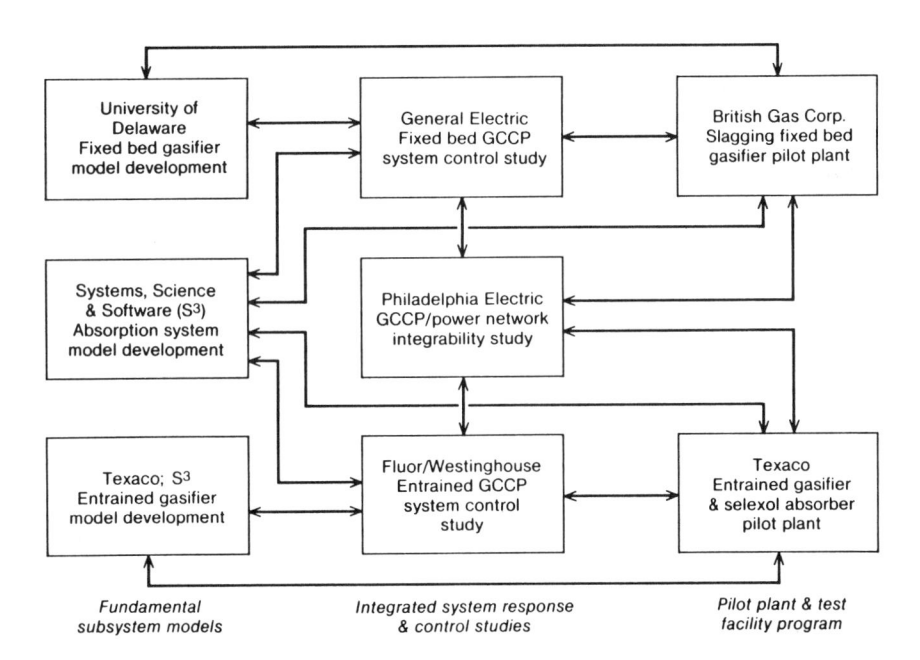

Figure 1. Gasification-Combined Cycle Power Plant Simulation and Control Projects.

This $300 million project will build a full-scale single-train process, under joint sponsorship of Bechtel, General Electric, EPRI, SCE, Texaco and others yet to be announced. It is not coincidental that this is the largest single EPRI project involvement to date (i.e., at $50 million). One of the important objectives of the Cool Water project is to verify experimentally the load-following capabilities of GCC power plants.

The demonstration plant flowsheet will evolve from an early conceptual design study for EPRI by R. M. Parsons Co., based on a conventional combined cycle plant (CCP) design with the combustion turbine exhausting to a heat recovery steam generator. This cycle contrasts sharply with the GCC plant built earlier by STEAG at Luenen, West Germany burning low Btu gas from a Lurgi gasifier in a supercharged boiler to produce steam before discharging to a Kraftwerk Union gas turbine. Cycle comparisons have been reported by Fluor Engineers & Constructors. Interestingly, the Luenen plant was planned as an intermediate load station suitable for daily startup and shutdown. Without benefit of prior dynamic analyses, the plant experienced a rather lengthy startup period.

POWER PLANT RESPONSE GUIDELINES

Large electric power plants have generally been designed for baseload service, however, it appears that future fossil power plants must be capable of cyclic operation. In prior years, as conventional plants aged and became expensive to operate, they were relegated to daily cycling with lower annual capacity factors. The age and design of such plants generally contribute to slower overall response rates.

It is difficult to set absolute requirements for the response rates of new power plants, however some guidelines have been outlined. Typical ranges of response for interconnected power system service are shown in the table below (which corresponds to Figure 2):

Table 1. Typical Ranges for Cyclic Operation of Power Plants

	Frequency Regulation		Tie-Line Thermal Backup		Daily Load Following	
Magnitude of Change % of rating	2	5	7	12	20	40
Response Rate Required % rating per min.	20	10	7	6	5	2
Time to Perform Change minutes	0.1	0.5	1	2	4	20

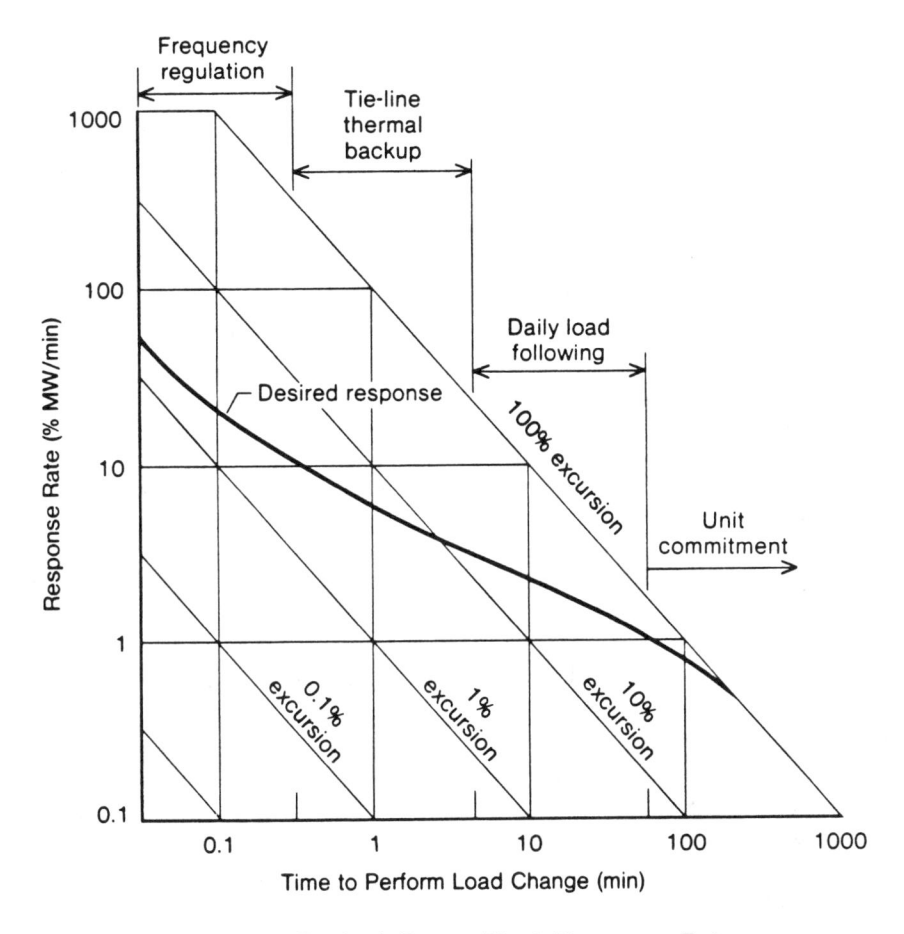

Figure 2. Desired Power Plant Response Rate.

Concisely, such guidelines require each unit to be: (1) independently stable under all forms of control, (2) controlled to within reasonable response limits, and (3) able to assume a proportionate share of load and/or frequency regulating duty. These requirements imply that even baseload units must be prepared to vary electrical output when necessary. Moreover, it appears that future units which can load-follow effectively will assume a larger share of the burden.

GCC PLANT SIMULATION

The dynamic response of GCC plants based on different gasifier technologies of specific interest for electric power generation were analyzed in separate simulation studies. Economic studies generally favor plants based on medium-Btu fuel gas from oxygen-blown, pressurized, slagging gasifiers of the entrained-flow (Texaco) type or the moving-bed (BGC/Lurgi) type.

Moving-bed GCC plant technology based on the Lurgi-type reactor has been simulated by General Electric, who initially examined air-blown, dry-ash operations. The entrained-flow GCC technology based on the Texaco-type partial oxidation reactor was simulated by Fluor with Westinghouse as subcontractor. In both cases, a full GCC plant simulation was developed for use with a power system network model, which generated load requirements for evaluating plant control strategies. In continuing efforts, GE will analyze dynamics and control of a moving-bed GCC plant based on a slagging version of the Lurgi gasifier developed by British Gas Corporation (BGC).

In a subsequent project, Philadelphia Electric is evaluating the insertion of GCC plants into a power system simulation of actual operations on the Pennsylvania-New Jersey-Maryland (PJM) interconnection. They formed simplified plant models, and verified them against the abovementioned GE and Fluor/Westinghouse models. They are examining effects caused by the manner of loading GCC plants onto the system in place of conventional fossil units. Preliminary results indicate the sensitivity of results to the assumed incremental cost curve for the plants, which remain to be accurately determined. Sequential loading of units in a given plant, or of multiple trains appears to offer some response advantages.

PLANT CONFIGURATION

For the simulation, the GCC plants were configured using the conventional combined cycle flowsheets from earlier economics studies by Fluor for EPRI. Therefore, certain design features may not be typical, such as the 2400°F gas turbine temperature in the entrained GCC study. However, in the absence of commercial plant designs, best engineering judgment was applied to represent the salient dynamic elements of a real plant.

In both the moving-bed and entrained GCC studies, the combined cycle process contains two gas turbines (GT), two heat recovery steam generators (HRSG), and a single steam turbine (ST). Each also simulated a single gas cleanup train, based on acid gas removal by physical absorption (Allied Chemical/Selexol type) to selectively remove hydrogen sulfide.

The moving-bed GCC plant generated low-Btu gas from four air-blown Lurgi type gasifiers with dry-ash bottom. The entrained GCC plant produced medium-Btu gas for two oxygen-blown Texaco-type gasifiers with slagging bottom. The oxygen plant for the latter study was not simulated in detail, but rates of oxygen supply and steam demand were varied to impact system dynamics realistically.

Plant integration schemes for both studies is shown in Figures 3-A and 3-B. Major differences are dictated by gasifier operating characteristics, which are outlined in Figures 4-A and 4-B.

Basically, the high temperatures in the entrained reactor consume all tars and oils, and produce considerable heat for high-pressure steam generation in a waste heat boiler. Lower temperature off-gases from the moving-bed reactor produce tar and oil, but generate only low-pressure jacket steam. Overall, the air-blown, moving-bed unit is a large steam consumer, while the entrained unit is a net steam producer. This directly affects fuel system integration, causing each plant to respond quite differently.

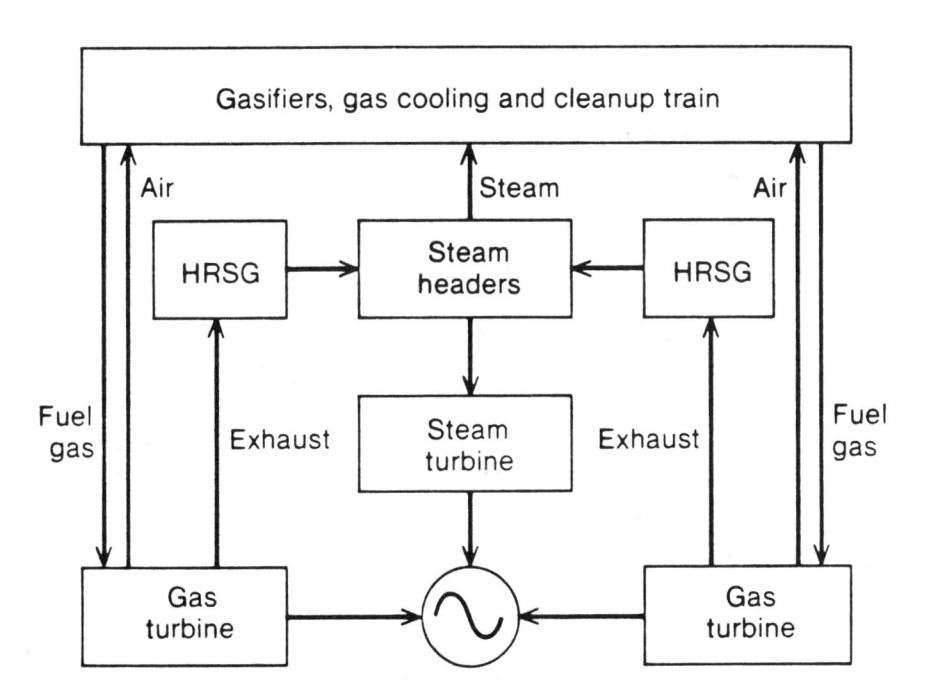

Figure 3a. Moving Bed GCC Plant Integration.

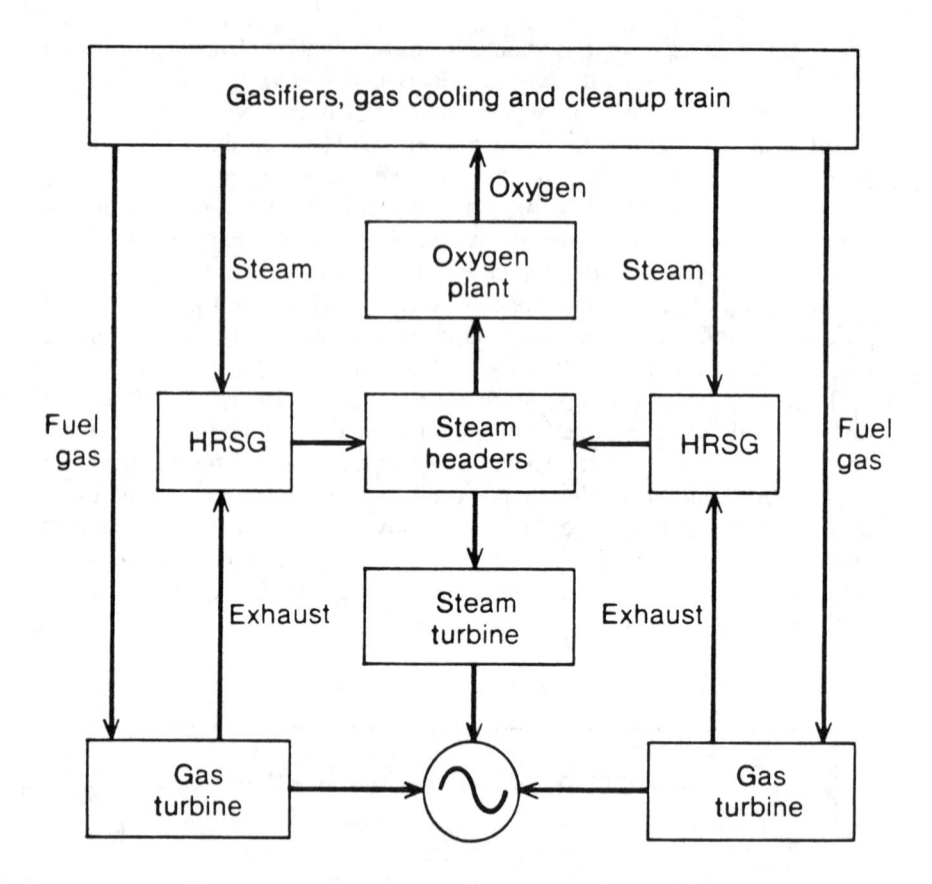

Figure 3b. Entrained Flow GCC Plant Integration.

The resulting differences in CCP characteristics are shown in Figure 5. In the moving-bed GCC plant, a large extraction steam flow is required for the gasifier, leading to a low steam turbine electrical output. Conversely, the entrained GCC plant offers a higher ST electrical output by admitting high pressure steam produced in the gasifier/ waste heat boiler.

RESULTS

Process Dynamics

The dynamic performance of the Texaco-type entrained gasifier was markedly different from the Lurgi-type moving-bed gasifier. The en-

Type	Lurgi	British gas	Texaco
Class & operation	Moving bed dry-ash	Moving bed slagging	Entrained flow slagging
Pressure (psi)	350	350	600
Coal feed	Dry/batch	Dry/batch	Continuous water slurry
Coal size	Lump	Lump	Pulverized
Gas feed	Air/steam	Oxygen/steam	oxygen
Residence time of solids	Long	Moderate	Very short
Gas throughput	Moderate	High	High
Product gas HHV (BTU/SCF, dry)	180	380	280
By-products	Sulfur, ammonia, tar & oil	Sulfur, ammonia, tar & oil	Sulfur
Off-gas temp. (°F)	1000° ±200	900° ±200	2500° ±200
Jacket steam	Yes	Yes	No

Figure 4a. Coal Gasification Reactor Characteristics.

trained reactor is capable of large step changes in throughput with relatively constant fuel gas temperature, composition, and heating value. Such results have been substantiated by pilot plant tests.

By comparison, fuel gas from the air-blown, Lurgi-type unit undergoes wide variation in heating value during transients, caused by fluctuating content of volatile hydrocarbons released from coal at the top of the reactor. For example, upon a load reduction at the gas turbine fuel valve, pressure control decreases the air and steam flows to the gasifier. This sudden loss of diluting flow from the bottom of the reaction bed, with constant release of volatiles at the top, causes a transient increase in gas heating value. This creates a positive feedback effect, generally hampering stable control.

Moving Bed Reactor	Entrained Flow Reactor
Coal flow and moisture	Slurry flow and water content
Caking properties and size	Coal particle size
Oxidant flow and temperature	Oxidant flow and temperature
Steam flow and temperature	
Maximum bed temperature	Reactor temperature
Coal volatiles release	
Bed pressure drop/flow distribution	Mixing (turndown effects)
Water balance	
Recycle tar, oil & dust	Recycle dust
Ash fusion temperature	Ash fusion temperature

Figure 4b. Important Factors Affecting Gasifier Operation.

Fuel gas pressure changes associated with pressurizing the coal feed lockhoppers (i.e., using fuel gas from the header) also caused variations in coal devolatilization rate and gas heating value. Such effects are not encountered with the entrained gasifier, since volatile compounds are not present in the product gas. Typical fuel gas compositions are shown in Figure 6.

Only the packed column absorber was simulated in these studies, and not the regenerator column or auxiliaries. A comprehensive, generalized simulation of a complete H_2S absorption systems (suitable for either chemical or physical processes) is being developed in another project by Systems, Science & Software to examine dynamic response in more detail.

In the entrained GCC study, the partially-simulated oxygen plant was permitted only a slow rate of change from baseload production. Assumed rates for these studies were typical of prior experience with high-purity oxygen plant designs (10–25%/hour). However, there are economic implications regarding the means of oxygen supply for the load-following abilities of a large oxygen plant, which led to specific studies with oxygen

	Moving bed GCCP	Entrained flow GCCP
Gas Turbines, rating	2 x 85 MW	2 x 113.1 MW
Combustor temp. (°F)	2000	2400
Pressure ratio	11	19
Exhaust temp. (°F)	1000	1150
HRSGs	2	2
	Economizer Evaporator superheater	Economizer evaporator superheater reheater gas heater
Steam Turbine, rating	1 x 35 MW	1 x 127.4 MW
High pressure (psi)	1265	1420
Temperature (°F)	900	900
Reheat temp. (°F)	—	1000
Process steam	Extraction flow	Admittance flow
Combined Cycle Net Power	205 MW	353.6 MW

Figure 5. Combined Cycle Plant Characteristics.

vendors. For example, the gasifier does not require high-purity (99+%) oxygen, therefore relaxation of the purity restriction can lead to improvements in operating flexibility and response rate.

One of the more important results of these simulation studies was the potentially large mismatch in rates of change for the gasification plant and the oxygen plant. Recent transient tests undertaken for EPRI by Air Products & Chemicals on a commercial (1000 tpd) oxygen plant have since dispelled any doubt about the ability of such plants to load-follow. In those tests, closed loop computer control was used effectively, applying feedforward/feedback algorithms to make rapid, well-balanced changes in load. Observed rates of change were consistent with highest expected rates of change for the GCC plant (e.g., up to 10%/minute).

	Gasifier Type/Oxidant		
Component	Lurgi/air	BGC/oxygen	Texaco/oxygen
CH_4	2.85	7.29	0.08
C_2H_4	0.03	0.22	—
C_2H_6	0.04	0.32	—
H_2	17.17	28.56	28.99
CO	11.94	54.51	42.45
CO_2	9.71	1.82	8.71
H_2S	0.52	1.29	1.01
COS	0.02	0.06	0.06
N_2	29.63	0.41	0.82
NH_3	0.35	0.85	—
H_2O	27.74	4.67	17.88
Total	100.00	100.00	100.00
Tar and oil, (lb/lb)	0.016	0.042	None

Note: The above results are based upon Illinois #6 coal.

Figure 6. Typical Gasifier Product Gas Composition.

System Control Analyses

Two general plant control strategies were evaluated, including a gas turbine-lead strategy and the alternative gasifier-lead mode. The turbine-lead mode represents a familiar power plant control strategy whereby the turbine fuel valve is first to respond to a changing load demand from the power system, and the fuel gas flow is regulated accordingly by pressure control of the gasifier feed streams. In the gasifier-lead mode, the gasifier feed valves respond first to a changing load demand, and the gas turbine fuel valve subsequently responds to pressure control. (See Figures 7 (A, B))

The results of each control mode for both simulation studies are shown in Figures 8 (A, B, C, D). In each case, the overall plant response is shown following a 20% decrease in electrical demand, imposed as a 4%/minute ramp change from the station controller. The unique operating differences between the moving-bed and entrained-flow fuel processes are apparent in their dynamic response.

The response of the moving-bed reactors was slower as expected, because of the larger reacting mass of coal, and more complex to control because of changes in product gas heating value. Again, by comparison, the response of the entrained-flow gasifier was rapid, exhibiting constant gas heating value, making it very compatible for closely-coupled operation with a gas turbine.

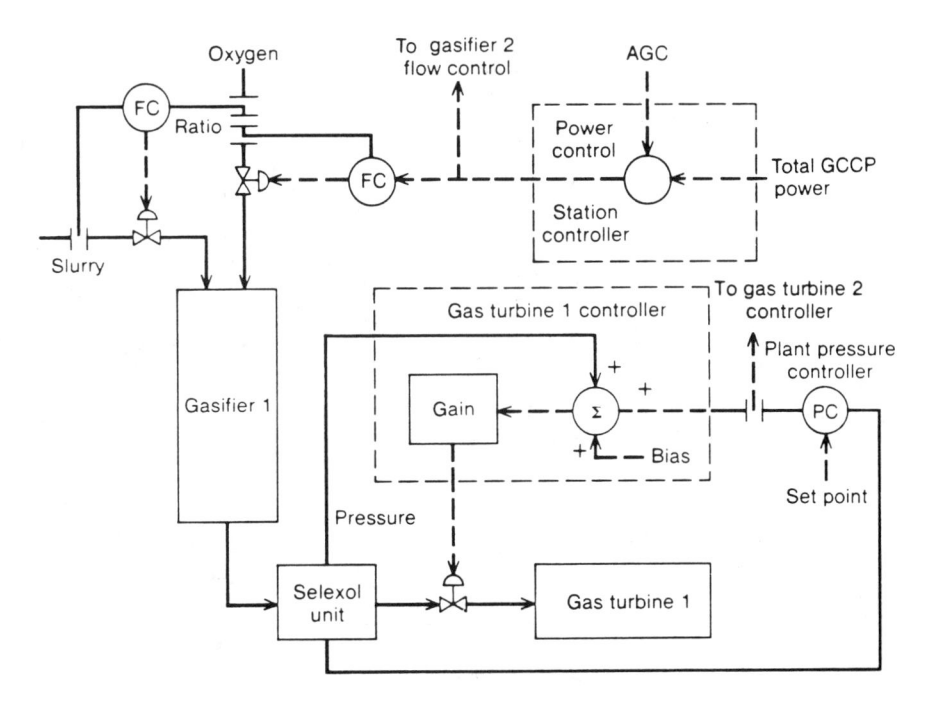

Figure 7a. Gasifier-Lead Control Diagram—Power Error to Gasifiers, Pressure Error to Gas Turbines.

428   Chemical Process Control II

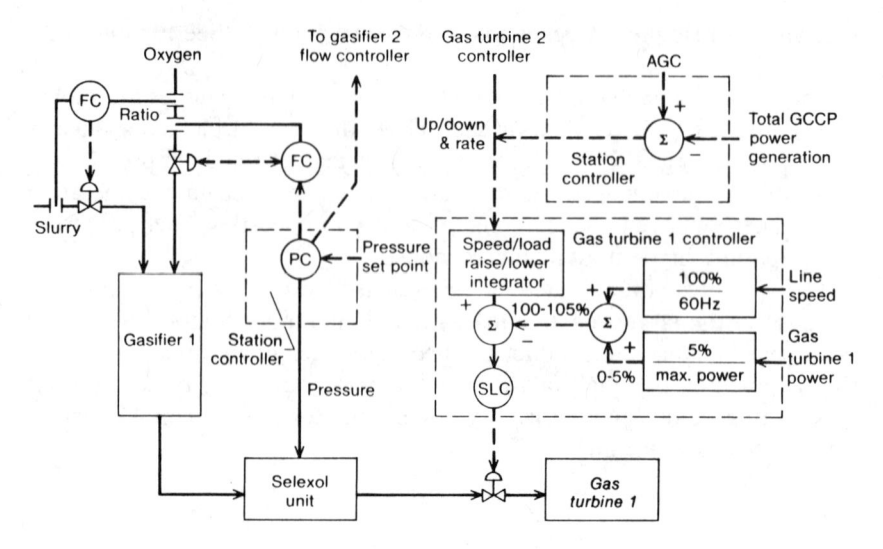

Figure 7b. Turbine-Lead Control Diagram—Power Error to Gas Turbines, Pressure Error to Gasifiers.

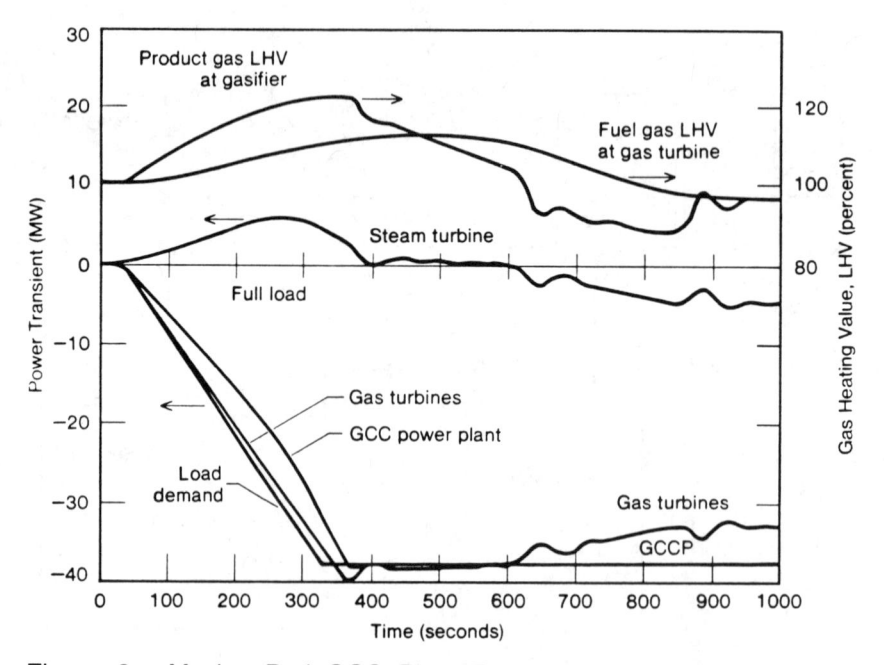

Figure 8a. Moving Bed GCC Plant Response—Turbine-Lead Control Mode—20% Decrease in Station Load at 4%/Minute Ramp Rate.

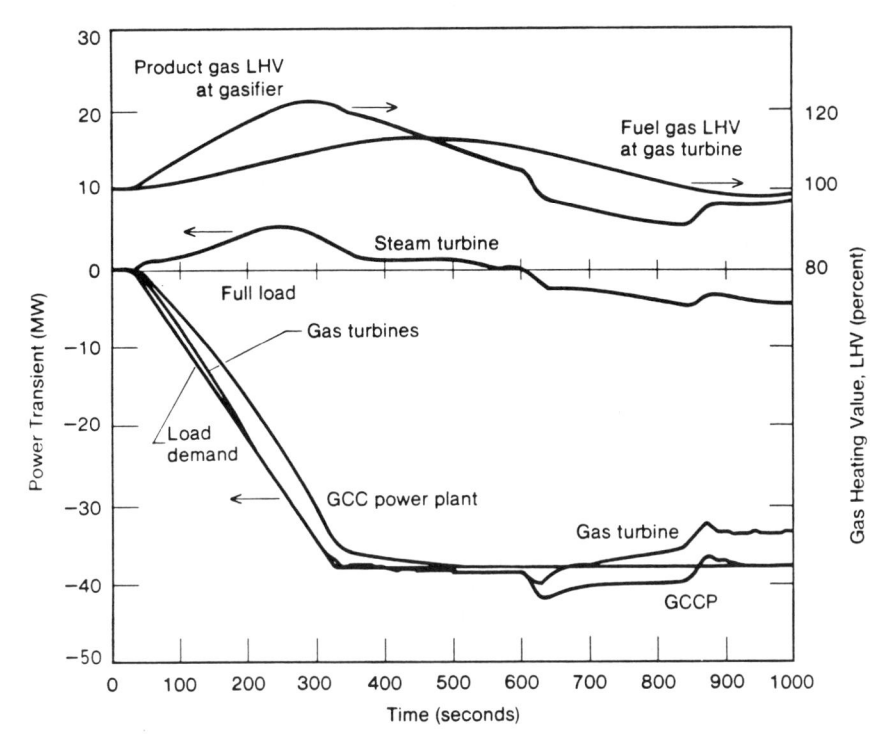

Figure 8b. Moving Bed GCC Plant Response—Gasifier Lead Control Mode—20% Decrease in Station Load at 4%/Minute Ramp Rate.

CONCLUSIONS

Simulated operation of the moving-bed GCC plant was characterized by fuel system transients as shown, which necessarily limited plant response to avoid serious upsets. Nevertheless, the plant was controlled satisfactorily to meet load-following requirements in the preferred turbine-lead control mode. The plant was less responsive when operated in the gasifier-lead mode, and experienced control difficulty making this strategy less acceptable.

The entrained-flow GCC plant demonstrated excellent capability for cyclic operation to meet power system needs. Both control strategies (i.e., turbine-lead and gasifier-lead) were quite successful in the closed

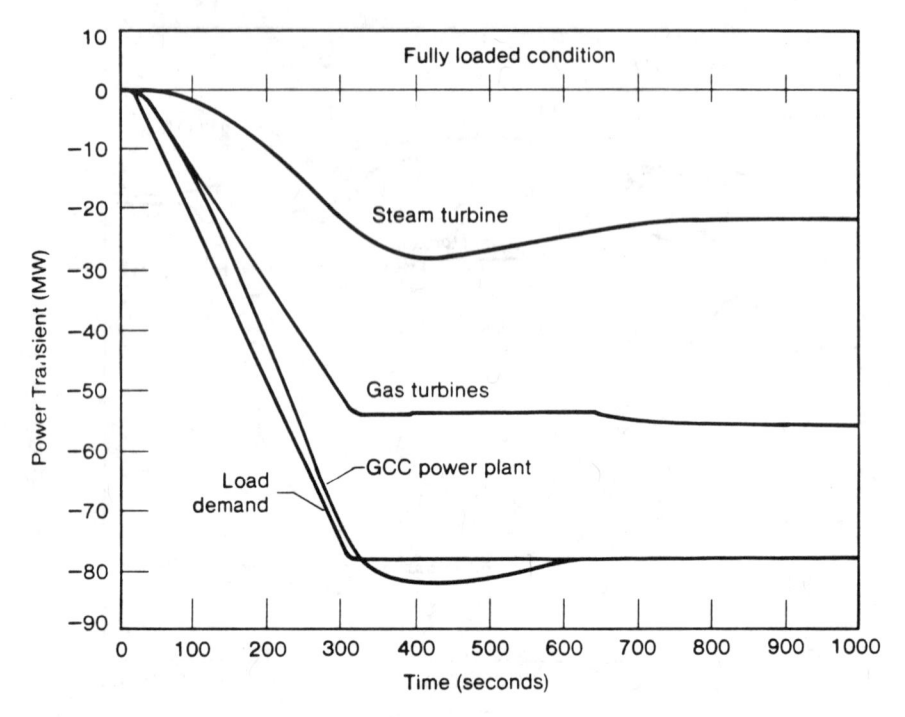

Figure 8c. Entrained Flow GCC Plant Response—Turbine-Lead Control Mode—20% Decrease in Station Load at 4%/Minute Ramp Rate.

loop control of the overall plant, both during normal transients and during emergency conditions following loss of key components.

Closed loop operation was successful in both plants studied, because the gas turbine control system effectively handled much of the burden. Perhaps the most significant result of these analyses was that both the moving-bed and entrained-flow GCC plants demonstrated capability for closely-coupled operation, without the need for intervening gas storage buffers or supplementary fuels. Despite some anomalous occurrences with fuel process integration, both plants were quite responsive, using familiar power plant control strategies.

Certain forms of coordinated control may offer further improvement, although the only attempt at such concepts in these studies was to provide tight pressure control at the H_2S absorber. However, the capability now exists to pursue such optimal control concepts.

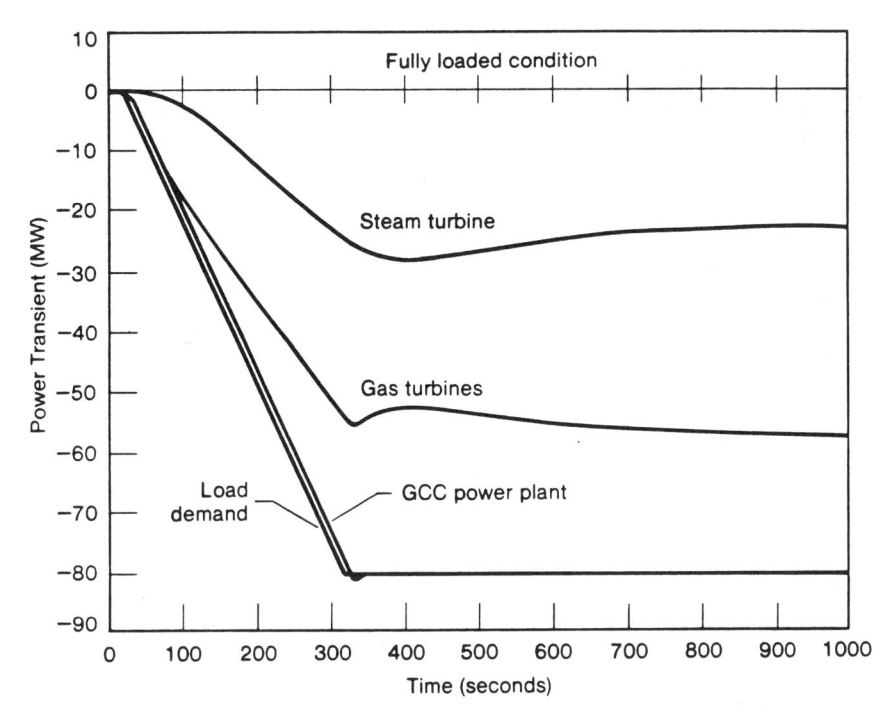

Figure 8d. Entrained Flow GCC Plant Response—Gasifier Lead Control Mode—20% Decrease in Station Load at 4%/Minute Ramp Rate.

There has been no prior experience with such GCC plants based on the conventional combined cycle design. Therefore, it is important to verify the general results of these analyses with data obtained on large-scale integrated plants. The Cool Water project should provide an excellent opportunity for such verification.

ACKNOWLEDGMENTS

Since the control analyses discussed in this paper represent the culmination of the efforts of many EPRI contractors, I would like to acknowledge the key personnel on these projects:

David Ahner and Anami Patel, General Electric

Stephen Smelser, Fluor and Milton Hobbs, Westinghouse
John McDonald and David Weber, Philadelphia Electric

Figures 1–8 for this paper are reprinted with permission from Volume 42, Proceedings of the American Power Conference where they originally appeared in conjunction with reference 14.

REFERENCES

[1] Foster-Pegg, R. W. "The Integration of Gasification with Combined Cycle Plants." Proceedings of 1979 JACC, Denver, June 1979.
[2] Kwatny, H. G. and McDonald, J. P. "Changing Electric Power Plant Control Requirements." Proceedings of 1979 JACC, Denver, June 1979.
[3] EPRI Report AF-642 "Economic Studies of Coal Gasification Combined Cycle Systems for Electric Power Generation." Palo Alto, Calif., Jan. 1978.
[4] O'Shea, T., Alpert, S., Holt, N., and Gluckman, M. "First U.S. Coal Gasification Combined Cycle Power Plant: SCE/Cool Water Project." Proceedings 7th Energy Technology Conf., Washington, D.C., Mar. 1980.
[5] EPRI Report AF-880 "Preliminary Design Study for an Integrated Texaco GCC Power Plant." Palo Alto, Calif., August 1978.
[6] Meyer-Kahrweg, H. "Operation and Control of the 170 MW Coal Pressurized GCC Plant at Luenen, West Germany." Proceedings 1979 JACC, Denver, June 1979.
[7] EPRI Report AF-1288 "Economic Evaluation of GCC Power Plants Based on STEAG Design and Comparison with U.S. Design." Palo Alto, Calif., Dec. 1979.
[8] EPRI Workshop Report WS-77-50 "Cycling Ability of Large Generating Units." Palo Alto, Calif., February 1978.
[9] EPRI Report EL-975 "Survey of Cyclic Load Capabilities of Fossil-Steam Generating Units." Palo Alto, Calif., February 1979.
[10] Ewart, D., Dawes, M., Schulz, R. and Brower, A. "Power Response Requirements for Electric Utility Generating Units." Proc. 1978 Am. Power Conf., Chicago, Apr 1979.
[11] Ahner, D., Patel, A. and Quentin, G. "Integrated GCC Plant Interaction and Control." ASME Paper 79-GT-60, 1979 Gas Turb. Conf., San Diego, Mar. 1979.
[12] Smelser, S. and Quentin, G. "Control of Combined Cycle Power Plants Based on Texaco Gasification." Presented 1979 JACC, Denver, June 1979.
[13] EPRI Report AP-1422 (3 vols.) "Entrained GCC Control Study." Palo Alto, July 1980.
[14] Quentin, G., Ahner, D., Patel, A., Smelser, S., McDonald, J., Weber, D. "Control Strategies for GCC Power Plants." 1980 Am. Power Conf., Chicago, Apr 1980.
[15] Chatterjee, N., Sorenson, J., Patrylak, A. and Quentin, G. "Dynamic Response Tests of a Large Commercial Oxygen Plant Under Closed Loop Computer Control." Presented to IFAC Workshop/1980 ISA Meeting, Houston, Texas, October 1980.

A GENERAL INDUSTRIAL ENERGY MANAGEMENT SYSTEM—FUNCTIONALITY AND ARCHITECTURE

C. Ronald Simpkins
Engineering Department
E. I. Du Pont de Nemours & Co.
Wilmington, Delaware 19898

ABSTRACT

Petrochemical plant sites usually contain multiple processes served by a central powerhouse. That powerhouse typically contains boilers and a mixture of compressors, turbines, refrigeration machines, and generators. Rising energy costs and more cost effective digital technology have created justification for computer-based Energy Management Systems that can reduce the cost of energy conversion and distribution by 3–5%. The functionality and architecture of a general industrial Energy Management System is discussed. Specific functions include optimum operation of individual equipment, load distribution, global optimization, user accountability, and energy accounting. Current vendor offerings are reviewed generically and future industry directions are suggested.

INTRODUCTION

Petrochemical-based processes are the backbone of much of our current chemical processing industry and are large energy users. These plants typically have multiple processes served by a relatively large capacity central powerhouse. This central powerhouse usually produces steam at several pressure levels, compressed air, refrigeration, and electricity. While the individual pieces of equipment are straightforward to model, operate, and control (compared to such things as distillation columns with chemical reactions), operating these pieces of equipment simultaneously presents the operator with a complex maze of choices. This complexity, combined with the generally large-scale powerhouse operation, provides an excellent opportunity to use computers to help save energy and operate the powerhouse less expensively. Traditional process control computer systems, combined with modeling and optimization packages, can be the building blocks of a standard Energy Management System.

WAYS TO SAVE ENERGY

There are several ways to save energy in a powerhouse by operating equipment more efficiently and better utilizing fuel sources.

First, there are savings from operating individual pieces of equipment more efficiently. These methods include not only the obvious, such as trimming excess oxygen on boilers, but also recognizing and correcting equipment malfunctions, such as high condenser head pressure in a refrigeration machine.

Second, a load can be distributed for minimum cost where it is shared among several pieces of equipment, for instance multiple boilers on a common header. This is possible because individual units generally have different overall efficiencies that vary with the load. To distribute the load, the amount of electric power to be generated or purchased must be determined. Whenever there is a choice of how to load equipment, there is an opportunity to save energy.

Additional energy savings result from properly choosing equipment to be turned on or off. The total load in a given service can sometimes be satisfied by operating less than the total number of pieces of equipment available.

Alternate fuels may be available, sometimes as wastes, which are essentially free. The efficiencies of boilers buring waste fuels rather than primary fuels will generally be different. Some fuels, such as waste gas, cannot be stored and the powerhouse operation must be adjusted to ensure their complete use.

HOW COMPUTERS CAN HELP

Powerhouses represent a mixture of new and old. Most are over 10 years old with a mixture of vintage and recent instrumentation that ranges from 30-year-old boiler control panels with no automatic controls to isolated installations of the new microprocessor instrumentation. In most cases, the controls are basic and provide only enough information to maintain operation. Recorders, indicators, and additional measurements for optimization or improved operation are minimal.

On-line process computer systems can help in three basic ways. They can help operators perform their current functions more effectively. They can provide the operator with additional, more understandable information to use in making decisions about operating parameters. They can help in analyzing equipment performance by providing logic that an experienced engineer would use in analyzing malfunctions.

Computers can provide new functions, such as advanced control schemes for optimization. Computers provide a level of optimization beyond what operators currently accomplish. For instance, to allocate a refrigeration load the operator can make only rough attempts to distribute the load and decide when to turn equipment on and off. With information calculated from on-line measurements made by a computer, these decisions can be substantially refined.

Finally, computers can provide "Process Management." The computer can generate a daily analysis of the operation, perform energy accounting, and improve the opportunity for management to assign accountability.

The biggest opportunities to save energy are in the process areas rather than in the energy generation area. Potential savings are estimated at 20% or more for utilization vs. 5% to 10% in the energy generation. However, while attacking problems in the powerhouse, tools and techniques are being developed that will be available for the more challenging problems in the process areas. This paper deals with concerns and questions for energy management systems in the energy generation and distribution area.

The need exists to save energy and reduce the cost of operating a powerhouse. A system that performs these functions is an Energy Management System (EMS).

NEED FOR EMS; LACK OF COMMERCIAL AVAILABILITY

Spurred by intense interest among both potential vendors and people at Du Pont plants, in early 1979 the Du Pont Engineering Department decided something needed to be done about energy management systems. The "Engineering Department Energy Management Systems Coordinating Committee" was assembled in March of 1979. It included personnel from many disciplines, including power, instrumentation, computer control, and optimization.

The specific responsibilities of the committee included:

- Exploring needs and opportunities for energy management systems to help meet the corporate energy conservation goal.
- Determining ability of vendors to provide systems to meet these needs.
- Reviewing projects.

The need for an EMS was determined by a survey of major corporate power facilities. The survey indicated that the Du Pont Company has more than 70 domestic manufacturing sites with an annual purchased fuel and power bill approaching one-half billion dollars. Of these sites, 16 account for more than 65 percent of the power bill. This survey, made by the Engineering Service Division Power Consulting Group, indicated that 5 percent of the total power bill could be saved by installing energy management systems. The survey also indicated that many plants have sufficiently complex power facilities to justify energy management systems with an optimizing capability.

Meanwhile, the committee surveyed three major instrument vendors and two companies whose main product was energy management and optimization systems. The survey revealed that several vendors could supply an energy management system for a specific plant application; however, none offered a standard system that could meet the requirements of more complex facilities without substantial customizing.

Because Du Pont computer groups have extensive backlogs of work and there is a shortage of trained computer professionals, the objective in the vendor survey was to find a vendor or vendors who could supply systems that could be implemented for a variety of applications for many plant sites.

The vendor survey indicated that the technology of energy management systems is still in the early stage of development. Several vendors recognize the energy management area as a potential new market for control systems but no vendor could provide a standard off-the-shelf system that could be installed for a set of applications without substantial customizing.

Several in-depth meetings with vendors revealed that all were doing custom programming, in many cases in low-level languages. They were using relatively small computers that limited the size of the application they could handle. Most were considering only partial, rather than global, optimization.

Because of these observations and previous experience with computer systems' vendors, the committee concluded that more Du Pont computer professionals would be required to install and maintain these vendor systems than to put together, install, and maintain our own systems.

The committee recommended and management approved a program which included the development of a corporate Standard Energy Management System.

STANDARD SYSTEMS

To eventually install several systems, it is cheaper from an overall cost standpoint to spend the initial money needed to develop a standard system.

What is a standard computer system? A standard computer system, by our definition, is one that does not require a computer programmer to adapt it to a new application. For instance, we expect this standard energy management system to be installed by site power engineers. Another feature of a standard system is that it can be supported and maintained at a central location.

Based on these two features, it is cheaper to develop a standard system. Figure 1 shows relative system initial software costs and is based on nonquantitative experience of people who build computer software systems. If we take the cost of installing a custom system as a basis, developing and installing the first standard system is expected to cost in the order of three times as much. However, once the standard system is available it can be installed on each subsequent application for about one-third the cost of the custom system.

In addition, computer programs require maintenance to eliminate malfunctions and to keep abreast of changes which will occur over the lifetime of an operating facility. Program maintenance is generally about 10 percent of the original operating cost per year. Maintaining a single standard energy management system as opposed to maintaining a number of custom systems requires fewer people.

On the basis of twenty custom systems, the cost for maintaining a standard system is expected to only be about one-tenth as much.

This concept of a standard system results from experience in several areas, not only in process control, but also in modeling and optimization.

Relative Systems Initial Cost

Each Custom System — 1X
First Standard System — 3X
Each Additional Standard System — 1/3X

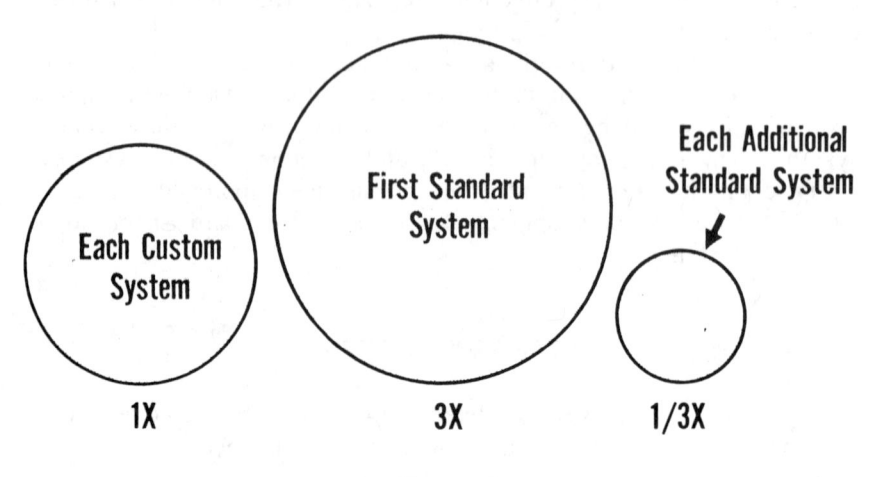

Figure 1.

In all three of these areas the viability of maintaining standard systems has been demonstrated with the resulting savings in computer personnel and program maintenance costs. These savings are real.

STANDARD SYSTEM ELEMENTS

Figure 2 shows the components of a standard energy management system. The foundation of this system is the combination of hardware, operating system and utilities or service programs.

The most important hardware consideration is whether the system is powerful enough to do the job.

An operating system is a set of programs that manage and control the computer resources. It is standard software and is always supplied by the vendor. Important considerations for the operating systems are:

- How friendly is the operating system to the user?
- What degree of protection does it supply for on-line development and testing?
- Are all required services provided?

System Structure

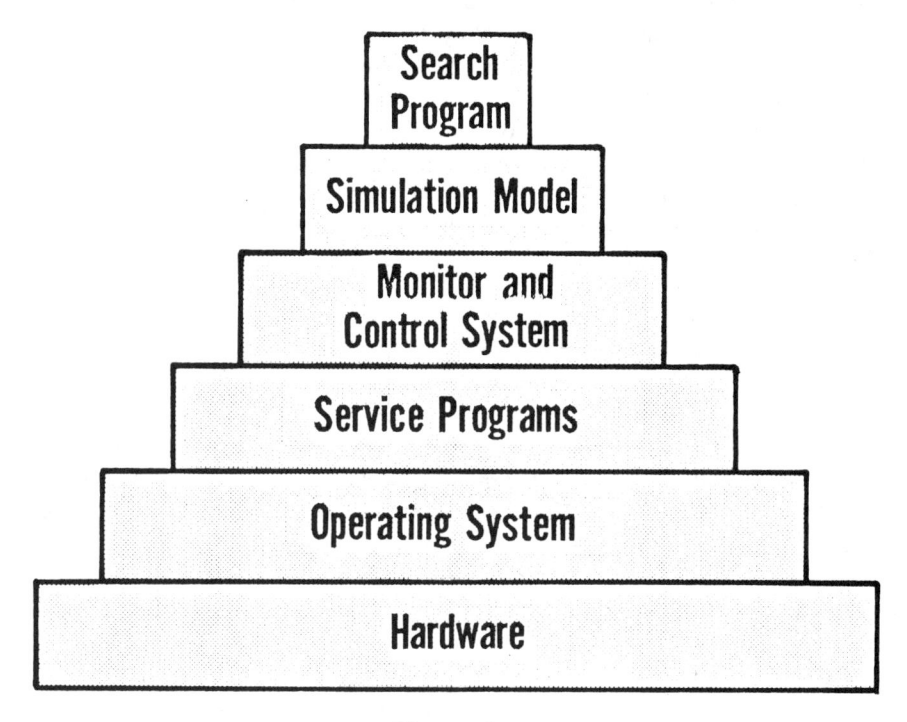

Figure 2.

```
        ┌─────────────────┐
        │  Search         │
        │  Program        │
      ┌─┴─────────────────┴─┐
      │   Simulation Model  │
    ┌─┴─────────────────────┴─┐
    │      Monitor and        │
    │     Control System      │
  ┌─┴─────────────────────────┴─┐
  │      Service Programs       │
┌─┴─────────────────────────────┴─┐
│        Operating System         │
├─────────────────────────────────┤
│            Hardware             │
└─────────────────────────────────┘
```

There should be a baseline operating system that can be generated in the field by the user. A feature that seems particularly important in optimizing is maximum program size. The great relaxation of program size requirements provided by the newer virtual-memory systems is particularly attractive. The user must also be able to easily and manageably make changes to programs and install them.

The utilities should be supported by and run under the operating system. They should include editors, assemblers, compilers, etc. System debugging tools are particularly important. Utilities should permit the user to examine programs and set break points to examine and modify memory programs and set break points to examine and modify memory while the program runs. File management services and communications with other computers are also important.

Other components of a standard system are a monitoring and control system, a process simulation model, and a search program.

The process monitoring control system provides the process interface and operator interface. It also supplies several basic functions, including conversion to engineering units, limit checks, alarm violation notification, logging of critical events, saving and retrieving historical data, and supervisory control. All functions should be supplied through a fill-in-the-blanks data setup. The system should be capable of interfacing to pneumatic, electronic analog, as well as the new microprocessor-based process instrumentation. This is important because in many cases the existing instrumentation does not have to be completely replaced to accomplish energy management goals.

The standard system also requires some form of process simulation model. The choices are to develop a hand-coded, customized set of equations as a model for each powerhouse or to use a general purpose modeling system.

Du Pont chose to use a general steady-state simulation system for its models. It consists of a series of programs or modules that represent the performance of various types of processing equipment. These equipment modules are set up with fill-in-the-blanks data. The system provides capability for doing simultaneous energy and material balances, vapor/liquid equilibrium, and chemical kinetics. This system was developed in-house, has been used for over 15 years for process modeling, and is now available for on-line use. New programs or computer programmers are generally not required.

Figure 3 is a representation of a typical powerhouse. The first step in using a general simulation model is to set up a single-line schematic diagram or a flow sheet. Each individual flow or stream and each individual equipment module are uniquely numbered.

Figure 4 shows an example of how an individual piece of equipment, in this case a refrigeration machine, might be set up. The identification of the streams entering and leaving the equipment and required operating parameters are specified.

The basic premise of this system's operation is that basic mechanistic equations with specified operating parameters are used to calculate output from given inputs. Logic is provided to recognize that outputs from a given piece of equipment may be inputs to subsequent pieces and for interactive calculations when necessary. However, by providing calculation options the same modules may be used to calculate efficiencies or to compare measured variables vs calculated variables as an indication of equipment malfunction.

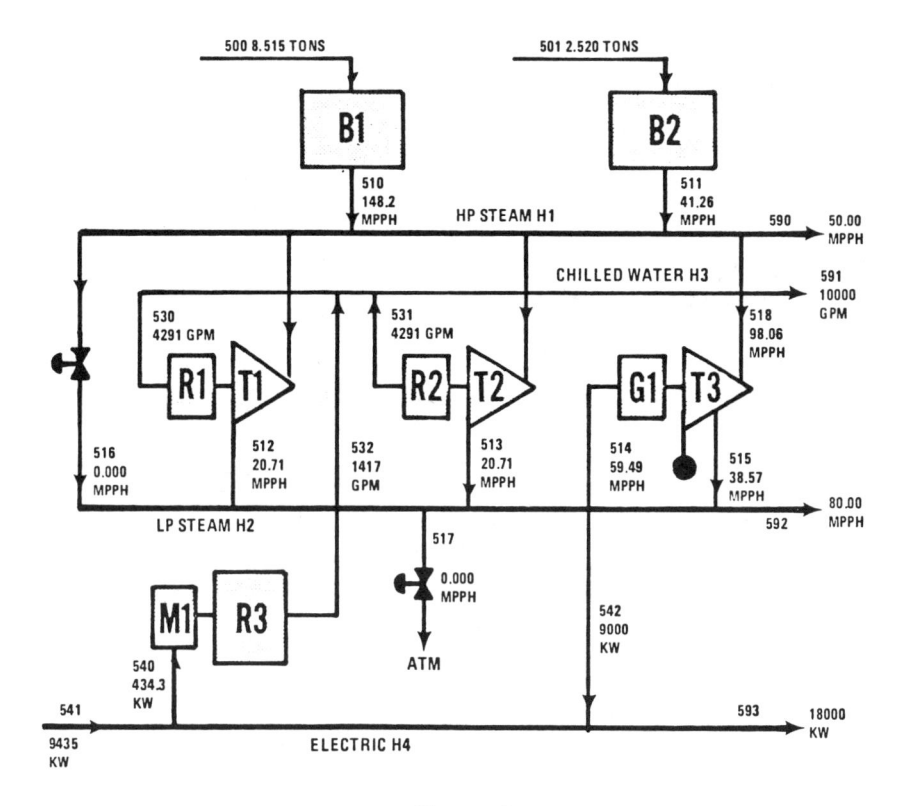

Figure 3.

Again, the reason that we use this system to model chemical processes and plan to use it in the energy management system is that is provides a means to develop and use the model for fewer manpower dollars and the people required need not be computer programmers.

The next layer of the system is the search program or the optimizer. The model is used to calculate a set of conditions that result from setting various operating parameters in the powerhouse. The optimizer manipulates inputs to the model so as to produce required utilities at minimal cost. The optimizer and the model may be separate but must pass information back and forth.

A number of optimization routines are available. These include not only linear programs but also some so-called "mixed integer" programs that allow on/off decisions to be made easily. Additionally there are routines such as a random search optimizer that is a superior technique for certain classes of nonlinear problems.

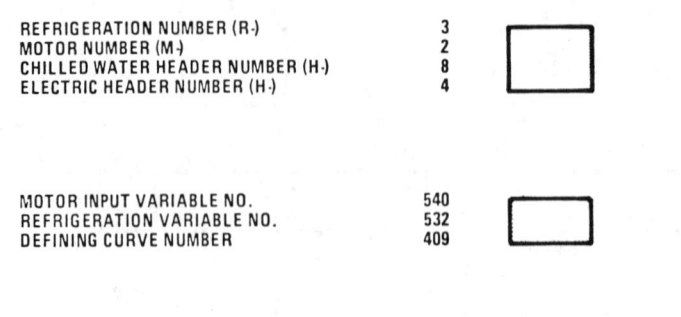

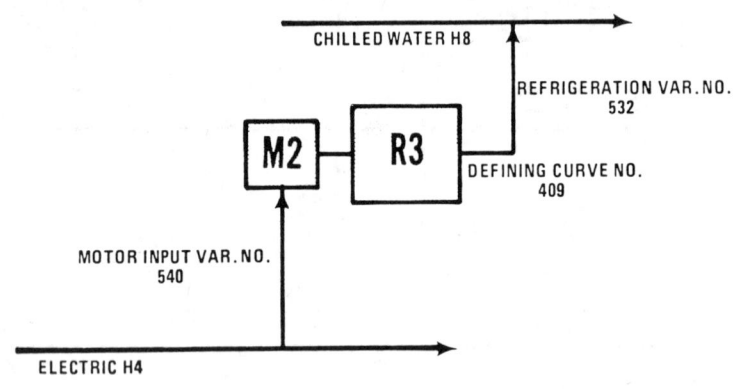

```
DESCRIPTION

REFRIGERATION NUMBER (R-)           3
MOTOR NUMBER (M-)                   2
CHILLED WATER HEADER NUMBER (H-)    8
ELECTRIC HEADER NUMBER (H-)         4


MOTOR INPUT VARIABLE NO.            540
REFRIGERATION VARIABLE NO.          532
DEFINING CURVE NUMBER               409
```

```
                    CHILLED WATER H8
                                          REFRIGERATION VAR. NO.
                                                   532
        ┌────┐   ┌────┐
        │ M2 │───│ R3 │      DEFINING CURVE NO.
        └────┘   └────┘              409
          ▲
    MOTOR INPUT VAR. NO.
            540

                    ELECTRIC H4
```

Figure 4.

THE SYSTEM'S FUNCTIONS

Now that some groundwork has been laid as to vendor offerings and system architecture, the functionality of a standard energy management system can be considered. "Exactly what is it that this system is to do?" The standard system is modular, meaning that any functions not needed for any given powerhouse, are not implemented. In some installations, the choices of how to operate equipment may be constrained such that there is no real opportunity for optimization. The modular concept of the standard energy management system makes it appropriate for use even in those cases.

Process Monitoring Control System General Functionality—The base process system should have the characteristics generally accepted in a modern process control system. The operating console should be designed for easy operator use, with function buttons and quick response for easy operator access to data. The console should use color dynamically to highlight unusual conditions.

Pictorial graphics should be available through fill-in-the-blanks data setup utilizing shape libraries. Variable color with changing shapes should represent changes in process conditions. Live data should be updated on the screen along with the pictorial graphics.

The capabilities are available in-house and in several commercially available systems. These required features are not unique to energy management systems.

General Model Functionality—The model is, in general, a mechanistic, nonlinear mathematical representation of the powerhouse. It must operate in several modes and must be able to combine measured and manually entered data. Default values should be included so the model will run even when pieces of data are missing. (Of course, the assumptions made in supplying the default values must be clearly flagged and presented with the results of the model calculation.)

The model should be as mechanistic as possible. Using basic heat and material balances is strongly preferred over a simple "output = efficiency × input" approach, even if some values must be manually input rather than measured.

The result of this mechanistic approach is a model that not only does a better job of representing the status of the equipment at a given set of operating conditions, but is less sensitive to changes in operating conditions. For example, the efficiency of a turbine is a function of the superheat of the steam to the turbine. If turbine efficiency is merely represented as "output = efficiency × input," then the efficiency would have to be adjusted to compensate for changes in steam superheat. On the other hand, a model that takes into account enthalpy and entropy of entering steam does not require parameter changes within the model to accommodate such a change.

Alternate modes of operation of the model must include:

- Given inputs, it should calculate outputs (the normal mode of operation of the optimizer).
- Given some measured variables, it should calculate other variables and make comparisons to identify bad measurements. This mode of operation is called data validation.
- Given certain measured variables, it should calculate parameters, such as heat transfer coefficients, used to identify malfunctioning equipment. This is called equipment analysis.

This general process monitoring control system and model functionality are required as part of the specific energy management functions now discussed.

Optimization—An energy management system is used for optimization on four levels:

- Optimizing Individual Pieces of Equipment—Boiler excess oxygen trim control is a good example. There are several methods of optimum excess oxygen trim currently under discussion. Using CO measurement, following a predefined curve, and using a "hill-climber," are all possibilities for specific cases. (The question of using CO in the flue gas to control is an extremely interesting argument but beyond the scope of this paper.) These advanced control schemes require instrumentation and control capability beyond what has classically been available through analog controls. In some cases these strategies can be implemented in the new microprocessor-based control systems, if used.
- Local Optimization—With individual pieces of equipment running as efficiently as possible, the next step is distributing the load among like pieces of equipment; for instance, the steam load among several boilers. Also, choices must be made where services can be provided in alternate ways, such as providing refrigeration with either electric- or steam-driven equipment. At this level, in general, only equipment that is turned on is considered in the optimization, and loadings are changed within allowable limits.

  Techniques for accomplishing this level of optimization generally involve using the nonlinear model to generate and pass coefficients to a linear programming package (LP). The LP manipulates operating conditions to provide desired services at a minimum cost, and compares those operating conditions with the ones used in the model to produce the coefficients. If they match, the problem is considered solved; if not, the conditions are fed back to the model to generate a new set of coefficients. In general, this convergence may take several passes between the model and the LP.
- Global Optimization—This level of optimization involves turning equipment on and off. For instance, if the desired load can be achieved by running only six of seven available refrigeration machines, sometimes shutting down one unit reduces costs.

  The techniques for handling global optimization problem are generally the same as local optimization, with an added level of complexity. Generally, penalties are assigned for starting up or shutting down a piece of equipment, and the operator can override the option of turning a piece of equipment on or off.

  These cases generally are handled by implementing a logic that runs alternate cases of the local optimization with various pieces of equipment on and off. Since the number of possible cases to be

evaluated can be very large (2^N, where N equals the number of pieces of equipment to be turned on or off), "branch and bound" logic must be added to restrict further evaluation of a logic-tree branch once it has been found to be more costly than a previously evaluated branch. This "branch and bound" logic is generally created only after exploring the behavior of the model and the sensitivity to various real operating parameters.
- "What If" Cases—Energy management systems should allow engineers to explore the effect of various changes in equipment configuration and operating conditions. This means that not only must operating parameters be accessible and easily changed, but also the model itself. It must be possible to run these studies while the system is on-line, performing its other normal functions. On the practical side, it is necessary to set up file structures so the parameters, model, and results from multiple runs are available and manageable.

The system should implement optimization results either through advisory messages or as control outputs to the front line instrument system. Results from the "What If" cases should be available on both the CRT and hard copy, in alphanumeric and graphical form. "X-Y" plotting should be available in addition to the usual "X-time" plotting feature found in most process control systems.

Several other functions are available to support optimization and to assist operations.

Data Validation (e.g. Boiler feed Water + blowdown = (?) steam)— This is useful not only to assure that the optimizer is working on reasonably good data, but also as an aid in pinpointing bad measurements.

Load Forecasting—Most plants have purchased electric power contracts that include both a demand and a peak period charge, To provide cost of purchased power for the optimizer, both a short- and long-term forecast of total load and purchased power must be produced. These forecasts must, in general, consider both production and seasonal variations.

Equipment Analysis— The model can be used to calculate and compare expected and measured variables, such as efficiencies or heat exchanger exit temperatures. A substantial deviation from expected conditions indicates equipment malfunction and often requires attention from supervision or the engineering staff for analysis. The experienced power engineer usually follows a logic-tree type analysis ("Is it this? No. Well, is it that?") in deciding what the problem really is. That logic-tree analysis can be built into the model using the calculation services available (e.g., to calculate the enthalpy of a stream, "CALL ENTHALPY"). This capability gives the operator access to an engineering analysis of equipment performance at any time.

Accounting and Reporting—This is a somewhat mundane area but an important one to the operating supervision and management of power facilities. And, in most cases, it is well justified. Data is usually available in an energy management system to produce the required reports of fuel consumption, energy generation, and "power for power," the energy used within the powerhouse. Daily "performance reports" are a great help in tracking day-to-day problems.

User Accountability—The energy management system can collect data and be used to develop standard usage curves that reflect energy consumed per unit of production. Since process monitoring usually improves utilization, these data can be an important tool to help operating areas use less energy. Additional transmitters at remote locations would be required for this function. However, availability of rented telephone lines and signal multiplexing techniques now make this an attractive possiblity.

CONCLUSIONS

Several points should be emphasized:

- Energy management systems are justified in power generation and distribution facilities.
- Optimizing is needed in most moderate to large petrochemical installations.
- A standard energy management system, one that can be installed with no or at least with a minimum of programming, is needed not only because of the shortage of computer professionals but also to minimize installation and long term support costs.
- Standard Energy Management Systems are not commercially available. Vendors should further consider this potential market.
- While this discussion concentrated on power generation and distribution facilities, the system described here has a broader application, beyond the powerhouse. The process monitoring control system, running on-line with a general purpose modeling system tied to an optimizer, is an extremely powerful tool with broad application in the process area. With the continuing downward spiral of computer price/performance ratios accelerated by the introduction of the new 32-bit process computers, this technological step is feasible and, in fact, is being taken.

ACKNOWLEDGMENTS

This paper in part reports work done and conclusions reached by the Du Pont Engineering Department Energy Management Systems Committee. The chairman of that committee is W. B. Johnson; membership during 1979 and the first half of 1980 included B. F. Coe and F. E. O'Toole, Instrumentation; B. Quillman, J. D. Robnett, and J. T. Kephart, Power; C. H. White, Optimization; and J. J. Riley and C. R. Simpkins, Process Computers. Many of the concepts in the paper were developed in joint discussions between W. B. Johnson, J. J. Riley and the author. Part of the material was presented in a talk, "Computers in the Energy Business" by W. B. Johnson at the Du Pont Energy Forum, March 13, 1980.

PLANT UTILITY MANAGEMENT

Richard A. Hanson
Section Manager
Application Engineering Dept.
Taylor Instrument Company
Rochester, New York

Utility Management has numerous meanings depending upon what your involvement is in the use of energy.
1. Building—working and living conditions.
2. Power Plant—the generation of energy in different forms.
3. Plantwide—the efficient generation of energy and the efficient use of energy by the different process areas.

There are a wide variety of users of different forms of energy and a number of different types of Energy Management Systems to deal with these variations. This accounts for the many suppliers of Energy Management Systems. This variety also means that there is not a single Energy Management System that will fit all applications. All Energy Management Systems monitor and control and/or manage the *"use"* of energy, not the actual energy itself. Therefore, a more appropriate name would be Energy Use Management. The energy use management is accomplished by controlling input and output conditions of the energy used and the products being produced. This energy may be in many different forms: oil, gas, coal, bark, waste, bagasse, and a multitude of other fuels.

This paper will address the Industrial Power Plant where the energy is converted from fossil fuels; such as, oil, gas, coal, bark, etc.; to steam, electricity, water, compressed air, to be used by other process areas in the plant to provide a final product.

The Power Plant is providing a "service" to the rest of the plant in the form of utilities. The more efficiently the powerhouse can provide that service, the better the results in the form of a direct reduction in plant operating costs. The utilities supplied must be controlled to a specific temperature, pressure, purity, voltage, etc. This requirement is the area where the "management" of the different control systems, supplying these utilities, can be improved. Improvement results from closer control of the different process parameters; i.e., temperature, pressure, flows; and the capability of feedforward signals from the users in the plant, before the impact is detected at the supply end.

To analyze the need for an Energy Use Management System, a utilities study needs to be performed on the system. From the results obtained, the capability to obtain an appreciable "Return on Investment" may be determined. The study may be a single, comprehensive investigation by an outside firm or a preliminary study done by "in-house" personnel to obtain an estimate of the potential savings available. Assuming that the results indicate a valid reason to proceed, the next step is to decide what areas of the powerhouse require attention.

ENERGY MANAGEMENT SYSTEM

An Energy (Use) Management System should cover the complete powerhouse operation. This can be separated into four main categories:

> Item 1: Steam Generation
> Item 2: Steam Distribution
> Item 3: Purchased Power
> Item 4: Management Services

A description of these categories follows:

1. *Steam Generation* ("Boiler Load Optimization")—is the efficient operation of the steam generators to supply the steam demanded by the process and the powerhouse at the needed pressures, temperatures, and purity.
2. *Steam Distribution*—is the allocation of steam for electrical generation and for process use. Turbine-Generators require steam at the proper temperature and pressure to meet the mill demands. They also supply process steam at the required pressure and temperature via their extraction stages. If an electrical load increase is required, this system would determine where the increased electricity should

come from due to economic considerations; that is, from the Turbine-Generators or from the purchased power.
3. *Purchased Power*—monitors the electrical loads in the plant and powerhouse and the electric supplies (Turbine-Generators and purchased power) to determine their respective needs and priorities. If the electrical load is reaching a limit on purchased power, this system would have the ability to automatically shut down loads (according to a predetermined schedule of priorities) to prevent exceeding the established demand limit set.
4. *Management Services*—would be the generation of logs, reports, displays, and energy balances in the plant and additional information on the power plant's operation.

There are other areas which can be addressed if it is deemed necessary or the pay back potential is there:

    Item 5: Water Management
    Item 6: Air Management
    Item 7: Fuels Management
    Item 8: Environmental Considerations

This paper will be limited to the first four areas of an Energy Management System, the details of which will be presented in more detail.

STEAM GENERATION

This area of the Energy Management System encompasses the generation of steam by the boilers and the pressure reducing station. The first step is to determine the present firing conditions and the boiler efficiency. This is accomplished by checking the calibration and installing a stack gas monitor for percent oxygen and opacity (as a minimum). A combustibles analyzer when firing Natural Gas or light fuel oils may also be required. In most medium to large industrial installations there are more than one steam pressure header; therefore, the possibility of supplying the lower steam header with steam from the higher steam header via the pressure reducing station has to be considered. This is based on economic consideration because the higher pressure boilers are usually newer; fire coal, waste product, heavy fuel oil, or a combination of these fuels; and are very often the more efficient units. Therefore, they may be the lowest cost energy supplier. The result of optimizing the Steam Generators is not necessarily to operate all boilers at their maximum efficiency, but to operate the boilers at the total optimum load. This results in the total steam being generated for all headers being supplied at least total cost to the plant. The following is an example of a typical situation:

Assume you have 4 boilers and need to supply 300K #/Hr to a high pressure header and 200K #/Hr to the low pressure header. How should the boiler loads be allocated to supply the required steam at least cost? Shown in Figure #1 are the efficiency curves for 4 boilers, 2 high pressure and 2 low pressure. First, for simplicity, we will assume all boilers are firing the same fuel; because of this situation, the efficiency curves can be directly converted to cost curves with the cost axis replacing the efficiency axis.

The following are some examples of a typical situation:

Present Operation—600 psi header
  Demand = 200,000 #/Hr
  Boiler #1—(100K#/Hr)(5.195$/K#)   519.5
  Boiler #2—(100#/Hr)(5.150$/K#)   515.0
                                         $1034.5 per Hour

Optimization Program—recalculate cost at new boiler loads
  Boiler #1—(80K#/Hr)(5.20$/K#)   416.0
  Boiler #2—(120K#/Hr)(5.14$/K#)   616.8
                                         $1032.8 per Hour

Potential Savings Per Year = $14,892

Present Operation—900 psi header
  Demand = 300,000 #/Hr
  Boiler #3—(150K#/Hr)(5.755$/K#)   863.25
  Boiler #4—(150K#/Hr)(5.715$/K#)   857.25
                                         $1721.50 per Hour

Optimization Program—recalculate cost at new boiler loads
  Boiler #3—(110K#/Hr)(5.775$/K#)   635.25
  Boiler #4—(190K#/Hr)(5.690$/K#)   1081.1
                                         $1716.35 per Hour

Potential Savings Per Year = $45,114
TOTAL YEARLY SAVINGS = $60,006

These examples demonstrate why you do not want to operate each boiler at its maximum efficiency to obtain least overall cost. These examples only give different operating levels and are not necessarily "optimum" level for the given boilers and loads.

With this time consuming problem to determine the optimum loading of each boiler for an infinite number of combinations of plant load, the use of a computer is essential. A computer with on-line facilities can readily digest all of the information and provide the operator with the necessary information to take corrective action. One method for analyzing this data is to use the Incremental Heat Rate calculation; this is the slope of the

"Input-Output" curve at any given load. The "Input-Output" curves are the efficiency curves, and the slope is the first derivative of the curve, that are shown in Figure 1. If the derivatives are taken, then Incremental Cost curves can be determined and can be defined by straight line as shown in Figure 2. Now let us consider two boilers with different efficiency curves. It can be proven that the least cost to supply a given load is where the incremental fuel costs are equal for each boiler.

If we take the graph given in Figure 2, the boilers with the higher dC/dL can be reduced by increasing the load on the second boiler, which has a lower dC/dL. This change of load will be beneficial until the values of dC/dL for both boilers are equal. This is now optimum (minimum) cost for a given load. This can then be extended to multiple boiler installations.

There is one set of curves for the high pressure header boilers and another set of curves for the low pressure boilers. The high pressure boilers can supply the low pressure header via the P.R.V. When the high pressure boilers provide low pressure steam, another set of incremental rate curves need to be generated.

It is from these incremental rate functions that the optimum load for all boilers can be obtained. If the demand rate to the process area is

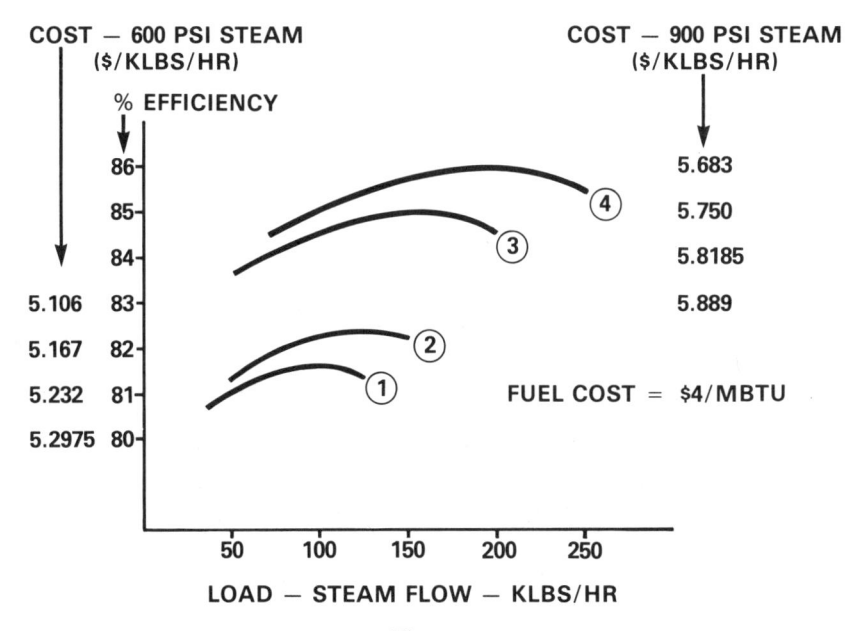

Figure 1.

INCREMENTAL RATE CURVES

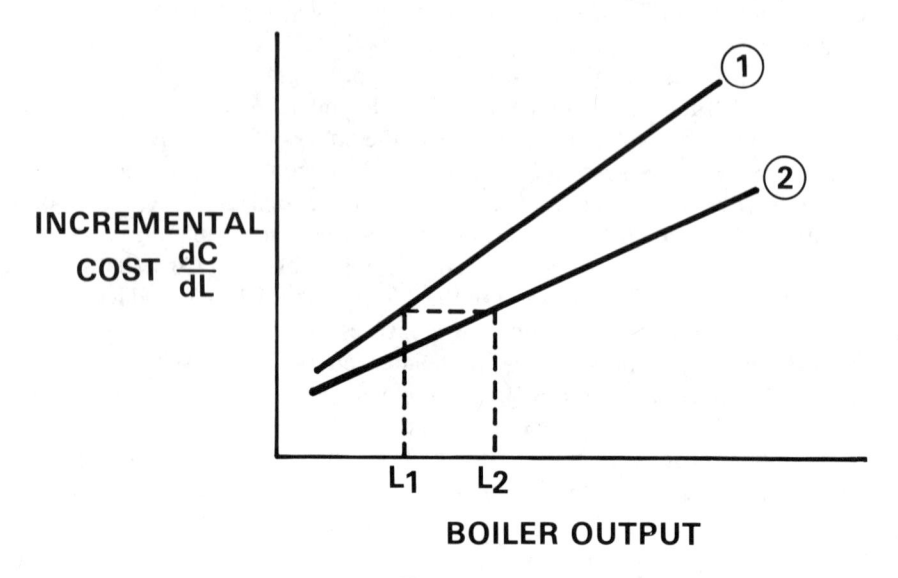

Figure 2.

measured and input into the system, then the minimum cost of the next unit of steam will be specified by the incremental rate function. This function will provide the value of load on all of the boilers to achieve the *minimum overall cost*.

STEAM DISTRIBUTION

If the powerhouse does not contain Turbine/Generators, the steam distribution is performed by the Optimization Program for the Steam Generation. If the system contains Turbine/Generators, the Steam Generators supply the steam to the steam headers as required by the process plant and the Turbine/Generators. Assuming there exist single or multiple extraction turbines, then this portion of the Energy Management System optimizes the distribution of Steam and Electricity to the plant, at the proper pressure, temperature, voltage, frequency, etc. The optimization program for this portion consists of monitoring the Turbine—throttle, extraction and exhaust; flows, temperature and pressures, along with the power output of the generator and the electric utility supply coming into the plant. The optimization program will then determine how the next unit of energy, that has been requested by the plant, will be supplied at least cost.

Assuming there has been an increase in plant Energy use, the optimization program would compare the cost to generate the additional power, via various turbine extraction rates and determine what the new turbine loads should be, to the purchase of additional power from the utility company. The cost of steam input to the T/G is obtained from the boiler load optimization program and it takes these cost values and compares them with the cost to internally generate the needed additional power.

This "cost to generate" is compared to the "cost to purchase" additional power from the utility company. Whichever total cost is lowest would be used to supply the energy to the plant. A detailed example of the Turbine/Generator evaluation is discussed in a paper entitled "Optimization in the Industrial Power Plant."[2]

PURCHASED POWER

Industries which purchase electricity from a utility are charged not only for the total electrical energy consumed, but also for the rate at which the electrical energy is consumed. The rate charge (or demand charge) is based on the amount of energy used during a demand interval divided by the demand interval period in hours. The demand interval is usually a 5, 15, or 30 minute time period synchronized by an on-site utility demand meter pulse. The demand meter keeps track of the energy rate of the highest consumption demand interval in a given month. In some installations, the demand period used for the demand charge is "floating." For example, a 15 minute demand period is made up of the 5 highest consecutive 3 minute intervals of the month. A "floating" demand interval usually results in significantly higher demand charges for the month. Typical demand charges range from 1.00 to 10.00 $/KW depending upon the utility providing electrical service. In addition, some utility tariff structures allow back charges for high consumption demand intervals occurring up to twelve months *prior* to the billing month. Depending upon the overall load factor of the plant (average power/maximum power), the demand charge may constitute as much as 50% of the total bill. The main function of an electrical demand control system is to limit peak electrical demand by shedding non-essential loads during periods of high consumption.

There are several methods of determining how much electrical load should be shed or how much can be added at a given moment. The method which best utilizes the capabilities of a digital computer and which best suits a fluctuating plant load situation is the "Forecasting Method." The forecasting method is actually an optimized combination of the "ideal rate" and "continuous integral" methods commonly used in less sophisticated systems incorporating programmable controllers or mechanical timing devices.

In the *Ideal Rate Method*—only the actual energy used up to a particular point in the demand interval is looked at. If this energy consumption is higher than a set maximum for that point in the demand interval, loads are shed. If the energy consumption is lower than a set minimum for the point in the demand interval, loads are turned back on.

The *Continuous Integral Method*—also known as "instantaneous rate," compares the rate of energy consumption at a given point in the demand interval to an instantaneous rate equal to the energy demand limit divided by the demand interval period. If the instantaneous rate is too high, loads are shed. If the rate is too low, loads are turned back on.

The *Forecasting (or "Predictive") Method*—utilizes both the energy consumed and the rate of energy consumption as well as the time remaining in the demand interval to determine load control requirements. In the forecasting method, the demand limit is fixed for a given demand interval, but the electrical energy rate is fluctuating due to random plant power demands. Ideally, the total energy consumed during the demand interval should be less than or equal to the demand limit (see Figure 3).

Results: The program will shed loads whose net power consumption is equal to the difference in slopes between "projected" and "desired" energy usage levels.

Therefore, an Electrical Demand control system will utilize the "floating forecasting method" with an automatically adjustable demand limit and user/utility defined demand interval. This method can use a utility meter supplied synchronization pulse for the demand intervals, if available, but it can also keep track of an internally supplied starting point. This feature can be used to advantage in areas where the utility is allowed a continuously floating demand interval, or when no synchronization pulse is provided.

In order to fully utilize the capabilities of the system and to eliminate undesirable off-hour power peaks, provisions are made to automatically adjust the demand limit depending upon the time of day, day of the week, or whatever parameter is deemed most useful.

PURCHASE POWER USAGE

Figure 3.

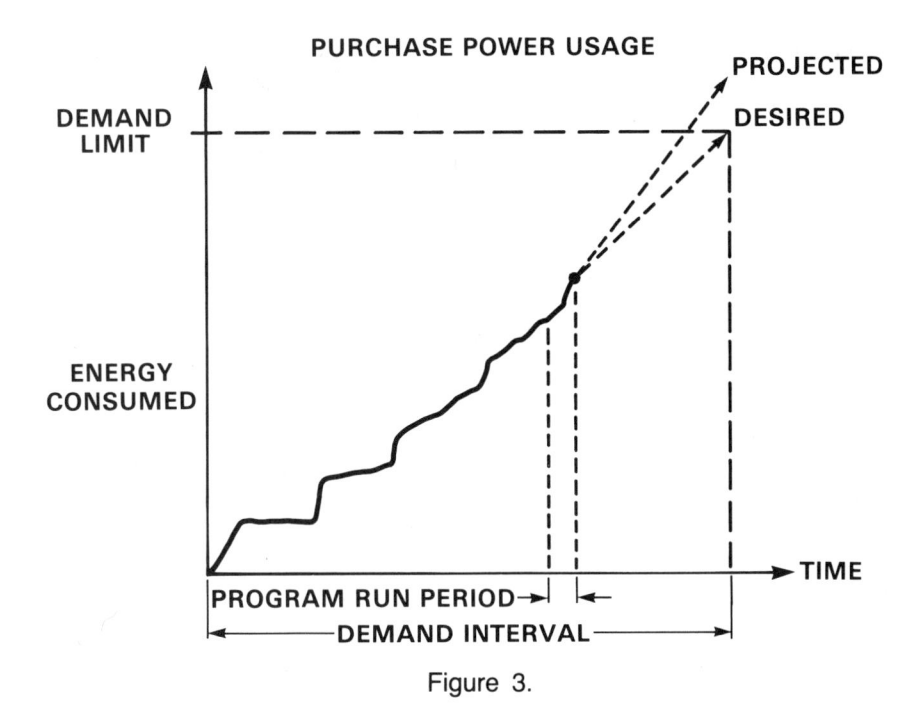

DEVICE PARAMETERS

Each sheddable electrical device has an associated:

> name,
> load priority,
> power rating,
> maximum off time,
> minimum off time,
> maximum on time,
> minimum on time,
> device timer.

These parameters are displayed and input via CRT and keyboard and can be altered on line at any time.

Every time the program is run, the field status of each device is checked and compared to its previous status. If the status has changed or if any of the device parameters have been reached, then the devices are put on various lists; "turn-on," "available-to-add," 'turn-off," "available-to-shed"; and arranged according to priority and power levels.

The program will shed or add loads, on a priority basis, whose net power consumption is equal to the difference in slope between the "projected" and "desired" energy usage rate. In addition to Demand control, the program will have the capability to start and stop devices by time of day, week, etc.

MANAGEMENT SERVICES

This part of the Energy Management System consists of Reports and Displays to help the operators and management personnel to operate the plant more efficiently. A color CRT is used as the main interface device for the operator to obtain information about the system's present (or past) operation. All information the computer senses or calculates is available to the operator through this terminal. The following are some examples of Displays and Reports which are available to the operator and/or management personnel.

Displays
    Status Display
    Trend Display
    Alarm Display
    Graphic Display

Reports
    Shift—operating variables, throughput variables and totals.
    Daily—operating variables, throughput variables and totals.
    Weekly (or) Monthly—throughput variables

OPERATOR DISPLAYS

The *Status Display* shows the current value of all monitored variables. The controller mode, present input value, setpoint, output, and alarm condition should be displayed for the operator. An example of this display is shown in Figure 4.

The *Alarm Display* shows any variables that are currently in alarm. Alarm limits are defined in the computer for each variable. If these limits are exceeded for a measurement, that variable will be displayed along with the limit and actual reading. Figure 5 is an example of this display.

The *Trend Display* shows graphically the performance of a measurement for a desired time base. The system saves samples for any defined variables at specified time intervals and displays it in trend record format on demand. The horizontal and vertical axis may be independently

```
              STATUS PAGE #1      BOILER #1        10/8/80    9:25

LINE   I.D.        FUNCTION              MEASURED       SETPT    OUTPUT  ALARM
  1   LIC-101   DRUM LEVEL              0.2   IN.        0.0     50.9 %
  2   FIC-102   STEAM FLOW             52.5   K#/H
  3   FIC-103   FEEDWATER FLOW         50.1   K#/H      50.1     65.2 %
  4   PIC-100   600# MASTER PRES.       600   PSI        600     38.9 %
  5   HIC-105   BOILER MASTER          38.9   %          6.2     45.1 %
  6   FIC-106   TOTAL FUEL CONT.       45.1   %         45.1     43.1 %
  7   FIC-107   AIR FLOW CONT.         45.0   %         45.1     46.1 %
  8   AIC-108   % OXYGEN CONT.          2.4   %          2.4     52.0 %
  9   PIC-104   FURNACE PRES. CONT.    -.11   IN.       -.10     60.5 %
 10
 11   PI-110    DRUM PRESSURE           596   PSI
 12   PI-111    PRES. - S.H. OUT        591   PSI
 13   TI-112    TEMP. - S.H. OUT        713   °F
 14   PI-113    OIL BURNER - PRES.       38   PSI
 15
```

Figure 4.

```
                         BOILER #1                   10/7/80       9:10
         ID         FUNCTION           ALARM      LIMIT       ACTUAL

  1   LIC-101    DRUM LEVEL             LOW       - 4          - 5
  2   PI-115     FEEDWATER PRESSURE     LOW        688          675
  3   AIC-108    % OXYGEN               HIGH       3.5          4.0
```

Figure 5.

adjusted to obtain a "blow-up" view of certain areas and variables. The operator also has the capability to display more than one display on the CRT at a time and to coordinate the time bases to analyze a system upset. Examples of these displays are shown in Figure 6.

REPORTS

The *Boiler Shift Report* details the boiler operation for a shift. The shift report may contain instantaneous values at the specified time (i.e., every hour or every two hours), or the variables can be averaged for this time period and display the averaged value. This report is automatically printed out at the end of each shift, or can be called up on demand and look back at the last eight hours of operation. Figure 7 is an example of this report.

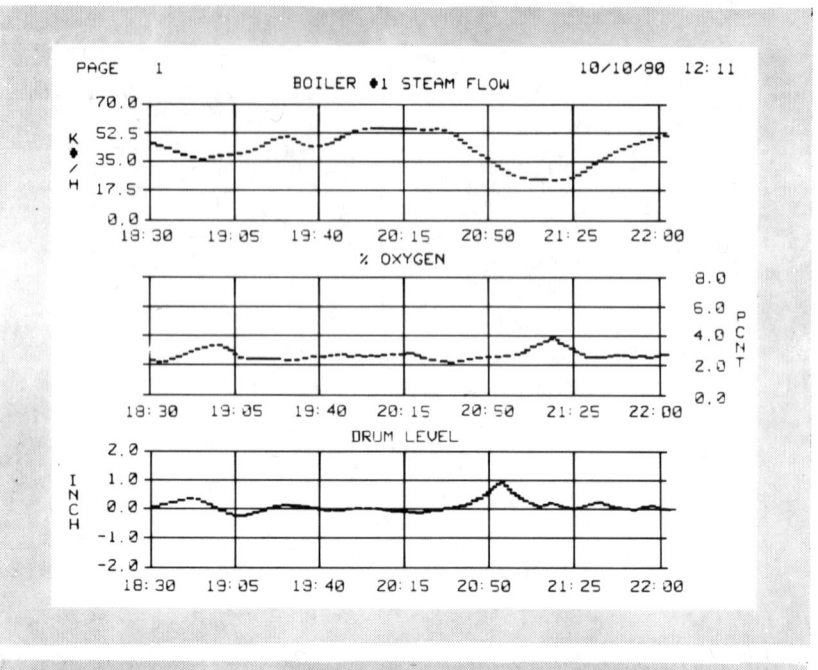

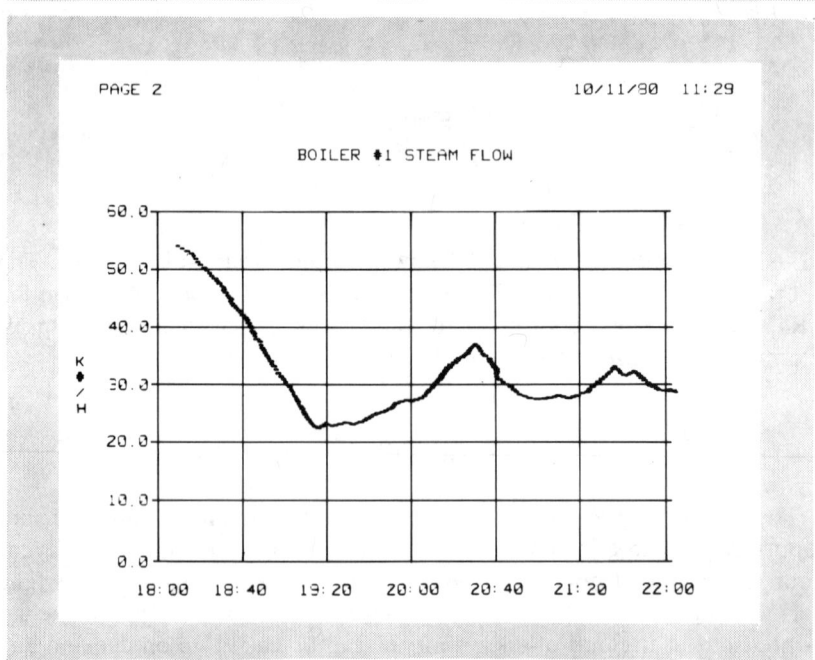

Figure 6.

R. A. Hanson 461

```
                        SHIFT REPORT - DAY    #1 BOILER    1400HRS  10/ 8/80

OPERATING VARIABLES       0700     0800     0900     1000     1100     1200     1300     1400
DRUM LEVEL      -INCH      0.4    - 0.1      0.7      1.2      0.8      1.9      0.0    - 1.9
STEAM FLOW      -K#/HR    69.0     65.0     64.0     59.0     68.0     57.0     53.0     45.0
FEEDWATR FLOW   -K#/HR    68.0     70.0     67.0     65.0     70.0     66.0     60.0     49.0
DRUM PRES.      -PSI     600.0    595.0    594.0    592.0    594.0    595.0    594.0    591.0
S.H. OUTLET PRES.-PSI    595.0    591.0    590.0    588.0    590.0    592.0    591.0    589.0
AIR FLOW        -%        62.1     61.0     59.5     55.2     61.8     53.5     50.2     45.3
OIL - FUEL FLOW -GPM      14.6     14.0     14.2     13.7     14.5     13.4     12.8     12.1
OIL - BURNER PRES.-PSI    50.0     47.0     48.0     45.0     50.0     44.0     41.0     37.0
ATM. STM. PRES. -PSI      75.2     71.8     73.2     70.6     74.0     69.2     67.0     63.4
% OXYGEN        -%         3.0      3.5      3.1      3.7      3.2      3.3      4.0      3.4

TEMPERATURES - DEG F
STEAM - S.H. OUTLET      736.0    713.8    714.4    712.6    719.2    712.6    714.4    717.4
AIR - A.H. INLET          84.0     86.0     88.0     90.0     94.0     96.0     98.0     98.0
AIR - A.H. OUTLET        271.0    268.0    274.0    274.0    278.0    274.0    274.0    270.0
GAS - A.H. INLET         544.0    536.0    535.0    532.0    545.0    538.0    534.0    529.0
GAS - A.H. OUTLET        391.0    386.0    387.0    386.0    396.0    390.0    387.0    382.0
FUEL OIL                 220.0    220.0    221.0    220.0    222.0    221.0    224.0    220.0

DRAFT PRESSURES - INCH
F.D. FAN DISCHARGE         6.0      6.0      6.6      6.6      5.9      5.8      5.9      6.3
WINDBOX                    3.0      4.3      4.2      4.3      4.4      2.6      2.1      2.4
FURNACE                  - .3     - .2     - .2     - .1     - .2     - .1     - .2     - .1
BOILER OUTLET            - 5.4    - 4.6    - 4.7    - 4.5    - 4.4    - 5.8    - 6.2    - 5.9
GAS - A.H. OUTLET        -10.0    -10.8    -10.1    -11.0    -11.2    -10.2    -10.9    -10.6
DUST COLLECTOR OUTLET    -12.1    -12.9    -11.9    -13.1    -12.9    -12.5    -12.0    -12.6
```

Figure 7.

The *Daily Report* gives the "Hi," "Lo," and "Average" values for each shift and for the day. This report contains the same values that are given in the "Shift Report" but, by presenting the "Average" and the "High" and "Low," it gives a representative operation of the boilers. An example of the report is shown in Figure 8.

The *Daily Throughput Variable Report* shows the totals for: steam generated, feedwater and fuels flows; other variables could also be displayed if desired. This report should provide the totals on a "Shift," "Daily" and "Weekly" (or Monthly) basis, and with the total power plant usage summarized. A sample of this type of report is shown in Figure 9.

GRAPHIC

The capability to generate a "Graphic Display" to help the operator visually "see" the total power plant operation is a necessity. The graphic for each power plant would be customized for that system, but some examples are given in Figures 10 and 11. With the information displayed as shown in Figure 10, the operator can see all of the pertinent boiler parameters and make decisions on the plant's operation; a further assistance to the operator would be the next display Figure 11. This display contains bar graphs for each boiler. The bar graph will show the present load on each boiler and display the operating range in which the boiler can swing under automatic operation.

462   Chemical Process Control II

```
                       DAILY REPORT      #1 BOILER       10/ 8/80

                          DAY SHIFT         EVE SHIFT         NYT SHIFT           DAILY
OPERATING VARIABLES     AVERAGE  HI   LO  AVERAGE  HI   LO  AVERAGE  HI   LO  AVERAGE  HI   LO
DRUM LEVEL      -INCH    -0.2  -1.3  -2.1    0.6   6.9  -7.1    1.1   5.8   -.9    1.0   6.9  -7.1
STEAM FLOW      -K#/HR   56.0  69.0  44.0   48.6  72.0  37.0   61.8  75.0  44.0   55.5  75.0  37.0
FEEDWATR FLOW   -K#/HR   64.3  73.0  43.0   52.1  69.0  25.0   60.4  69.0  53.0   58.9  69.0  25.0
DRUM PRES.      -PSI    595.5 601.0 591.0  594.8 614.0 581.0  595.9 614.0 573.0  594.4 614.0 573.0
S.H. OUTLET PRES.-PSI   590.2 595.0 588.0  590.8 606.0 579.0  591.6 606.0 572.0  590.9 606.0 572.0
AIR FLOW        -%       51.8  65.3  43.2   48.6  69.4  44.2   59.6  71.0  47.5   54.4  71.0  43.2
OIL - FUEL FLOW -GPM     13.1  14.6  12.0   12.3  14.5  10.7   13.9  14.4  13.5   13.1  14.6  10.7
OIL - BURNER PRES.-PSI   37.1  38.0  37.0   43.3  50.0  34.0   46.3  49.0  44.0   42.2  50.0  34.0
ATM. STM. PRES. -PSI     63.5  63.6  63.4   68.6  75.0  60.2   71.2  73.4  69.4   67.8  75.0  60.2
% OXYGEN        -%        3.4   3.5   3.3    3.3   3.7   3.1    3.1   3.6   2.5    3.2   4.0   3.5

TEMPERATURES - DEG F
STEAM - S.H. OUTLET     718.9 740.4 712.4  712.5 743.8 698.8  713.1 749.2 695.2  714.5 749.2 695.2
AIR - A.H. INLET         91.0  98.0  84.0   97.9 103.0  93.0   90.5  94.0  88.0   95.5 103.0  84.0
AIR - A.H. OUTLET       274.5 278.0 268.0  268.3 278.0 264.0  251.1 268.0 246.0  263.3 278.0 246.0
GAS - A.H. INLET        535.2 545.0 529.0  523.8 532.0 509.0  486.5 534.0 285.0  513.2 544.0 285.0
GAS - A.H. OUTLET       387.5 396.0 382.0  378.2 394.0 369.0  421.1 614.0 378.0  393.9 614.0 369.0
FUEL OIL                220.0 224.0 218.0  222.0 225.0 217.0  221.0 228.0 215.0  222.0 226.0 217.0

DRAFT PRESSURES - INCH
F.D. FAN DISCHARGE        5.9   6.9   5.8    5.9   7.1   5.4    6.1   7.2   5.0    6.0   7.2   5.0
WINDBOX                   3.7   4.4   2.0    3.6   4.3   1.7    3.8   4.0   2.7    3.8   4.4   1.7
FURNACE                   -.1    .0   -.4    -.1    .0   -.4    -.2    .0   -.5    -.1    .0   -.5
BOILER OUTLET            -5.1  -4.1  -6.4   -4.8  -4.2  -6.5   -4.9  -4.3  -6.7   -5.0  -4.1  -6.7
GAS - A.H. OUTLET       -10.7 -10.0 -11.2  -10.6 -10.1 -12.0  -10.2 -10.0 -11.0  -10.4 -10.0 -12.0
DUST COLLECTOR OUTLET   -12.5 -12.0 -13.1  -12.3 -12.2 -13.5  -12.2 -12.0 -12.7  -12.4 -12.0 -13.5
```

Figure 8.

```
                DAILY SUMMARY REPORT - THROUGHPUT VARIABLES      10/ 8/80

                                           DAY       EVE       NYT         TOTALS
                                          SHIFT     SHIFT     SHIFT     DAILY      WEEK
NO.1 BOILER - STEAM GENERATED - K#         454.0     384.0     474.0    1312.0     3391.0
            - FEEDWATER       - K#         467.0     404.0     493.0    1362.0     3517.0
            - OIL             -GAL        3789.0    3210.0    3917.0   10916.0    78056.0
            - OIL    -#STM/GAL OIL         119.8     119.6     121.0     120.1     ------
            - % O2 (AVERAGE)  - %            3.3       3.4       3.1       3.2       ---

NO.2 BOILER - STEAM GENERATED - K#           0.0       0.0       0.0       0.0       0.0
            - FEEDWATER       - K#           0.0       0.0       0.0       0.0       0.0
            - OIL             -GAL           0.0       0.0       0.0       0.0       0.0
            - OIL    -#STM/GAL OIL           0.0       0.0       0.0       0.0       0.0
            - GAS             -KCF           0.0       0.0       0.0       0.0       0.0
            - GAS    -K# STM/KCF GAS         0.0       0.0       0.0       0.0       0.0
            - % O2 (AVERAGE)  - %            0.0       0.0       0.0       0.0       0.0

NO.3 BOILER - STEAM GENERATED - K#        1319.0    1156.0    1118.0    3593.0    12376.0
            - FEEDWATER       - K#        1327.0    1191.0    1146.0    3664.0    12577.0
            - OIL             -GAL       10850.0    9470.0    9180.0   29500.0   102780.0
            - OIL    -#STM/GAL OIL         121.5     122.1     121.8     121.8
            - GAS             -KCF           0.0       0.0       0.0       0.0       0.0
            - GAS    -K# STM/KCF GAS         0.0       0.0       0.0       0.0       0.0
            - % O2 (AVERAGE)  - %            3.1       3.2       3.0       3.1

NO.4 BOILER - STEAM GENERATED - K#        1511.0    1599.0    1635.0    4745.0    31944.0
            - FEEDWATER       - K#        1454.0    1537.0    1573.0    4564.0    32645.0
            - OIL             -GAL           0.0       0.0       0.0       0.0       0.0
            - OIL    -#STM/GAL OIL           0.0       0.0       0.0       0.0       0.0
            - GAS             -KCF        1848.0    1950.0    2010.0    5808.0    39350.0
            - GAS    -K# STM/KCF GAS       817.0     820.0     817.0     818.0       0.0
            - % O2 (AVERAGE)  - %            2.4       2.5       2.3       2.4

NO.5 BOILER - TOTAL STEAM GEN - K#        2698.0    2730.0    2871.0    8299.0    49590.0
            - FEEDWATER       - K#        2806.0    2815.0    2963.0    8584.0    51817.0
            - BARK - STEAM    - K#        1454.0    1193.0    1377.0    4024.0    25437.0
            - OIL             -GAL       10770.0   13390.0   13010.0   37170.0   225740.0
            - OIL    -#STM/GAL OIL         115.5     114.8     114.8     115.0
            - % O2 (AVERAGE)  - %            4.5       3.8       4.3       4.1

                              ------------TOTALS (COMPUTED)------------
ALL BOILERS - STEAM GENERATED - K#        5982.0    5869.0    6098.0   17949.0    97301.0
            - FEEDWATER       - K#        6054.0    5945.0    6145.0   18174.0   100556.0
            - OIL             -GAL      25409.0   26070.0   26107.0   77586.0   406576.0
            - GAS             -KCF        1848.0    1950.0    2010.0    5808.0    39350.0
```

Figure 9.

R. A. Hanson 463

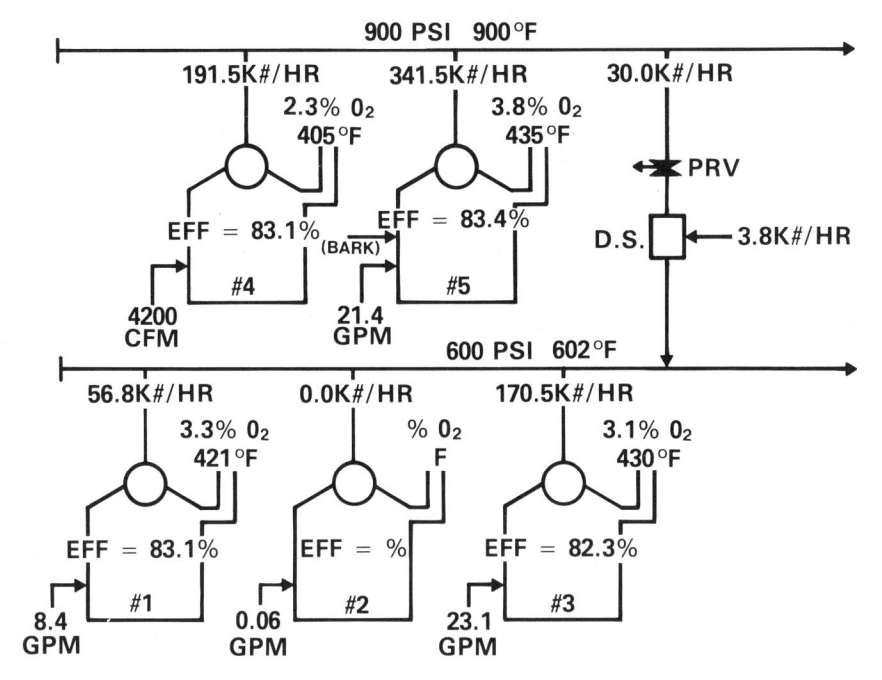

900 PSI 900°F

191.5K#/HR 2.3% O₂ 405°F EFF = 83.1% #4 4200 CFM

341.5K#/HR 3.8% O₂ 435°F EFF = 83.4% (BARK) #5 21.4 GPM

30.0K#/HR PRV D.S. ← 3.8K#/HR

600 PSI 602°F

56.8K#/HR 3.3% O₂ 421°F EFF = 83.1% #1 8.4 GPM

0.0K#/HR % O₂ °F EFF = % #2 0.06 GPM

170.5K#/HR 3.1% O₂ 430°F EFF = 82.3% #3 23.1 GPM

Figure 10.

BOILER — STEAM FLOWS

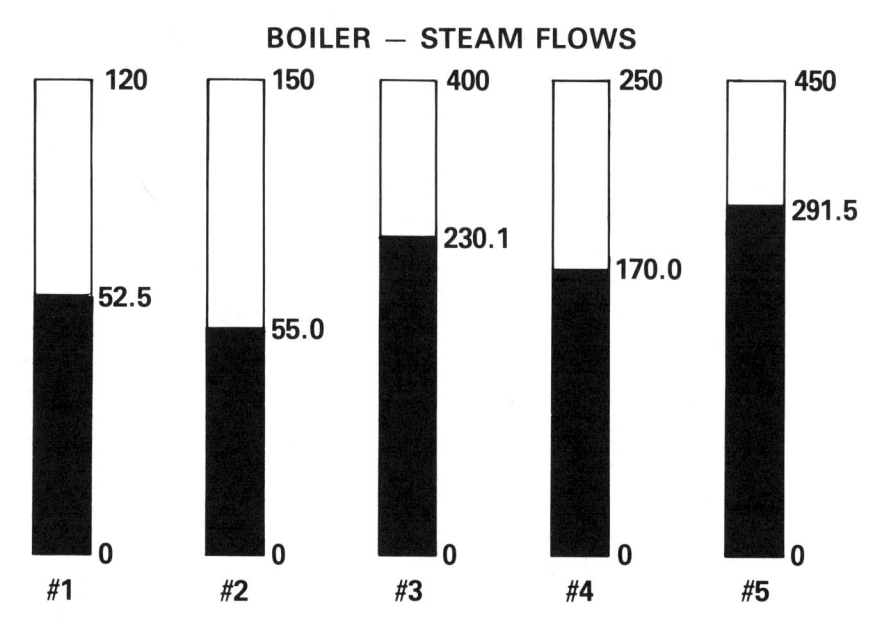

#1: 120 / 52.5 / 0
#2: 150 / 55.0 / 0
#3: 400 / 230.1 / 0
#4: 250 / 170.0 / 0
#5: 450 / 291.5 / 0

Figure 11.

CONCLUSIONS

This paper has presented the main areas of a Utilities Management System that should be investigated and implemented; that is: Steam Generation, Steam Distribution, Purchase Power, Management Services. There are other areas as described earlier; Water, Air, Fuels and Environmental Management; and the list can be expanded for particular installations. There are also numerous feedforward capabilities that can now be implemented. The main would be remote sensing of the steam load (increase or decrease) and feedforward to the combustion control system. The boiler firing rates can then be changed before the header pressure changes, resulting in closer control of header pressure and more direct boiler optimization calculation.

REFERENCES

[1] W. A. Vopat and B. G. A. Skrotzki, "Power Station Engineering and Economy," McGraw-Hill, New York, 1960.
[2] R. A. Hanson, "Optimization in the Industrial Power Plant," paper given at T.A.P.P.I. Engineering Conference, San Francisco, 1978.

Design of Control Systems
For Integrated Chemical Plants

INTEGRATED PLANT CONTROL: A SOLUTION AT HAND OR A RESEARCH TOPIC FOR THE NEXT DECADE?

Manfred Morari
Chemical Engineering Department
University of Wisconsin
Madison, WI 53706

ABSTRACT

The applicability of currently available controller design and tuning techniques to the control of integrated plants is investigated. Deficiencies from a theoretical and a practical point of view are pointed out and directions for future research are given.

INTRODUCTION

The textbooks, courses, research papers, in short most of the educational and research efforts in the area of process control revolve around "small" systems (involving a maximum of three input and three output variables). If the dynamic and control properties of physical systems are examined the attention is generally restricted to one or maximally two

---

The U.S. Government retains a nonexclusive, royalty-free license to the use of this paper.

coupled unit operations (reactor, distillation column, etc.) which have been carefully modeled, whose parameters are known with high accuracy, and whose control structure (the measurements and manipulated variables) has been selected a priori in an ad hoc manner. Chemical plants consist of many more than two unit operations. Moreover, economic necessities have forced the designer to introduce tighter coupling between the units through heat integration, recycles and the removal of intermediate storage units, thus improving raw material and energy efficiency. Therefore, it is usually not possible to look at a plant as an aggregate of essentially independent small groups of unit operations and "small system methods" are unlikely to be successful.

Industry noticed early that a control strategy of using individual control loops to force variables to their steady state operating levels as fast as possible usually failed because of the interactions between adjacent loops. An alternate strategy termed "set-point environmental control," required flows, temperatures, pressures, etc. in the plant to be kept constant everywhere and thus to guarantee a constant production rate and quality. These requirements can be contradictory, however, because it is impossible to manipulate a flowrate to compensate for a concentration disturbance in a unit and at the same time to keep the flowrate through the unit constant. In order to resolve this conflict Buckley (1964) introduced the concept of "dynamic process control." He suggested to distinguish between material balance and product quality controllers. Whereas material balance controllers act as low pass filters thereby taking maximum advantage of the storage capabilities in the plant, product quality controllers act as high pass filters. Thus, by choosing storage capacities and controller gains appropriately, the break frequencies of the two control systems can be placed sufficiently far apart to avoid interaction. With other words, in order to achieve decoupling, we pay for poor quality control by having to put in large storage tanks. This ingenious policy of the optimal compromise has found wide industrial acceptance and has been able to overcome many problems.

Mass flow and energy integration and the trend to reduce the size of storage tanks to keep inventory costs down have yielded many processes multivariable in nature with complex interactions between the variables. The concept of decoupling material balance and product quality loops is insufficient to deal with this new situation. Since Buckley's pioneering ideas almost twenty years ago, control theory has undergone revolutionary changes. We have learned about optimal state estimation and optimal control in the presence of noise (the Linear Quadratic Gaussian problem), about pole placement to achieve a desired speed of response and adaptive control in a stochastic and a deterministic environment, just to name a few which would be suspected to have a decisive impact on the process control practice. The goal of this paper is an attempted analysis on why the influence of the new theoretical developments has been so small and in particular why these and other modern techniques have left "plant control" mostly unchanged for twenty years.

The four main reasons are believed to be:

1. *Large Scale System Aspects.*

The computing power generally available today is rarely a restricting factor. The successful implementation of a plant control system requires a prior structuring of the control task and the proper selection of the manipulated and measured variables, however and this problem is far from trivial for a large chemical plant. There is very little theory available to help the designer with these decisions, which can be of crucial importance for the success or failure of a plant control system.

2. *Sensitivity*

Most control theory is derived under a set of idealizing assumptions, which are not satisfied in practice (e.g. linearity, perfect system and disturbance model). Through the standard tuning procedures a certain (unknown) degree of robustness against modelling errors is achieved implicitly. Sensitivity considerations should be a primary and explicit goal for tuning controllers and structuring control tasks. The presently available sensitivity theory for multivariable systems is impractical as a design tool.

3. *Fundamental Limitations to Control Quality*

For linear systems with perfectly known models perfect control can not be achieved (not even in a limiting sense) in the presence of limitations like RHP zeroes. These limitations cannot be removed by a controller, but are inherent in the system and could therefore form a criterion for selecting manipulated and measured variables, decoupling control loops and even judging different system designs from an operability point of view. This very simple concept has not been explored in any depth.

4. *Education*

The widely acclaimed theory-practice gap is often attributed to ignorance on part of the practitioner. While there may be some truth in this analysis, there is just as much fault on the part of the theoretician who is either solving irrelevant problems or presenting the solutions to important problems in obscure formulations. As G. E. P. Box once remarked, the relationship between theory and practice is comparable to that between a man and woman: without their interaction there is no life. Though the education problem would certainly be worth discussing, it is considered to be beyond the scope of this analysis.

It is obvious that plant control is a very complex problem. In our search for answers we should learn from the experience other disciplines have had in the solution of complex problems. Process synthesis, the development of new chemical processes, was first approached via fancy optimization techniques. The improvements over designs carried out by conventional methods were often marginal, the value of the techniques was not convincing and they found only limited acceptance in industry. Recently, for example in the area of heat exchanger network synthesis, optimization algorithms were quite successfully replaced by simple structural and physical arguments, derived from a thorough understanding of the specific problem (Linnhoff, Flower, 1978, Flower, Linnhoff, 1980). Industrial acceptance of the new methods is spreading fast, because of their simplicity, effectiveness and practical appeal. Following the same promising philosophy, optimization, optimal control and related techniques will play a minor role in the subsequent discussions. We will show that many of the problems encountered in plant control could be solved in a superior manner if the deficiencies 1–3 were removed. Consequently, research done in these three areas should be of great interest to the practitioner and should offer ample playground for the theoretician.

A general definition of plant control will be followed by a discussion of some of the current approaches and their shortcomings illustrated by examples. Finally some promising research directions and results which have appeared recently will be reviewed.

A GENERAL DEFINITION OF PLANT CONTROL

The following systematic treatment expands on that given by Morari et al. (1980). Similar approaches have been used by Govind and Powers (1976), Umeda et al. (1978) and Shinnar (1980).

The problem of plant control can be divided into five parts:
1. Formulation of the control objectives
2. Selection of the variables to be controlled
3. Selection of the variables to be measured
4. Selection of the variable to be manipulated
5. Design of the interconnection structure between measured and manipulated variables.

As a *control objective* we require in mathematical terms the variations of the state variables $x(t)$, the manipulated variables $u'(t)$ and the disturbances $d(t)$ to be in some acceptable "operating window" of the function space $X \times U' \times D$, i.e.

$$(x(t), u'(t), d(t)) \in O \subset X \times U' \times D \qquad (1)$$

This complex expression includes implicitly the physical system, i.e. the relationship between u', d and x, as well as product specifications and operational constraints. The physical system is assumed to be fixed for the moment, effects of design on control are going to be discussed later in this paper. The set of feasible x(t) and u'(t) for a given disturbance d(t) could have been determined on the basis of the optimization of some economic objective. While such a clear formulation of the goals would be desirable, it is usually infeasible because it is too complex and requires a detailed process model and good, a priori information on the disturbances. More often the control objectives are expressed in linguistic terms or are just a general design philosophy which is never formulated explicitly: "Satisfy product specifications and operational constraints, guarantee safe and reliable operation and be as economical as possible." The selection of the controlled, the manipulated, the measured variables and the interconnection structure is then made in an evolutionary and iterative manner along these guidelines.

It is rarely feasible or necessary to control all the state variables x(t), but generally a set of *controlled variables* y(t) is selected whose evolution with time is to be controlled through the action of some of the available *manipulated variables* u(t). This should be done in a manner such that a satisfaction of the control objectives is implied:
Select y(t), u(t) such that

$$(y(t), u(t), d(t)) \epsilon O^* \subset X \times U \times D \Rightarrow (x(t), u'(t), d(t)) \epsilon O \quad (2)$$

The *measurements* m are selected to give maximum direct or indirect information on the controlled variables y. Often a subset of the disturbances d_m is also measured and used with m to obtain estimates $\hat{y}$ of y

$$\hat{y}(t) = \hat{y}(m(t), d_m(t), u(t)) \quad (3)$$

Finally a decision is made on the *interconnection structure*, i.e. how measured and manipulated variables are to be paired (feedforward, feedback, SISO, SIMO, MIMO, etc) and what control laws are to be used. We are looking for the functional dependence of u on the measurements such that a satisfaction of the control objectives is implied:

$$u(t) = u(y(t), d_m(t)) = u(m(t), d_m(t)) \Rightarrow (y(t), u(t), d(t)) \epsilon O^* \quad (4)$$

Most current theory is obsessed with this last part, the control law, probably because it can be expressed in mathematical terms most conveniently. It is our conjecture that the earlier decisions during the selection of the manipulated, the measured and the controlled variables, for which no theoretical guidelines exist, are at least as important if not more important than an "optimal" control law. We will discuss examples where incorrect selections can make good control difficult or even impossible.

SOME CURRENT APPROACHES TOWARD A SOLUTION OF THE PROBLEM OF PLANT CONTROL

In view of the fact that plants of realistic size can easily involve hundreds of measurable and manipulatable variables, a global solution of the plant control problem without simplifications is impossible. It is almost a natural tendency to attempt to divide the overall problem into subproblems which are more amenable to currently available solution techniques, an idea Descartes suggested several hundred years ago:

> "Divide each problem that you examine into as many parts as you can and as you need to solve them more easily"
>
> Descartes, Discours de la Methode

A rigorous theoretical treatment of the different decomposition schemes in the context of control and optimization can be found in Mesarovic et al. (1970a) and in a collection of papers edited by Wismer (1971). Qualitative introductory outlines are given by Lefkowitz (1966, 1975) and Mesarovic (1970b).

We can distinguish three approaches to make the problem tractable which are currently in use:

1. *Multilayer (vertical) decomposition*

The idea is illustrated in Fig. 1. The sets of disturbances, manipulated and measured variables are divided into two or more sets. Rules for this division are not well defined but it is generally assumed that the manipulated variables in the first layer can stabilize the system against "fast" disturbances by employing high frequency measurements and control actions. At the second layer the system is regarded to be at a quasisteady state. As a consequence set point changes and direct changes of the manipulated variables in this layer can be computed on the basis of a steady state model. For the control problem 2–6 the following structure results:

Frequency of Adjustment of u	Objective	Control Law
low	$(y_2, y_{1s}, u_2, d) \epsilon O_2$	$\begin{pmatrix} u_2 \\ y_{1s} \end{pmatrix} = f_2(m_2, d_{m2})$
high	$(y_1, y_{1s}, u_1, d) \epsilon O_1$	$u_1 = f_1(m_1, d_{m1})$

where the subscript 2 refers to variables which are measured or adjusted in the second (low frequency) layer and y_{1s} are the set points determined in layer 2 and used in layer 1. More than two adjstment frequencies could be used giving rise to more than two layers. The upper layers can also serve some optimizing or adaptive function as suggested in Fig. 1. It is important to realize that the simplification resulting from the multi-layer structure is not so much due to the different adjustment frequencies of the manipulated variables as to a change in the modeling requirements and the smaller number of variables in each layers. Instead of the nonlinear dynamic model needed in the global approach, a linear dynamic model and a nonlinear steady state model is often sufficient for the two layer structure. Because only a subset of the variables is considered in each layer, the selection of the control structure is simplified.

2. *Multiechelon (horizontal) Decomposition*

We divide the overall system into a set of interacting subsystems, in general consisting of groups of unit operations. This separation is meaningful if the interactions between the subsystems are relatively weak so that the subsystems can be controlled quite independently and the demands on the "coordinator" (see Fig. 2) are minor. The multiechelon decomposition can be applied by itself or in conjunction with the

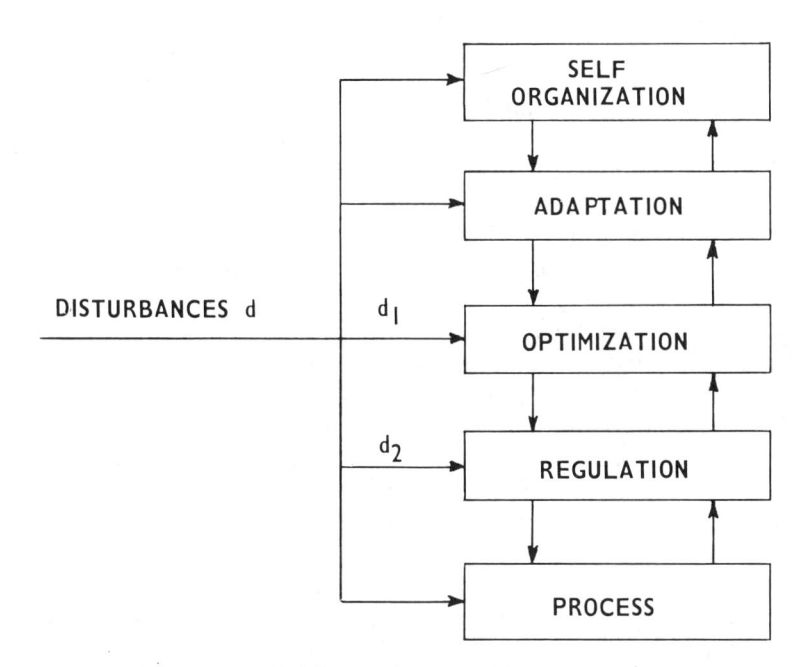

Figure 1. Multilayer Process Decomposition.

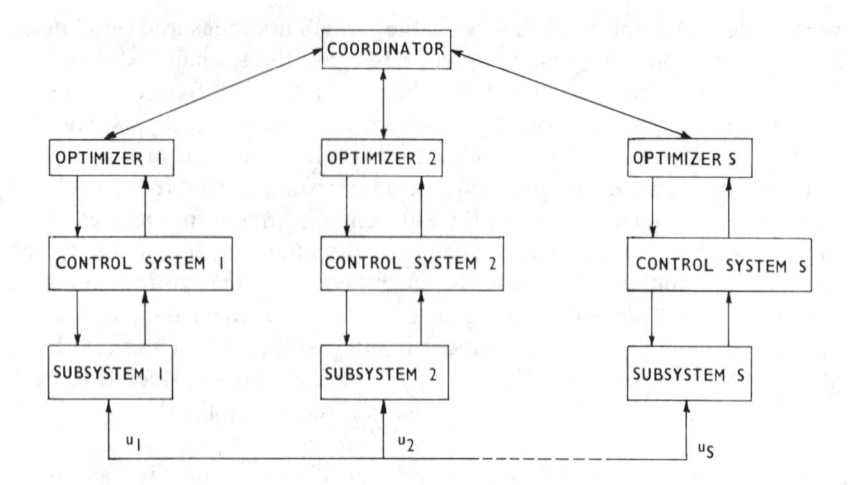

Figure 2. Multiechelon Process Decompsition.

multilayer decomposition. In the latter case each layer can be decomposed horizontally. A decomposition of the first layer corresponds to a decoupling of sets of control loops. In the second layer the optimization tasks for the different subsystems would be carried out separately and only periodically coordinated by the coordinator. Again, because of the reduction in the number of variables considered in each subsystem the synthesis of the control structure becomes easier.

3. *Evolutionary Approach*

A path followed widely by industrial practitioners is to synthesize first a preliminary control structure according to rules of thumb (heuristics) established by experience. (Govind and Powers (1976) have collected quite a number of these heuristics through discussions with control engineers in industry.) In a series of subsequent steps the features of the preliminary structure are analyzed and improved progressively. Douglas (1980) has devised rules to guide the initial steps. He uses simplified process models and economic criteria to identify the "key" disturbances and control variables and shows how to obtain estimates of the economic incentives for installing simple SISO and optimal multivariable controllers respectively. While the multiechelon and the multilayer decomposition can be used to aid the design as well as the implementation, the evolutionary approach is mostly a design tool. It can be and generally is applied in conjunction with the other two decomposition methods. Thus, for example, the regulatory and the optimizing control layer could be developed separately in an evolutionary manner. Even the decomposition itself could be created through an evolutionary process.

Though decomposition appears to be the logical approach to the analysis, design, and implementation of plant control systems, and though it is widely used in an ad hoc fashion, guidelines for carrying out this division are virtually nonexistent despite their obvious importance. In the 17th century Leibniz raised concerns similar to ours about Descartes' rule quoted above:

> "This rule of Descartes is of little use as long as the art of dividing remains unexplained. By dividing his problem into unsuitable parts, the unexperienced problem solver may increase his difficulty"
>
> Leibniz, Philosophische Schriften

When dividing the synthesis problem or when decomposing the control tasks for implementation the designer should follow two guidelines:

1. The decomposition and the accompanying simplifications should leave the system operability and in particular the system stabilizability unimpaired. The properties of controllability and operability should be preserved.

2. A control task should be implemented in distributed and/or hierarchical fashion if this decreases the complexity of the necessary hardware and software or if there exists a hardware/software tradeoff situation which is favorable in terms of economics, reliability, transparency for the operator, etc. Though the validity of these rules is too obvious to merit a discussion it should be emphasized that very little progress has been made in their further development in a specific and quantitative direction. The following examples are meant to illustrate this point, to identify it as an important research area and to relate decomposition to the issues of "sensitivity" and "limitations to control quality."

PRACTICAL ASPECTS OF DECOMPOSITON

The Fluid Cat Cracker

Let us take a fresh look at Kurihara's fluid cat cracker (FCC), (Kurihara, 1967, Gould et al., 1970) which has been a popular example system during the past ten years (Schuldt, Smith, 1971, Lee, Weekman, 1976, Arkun, 1979, Shinnar, 1980). A typical scheme with a conventional control structure is shown in Fig. 3. Gas oil feed is mixed with hot catalyst from the regenerator. During the endothermic cracking reaction in the riser and the reactor carbonaceous material deposits on the catalyst and deactivates it. In the regenerator the material is burned off and the catalyst is heated up through the heat of combustion. The overall control

476   Chemical Process Control II

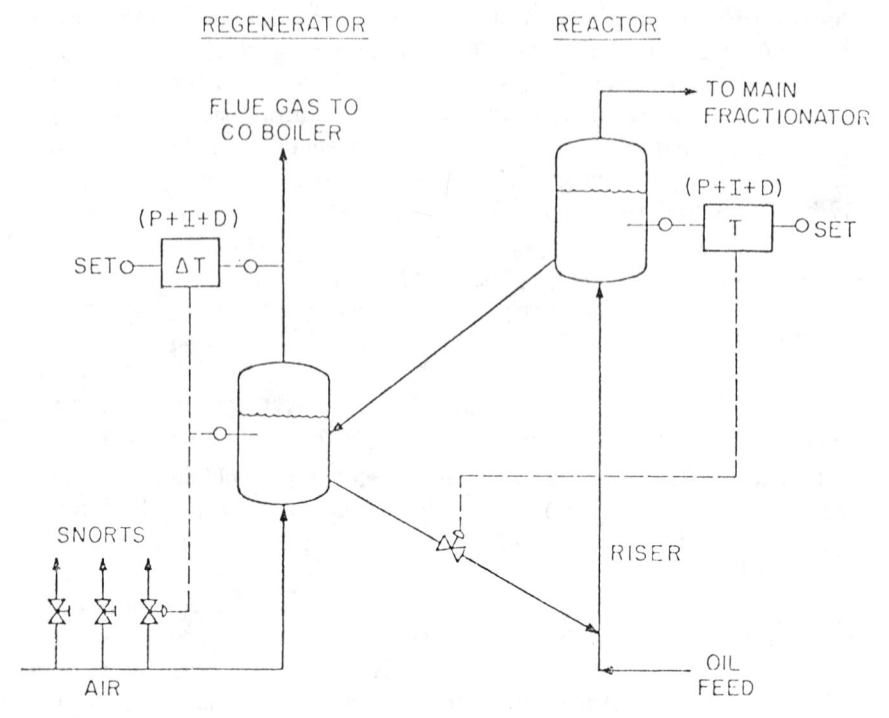

Figure 3. Conventional FCC Control.

objective is to achieve an economically attractive conversion of gas oil. These economics can be quite complex relating to the yield and octane number of gasoline, and properties and amount of fuel oil, for example. The air flowrate to the regenerator and the catalyst recirculation rate are usually used as manipulated variables, feed preheating would be another one. There is a variety of constraints which can become active depending on the design parameters, the catalyst used and the feedstock: Temperature in the reactor and the regenerator, catalyst recirculation rate, air flowrate, and oxygen content in flue gas from the regenerator are the most important ones. As "fast" measurements only temperatures at different places in the system are available. At longer intervals the analyzer measurements from the fractionator succeeding the unit could be used. A more detailed discussion of all the options is given by Shinnar (1980). In short, there is quite a large number of variables available, the definition of the layers, the selection of the proper measurements and control loop pairings, and the decoupling of the loops are clearly nontrivial tasks.

In the dynamic control layer of the conventional control scheme the flue gas oxygen content (or the amount of afterburning as measured by T) is controlled by the air rate in the regenerator and the reactor temperature is controlled by the catalyst circulation rate. This structure was probably arrived at in an evolutionary fashion. It assumes that for some disturbance range all the process variables can be held within the constraints by these two dynamic controllers whose setpoints are adjusted according to some economic criteria. This scheme makes sense if the reactor temperature is the key factor affecting the conversion and if we know from experience or modelling that the (uncontrolled) regenerator temperature will stay within bounds.

Because of its superior dynamic characteristics, Kurihara (1967) suggested an alternate control scheme (Fig. 4) where the regenerator temperature is controlled by the air flow rate and the afterburning by the catalyst recirculation rate. Operator resistance against the new scheme (Lee, Weekman, 1976) is only a secondary effect. The basic flaw here is that as a result of the narrow emphasis on the dynamics, the overall control goal was neglected, the key economic variable, reactor temperature, was not included in the control scheme. This was corrected through the cascade control scheme by Lee and Weekman (1976) which has also a good dynamic performance (Fig. 5). Shinnar (1980) has further improved the control by adding fuel preheat control and regenerator cooling coils.

Kurihara arrived at his control scheme after a very detailed economic and dynamic analysis of a particular FCC with specific design parameters. Many other people have followed and have designed superior control schemes for the "Kurihara-FCC." It is questionable, however, if any of these imaginative variations helped to solve the *real* control problem. It appears to be a well accepted fact that of FCC's in general the economically optimal operating point tends to lie at the intersection of constraints and that the set of active constraints changes depending on the disturbances (Lee, Weekman, 1976, Prett, Gillette, 1979, Arkun, 1979). Therefore, in general, the conventional control scheme will be most appropriate if the flue gas oxygen content and the reactor temperature are the active constraints (they can be controlled directly and tightly), the Kurihara scheme when the regenerator instead of the reactor temperature constraint is active. Economically optimal operation would require a switching between different single loop control structures. The development of an adaptive control scheme robust enough to cope with the switching was suggested for this purpose. This is likely going to be the wrong approach. For the FCC we have every indication that the regulatory and the optimizing constraint control tasks on one hand and the

478   Chemical Process Control II

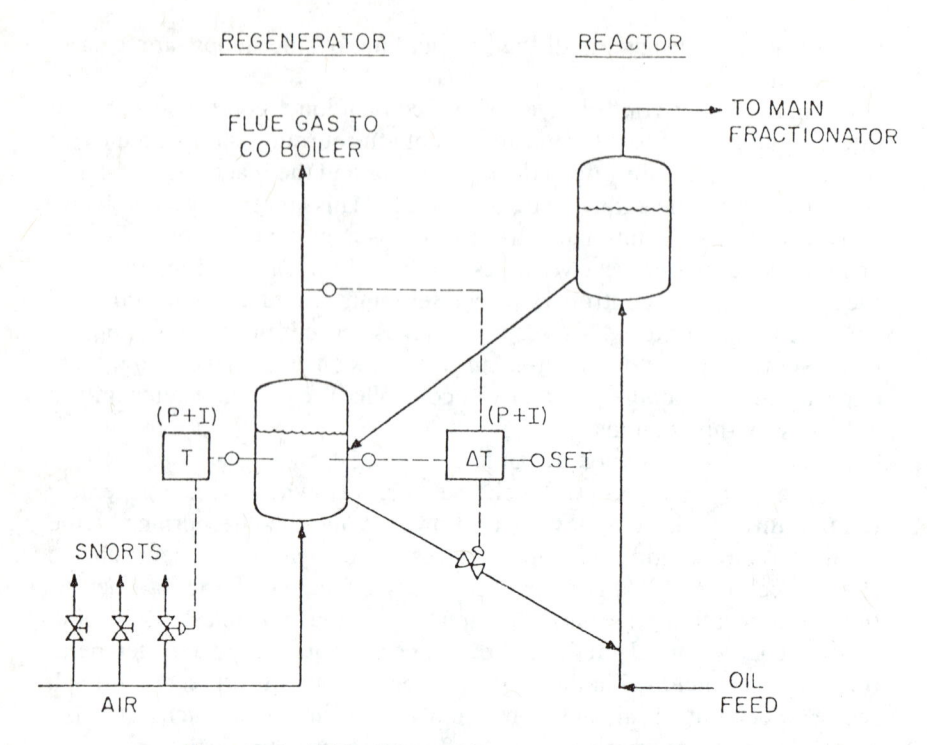

Figure 4. FCC Control Proposed by Kurihara.

different SISO control loops on the other are tightly coupled and that a *decomposition* is inappropriate. All the constrained and measured variables and all the manipulated variables should be considered simultaneously to arrive at the appropriate control action. Using computer control, this task is not only feasible but also practical in an industrial environment as was demonstrated by Shell Oil Co. (Prett, Gillette, 1979).

A further question of practical importance is how these control schemes behave in the face of process nonlinearities which were not considered explicitly in any of the design techniques. *Sensitivity* studies, either of a theoretical type or through simulation are not available for any one of them.

Chemical Reactors

It is well known that chemical reactors can exhibit all kinds of exotic and potentially dangerous behavior like multiple steady states, instabili-

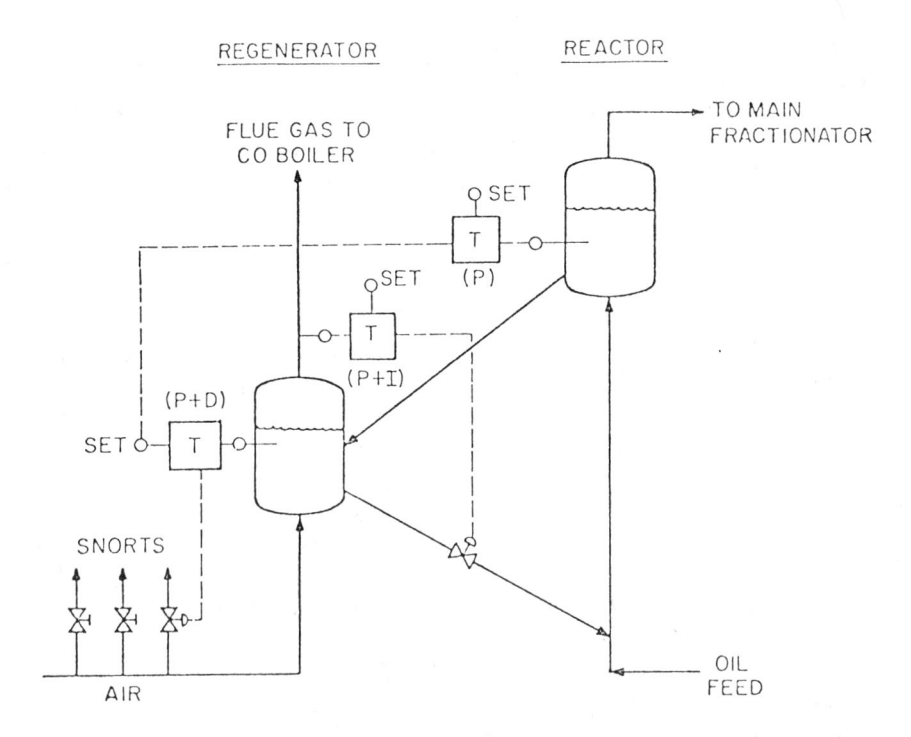

Figure 5. FCC Control Proposed by Lee & Weekman.

ties and limit cycles. In most circumstances any variations in outlet specifications are undesirable. Usually temperature control is employed in the first layer to stabilize the reactor and the temperature setpoints are corrected in the second layer on the basis of concentration measurements to compensate for drifts. This decomposition appears to work well: For example, the Aris-Amundson reactor (1958) can be stabilized through a proportional controller using temperature measurements, it cannot be stabilized on the basis of composition measurements alone. The same can usually be said about exothermic tubular reactors. Certain types of polymerization reactors, however, suffer from multiple steady states even under isothermal conditions (Schmidt, Ray, 1980) and the problem becomes several orders of magnitude more complex when thermal effects are also included (Jaisinghari, Ray, 1977). It is quite clear that temperature control will be insufficient for stabilization here and the proper decomposition of the control layers which guarantees safe and smooth operation is not at all obvious. What is the best way to handle a reactor with parallel endothermic reactions, which is also known to be able to exhibit multiplicity? The development of criteria for the decomposition of the control layers for chemical reactors, which can be implemented with the available measurement technology seems to be an important unsolved problem.

Model Simplification vs. Control System Sensitivity

As mentioned above, a further reason for the multilayer decomposition is the ability to use models of different complexity in the different layers. For the first (dynamic) layer a simple linear model, e.g. first or second order with dead time, is often viewed as sufficient. But what does "sufficient" mean? We would like to have a quantitative measure for the degradation of the performance as a function of the model inaccuracies so that we can make a rational decision on the modelling and/or identification effort which is justifiable in a particular situation. Though many books and papers deal with the subject of control system *sensitivity*, the results are rarely useful as a design technique. Maximum insight is obtained in the frequency domain (Horowitz, Shaked, 1975), analyses in the state space tend to obscure the relevant issues and become very complex in the presence of time delays. In this spirit Shinnar and his students (Kestenbaum et al., 1976, Palmor, Shinnar, 1979) have performed an extensive set of case studies. For continuous systems they come out quite strongly in favor of PID against optimal control. The comparison is not quite fair because the optimal controller used was derived for servo- and not for regulatory control. This difference is very critical for systems with time delay (Garcia, Morari, 1981). For the discrete case a variety of controllers with favorable sensitivity characteristics was suggested depending on the structure of the system and noise models. There are no useful studies of this type available for multivariable systems.

From application and simulation studies the newly proposed techniques of Model Algorithmic Control (MAC) (Richalet et al., 1978, Mehra, et al., 1980) and Dynamic Matrix Control (DMC) (Cutler, Ramaker, 1979) seem to hold a very promising future. Both techniques were developed by or in close cooperation with industry and display a very robust behavior. Final judgement on their qualities, however, has to be reserved pending a thorough theoretical analysis. Some structural similarities between DMC, MAC and linear quadratic (LQ) control will be discussed later in the paper. It is very likely that tuning methods for these controllers can be developed which compensate for model-reality differences in a direct manner and consequently allow one to establish a straight forward relationship between model inaccuracies and deterioration in control quality.

Decoupling of Control Loops

All the examples discussed so far dealt with the aspects of multilayer (vertical) decomposition. The motivation for multiechelon (horizontal) decomposition is to split the control or optimization task into subtasks by considering only the dominant interactions between groups of input and output variables. Decomposition carried to the extreme, namely fragmenting the regulatory task into SISO regulatory loops was studied quite extensively in the process control literature. A series of interaction indices was defined over the years, of which the Relative Gain Array (RGA) (Bristol, 1966) has found widespread use. Tung and Edgar (1981) have derived a dynamic version of the RGA. With the exception of Rosenbrock's (1974) diagonal dominance criterion all the different interaction measures provide only an approximate check of the severity of the interactions but have no direct implications for controller tuning or performance. Often it is not possible to form SISO pairs with weak interactions between all the pairs, but it would probably be possible to form groups of variables such that the interactions between the groups are weak. The identification of interacting and non-interacting groups of variables would form a rational basis for structuring plant control system but none of the currently available interaction measures is suitable for that purpose. Such a tool would also significantly enhance the power of graphical multivariable controller design techniques like the Inverse Nyquist Array, which become quite impractical for systems with more than three inputs and outputs.

We can also provide special control action to decouple control loops artificially. Almost since the advent of noninteracting or decoupling control (Boksenbom, Hood, 1949), the first multivariable control method suggested, there has been disagreement about the merits of decoupling. Its value for servo control and the simplifications for controller tuning arising from it are unquestioned. Sensitivity problems can arise because of the necessary inversion (Jafarey, McAvoy, 1978), but they are usually easy to detect. There is, however, a strong indication that the regulatory qualities of the controller, i.e. its ability to suppress disturbances, can deteriorate when decoupling control is used instead of two SISO loops or other multivariable control schemes. Niederlinski (1971) for example, demonstrates on the Rosenbrock distillation column that complete decoupling will yield inferior two-composition regulatory control. Despite that fact decoupling is the standard industrial solution to multivariable control problems. It would probably be used more selectively if there were some theoretical understanding of the consequences of decoupling and if one did not have to rely solely on simulation and case studies. Bristol's "pinned zero" concept (1981) promises an answer to this question and should be further explored. It postulates that certain RHP zeros, the "pinned zeroes," are inherent in the transfer matrix model and cannot be removed but only shifted around. Sometimes decoupling duplicates these zeroes and therefore degrades the quality of control.

This brings up the last one of the three areas of emphasis, the *fundamental limits to control quality*. At present, when different decoupling techniques are compared, when the (dis) advantages of grouping variables into multivariable control blocks are judged, the decision on the superior scheme hinges crucially on the employed controller tuning method. There seems to be no guarantee that with more careful tuning decoupling might not indeed be preferable for the Rosenbrock column, contrary to Niederlinski's claims. Thus it would be desirable to have an absolute bound on the achievable control quality which can easily be calculated and against which the different control schemes can be judged. Perfect control is possible by inverting the transfer function. Nonminimum phase behavior, i.e. time delays and RHP zeros, is the only fundamental limitation to perfect control because it makes this inversion, even in an approximate sense, impossible. Therefore, we postulate that a first evaluation of the "controllability" of a system via a certain control scheme can and should be made solely on the way it handles nonminimum phase elements. In this manner Bristol (1980) explained Niederlinski's result: The decoupler generates an additional zero in the RHP. With some refinement, the concept of the "fundamental limits to control quality" should allow to base a decision on control quality not on vaguely defined procedures but on exact rational criteria.

SOME PROMISING NEW DIRECTIONS

In our discussion so far the emphasis has been on an analysis of the shortcomings of currently practiced techniques. But the situation is not all bleak. A shift in research directions has become recognizable recently which holds significant potential. Some of them, like the study of zeroes, has been mentioned above. In the remaining part of the paper we will focus on three of those new topics. They were selected mainly because they are also discussed in other papers at this conference and because we would like to look at them in the framework of plant control.

MEASUREMENT SELECTION AND ESTIMATION

Plant Data Estimation and Adjustment

In a large plant hundreds of temperature, flowrates and composition measurements are available from different locations at periodic intervals.

Exploiting temporal and spatial redundancy, i.e. repeated measurements and measurements related through mass and/or enthalpy balances, we seek to solve the following three problems:

—reconciliation: correct measurements to satisfy mass and enthalpy balances
—coaptation: estimate unmeasured flowrates, compositions, etc. from measured ones
—fault detection and rectification: detect faulty instruments and leaks and correct the measurement errors arising from them

The first in-depth analysis of the problem is probably due to Vaclavek and his students. Mah, Stephanopoulos and their students refined and significantly improved the early techniques. All the early work up to 1975 is covered in the excellent review by Hlavacek (1977). Most of this work has concentrated on structural aspects: Due to the specific structure of the mass and enthalpy balances certain measurements can be corrected, certain variables estimated and others cannot. Typical goals could be to select the minimum number of measurements such that a set of desired variables can be estimated or to select those measurements which should be corrected (by invoking mass and enthalpy balances) to improve the accuracy of a set of desired measured or unmeasured variables. Vaclavek (1969), Vaclavek & Loucka (1976) and Mah (Mah et al., 1976, Stanley, Mah, 1977) follow a graph oriented approach, Stephanopoulos' (Romagnoli, Stephanopoulos 1980, 1981) is equation oriented. Recently Stanley and Mah (1981) have employed the concepts of structural observability for the classification of the measurements. The work can be regarded as a link between the graphical and the equation oriented procedure. The method by Romagnoli and Stephanopoulos (1980) is probably most flexible for extracting different kinds of structural information.

By using an extended Kalman filter for the quasi steady state mass and enthalpy balances, Stanley and Mah (1977) take advantage of repeated measurements in an efficient easily implementable manner. Their technique for fault detection (1976) is superior to that by Almasy & Sztano (1975) because it stands on rigorous statistical grounds and moreover under a few reasonable assumptions it has the ability to distinguish between leaks and measurement biases. In all these works on the synthesis and analysis of measurement structures special consideration is given to the large scale nature of the problem. The first step is usually a simplification of the problem by merging or decomposition.

Contrary to the structural questions much less progress has been made on the statistical aspects. The reconciliation problem can be formulated as a constrained least squares problem. If only overall mass balances are corrected, the constraints are linear and efficient decomposition techniques have been adapted by Romagnoli & Stephanopoulos (1981) for the solution. If component and enthalpy balances are also of interest, the problem becomes nonlinear (bilinear) and the solution of a problem of realistic size is infeasible with currently available techniques. That is where future research efforts should be directed. Further on it would be useful to select measurements not only on structural but also on statistical grounds. No work in the open literature has dealt with this problem.

State Estimation in the Presence of Noise and Modelling Errors

Measurements can be used to obtain estimates of measured and unmeasured process states. These estimates are used in turn for monitoring and/or controlling the process. The estimates can be based on a process model and are therefore subject to errors from three possible sources:

—model-reality differences
—measurement noise
—state excitation noise (= input disturbances)

Model-Reality Differences—The only measurement selection criterion which aims to minimize the effect of modelling errors on the state estimates is due to Weber and Brosilow (1972), who used it for finding suitable temperature measurements for the composition control of a distillation tower. It is interesting to note that increasing the number of measurements generally increases the sensitivity of the estimates to modelling errors.

Measurement Noise—If all other error sources are neglected certain properties of the observability matrix can be used directly for measurement (Lückel, Müller, 1975, Mehra, 1976). In process control this situation is rarely relevant, however.

State Excitation and Measurement Noise—The general approach is to use a Kalman filter for the estimation of the states. The solution of the Riccati equation, which is needed to determine the filter gain, provides the error covariance matrix of the state estimates. Some scalar measure of the steady state value of this matrix (usually the trace) is used as a measurement selection criterion. White noise is usually assumed but the extension to colored stationary noise is trivially accomplished by augmenting the state equations with the noise model. This approach was theoretically refined and applied to tubular reactors by Kumar & Seinfeld (1978a & b) and also by Harris et al. (1980). Neglecting dynamics a simpler approach was developed by Alvarez et al. (1981). It was also suggested to allow changes in the set of measurements during the time evolution of the process in order to improve the estimates (variable measurement structures). The computational procedures suggested by Athans (1972), Herring and Melsa (1974) for linear systems for this purpose are quite involved. But, as is well known, the error covariance matrix is independent of the actual values of the measurements, and therefore the optimal sequence of measurements can be precomputed off-line. Alvarez et al. (1981) developed quite an efficient procedure to vary the measurement set on-line for state estimation of non-linear systems. They use an extended Kalman filter and determine the best measurement to be used at the next time step by predicting the error covariance for different measurement structures. Based on the results of their simulation study of a tubular reactor it can be concluded that variable measurement structures will only be justifiable in special circumstances because the reduction in error over that obtained by a fixed measurement structure is often quite small.

Usually the state estimates are to be used in a controller and therefore a complete study should evaluate the control performance resulting from the selected measurement set. Mellefont and Sargent have tackled this problem theoretically (1977) and applied their method to a distillation column (1978). The results of this case study agree qualitatively with those obtained by Shunta & Luyben (1971) using completely different arguments.

State Excitation Noise—We believe that in process control unmeasured input disturbances will generally have the most serious effect on the state estimates. They are usually "long lasting" and are best described by a nonstationary noise model. The simplest model of this type is the Wiener Process. Based on this realistic noise model measurement selection criteria for linear lumped (Joseph, Brosilow, 1978, Morari, Stephanopoulos, 1980a,b) and distributed parameter systems (Morari, O'Dowd, 1980) and nonlinear lumped parameter systems (Morari, Fung, 1981) were developed. Simulation studies of a distillation column and experimental investigations of a heated steel ingot demonstrated that the estimates can be significantly improved by choosing the correct set of measuremnts. Though all the example systems were openloop, it can be shown rigorously that the estimation improvements yield equivalent off-set reductions, when feedback control is employed.

INTERNAL MODEL CONTROL (IMC)

Let us suppose that the SISO plant with the transfer function $z^{-T}G(z)$ ($G(z)$ does not contain any time delays and is assumed to be minimum phase for this discussion) is controlled by a controller with the structure shown in Fig. 6. $z^{-\tilde{T}}\tilde{G}(z)$ is the transfer function of the plant model.

486   Chemical Process Control II

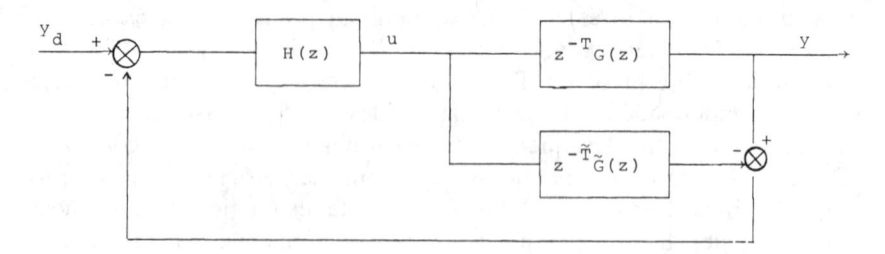

Figure 6. Structure of the Internal Model Controller (IMC).

Influenced by a paper by Brosilow (1979), Garcia and Morari (1981) showed that this structure has a variety of interesting properties and these results will be summarized in the following. Knowingly or unknowingly this structure is used for Smith predictors, Brosilow's inferential control (1979), Model Algorithmic Control (MAC) (Richalet et al. 1978, Mehra et al., 1980) and Dynamic Matrix Control (DMC) (Cutler, Ramaker, 1979). It can be shown easily that

—there will be no offset for $H(1) = \tilde{G}(1)^{-1}$
—there will be "perfect" control for $\tilde{T}=T$, $\tilde{G}(z)=G(z)$, $H(z)=G(z)^{-1}$

In the usual feedback formulation we have to vary a set of parameters to achieve a complex compromise simultaneously between stability, quality of response and modelling sensitivity. Here on the contrary, we start out with perfect control and then move away from this ideal in a well defined two step procedure. First we take care of the "fundamental limitations to control quality," the RHP zeroes, which make an inversion impossible. Then we design a filter to solve the "sensitivity" problem. The current design techniques for H(s) are not as clear cut but the possibility is there.

Let us assume that the process is described by an impulse response model truncated after n time steps. H(z) is chosen such that u(k) minimizes

$$\sum_{k=1}^{p} (y(k + T) - y_d(k + T))^2 + wu^2(k)$$

where $y_d(k)$ is the desired value of the output and w some scalar weight. Over the optimization horizon p, u is allowed to vary for the first m intervals only, afterwards it remains constant. At each step only the first computed u is implemented, then the horizon is moved one step and the same optimization is performed again for the next step. Using the impulse response model this yields a control law which is easily determined and implemented.

When p=m=n, w=0, MAC results and $H(z)=\tilde{G}(z)^{-1}$. We can relax the requirement of an exact inverse by setting m < n=p, as is done for DMC. This algorithm is stable for all systems when w is sufficiently large and stable for a wide class of systems even when w=0 as long as m is sufficiently small (Garcia, Morari, 1981). Simulation studies and applications in practice have shown these controllers to be superior to the standard PID in terms of quality of response and robustness. Constraints on the controlled variables can be easily incorporated in the least squares formulation and the extension to multivariable systems is conceptually straight forward. Unfortunately sensitivity does not appear as an explicit tuning criterion here but the structure offers the potential and a research effort in this direction seems certainly worthwhile.

THE INTERDEPENDENCE OF PROCESS DESIGN AND PROCESS CONTROL

In recent years, rising raw material and energy costs have brought about a change in the philosophy of plant design. Economic necessity has forced designers to strive for more efficient utilization of natural resources. Case studies demonstrate the impressive improvements obtainable by the new approach. The weakness in the improvements in design efficiency is that they usually lead to stronger interactions between the different units in a plant, which can result in operational difficulties.

Some imaginative schemes for resource conservation have not been applied in practice because they do not satisfy the basic criteria put forward by process engineers. In practical operation design efficiency is a secondary consideration, while the primary concern is that the process should remain sufficiently flexible, operable and controllable, attributes which we call resilience. The main process should be served by the resource management system, rather than be subjugated by it.

To make the new design techniques practicable, the steady state, essentially economic, aspects and the operational aspects should be considered simultaneously, rather than sequentially as is done now. Progress in this direction, particularly the development of suitable design tools, is the objective of an extensive investigation at the University of Wisconsin (Lenhoff & Morari, 1981, Marselle et al., 1981).

We can distinguish between dynamic resilience and steady state resilience. By dynamic resilience we mean the ability of a process to cope safely and efficiently with process disturbances and changes in operating conditions not only as far as the steady states are concerned but also during the transient introduced by these inputs. In some cases the dynamics are not critical and of minor interest only. Then a consideration of the steady states suffices and we speak of steady state resilience. Two examples will be used to illustrate these two concepts and to show how they can be incorporated into the usual design procedures. Thermally coupled distilliation columns will be designed to achieve not only energy efficiency but also dynamic resilience. The dynamics of heat exchanger networks are generally fast compared to other processing units and can be safely neglected. Therefore steady state resilience will be used as a design criterion.

SYNTHESIS OF RESILIENT ENERGY MANAGEMENT SYSTEMS

The standard heat exchanger network problem stated in 1969 (Masso, Rudd, 1969) is:

Given n_h hot streams to be cooled and n_c cold streams to be heated from specified supply temperatures to specified target temperatures, design the network of heat exchangers, heaters and coolers accomplishing this task at the least cost.

This problem has attracted strong interest because heat exchanger networks are an important industrial energy management tool for which only empirical design methods existed. After over a decade of research and scores of publication some fundamental advances have been achieved mainly through the work by Linnhoff and Flower (1978, 1980).

Though recent reports indicate that these synthesis techniques find some use in an industrial environment, case studies (Marselle et al. (1981)) indicate that often the resulting networks do not satisfy the basic resilience criteria put forward by the practicing process engineer. Marselle (1980) developed a synthesis technique which pays special attention to aspects of resiliency. It is achieved as the result of two interdependent synthesis tasks:

1. A network structure is found which can handle fluctuations of the inlet conditions (temperatures, flowrates) with maximum energy efficiency.
2. The regulatory and optimizing control structure is synthesized which keeps the network outlet temperatures at their setpoints or within constraints and minimizes the utility consumption.

A series of fundamental properties of resilient heat exchanger networks was discovered. As a result a simple on-line model-free optimizing controller was developed, which guarantees minimum energy consumption for a given network and specified outlet conditions, when the inlet conditions vary. In Fig. 7 two alternate control structures are shown for a six stream problem which were synthesized by the new method, the three

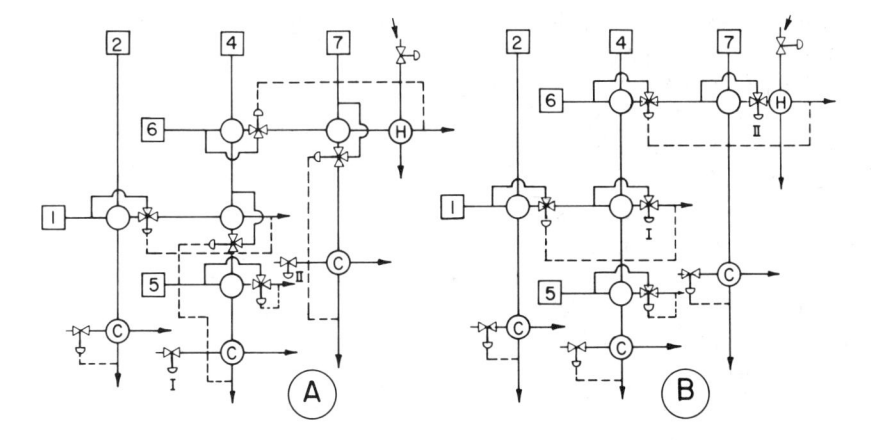

Figure 7. Two Examples of Control Structures Synthesized for a Network with Three Hot Streams (2, 4, 7) and Three Cold Streams (1, 5, 6).

valves which are not connected to any specific control loops are used in the optimizing controller to minimize the steam consumption of the network.

Many questions are still unresolved, however. For example, no efficient methods to size the exchangers for resiliency are available and stream splits are excluded. Because of the dependence of the sizes on the selected control structure and vice versa, the sizing problem forms a formidable challenge.

SYNTHESIS OF THERMALLY COUPLED DISTILLATION SYSTEMS UNDER CONSIDERATION OF RESILIENCE

Distillation is one of the most widely used separation processes, especially in the petro-chemical industry, and because it is very energy intensive, it presents itself as a prime candidate for improvement in energy efficiency (King, 1971, Mix et al. 1978).

Heat integration achieves savings in energy consumption by an appropriate matching of energy sources and sinks (reboilers and condensers) in a network of distillation columns. In recent years a number of techniques has been developed for the steady-state design of distillation networks with heat integration (Rathore et al. 1974, Siirola, 1978, Morari, Faith, 1980) and it has been noted by Siirola that a suboptimal steady-state design may be more favorable than an optimal one from a dynamic

490 Chemical Process Control II

point of view. However, little (Tyreus, Luyben, 1975) has been published on solving the dynamic problems.

Lenhoff (1979) used the integral square error as an indicator of operability, a "Dynamic Performance Index" (DPI), and developed an efficient decomposition method to find lower bounds on the minimum integral square error for a general system consisting of interconnected subsystems. This allows him to discard alternatives which are not attractive from a dynamic point of view without a detailed evaluation. The method was employed to investigate the operability properties of systems of coupled distillation columns used to separate a mixture of 70% methanol/30% water. In addition to the design structure (Fig. 8) the number of stages and the control structure (manipulated variables) were varied, giving rise to about 50 different systems. A vector objective function was optimized. One component is the steady state energy requirement, the Economic Performance Index (EPI), the other one the DPI. The paretooptimal set is shown in Fig. 9. Design structure A is seen to be more difficult to control than C, but to be more energy efficient. The final decision on the tradeoff will depend on the plant environment, i.e. steam availability and the expected disturbances.

Unfortunately this design by optimization does not provide any physical insight into why certain structures are more difficult to control than others. A very interesting approach appears to compare the "funda-

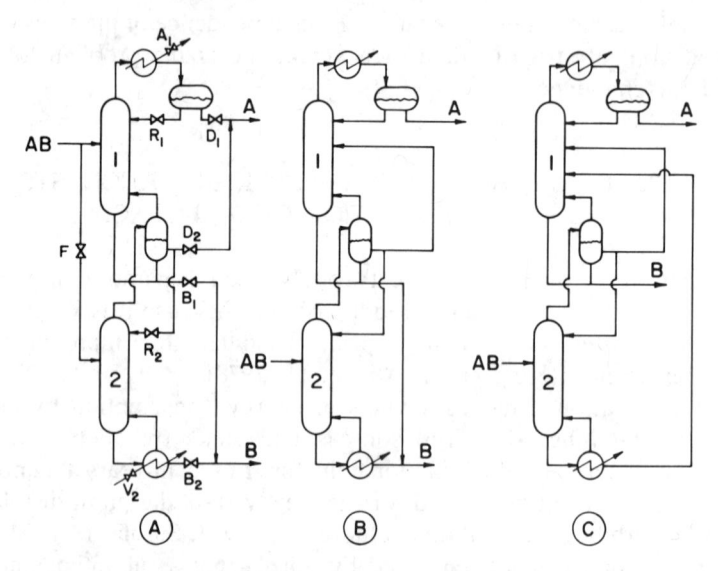

Figure 8. Three Configurations for Binary Distillation with Heat-integration.

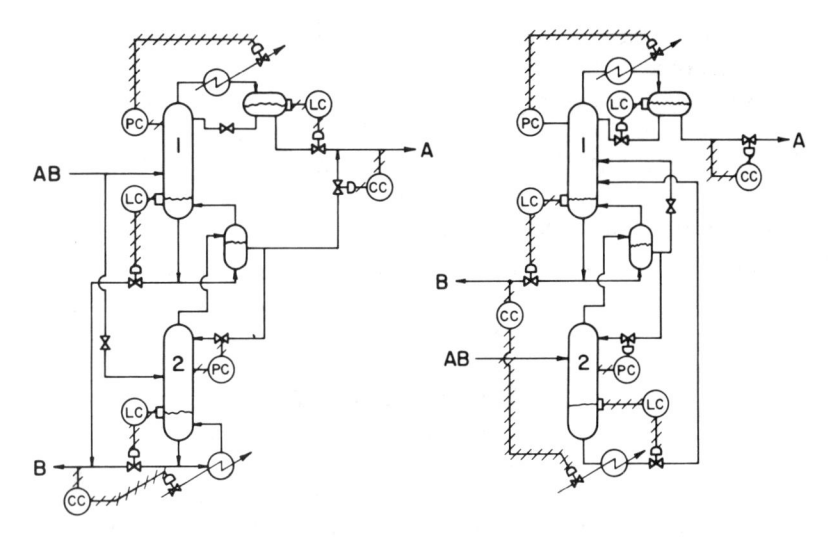

Figure 9. The Three Noninferior Designs: 29 (column 1: 20 stages, column 2: 25 stages), 41 (column 1: 20 stages, column 2: 25 stages) and 46 (column 1: 21 stages, column 2: 20 stages).

mental limitations of control quality" for the various designs. These limitations will effect operability significantly and cannot be removed by any type of control system. Therefore they are the most basic and most important operability criteria for judging different designs. A preliminary check of the distillation column examples showed RHP zeros for most of the "bad" designs and further investigations along this promising direction are currently underway.

CONCLUSIONS

Through a series of theoretical arguments and practical examples we have demonstrated that the control of integrated plants can offer challenges which cannot be met by the currently available design methodologies and implementation techniques developed to deal with "small systems." In particular the structuring of the control tasks, the selection of the measured and manipulated variables and the interdependencies between process design and control can be demanding problems while they are usually trivial for single units operations. We propose to focus future work on three areas found of central importance in our analysis:

1. The large scale system aspect of the problem, in particular the development of decomposition criteria
2. Control system sensitivity, presently not occupying the central role it deserves as a synthesis tool
3. The concept of the "fundamental limitations to control quality" could play an important part in decomposition, controller tuning and as a screening criterion for alternate design and control structures.

While progress on these three questions will certainly not solve all control problems, it should bring us closer to the ultimate goal of plant control.

ACKNOWLEDGMENT

Financial support from the National Science Foundation (ENG-7906353) and the Department of Energy (DOE contract DE-AC02-80ER10645) is gratefully acknowledged.

REFERENCES

[1] Almasy, G. A. and T. Sztanò, "Checking and Correction of Measurements on the Basis of Linear System Model," Prob. Contr. Inf. Theor., 4(1):57 (1975).
[2] Alvarez, J., J. A. Romagnoli, G. Stephanopoulos, "Variable Measurement Structure for the Control of a Tubular Reactor," submitted for publication (1981).
[3] Aris, R. and N. R. Amundson, "An Analysis of Chemical Reactor Stability and Control," Chem. Engr. Sci., 7:121 (1958).
[4] Arkun, Y., "Design of Steady State Optimizing Control Structures for Chemical Processes," Ph.D. Thesis, Univ. of Minnesota, Minneapolis (1979).
[5] Athans, M., "On the Determination of Optimal Costly Measurement Strategies for Linear Stochastic Systems," Automatica, 8:397 (1972).
[6] Boksenbom, A. S. and R. Hood, "General Algebraic Method Applied to Control Analysis of Complex Engine Types," Nat. Adv. Comm. for Aeronautics, Washington, D.C., Report NCA-TR-980 (1949).
[7] Bristol, E. H., "On a Measure of Interaction for Multivariable Process Control," IEEE Trans. Aut. Contr., AC-11, 133 (1966).
[8] Bristol, E. H., Seminar at the University of Wisconsin—Madison (1980).
[9] Bristol, E. H., "A New Process Interaction Concept: Pinned Zeros," submitted for publication (1981).
[10] Brosilow, C. B., "The Structure and Design of Smith Predictors from the Viewpoint of Inferential Control," Joint Aut. Contr. Conf., Denver (1979).
[11] Buckley, P. S., Techniques of Process Control, Wiley & Sons, New York (1964).
[12] Cutler, C. R. and B. L. Ramaker, "Dynamic Matrix Control—A Computer Control Algorithm," AIChE 86th Nat. Mtg. (April 1979).
[13] Douglas, J. M., "A Preliminary Design Procedure for Control Systems for Complete Chemical Plants," AIChE 73rd Annual Mtg., Chicago, paper 5a (1980).

[14] Flower, J. R. and B. Linnhoff, "A Thermodynamic-Combinatorial Approach to the Design of Optimum Heat Exchanger Networks," AIChE J., 26:1 (1980).
[15] Garcia, C. E. and M. Morari, "Internal Model Control," submitted for publication (1981).
[16] Gould, L. A., L. B. Evans, and H. Kurihara, "Optimal Control of Fluid Catalytic Cracking Processes," Automatica, 6:695 (1970).
[17] Govind, R., and G. J. Powers, "Control System Synthesis Strategies," AIChE 82nd Nat. Mtg., Atlantic City, NJ. (1976).
[18] Harris, T. J., J. D. Wright and J. F. MacGregor, "Optimal Sensor Location with an Application to a Packed Bed Tubular Reactor," AIChE J., 26:910 (1980).
[19] Herring, K. D., and J. L. Melsa, "Optimum Measurements for Estimation," IEEE Trans. Autom. Contr., AC-19, 264 (1974).
[20] Hlavacek, V., "Analysis of Complex Plant-Steady State and Transient Behavior," Comp. & Chem. Eng., 1:75 (1977).
[21] Horowitz, I. M. and U. Shaked, "Superiority of Transfer Function Over State Variable Methods in Linear Time-Invariant Feedback System Design," IEEE Trans. Autom. Contr. AC-20, 84 (1975).
[22] Jafarey, A. and T. J. McAvoy, "Degeneracy of Decoupling in Distillation Columns," IEC Proc. Des. Dev., 17:685 (1978).
[23] Jaisinghari, R. and W. H. Ray, "On the Dynamic Behavior of a Class of Homogenous Continuous Stirred Tank Polymerization Reactors," Chem. Eng. Sci., 32:811 (1977).
[24] Joseph, B. and C. B. Brosilow, "Inferential Control of Processes," AIChE J., 24:485 (1978).
[25] Kestenbaum, A., R. Shinnar, and R. E. Thau, "Design Concepts for Process Control," Ind. Eng. Chem., Proc. Des. Dev., 15:2 (1976).
[26] Kumar, S. and J. H. Seinfeld, "Optimal Location of Measurement in Tubular Reactors," Chem. Eng. Sci., 33:1507 (1978a).
[27] Kumar, S. and J. H. Seinfeld, "Optimal Location of Measurements for Distributed Parameter Estimation," IEEE Trans. Autom. Contr. AC-23, 690 (1978b).
[28] Kurihara, H., "Optimal Control of Fluid Catalytic Cracking Processes," Sc.D. Thesis, MIT, Cambridge (1967).
[29] Lee, W., and V. W. Weekman, Jr., "Advanced Control Practice in the Chemical Industry," AIChE J., 22:27 (1976).
[30] Lefkowitz, I., "Multilevel Approach Applied to Control System Design," ASME Trans. J., Basic Eng., 392 (June, 1966).
[31] Lefkowitz, I., "Systems Control of Chemical and Related Process Systems," Proc. 6th IFAC Cong., Boston, 38.2 (1975).
[32] Lenhoff, A. M., M. S. Thesis, Chemical Engineering Dept., University of Wisconsin (1979).
[33] Lenhoff, A. M., and M. Morari, "Design of Resilient Processing Plants I., Process Design under Consideration of Dynamic Aspects," Chem. Eng. Sci. (1981).
[34] Linnhoff, B. and J. R. Flower, "Synthesis of Heat Exchanger Networks," Parts I and II, AIChE J., 24:633 (1978).
[35] Luckel, J., and P. C. Muller, "Analyse von Steuerbarkeits- Beobachtbarkeits- und Storbarkeitsstrukturen linearer zeitinvarianter Systeme," Regelungstechnik, Heft 5:163 (1975).
[36] Mah, R. S., G. M. Stanley, D. M. Downing, "Reconcilation and Rectification of Process Flow and Inventory Data," IEC Proc. Des. Dev., 15:175 (1976).
[37] Marselle, D. F., M.S. Thesis, Chemical Engineering Department, University of Wisconsin (1980).
[38] Marselle, D. F., M. Morari, and D. F. Rudd, "Design and Resilient Processing Plants II., Design and Control of Energy Management Systems," Chem. Eng. Sci. (1981).

[39] Masso, A. H. and D. F. Rudd, "The Synthesis of System Design. II, Heuristic Structuring," AIChE J., 15:10 (1969).
[40] Mehra, R. K., "Optimization of Measurement Schedules and Sensor Design for Linear Dynamic Systems," IEEE Trans. Autom. Contr., AC-21, 55 (1976).
[41] Mehra R. K., R. Rouhani, L. Praly, "New Theoretical Developments in Multivariable Predictive Algorithmic Control," Proc. Joint Autom. Contr. Conf., San Francisco (1980).
[42] Mellefont, D. J., and R. W. H. Sargent, "Optimal Measurement Policies for Control Purposes," Int. J. Contr., 26:595 (1977).
[43] Mellefont, D. J., and R. W. Sargent, "Selection of Measurements for Optimal Feedback Control," IEC Proc. Des. Dev., 17:549 (1978).
[44] Mesarovic, M., D. Macko, and Y. Takahara, "Theory of Hierarchical Multilevel Systems," Academic Press (1970a).
[45] Mesarovic, M., Multilevel Systems and Concepts in Process Control, IEEE Proc., 58:111 (1970b).
[46] Mix, T. W., J. S. Dweck, M. Weinberg and R. C. Armstrong, "Energy Conservation in Distillation," Chem. Eng. Progr., 74:49 (April 1978).
[47] Morari, M., Y. Arkun and G. Stephanopoulos, "Studies in the Synthesis of Control Structures for Chemical Processes," Parts I, II and III, AIChE J., 26:220 (1980a).
[48] Morari, M., D. C. Faith III, "The Synthesis of Distillation Trains with Heat Integration," AIChE J., 26:916 (1980).
[49] Morari, M. and A. K. W. Fung, "Nonlinear Inferential Control," Comp. & Chem. Engr. (1981).
[50] Morari, M. and M. J. O'Dowd, "Optimal Sensor Location in the Presence of Nonstationary Noise," Automatica, 16:463 (1980).
[51] Morari, M., and G. Stephanopoulos, "Minimizing Unobservability in Inferential Control Schemes," Int. J. Contr., 31:367 (1980b).
[52] Niederlinski, A., "Two-Variable Distillation Control: Decouple or Not Decouple," AIChE J., 17:1261 (1971).
[53] Palmor, Z. J. and R. Shinnar, "Design of Sampled Data Controllers," IEC Proc. Des. Dev., 18:8 (1979).
[54] Prett, D. M. and R. D. Gillette, "Optimization and Constrained Multivariable Control of a Catalytic Cracking Unit," AIChE 86th Nat. Mtg., (April 1979).
[55] Rathore, R. N. S., K. A. Van Wormer and G. J. Powers, "Synthesis of Distillation Systems with Energy Integration," AIChE J., 20:491 and 940 (1974).
[56] Richalet J., A. Rault, J. L. Testud and J. Papon, "Model Predictive Heuristic Control: Applications to Industrial Processes," Automatica, 14:413 (1978).
[57] Romagnoli, J. A. and G. Stephanopoulos, "On the Rectification of Measurement Errors for Complex Chemical Plants," Chem. Eng. Sci., 35:1067 (1980).
[58] Romagnoli, J. A. and G. Stephanopoulos, "Rectification of Process Measurement Data in the Presence of Gross Errors," Chem. Eng. Sci., 36 (1981).
[59] Rosenbrock, H. H., "Computer-Aided Control System Design," Academic Press (1974).
[60] Schmidt, A. D. and W. H. Ray, "The Dynamic Behavior of Continuous Polymerization Reactors," Chem. Eng. Sci. (1981).
[61] Schuldt, S. B., and F. B. Smith, Jr., "An Application of Quadratic Performance Synthesis Techniques to a Fluid Cat Cracker," Proc. Joint Autom. Contr. Conf., 270 (1971).
[62] Shunta, J. P. and W. L. Luyben, "Dynamic Effects of Temperature Control Tray Location in Distillation Columns," AIChE J., 17:92 (1971).
[63] Siirola, J. J., "Progress Toward the Synthesis of Heat Integrated Distillation Schemes," paper 47a, 85th Nat. AIChE Mtg., Philadelphia (1978).

[64] Shinnar, R., "Chemical Reactor Modelling for Purposes of Controller Design," submitted for publication (1980).
[65] Stanley, G. M. and R. S. H. Mah, "Estimation of Flows and Temperatures in Process Networks," AIChE J., 23:642 (1977).
[66] Stanley, G. M. and R. S. H. Mah, "Observability and Redundancy in Process Data Estimation," Chem. Eng. Sci., 36, (1981)
[67] Tung, L. S. and T. F. Edgar, "Analysis of Control-Output Interactions in Dynamic Systems," AIChE J., (1981).
[68] Tyreus, B. D. and W. L. Luyben, "Dynamics and Control of Heat-Integrated Distillation Columns," 68th Annual AIChE Mtg., Los Angeles (1975).
[69] Umeda, T., T. Kuriyama and A. Ichikawa, "A Logical Structure for Process Control System Synthesis," Proc. IFAC Congress, Helsinki, (1978).
[70] Vaclavek, V., "Optimal Choice of the Balance Measurements in Complicated Chemical Engineering Systems," Chem. Eng. Sci., 24:947 (1969).
[71] Vaclavek, V. and M. Loucka, "Selection of Measurements Necessary to Achieve Multicomponent Mass Balances in Chemical Plant," Chem. Eng. Sci., 31:1199 (1976).
[72] Weber, R. and C. Brosilow, "The Use of Secondary Measurements to Improve Control," AIChE J., 18:614 (1972).
[73] Wismer, D. A. (Ed.), "Optimization Methods for Large-Scale Systems," McGraw Hill, New York (1971).

PROCESS OPERABILITY AND CONTROL OF PRELIMINARY DESIGNS

J. M. Douglas
Department of Chemical Engineering
University of Massachusetts
Amherst, MA 01003

ABSTRACT

A procedure is outlined for evaluating the steady-state operability and control of complete chemical plants at the preliminary design stage. The purpose of the analysis is to estimate the design modifications, i.e., additional equipment and capacity, that are needed to control a plant. These costs should be included in a comparison of process alternatives.

The method is based on short-cut design procedures, which normally provide sufficient accuracy for screening purposes at the preliminary design stage. Some potential operability problems can be identified and first estimates of control structures can be determined. However, the analysis represents only the first step in the development of control systems, i.e., some examples are given where the steady-state controllers would not provide satisfactory dynamic operation.

INTRODUCTION

In current practice most control problems are identified only after the plant has been built and put into operation. Some attempts are made to ensure that the final design has sufficient flexibility to handle most of the disturbances and some controllers are installed based on experience. Also, consideration is given to the possibility of installing additional controllers after the plant start-up, if they are needed.

This "ad hoc" approach to process control normally proved to be satisfactory in the past because there are not many process interactions in most plants. That is, there may be a liquid and/or a gas recycle loop for unconverted raw materials, as well as one or more feed-effluent heat exchangers around reactors or distillation columns to provide energy integration. Normally, there are control points available within each of these recycle loops that can be used to prevent disturbances from propagating around the loops. Thus, in most cases potential interactions can be nullified fairly simply, and most control studies focus on the control of individual process units.

For cases where the process interactions caused unexpected control problems, trial-and-error simulation studies on large analog computers were used to develop improved control systems for the complete plant. Some early studies of this type were published by Bode and Bakanowski [2] for a nitric acid plant and by Bettes and Wright [1] for a solvent dewaxing plant. Simulation studies are still used to study control, although current practice uses digital computers and often includes process optimization; for example see Shah and Stillman's [10] analysis of a methanol plant and Weisenfelder, Fritz, and Thompson's [12] analysis of an ammonia plant.

In recent years there has been an attempt to provide a mathematical formalism for designing control systems for large scale systems, such as chemical plants. Most of this theory describes multilevel, or hierarchical, control systems, which were introduced by Mesarovic [5]. Some recent improvements of the theory and some applications to chemical processing problems have been presented by Morari, Arkun, and Stephanopoulos [7, 8, 9]. There have also been some studies where simplified design equations have been used to develop improved controllers for trains of units; for example see Fauth and Shinskey's [4] analysis of multiple distillation columns.

Both the simulation studies and the hierarchical control analysis require that a fixed flow sheet is available. Also, both require a substantial investment of engineering effort. Thus, it seems as if these procedures are most useful for the final design of a control system, rather than as screening procedures to decide which flow sheet and its control systems might lead to the best process.

PRELIMINARY EVALUATION OF PROCESS OPERABILITY AND CONTROL

The current trend of increasing energy prices will eventually lead to increased energy integration in plants, and this energy integration will be accomplished by significantly increasing the amount of process interactions [6]. There is little experience with the operability and control of tightly coupled units, but we anticipate that we will encounter many more problems. As we consider various process alternatives having different extents of energy integration, we expect that the cost required to obtain satisfactory control of the plant will change. Hence, the cost of the control system must be included as part of the total processing costs before we can decide which process alternative is the least expensive.

Normally when we screen numerous process alternatives we use approximate design procedures. The goal of this screening activity is to get quick solutions that will allow us to discard the least promising alternatives, rather than to obtain rigorous design calculations. Thus, we want to develop procedures for studying operability and control at this preliminary design stage of the development of a project.

In this paper we consider only some of the steady-state aspects of process operability and control. We assume that the proposed control systems for a complete plant must be able to produce at least a satisfactory steady-state operation of the plant or they will be dropped from further consideration. Moreover, we want to use the same type of short-cut procedures to assess these controllers that we use for the preliminary design calculations.

ASSUMPTIONS REQUIRED FOR A PRELIMINARY PROCESS DESIGN

The preliminary design of a new process usually is based on information concerning the reactions, the reaction temperature and pressure, the desired production rate, and the desired product purity. In order to develop a basecase design using only this information it is necessary to introduce numerous assumptions. These assumptions can be broken down into three categories.

1. Assumptions that tie the process to its environment.
2. Assumptions that help fix the process flow sheet.
3. Assumptions that help fix initial values of some of the process variables.

Of course, the final design we obtain depends on the assumptions we introduce. Moreover, as we change these assumptions, we encounter different kinds of problems. Thus, it is helpful to consider each of these categories in more detail.

1. Assumptions that tie the process to its environment—
  Process operability and control problems

Some of the assumptions that fall into this category are the production rate; the desired product qualities; the flow rates, composition, temperature, and pressure of the raw material streams; and the flow rates, composition, pressure, and temperature of the utility streams, i.e., fuel, cooling water, steam, etc. For the base-case design, or even an optimum design, we choose constant values for these quantities. In some cases we might choose the largest, or the smallest, expected value, in order to obtain a conservative design. In other cases we base the design on the average value.

Changes in these assumptions correspond to disturbances entering the process. Thus, changes in these assumptions require a consideration of *process operability and control*. We also encounter control problems when there are process constraints that must be satisfied. For example, the hydrogen to aromatics ratio, as well as the hydrogen composition or partial pressure, might have minimum allowable values in the reactor for some processes to prevent excessive coke formation. These constraints must be satisfied even if disturbances upset the process operation.

2. Assumptions that help fix the process flow sheet—
  Process alternatives

Some assumptions that help to fix the process flow sheet are the choice of a separation system (sequences of distillation columns or extraction vs. distillation), the desirable amount of energy integration (including feed-effluent exchangers around reactors or distillation columns, thermally coupled distillation columns, etc.), a solvent recovery system (absorption, adsorption, refrigeration, etc.), the location of fresh feed and/or purge streams, etc. As we change these assumptions, we generate *process alternatives*. There is an optimum design (assuming no disturbances) for each process alternative.

3. Assumptions that help fix the values of some of the process variables—
  Process optimization

Some assumptions that fall into this category might include the reactor conversion, the composition of inerts in recycle streams with purge flows, the pressure level of gas recycle streams, the fractional recovery of energy, approach temperatures in feed-effluent exchangers or waste-heat boilers, fractional recoveries of raw materials and/or byproduct streams, reflux ratios in distillation columns, liquid flow rates in gas absorbers, etc. Initial estimates of most of these variables are chosen to correspond to common rules of thumb, e.g., 99% recoveries of valuable materials, R/R_m = 1.2 for distillation columns, L/mG = 1.4 for gas absorbers, 60°F approach temperatures in waste-heat boilers, etc. These rules of thumb represent the solutions of optimization problems for cases where the optimum design conditions normally are insensitive to the process and cost parameters.

If we change these assumptions (assuming no disturbances), we encounter *process optimization* problems. For cases where variables cannot be estimated using rules of thumb, i.e., reactor conversion or inert composition in gas recycle loops, it is common practice to use a series of case studies to estimate the optimum design conditions. Similarly, case studies can be used to determine the sensitivity of the design to changes in the rule of thumb guidelines.

STEADY-STATE OPERABILITY AND CONTROL

As we sketch a process flow diagram corresponding to a particular process alternative, we can identify where the disturbances will enter the process. Also, the production rate, the desired product purity, and the process constraints are apparent from the definition of the design problem. We require that the process design is sufficiently flexible that it can tolerate changes in the production rate as the market demand changes and that it can tolerate disturbances entering the process.

Initially, we consider only the low frequency disturbances that correspond to new steady-state operating levels, both because they are simplest to analyze and they provide the limiting behavior of any dynamic control system. Our goal is to make a preliminary assessment of the potential operability and control problems that are associated with each process alternative in order to estimate the incremental design costs required to solve these problems. Thus, the selection of the "best" alternative will be based on more complete information.

PRODUCT QUALITY

Very simple economic calculations can be used to demonstrate that it is essential to have the product quality meet the specifications for virtually all of the operating period. In most cases, profits are about 5% of sales, and if sales are lost because the product must be reprocessed the process profitability will be severely affected. Thus, the first step in our operability and control analysis is to consider factors that affect product quality.

For most processes, the product streams are purified in distillation columns. Experience indicates that often we can maintain the top and bottom compositions of these columns essentially constant by varying the distillate to feed ratio to satisfy the column material balance and by maintaining the vapor to feed ratio constant, see Shinskey [11] or Douglas, Jafarey and Seeman [3]. However, the satisfactory performance of a control system of this type depends on the magnitude of the disturbances and the flexibility of the column design. Thus, we need to estimate the maximum values of the disturbances and to ensure that the column condenser, reboiler, reflux drum, etc. are sufficiently large that they can accommodate the disturbances.

The disturbances that have the greatest effect on the product distillation column normally correspond to changing operating conditions in a flash drum and in the reactor. Since reactor disturbances will also affect the production rate, we initially consider only disturbances in the operating conditions in the flash drum. We anticipate that in some cases a consideration of these disturbances will require a modification of the base-case (or optimum design). We will treat reactor disturbances and production rate changes later in this paper.

FLASH DRUM DISTURBANCES

There are numerous processes where we partially condense the effluent stream from a reactor and then separate the vapor and liquid streams in a flash drum, see Fig. 1. The vapor stream leaving the flash often is recycled, although if inert materials are present in the gas some of the vapor leaves in a purge stream. The flash drum normally is designed to operate at 100°F because we prefer to use cooling water, recirculated through cooling towers, as a cooling medium for the condenser. We assume that cooling water is available at 90°F on a hot summer day, so that a $\Delta T = 10°F$ is maintained at the condenser outlet.

Thus, we design the condenser on a conservative basis to ensure that there will be an adequate heat transfer area on a hot summer day. However, during most of the year the cooling water temperature will be less than 90°F, and the flash drum will operate at a lower temperature. It is always desirable to operate the flash drum at as low a temperature as possible because the amount of product lost in the purge stream will

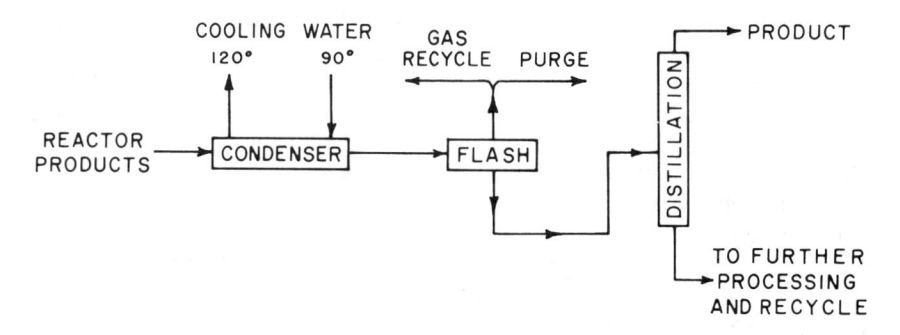

Figure 1. Potential Disturbances to Separation System.

decrease. Therefore, as the cooling water temperature drops, the amount of liquid leaving the flash drum and entering the distillation column will increase. Similarly, the feed composition to the column will change.

We want to estimate the magnitude of these disturbances to the column so that we can check to see if our column design is satisfactory. For cases where we obtain relatively sharp splits in the flash drum at the normal design conditions, we can estimate the component flows in the flash liquid using the expressions

$$\ell_i = f_i \frac{f_L}{K_i f_G}, \text{ for components lighter than the product} \quad (1)$$

$$\ell_j = f_j \left(1 - \frac{K_j f_G}{f_L}\right), \text{ for the product and heavier components} \quad (2)$$

where

f_G = total flow rate of all components lighter than the product entering the flash drum

f_L = total flow rate of product and heavier components entering the flash drum

The derivations of these equations are given in Appendix A. More exact solutions can be obtained by solving the flash equations, but the expressions above often are adequate for preliminary design and control studies (3–5% error).

Potential Operability Constraint for a Material Balance Controller

From Eq. (1) we see that the flow rates of the light components in the flash liquid will increase as the flash drum temperature, and therefore K_i, decreases. These light components will leave the distillation column with the product stream. In some cases the flow rates of these light components might become sufficiently large that it is not possible to achieve the desired product specifications.

In particular, if x_D = product purity specification, $\Sigma \ell_i$ = total flow of components in the flash liquid lighter than the product, $(1 - \epsilon_p)\ell_p$ = fraction of the product recovered overhead in the distillation column, and $\epsilon_{HK}\ell_{HK}$ = fraction of the heavy key recovered overhead in the column, then we require that

$$\frac{(1 - \epsilon_p)\ell_p}{\Sigma \ell_i + (1 - \epsilon_p)\ell_p + \epsilon_{HK}\ell_{HK}} \geq x_D \tag{3}$$

or

$$\ell_p \geq \left(\frac{x_D}{1 - x_D}\right)\left(\frac{\Sigma \ell_i + \epsilon_{HK}\ell_{HK}}{1 - \epsilon_p}\right) \tag{4}$$

in order to meet the product purity specification. If this restriction cannot be satisfied, then it will be necessary either to install an additional distillation column before the product column or to add a pasteurization section at the top of the product column. A discussion of the multiple column problem will be presented later.

If the restriction given by (4) is always satisfied, the material balance controller sets the distillate rate so that

$$D = \frac{\ell_p - x_w (\Sigma \ell_i + \Sigma \ell_j)}{x_D - x_w} \tag{5}$$

where x_D = product specification and x_w is the product composition in the bottoms corresponding to the base-case (or optimum) design conditions.

Potential Operability Constraints for V/F Controller

If we desire to make V/F = constant, or $V/(\Sigma \ell_i + \Sigma \ell_j)$ = constant, the reboiler and the condenser sizes must be adequate to handle the increased vapor load. Similarly, the column diameter must be large enough so that flooding does not occur when the flash drum temperature decreases.

Thus, in some cases it might be necessary to modify the base-case (optimum design).

Energy Integration Limitations

There is a general trend to increase the amount of energy integration in processes, and in some cases it might be advantageous to supply the heat to the product column reboiler by cooling the hot reactor effluent, for example. For tightly integrated plants the heat supplied at the base-case (or optimum) design conditions might just match the base-case load on the reboiler. Operability problems will be encountered for these cases when disturbances enter the column, so that it might be necessary to install an auxiliary reboiler for control purposes.

Sequence of Distillation Columns

In many processes there are light ends that must be removed before the product is purified. Similarly, there are heavy byproducts that are removed, see Fig. 2. Of course, in cases where the primary reactant is lighter than the product, the overhead streams from the second and third columns will be reversed.

The analysis of separation sequences of this type is esentially the same as the example discussed previously. Changes in the cooling water temperature will cause the composition and the flow rate of the flash liquid to vary. Eqs. (1) and (2) can be used to estimate the component flows in the flash liquid, except that Eq. (1) applies for all components that leave primarily with the vapor stream and Eq. (2) applies for all components that leave primarily with the liquid. The derivation in Appendix A details the assumptions used in the analysis.

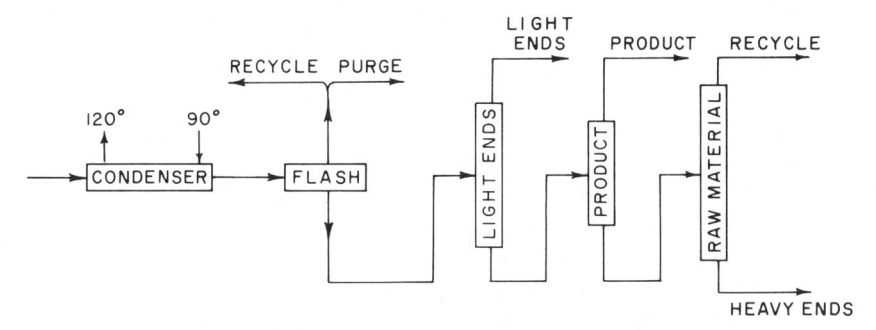

Figure 2. Sequence of Distillation Columns.

The product quality constraint given by Eq. (3) also must be modified. If we let $\epsilon_{LLK}\ell_{LLK}$ be the heaviest component in the light ends that exit at the bottom of the light ends tower, $\epsilon_{1p}\ell_p$ be the product leaving at the top of the first tower, and $(1 - \epsilon_{2p})\ell_p$ be the product leaving with the bottoms of the product tower, then the appropriate expression becomes

$$\frac{(1 - \epsilon_{1p})(1 - \epsilon_{2p})\ell_p}{\epsilon_{LLK}\ell_{LLK} + (1 - \epsilon_{1p})(1 - \epsilon_{2p})\ell_p + \epsilon_{HK}\ell_{HK}} \geq x_D \quad (6)$$

or

$$\ell_p \geq \left(\frac{x_D}{1 - x_D}\right)\left(\frac{\epsilon_{LLK}\ell_{LLK} + \epsilon_{HK}\ell_{HK}}{(1 - \epsilon_{1p})(1 - \epsilon_{2p})}\right) \quad (7)$$

Normally, there are no product specifications for the other overhead or bottoms streams. The preliminary design is usually based on the rule of thumb of a 99%, or greater, recovery of the key components in each column. However, this rule of thumb represents an approximate solution of a series of optimization problems. That is,

1. The amount of product leaving with the light ends represents a tradeoff between the value of the lost product and the incremental cost of adding trays and increasing the reflux ratio in the top of the light ends column.
2. As the amount of light ends leaving with the bottom stream from the first tower increases, the composition of the heavy key in the product stream must decrease so that the product specifications can be met. Thus, the fractional recovery of the light ends overhead in the first tower represents an economic balance between incremental trays in the stripping section of the first tower and the incremental trays and increased reflux ratio of the rectifying section of the product tower.
3. We can assume that the product leaving with the bottoms from the product recovery column will be recycled through the plant with the unconverted raw materials. Thus, the fraction of product recovered overhead in the product recovery column involves a tradeoff between incremental trays in the stripping section and the costs of oversizing all of the equipment in the recycle loop to handle the incremental product flow rate.
4. The fractional recovery of the unconverted raw material overhead in the last column involves an economic balance between the value of this material lost in the heavy ends stream and the incremental cost

of additional plates in the stripping section of the raw materials recovery column.
5. The fractional recovery of heavy ends in the bottoms stream of the final column involves a tradeoff between incremental plates and increased reflux in the rectifying section and the cost of oversizing all of the equipment in the recycle loop because of the incremental flow of heavy ends.

Thus, the control of the first and third columns in the distillation train does not contain any product quality operability restrictions such as Eq. (7). Instead, the decision to install controllers is based only on economic considerations. Other column configurations exhibit similar economic tradeoffs.

As the cooling water and flash drum temperatures decrease, the flow rate of product and unconverted reactant in the flash liquid will increase. If there is no control system on the first column, we expect that more product and unconverted reactant will be lost from the system with the light ends. We can use the appropriate versions of Eqs. (1) and (2) to estimate the increased feed rate to the distillation columns, and then we can use the column material balances along with the approximate design equation (3)

$$N = \frac{\ln [(x_D/x_w)_{LK} (x_w/x_D)_{HK}]}{\ln (\alpha/\sqrt{1 + 1/Rx_F})} \quad (8)$$

to estimate the stream flows leaving at the top and bottom. With this procedure, or with more rigorous numerical calculations, we can calculate the loss of product from the top of the column. From this result, we can calculate the economic incentive for installing a control system.

A similar series of calculations can be performed for the last column in the separation sequence. That is, we can calculate the incremental operating costs associated with recycling heavy ends and the value of the unconverted reactant lost from the bottom of the column if we don't add a control system. This result provides an estimate of the economic incentive for installing a controller on this last column.

REACTOR DISTURBANCES

If we consider a very simple case of a first-order reaction in an isothermal reactor, the design equation is

$$V_R = \frac{F \ln [1/(1 - x)]}{k\rho} \quad (9)$$

From this expression we see that for low frequency disturbances we expect the reactor conversion will change if the feed rate to the reactor or the reactor temperature changes. The flow rate to the reactor is affected by the flow rates of the raw material streams, as well as both the gas and liquid recycle flows. Similarly, the reactor temperature (or inlet temperature for an adiabatic reactor) depends on the temperatures of the fresh feed and/or makeup streams as well as the recycle streams.

For a constant production rate, the flow rates of the raw material streams can be maintained constant by using flow controllers. The flow rates of gas recycle streams depend to a slight extent on the operating temperature of the flash drum, but normally these fluctuations can be neglected for first estimates, i.e., whenever the base-case gas recycle flow rate is large and the split in the flash drum is fairly clean, see Appendix A, then the flow is essentially constant. Liquid recycle streams usually come from distillation columns, so that the flow rate is determined by the factors discussed above.

Reactor temperature changes might also depend on fluctuations in the fresh feed and recycle streams. The temperatures of fresh feed streams supplied by feed storage tanks will fluctuate with ambient conditions, while the temperatures of gas recycle streams depend on the flash drum temperatures. Liquid recycle streams usually remain at a fairly constant temperature. If the gas recycle rates are large as compared to the fresh feed streams, temperature variations primarily depend on cooling water temperature, whereas in the opposite case ambient changes are more imporant. Often, the reactor is preceded by a feed-effluent heat exchanger and a furnace, see Fig. 3. If the heat duty supplied by the furnace is maintained constant, the temperature changes in the fresh feed and recycle streams will cause the reactor temperature to fluctuate. Changes in the heating value of the fuel will also produce variations in the reactor temperature.

From Eq. (9) we see that increases in the reactor temperature, and therefore the reaction rate constant k, will cause the conversion to increase. For complex reactions, higher conversions almost always correspond to lower selectivities (smaller fractions of the limiting reaction going to form the desired product) and a greater production of by-products. Selectivity vs. conversion plots usually have the shape shown in Fig. 4, and for a large number of reactions the selectivity is essentially independent of temperature.

Effects of Disturbances

The conversion and selectivity changes in the reactor caused by the temperature fluctuations will affect the feed rates to the flash drum.

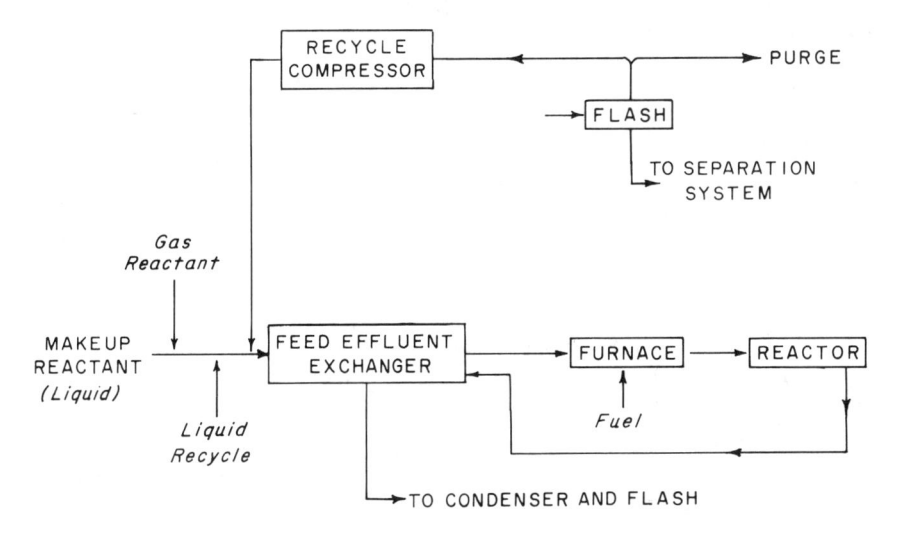

Figure 3. Reactor Flow Sheet.

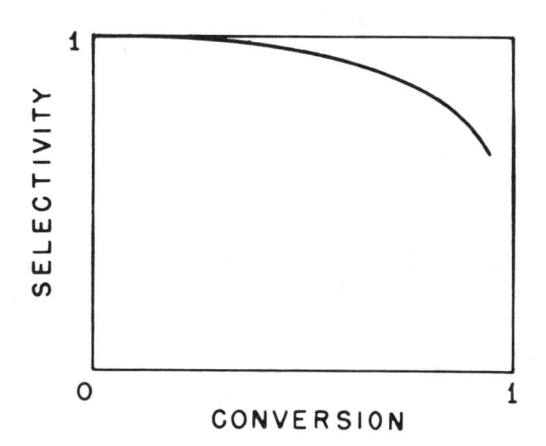

Figure 4. Selectivity vs. Conversion.

Furthermore, from Eq. (1) we see that the composition and flow rate of the flash liquid will vary. Procedures for estimating the effect of these changes on the separation system and column control systems were discussed previously.

510   Chemical Process Control II

Temperature Control of the Reactor

For the flow sheet shown in Fig. 3, we could maintain the reactor temperature constant simply by changing the heat input to the furnace. Fluctuations in the heating value of the fuel can be compensated in the same way. If we use this approach, then the size of the furnace must be designed to handle the largest possible disturbances.

PRODUCTION RATE CHANGES

Experience indicates that the total U.S. production varies from month to month, see Table 1 for the production of benzene, and we expect any of our processes to follow this same general pattern. Hence, we must be able to change the gas and liquid flows everywhere in the process as the production rate varies. We can use computer-aided design programs to solve the material balance problem fairly rigorously (the specifications of the fractional recoveries in the distillation columns are chosen to correspond to rules of thumb, so that the exact solutions will depend on these assumptions), but for first estimates we can obtain approximate solutions fairly simply. The approximate solutions have the advantage that we can understand the cause-and-effect relationships more easily. Of course, once we select the best flow sheet, we would eventually calculate the exact material balances that correspond to the optimum recoveries in the separation system.

Simplifying Assumptions

The process material balances depend on the reaction stoichiometry. However, some of the features of the material balancing problem are quite general. In order to illustrate the concepts, we consider a process for the hydrodealkylation of toluene to produce benzene, see Fig. 5, where the reactions are

$$\text{Toluene} + \text{Hydrogen} \rightarrow \text{Benzene} + \text{Methane} \tag{10}$$

$$2\ \text{Benzene} \rightarrow \text{Diphenyl} + \text{Hydrogen} \tag{11}$$

Table 1. Monthly Benzene Production Rate for 1977.

June	July	Aug	Sept	Oct	Nov	Dec
896.8	917.8	949.5	914.6	828.1	920.1	976.0

HYDRODEALKYLATION OF TOLUENE

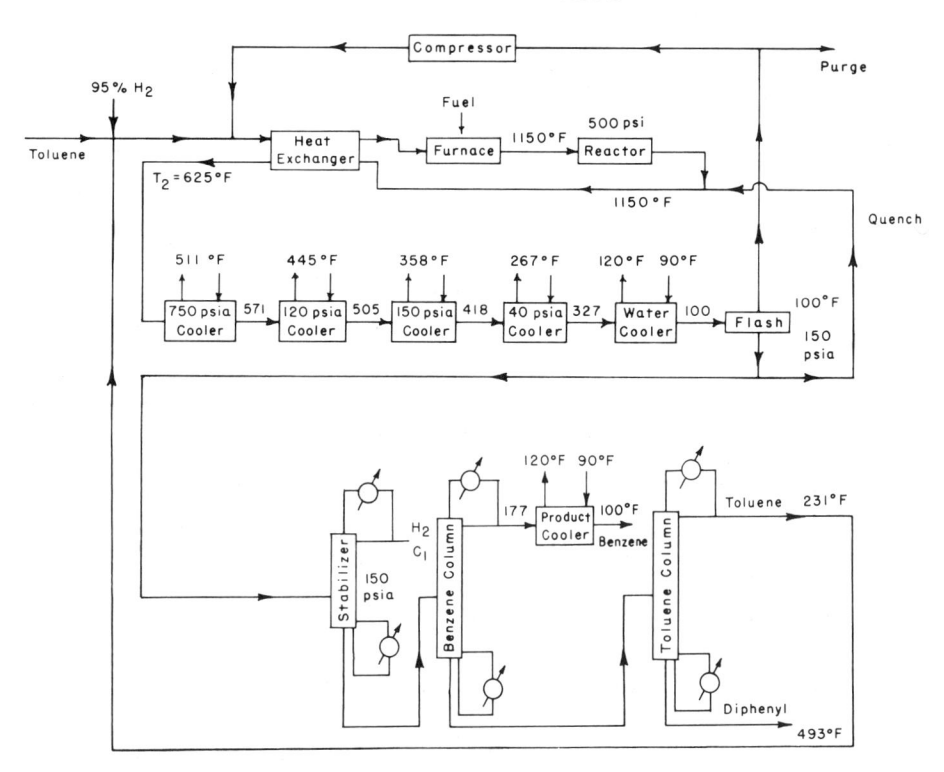

Figure 5. Flow Sheet Hydrodealkylation of Toluene.

The assumptions used to develop a base-case design are given in Table 2.

The reaction takes place at high temperature (1150 to 1300°F) and high pressure (500 psia), and at these conditions there is a tendency for aromatics to produce coke. In order to minimize coke production, the hydrogen to aromatics ratio at the reactor inlet is maintained at 5/1 and the hydrogen recycle composition must be greater or equal to 0.4. Thus, the prevention of coking imposes some operating constraints on the process. There are a large number of processes where this same type of behavior is encountered.

If we desire a very high purity benzene product, we install a stabilizer column before the product column to remove the small amounts of hydrogen and methane that dissolve in the flash liquid. The flash drum should give a fairly sharp split between methane and benzene (an ammonia synthesis process exhibits this same behavior), so that we can

512   Chemical Process Control II

Table 2. Design Assumptions.
___

Process Operability and Process Constraints
  1. Use all available toluene as feed.
  2. The feed is pure toluene.
  3. The feed temperature is 77°F and the pressure is 1 atm.
  4. The composition of the makeup gas stream is 95% H_2, 5% C_1.
  5. The pressure of the makeup gas stream is 550 psig and 100°F.
  6. The heating value of the fuel for the furnace is constant.
  7. Cooling water is availble from the cooling towers at 90°F and must be returned at less than 120°F.
  8. The water fed to each of the waste heat boilers is at its boiling point.
  9. The steam supplied to the reboilers of the distillation columns is at constant temperature and pressure.
  10. No heat losses to the atmosphere.
  11. The toluene recycle is pure toluene at its boiling point.
  12. The hydrogen to aromatics ratio at the reactor inlet is 5/1.
  13. The hydrogen composition at the reactor inlet must be greater than or equal to 0.4.
  14. The reactor effluent should be quenched to 1150°F.
  15. The desired benzene product composition is 99.99%.
  16. No energy losses from equipment to the environment.

Process Alternatives
  1. The aromatics losses in the purge stream and the stabilizer overhead are insignificant.
  2. Distillation is the cheapest way of separating the products.
  3. A stabilizer is required to achieve the desired benzene product purity.
  4. We want to recover diphenyl.
  5. We want to use three separate distillation columns, instead of columns with side streams.
  6. The heat recovery in the feed-effluent exchanger is 50%, and we want to make as much steam as possible at the highest pressure possible. The steam lines in the plant have pressures of 750, 420, 150, and 40 psia.
  7. Flash liquid is used as a quench stream, instead of cold hydrogen.
  8. Energy integration schemes for the distillation columns can not be justified.

Table 2. Cont'd.

Process Optimization
1. The reactor conversion is 0.75
2. The hydrogen recycle composition is 0.4.
3. The fractional heat recovery in the feed-effluent exchanger is 0.5.
4. The approach temperature in each waste-heat boiler is 60°F.
5. The recycle ratios in the distillation columns are 20% greater than the minimum.
6. The fractional recovery of toluene and diphenyl in the toluene column is 99%.
7. The fractional recovery of benzene in the benzene column is 99%.
8. The reactor pressure is 500 psig.

assume that the gas and liquid recycle streams are essentially independent of one another.

A design rule of thumb indicates that we want to recover 99%, or more, of all valuable materials. Thus, as a first estimate of the material balances, we introduce little error if we assume complete recoveries. That is, a one or two percent error in the process flows normally will not have a significant effect on the capital or operating costs of the various process units.

Production Rate

With the simplifying assumptions cited above, an overall material balance for the aromatics in the plant gives the production rate in terms of the fresh feed rate of reactant, F_F, and the selectivity, S,

$$\text{Benzene produced} = P_B = F_F S \qquad (12)$$

There are a very large number of processes where this same result is approximately valid, although often it is necessary to include a stoichiometric coefficient. From Fig. 4 we found that the selectivity usually depends on the reactor conversion. Thus, we see that we can change the production rate either by changing the fresh feed rate of reactant or by changing the reactor conversion (which depends on the reactor flow rate and temperature, see Eq. (9)).

Economic considerations dictate that most chemical processes are designed for high selectivities, i.e., greater than 95%, so that it is difficult to increase the production rate very much by decreasing the conversion, when the fresh feed rate is maintained constant. Moreover, if we decrease the production rate by increasing the conversion, we convert valuable raw materials into undesirable byproducts or fuel, which is economically disadvantageous.

Reactor Control System

From the discussion above we see that it is desirable to vary the fresh feed rate in order to change the production rate and that we can use a proportional controller, provided that we keep the reactor conversion constant. Furthermore, from Eq. (9) we find that we should adjust the reactor temperature (actually, the heat input to the furnace) so that

$$k\rho/F = \text{constant} \tag{13}$$

The appropriate expression for adiabatic reactors or more complicated kinetics is more difficult to solve, but on-line computer control systems could easily handle these problems.

In most commercial processes the reactor temperature is controlled, rather than the conversion. According to Eq. (9), with a constant reactor temperature we find that

$$F \ln [1/(1 - x)] = \text{constant} \tag{14}$$

Thus, if the flow rate through the reactor, F, varies when the production rate is changed, the conversion also will change. Furthermore, from Fig. 4, we see that the selectivity will change, and that more byproducts might be formed. We will describe a procedure for estimating the economic advantages of a reactor conversion controller, as compared to a temperature controller, after we have developed a procedure for estimating the reactor flow rate.

Changing the Makeup Hydrogen and Purge Rates

As the production rate changes, we expect that the makeup gas flow and the purge rate will vary. Thus, we need to understand the cause and effect relationships between these flows. An overall balance for the gaseous components must include the reaction stoichiometry. That is, the hydrogen and methane flows entering and leaving the system depend on the production and consumption of these components. A total mass balance relates the makeup gas and purge rates, F_G and P_G, to the production of methane and consumption of hydrogen, ΔC_1 and ΔH_2.

$$F_G + \Delta C_1 = P_G + \Delta H_2 \tag{15}$$

For processes with high selectivities, greater than 95%, we introduce only small errors in the flows in the gas recycle loop if we base the material balances on only the primary reaction, Eq. (10). From this expression we see that $\Delta C_1 = \Delta H_2$, and so Eq. (15) gives

$$F_G = P_G \tag{16}$$

Thus, the makeup gas and purge flow rates must be equal to one another.

Makeup Hydrogen Composition Disturbances and Recycle Composition Control

An overall component balance for hydrogen, assuming that the makeup gas stream contains 95% H_2 and letting the recycle gas composition be y_{H_2}, gives

$$0.95\, F_G = \Delta H_2 + y_{H_2} P_G \tag{17}$$

From Eq. (10) we see that the hydrogen consumed must be approximately the same as the toluene fresh feed rate, $\Delta H_2 = F_F$. Thus, we can write Eq. (17) as

$$F_G = \frac{F_F}{0.95 - y_{H_2}} = \frac{F_F}{y_f - y_{H_2}} \tag{18}$$

As the production rate changes, we change the toluene fresh feed rate, F_F. If our base-case design corresponds to the minimum allowable recycle gas composition, $y_{H_2} = 0.4$, and the feed composition of hydrogen, y_f, fluctuates (because of disturbances in our hydrogen production facility), Eq. (18) indicates how we should manipulate the makeup gas flow rate. Hence, the overall material balance indicates the appropriate control action for satisfying one of the process constraints.

Changing the Gas and Liquid Recycle Rates

Production rate changes will also require that we change the gas and liquid recycle flows. Since the hydrogen to aromatics ratio at the reactor inlet is specified to be 5/1, we initially consider the liquid recycle. Again, we introduce the simplifying assumption of no losses of valuable materials, so that a material balance at the toluene fresh feed and recycle mixing point gives

$$F_T = F_F + F_T (1 - x) \tag{19}$$

516   Chemical Process Control II

$$F_T = F_F/x \tag{20}$$

This same result is obtained for a very large number of processes. If we maintain the conversion, x, constant as we change the toluene fresh feed rate, F_F, proportional to the desired change in the production rate, we can use Eq. (20) to determine the change in the total toluene flow to the reactor.

The requirement of a hydrogen to toluene ratio of 5/1 at the reactor inlet leads to the result

$$5\,F_T = 0.95\,F_G + y_{H_2}\,R_G \tag{21}$$

After substituting Eqs. (18) and (20) this can be written as

$$R_G = \frac{F_F}{y_{H_2}}\left(\frac{5}{x} - \frac{0.95}{0.95 - y_{H_2}}\right) = \frac{F_F}{y_{H_2}}\left(\frac{5}{x} - \frac{y_f}{y_f - y_{H_2}}\right) \tag{22}$$

Thus, we can vary the recycle gas rate, R_G, to compensate for changes in the production rate (proportional to F_F) and disturbances in the hydrogen feed composition, y_f, while maintaining a 5/1 hydrogen to aromatics ratio at the reactor inlet. With this procedure we can satisfy the second process constraint.

Reactor Conversion Control vs. Reactor Temperature Control

A constant conversion controller, see Eq. (13), requires that

$$k\rho = F\cdot\text{const} \tag{23}$$

where

$$F = F_T + F_G + R_G \tag{24}$$

or

$$F = F_F\left[\frac{1}{x}\left(1 + \frac{5}{y_{H_2}}\right) - \frac{1}{y_{H_2}}\right] \tag{25}$$

It is interesting to note that this total flow is independent of fluctuations in the inlet gas composition.

If we maintain the conversion and y_{H_2} constant, we see that the total flow through the reactor is proportional to the fresh feed rate of toluene, which is proportional to the production rate, see Eq. (12). Thus, from Eq.

(23) we can calculate how the reactor temperature must change with production rate, and then we can use a heat balance to calculate how the furnace heat duty must be changed. Similarly, we can calculate the fuel costs for production rate variations comparable to those given in Table 1.

For a constant temperature controller, the conversion would vary according to the relationship

$$F \ln [1/(1 - x)] = \text{constant} = K \tag{14}$$

or, after substituting Eq. (25),

$$F_F \left[\frac{1}{x} \left(1 + \frac{5}{y_{H_2}} \right) - \frac{1}{y_{H_2}} \right] \ln \left(\frac{1}{1-x} \right) = K \tag{26}$$

Now we can use this result to eliminate F_F from Eq. (12), so that for a fixed production rate, P_B, we must have

$$P_B = \frac{K\,S}{\left[\frac{1}{x} \left(1 + \frac{5}{y_{H_2}} \right) - \frac{1}{y_{H_2}} \right] \left(\ln \frac{1}{1-x} \right)} \tag{27}$$

If we correlate the selectivity vs. conversion data shown in Fig. 4 and eliminate S, we obtain an expression that is in terms of the unknown conversion only. Once we solve this equation we can determine all of the process flows, and then the fuel costs corresponding to various production rate changes. In particular, we can use Eq. (26) to find F_F, Eq. (22) to find R_G, and we can use the expression

$$P_D = (1 - S)F_F/2 \tag{28}$$

to estimate the production rate of diphenyl.

The costs of toluene reactant, the diphenyl production, the fuel cost for the furnace and the cooling costs of the reactor products are expected to have the greatest effect on the difference in operating costs between the two control policies. This cost difference must then be compared to the costs of instruments and to other hardware required to implement the two policies.

OTHER CONTROL SYSTEMS

In our discussion of product quality and production rate control we have considered disturbances in the compositions and temperatures of the raw material streams, the heating value of the fuel to the furnace, the cooling water temperature to the partial condenser, and the recycle streams. Furthermore, we discussed control systems for the separation system and the reactor. The only other sources of disturbances for the flow sheet shown in Fig. 5 are the steam to the reboilers and the feed water for the waste-heat boilers.

Reboilers—Reboiler control will normally be part of the distillation column control system. If we control the column in part by maintaining V/F = constant, then the reboiler capacity and steam supply must be adequate to supply the necessary vapor rate, V. Moreover, if the quality of the steam (or any other heating medium) supplied to the reboiler fluctuates with time, we can merely supply more, or less, steam to obtain the desired vapor rate.

Waste-Heat Boilers—An energy balance indicates that the steam production in a waste-heat boiler is directly proportional to the flow rate of the process stream through this unit. Thus, as the production rate changes we can change the flow rate of boiler feed water to the unit. On the other hand, if we maintain the water flow rate constant, then a lower quality steam will be produced. However, the steam can be flashed away and the water recirculated to the unit, so that no controller is needed.

DISCUSSION OF STEADY-STATE CONTROL SYSTEMS

A consideration of the steady-state operability and control of a process merely represents a first step in the synthesis of the control systems for a complete chemical plant. The importance of the process dynamics is easy to see by considering the response of the gas and liquid recycle flows to changes in the production rate. That is, at steady-state conditions both the gas and liquid recycle flows must be decreased as the production rate decreases, see Eqs. (12), (20), and (22), but we expect that the gas flow rate will change much faster than the liquid flow (i.e., less holdup). Thus, unless dynamic compensation is included, the hydrogen/aromatic ratio at the reactor will be below 5/1 for some period of time, and during this interval reactor coking might lead to severe operability problems. Dynamic compensation also is necessary for the column control systems, because we expect that the response to changes in the vapor rate will be faster than the response to changes in the distillate flow rate.

However, the steady-state analysis is useful to determine whether or not additional process units are needed to ensure operability of the plant, see the discussion of energy integration limitations for an example, and to estimate the incremental capacity of units which is needed for control purposes. Both the incremental capacity and the additional units add to the design costs, and these additional costs need to be included when process alternatives are compared. The steady-state analysis also provides a first estimate of the appropriate control structures for the plant, see Table 3.

Short-cut calculation procedures normally are adequate for preliminary designs when numerous process alternatives are being considered. Similarly, often they are adequate to decide whether reactor conversion control might be more profitable than reactor temperature control. Of course, once the best alternative is selected, rigorous steady-state and dynamic simulation studies would be undertaken. Studies of this type are especially needed before an unconventional control system, such as reactor conversion control, is implemented.

ACKNOWLEDGMENT

The author is grateful to the National Science Foundation for partial support of this research under grant NSF CPE 7818041.

Table 3. Summary of Control Systems.

1. Change toluene fresh feed rate to compensate for desired changes in the production rate of benzene.
2. Use a ratio control system to change the recycle gas flow rate proportional to the toluene fresh feed rate.
3. Use a feedback control system to maintain the recycle gas composition constant by manipulating the makeup gas rate.
4. Adjust the purge rate, if necessary, so that it is equal to the makeup gas rate.
5. Adjust the quench flow rate to the reactor so that it is a fixed fraction of the flash liquid flow.
6. Manipulate the distillate to feed ratio of the benzene column to maintain the material balance and keep V/F constant.
7. Use the same distillation column control systems for the other two columns.
8. Install a furnace control system to maintain the reactor conversion constant by compensating for changes in the reactor flow rate, the inlet temperatures to the reactor, and the heating value of the fuel.
9. No controllers are needed for the partial condenser or the waste-heat boilers.

NOMENCLATURE

ΔC_1	= methane produced by reaction, mol/hr
D	= distillate flow
F	= feed rate to flash drum, reactor, or column, mol/hr
F_F	= fresh feed rate of toluene, mol/hr
F_G	= makeup gas rate
F_T	= toluene flow to reactor, mol/hr
f_G	= vapor flow, mol/hr
f_i, f_j	= feed rate of component i or j, mol/hr
f_L	= liquid flow, mol/hr
G	= gas flow to absorber
ΔH_2	= hydrogen consumed by reaction, mol/hr
K	= constant, see Eq. (14)
K_i, K_j	= K value for component i or j
k	= reaction rate constant
L	= liquid flow from flash drum, mol/hr, or liquid flow to gas absorber
ℓ_i, ℓ_j	= liquid flow of component i or j, mol/hr
m	= slope of equilibrium line
N	= number of plates
P_B	= production rate
P_D	= production rate of diphenyl
P_G	= purge rate, mol/hr
R	= reflux ratio
R_G	= recycle gas rate, mol/hr
S	= selectivity, fraction of toluene which goes to benzene
V	= vapor rate from flash drum or column, mol/hr
V_R	= reactor volume, ft^3
v_i, v_j	= vapor flow of component i or j, mol/hr
x	= conversion
x_D, x_F, x_W	= mole fraction of product
x_i	= liquid composition for component i
y_f	= feed composition of hydrogen
y_{H_2}	= recycle composition of hydrogen
y_i	= vapor composition of component i
z_i	= feed composition for component i

Subscripts

D	= distillate
F	= feed
HK	= heavy key

i   = components with $K_i > 10$
j   = components with $K_j < 0.1$
LLK   = lighter than light key
m   = minimum
P   = product
w   = bottoms

Greek

α = relative volatility
ϵ_p, ϵ_{HK}, ϵ_{1p}, ϵ_{2p} = fractional loss
ρ = density, mol/ft^3

REFERENCES

[1] Bettes, R. S. and L. T. Wright, Oil and Gas Journal, 58(17):202, (1960).
[2] Bode, R. N. and V. J. Bakanowski, Engineering Journal, 43(12):2, (1960).
[3] Douglas, J. M., A. Jafarey, and R. Seemann, Ind. Eng. Chem. Process Des. Dev., 18:203, (1978).
[4] Fauth, G. F. and F. G. Shinskey, Chem. Eng. Prog., 71(6), (1975).
[5] Mesarovic, M., IEEE Proc., 58:111, (1970).
[6] Linnhoff, B., "A Thermodynamic Approach to Practical Process Network Design," Paper 28B Presented at the 72nd Annual AIChE Meeting, San Francisco, Ca., Nov. 25–29, 1979.
[7] Morari, M., Y. Arkun, and G. Stephanopoulos, "An Integrated Approach to the Synthesis of Process Control Structures," JACC Proceedings, Phila., Pa., 1978.
[8] Morari, M., Y. Arkun, and G. Stephanopoulos, AIChEJ, 26:220, (1980).
[9] Morari, M. and G. Stephanopoulos, AIChEJ, 26:232, 247, (1980).
[10] Shah, M. J. and R. E. Stillman, Ind. Eng. Chem., 62(12):59, (1970).
[11] Shinskey, F. G., "Process Control Systems," Ch. 11, McGraw-Hill, New York, N. Y., 1967.
[12] Weisenfelder, A. J., J. C. Fritz, and W. G. Thompson, ISA 3.3–3.66, Oct. 1966.

APPENDIX A

APPROXIMATE FLASH CALCULATIONS

An overall material balance gives

$$F = V + L \tag{A-1}$$

522   Chemical Process Control II

and an component balance gives

$$Fz_i = Vy_i + Lx_i \quad \text{(A-2)}$$

The equilibrium relationship is

$$y_i = K_i x_i \quad \text{(A-3)}$$

We can combine these expressions to obtain

$$y_i = \frac{z_i}{\dfrac{V}{F} + \left(1 - \dfrac{V}{F}\right)\dfrac{1}{K_i}} \quad \text{(A-4)}$$

Then,

$$\text{If } K_i \gg 1, \quad Vy_i = Fz_i \quad \text{(A-5)}$$

so that this component is only very slightly soluble in the liquid.
  We can also combine Eqs. (A-1) through (A-3) to obtain

$$x_i = \frac{z_i}{(K_i - 1)\dfrac{V}{F} + 1} \quad \text{(A-6)}$$

Then,

$$\text{If } K_i \ll 1, \quad Lx_i = Fz_i \quad \text{(A-7)}$$

so that this component is essentially nonvolatile.
  For the case of sharp separations in a flash drum, i.e., whenever $K_{LK} \gg 10$ and $K_{HK} \ll 0.1$, the order of magnitude arguments above indicate that a reasonable estimate of the vapor flow will be the sum of the feed rates of the light key and all lighter components. Similarly, the liquid flow will be approximately equal to the sum of the feed rates of the heavy key and all heavier components.

$$V \simeq \sum_i f_i = f_G, \text{ for components where } K_i > 10;$$
$$L \simeq \sum_j f_j = f_L, \text{ for components where } K_j < 0.1. \quad \text{(A-8)}$$

The analysis above corresponds to a perfect split. However, we also want to estimate the small amounts of essentially noncondensible materials that dissolve in the liquid and the small amounts of essentially nonvolatile materials that vaporize. Since we know that these flows are small, we introduce only small errors if we merely superimpose the equilibrium relationships on the conditions for perfect separation. That is, we expect that

$$y_i = \frac{f_i}{f_G} \tag{A-9}$$

and we require that

$$x_i = y_i/K_i \tag{A-10}$$

Hence, the liquid flow rate of component i will be, approximately,

$$\ell_i = Lx_i = f_L x_i \tag{A-11}$$

Then, substituting Eqs. (A-8), (A-9), and (A-10), we obtain

$$\ell_i = \frac{f_i\, f_L}{K_i\, f_G} \tag{A-12}$$

Now we correct the vapor flow of component i for the amount lost in the liquid

$$v_i = f_i - \ell_i = f_i \left[1 - \frac{f_L}{K_i\, f_G} \right] \tag{A-13}$$

The results for the other components can be derived in a similar manner and they are

$$v_i = f_i \left[1 - \frac{f_L}{K_i\, f_G} \right], \quad \ell_i = \frac{f_i\, f_L}{K_i\, f_G} \tag{A-14}$$

$$v_j = \frac{K_j\, f_j\, f_G}{f_L}, \quad \ell_j = f_j \left[1 - \frac{K_j\, f_G}{f_L} \right] \tag{A-15}$$

A comparison between the approximate calculations and an exact solution for an example of the hydrodealkylation of toluene is shown in Table A-1.

Table A-1. Exact vs. Approximate Flash Calculations.

Component	f_i	K_i	Exact v_i	Exact ℓ_i	Approximate v_i	Approximate ℓ_i
H_2	1547	86.1	1545	2	1545	2
C_1	2321	12.4	2305	16	2304	17
	f_j	K_j	v_j	ℓ_j	v_j	ℓ_j
B	265	0.013	34	231	37	228
T	91	0.005	5	86	5	86
D	4	0.0001	0	4	0	4
	4228		3890	339	3891	337

DESIGN AND ANALYSIS OF PROCESS PERFORMANCE MONITORING SYSTEMS

Richard S. H. Mah
Northwestern University
Evanston, Illinois 60201

ABSTRACT

Process data reconciliation, variable and measurement classification, gross error detection and identification, and measurement placement form the basis for the design and analysis of process performance monitoring systems. Research progress in these areas is reviewed in this paper.

INTRODUCTION

Process data is the foundation upon which all control and evaluation of process performance are based. These functions include production planning, operation scheduling as well as the more familiar process control. They differ in time scale, ranging from minutes and hours in process control to months and quarters in production planning, and they differ in scope, ranging from a process stream or unit to a multi-refinery network. Such functions have already been alluded to in another paper presented at this conference (Fisher 1981). In this review we shall focus on the various questions which surround the design and analysis of a system which gathers, transforms and supplies process data for the various performance control and evaluation functions.

In a broad sense the problems that we address in this paper are as old as the process industry itself. When process measurements and data acquisition were regularly carried out by plant operators and engineers, the accuracy and consistency of the data were checked on the basis of their knowledge and experience of the process. Now that these functions are performed automatically by computers, a far greater volume of process data must be checked and processed without the benefit of human vigilance and intervention. Moreover, our concern with process efficiency leads us in the direction of highly integrated plants and immense scale of operations so that even a small error may have significant impact and broad ramifications. Consequently there is a need to develop a rational and systematic basis for checking and treating process data which underlie a process performance monitoring system.

Let us briefly summarize the key features of the process data problems. First, all measurements are subject to errors. These errors corrupt individual measurements and cause the measured values collectively to be inconsistent in the sense of discrepancies in enthalpy and material balance closure. They fall into two categories: Random errors which are commonly assumed to be independently and normally distributed with zero mean, and gross errors which are caused by non-random events such as instrument biases, malfunctioning measuring devices, incomplete or inaccurate process models. Second, not all process variables are measured for reasons of cost, inconvenience or technical infeasibility. Third, there is a data redundancy in the sense that there are more measurements (or data) available than needed if the measurements were not subject to errors. This redundancy which is brought about as a result of conservation relationship and the interconnectedness of a process network will be referred to as spatial redundancy. Fourth, with the data sampling and recording techniques now available, it is not uncommon to find process data being sampled continually at regular time intervals of 1 to 5 minutes. There is, therefore, also a data redundancy in the sense that more measurements are available than needed, if the process conditions were truly at a steady state. We shall refer to this as temporal redundancy.

Given the overall objective of developing a basis for improving the consistency and accuracy of process data, we can formulate the following subsidiary questions:

> How should the data be reconciled for consistency? (The formulation of data reconciliation problem).

How would the nature of the constraints influence the methods of solution?

Which measurements can be omitted without affecting our ability to estimate the values of the variables?

How do we decouple the reconciliation problem from the problem of estimating unmeasured variables? And more generally how can the computation be reduced through structural decomposition?

How should the structural (global) and numerical (local) effects be treated?

How can the presence of gross errors be detected? How can the error sources be identified?

How can the treatment be extended to dynamic and quasi-steady state systems?

What is the most effective placement of measurements and how should one design a performance monitoring system?

These questions are treated in the publications reviewed in this paper. To avoid excessive overlaps the questions will not be separately discussed, and the coverage of topics will be somewhat uneven, which reflects in part the state of the literature and in part our desire to devote the limited time and space to the most significant developments in the open literature. A recent review of Hlavacek (1977) listed most publications up to 1975 and gave an extensive coverage of Czech and other less accessible publications. Since many of these publications are still inaccessible to us, we will not list them in our bibliography, but, instead, will refer interested readers to Hlavacek's article (1977).

RECONCILIATION OF PROCESS DATA SUBJECT TO RANDOM ERRORS

Representation of chemical processes by graphs was studied by Kafarov et al. (1970a,b) and Korach and Hasko (1968). The underlying structure of a process may be described by a directed graph called a process graph (Mah et al. 1976). Its arcs generally correspond to the process streams whose directions are usually determined by the processing requirements. Its nodes generally correspond to the units, tanks and junctions in the process flow sheet. But since, at any given instant of time, only a subset of process units and tanks may be activated, the process graph may only contain a subset of such nodes. The process graph always contains an environment node. The process receives its feed (including utilities) from the "environment" and supplies its products to the "environment." The environment node may thus be perceived as the complement of the process.

With the inclusion of the environment node, the process graph is always cyclic. It is amenable to all the usual algebraic representations of a diagraph (Deo 1974). In particular we will make frequent use of a n x m incidence matrix **A** whose rows correspond to the nodes (excluding the environment node) and whose columns correspond to the arcs of the process graph. Let x_j be the measured flow rate, $\hat{x}_j$ the estimated flow rate and $\bar{x}_j$ the true flow rate of stream j. Then the adjustment δ_j and the measurement error ϵ_j may be defined as:

$$\delta_j = \hat{x}_j - x_j \tag{1}$$

$$\epsilon_j = x_j - \bar{x}_j \tag{2}$$

Exactly how $\hat{x}_j$ is estimated and what adjustment δ_j is made will, of course, depend on our assumptions on error statistics. If the measurement errors are normally distributed with zero mean and positive definite covariance matrix **Q**, then the least-squares estimation is also the maximum likelihood and the minimum variance unbiased estimation, which may be formulated as

$$\min_{\hat{x}} \delta^T Q^{-1} \delta \tag{3}$$

subject to the conservation constraints

$$A \hat{x} = 0 \tag{4}$$

The solution to this problem was given by Kuehn and Davidson (1961) as follows:

$$\hat{x} = x - Q A^T (A Q A^T)^{-1} A x \tag{5}$$

Subsequently, the same problem formulation and solution technique have been applied by Swenker (1964), Ripps (1965), Hoffman (1966), Vaclavek (1969a, b, c and 1972), Meharg (1972) and Murthy (1973) among others in different contexts. Computer simulations indicating substantial reduction in total absolute errors were reported by Mah et al. (1976). The effect of inventory changes in a quasi steady state process was studied by Vaclavek (1974b) and Mah et al. (1976).

In-plant experience of reconciling flow data using a linear objective function was discussed by Smith et al. (1969). Their approach appears to be motivated by the efficiency and availability of the out-of-kilter algorithm. No statistical justification was given for their selection of the objective function. Similar studies were carried out by Mathiesen (1974). Since the statistical assumptions are not obvious when gross errors are present, it was suggested (Hlavacek 1977) that one might just as well minimize the sum of weighted absolute errors; linear programming codes are at least efficient, readily available and can handle inequality constraints. Upon close examination this argument appears spurious. It would be much preferable to detect and identify gross errors and to eliminate their effects before data reconciliation. (Ripps 1965, Almasy and Sztano 1975 and Mah et al. 1976).

In the foregoing discussion x_j may be enthalpy, total mass or component mass flow rates. So long as these quantities are measured directly, the constraints (4) will be linear. But in practice it is more common for the temperature or the concentration of species to be measured along with the mass flow rate. Products of the variables will then appear in enthalpy and component balances, making these constraints bilinear rather than linear. If some of these variables are measured indirectly by other physical properties, e.g., concentration by density, pH or thermal conductivity, the constraints might even be nonlinear. Data reconciliation subject to nonlinear constraints may be carried out by nonlinear programming methods, direct algebraic substitution, the use of Lagrange multipliers among other methods, which have been reviewed by Hlavacek (1977).

It should be pointed out that the constraints in process data reconciliation are not always related to the structure of a process flow sheet. Reactor data reconciliation was studied by Vaclavek (1969c), Umeda et al. (1971), Murthy (1973, 1974) and Madron et al. (1977). In Vaclavek's investigations (1968, 1969c), reactants and products of a chemical reaction were represented by fictitious streams whose flow rates were made proportional to the extent of the reaction. This representation gives a chemical reaction the appearance of a network constraint but not all of its desirable attributes. Murthy first (1973) used the conservation of atomic species as constraints and later (1974) showed the advantages of using stoichiometric information when the reactions were known.

In practice both network and stoichiometric constraints may be present in a given application (see, for example, Duyfjes and Swenker 1976).

Up to this point we have made the tacit assumptions that the process operates under steady state conditions and that only a single set of measurements is available for data reconciliation. In reality, neither of these simplifications alway applies. At a given instant, a "snapshot" of the conditions of a continuously operated process may appear to be steady state, but over a longer period the successive "snapshots" of conditions will almost certainly reveal changing process conditions.

Continual and regular data sampling provides us with additional constraints of "temporal redundancy" which should be of assistance in data reconciliation. Not surprisingly the least-squares estimation in this case turns out to be an application of the discrete Kalman filter.

Examining data logged during a 2-month period on a crude oil distillation unit Stanley and Mah (1977) discovered that the unit was essentially at steady state except for slow drift or occasional sharp transition between steady states. They proposed a quasi steady state (QSS) model which predicted no change in the values of the state variables in the short run. The process noise covariance matrix was used as a filter design parameter. It was found possible to tune the QSS estimator to give rapid and stable response to sudden changes and at the same time reduce estimation errors by up to 70%. In almost all instances simulated, spatial redundancy contributed significantly to the overall error reduction. The computational requirements of the estimator were well within the capabilities of a process computer.

Data reconciliation for linear dynamic systems was treated by Gertler and Almasy (1973). They showed that the dynamic equivalent of constraints (4) could be derived in two different ways. The first approach starts from the state space description of the system

$$\dot{x}(t) = A\,x\,(t) + B\,u\,(t) \tag{6}$$

$$y(t) = C\,x\,(t) + D\,u\,(t) \tag{7}$$

which, after discretizing, z-transformation and expansion as a fraction of polynomials in z^{-1}, yields a sampled input-output relation that describes all the system outputs as explicit sampled data functions of the system inputs only. The second approach builds up the system description as a network of two types of units: dynamic units with single input and output and nodes with multiple inputs and outputs but without dynamics. Because approximation of the inter-sampling course is applied to the input of each dynamic unit instead of the system input, this description is somewhat less accurate than the first. Both approaches lead to affine constraints which are treated by Stanley and Mah (1981a). A previously developed estimation procedure (Almasy et al. 1969) may be applied to this problem.

The problem of least squares parameter estimation for steady state and linear dynamic models subject to liner conservation constraints was treated by Almasy and Sztaro (1978) and Almasy (1980).

ESTIMATION OF UNMEASURED VARIABLES

The least-squares formulation of Eq. (3) and (4) applies to the data reconciliation problem for which every state variable is measured. In most industrial processes because of cost, inconvenience or technical infeasibility not all of the variables are measured. But we may want to estimate the unmeasured process variables in terms of measured ones, and at the same time, subject the measured values to data reconciliation. How are the two problems related and under what conditions can the unmeasured variables be estimated?

The basic graph-theoretic results for mass flow networks (linear balances) may be stated as follows: (1) Reconciliation with missing measurements can be resolved into two disjoint problems: reconciliation on a graph which is formed by pairwise aggregation of the nodes linked by arcs of unmeasured flows, and the estimation of unmeasured flows in the tree arcs of the process graph. (2) Missing flow measurements can be determined uniquely, if and only if the unmeasured arcs form an acyclic graph (i.e. trees).

The use of node aggregation and "reduced balance scheme" was first pointed out by Vaclavek (1969), but a rigorous proof of the above two results was given by Mah et al. (1976).

For multi-component flow networks and bilinear constraints Vaclavek and Loucka (1976) proposed an algorithm (CLASS) for classifying measured variables into "overdetermined" and "not overdetermined" and unmeasured variables into "determinable" and "indeterminable." The theoretical basis of this algorithm was apparently published in less accessible literature (Vaclavek et al. 1972b and 1975b). Recently Stephanopoulos and Romagnoli (1980a) suggested that such a classification algorithm might be constructed on the basis of an output assignment algorithm (Stadtherr et al. 1974). If an unmeasured variable is assigned as the output to one of the contraining equations, then it is determinable. Otherwise, it is indeterminable. The classification of measured variables is less direct and makes use of the characteristics of unassigned equations (whether they contain only measured variables, unmeasured but determinable variables or unmeasured and indeterminable variables) and of disjoint subsystems (whether all external variables of a disjoint subsystem are measured). The classification can then be used as a basis for constructing a "reduced balance scheme," i.e. identifying constraints containing only measured variables.

OBSERVABILITY AND REDUNDANCY

The variable classification discussed above takes into account only structural information of the constraints. Stanley and Mah (1977, 1981a) showed how the classification can be made more rigorous and appropriate by formally and precisely defining the concepts of observability and redundancy. These concepts provide a link to the concepts of observability and controllability in dynamic system description and of identifiability in linear least-squares estimation, they are useful in differentiating local and global properties, and most importantly, they allow us to characterize the performance of process data estimators and to determine the feasibility and implications of problem decomposition. We shall give a brief description of these concepts and the major results.

A steady state system of n state variables $\mathbf{x}$ and ℓ measurements $\mathbf{y}$ may be characterized by the triplet $(S,\mathbf{h},V)$, where $S \subset R^n$ is the feasible set, $\mathbf{h}$ is the measurement function and V the set from which a particular value of "measurement noise" may be obtained. For instance,

$$S = [\mathbf{x} : \mathbf{g}(\mathbf{x}) = \mathbf{0}, \mathbf{x} \epsilon K] \tag{8}$$

where $\mathbf{g}$ is a set of p nonlinear constraints and K is the union of a finite number of convex sets. Loosely speaking, if we can determine x_i by taking a set of measurements $\mathbf{y}^o$, where

$$\mathbf{y} = \mathbf{h}(\mathbf{x}) + \mathbf{v}, \quad \mathbf{v} \epsilon V, \quad \mathbf{x} \epsilon \bar{S} \tag{9}$$

the $\mathbf{x}$ is observable.

Because $\mathbf{g}$ and $\mathbf{h}$ may be nonlinear, observability may depend on the numerical values of $\mathbf{x}$. For an index set I, $\mathbf{x}_I$ is said to be locally unobservable at $\mathbf{x}^o$ if there exists a sequence $[\mathbf{x}^k]_{k=1}^{\infty}$ such that

$$\mathbf{x}^k \to \mathbf{x}^o, \text{ as } k \to \infty \tag{10}$$

$$\mathbf{x}^k \epsilon S \tag{11}$$

$$\mathbf{h}(\mathbf{x}^k) = \mathbf{h}(\mathbf{x}^o) \quad \text{for all k} \tag{12}$$

$$\delta \mathbf{x}_I^k = \mathbf{x}_I^k - \mathbf{x}_I^o = \mathbf{0} \tag{13}$$

Notice that the local unobservability is defined as a property of the deterministic system $(S,\mathbf{h},0)$ for a subset of state variables $\mathbf{x}_I$ at a point $\mathbf{x}^o$. Starting with this definition similar definitions can be developed for local observability at a point, local unobservability and observability on a set

$S_1 \subset \bar{S}$, global unobservability and observability, and system unobservability and observability (when $I = \{1, 2, \ldots n\}$).

The relationships between these observabilities and their counterparts for linear and nonlinear dynamic systems were discussed by Stanley and Mah (1977, 1981a). The most notable features for the steady state systems are that the applications often require the observability of individual variables rather than that of the system as a whole, and that many questions of practical interest can be adequately answered without resorting to global observability. For these reasons steady state observabilities were defined in terms of local unobservability for a subset of state variables. The definitions also permit ready extensions to their counterparts in the redundancy of measurements. A measurement is redundant, if its removal causes no loss of observability.

Based on these definitions it is possible to establish sufficient conditions for local observability, local unobservability (Stanley and Mah 1981a Theorems 1–5). Stanley and Mah (1981a) also showed how local observability is directly related to the local uniqueness of the solution to the measurement equations and how local unobservability leads to estimator failure for both steady state (Theorems 7 and 8) and quasi steady state estimation (Theorems 10–14) and how the decomposition results for mass flow networks (Mah et al. 1976) could be generalized (Theorems 9 and 15).

With these theoretical constructs the key to the design and analysis of a performance monitoring system becomes the classification of observability and redundancy of each variable in the process network. Efficient classification algorithms which exploit the structural characteristics of a process network were developed by Stanley and Mah (1977, 1981b). The algorithms make use of properties of biconnected components, feasible unmeasurable perturbations and perturbation subgraphs.

A graph is separable or 1-connected if it contains a cut-node. Now suppose we split a cut-node into two nodes to produce two disjoint subgraphs and let us refer to this operation as splitting. If we repeat this operation until all subgraphs are non-separable, then the resulting subgraphs are called biconnected components. The biconnected components have a number of useful graph-theoretical properties (Aho et al. 1974 and Tarjan 1972), but of far more direct interest to us are the properties of the flow network associated with the biconnected components. For instance, no net mass flow can enter or leave a biconnected component of a process flow network. The δx^k in Eq. (13) are called feasible unmeasurable perturbations about x^o. The perturbations δx^k are "feasible" because the perturbed variables $x^k = x^o + \delta x^k$ are feasible, and "unmeasurable" because by Eq. (12) the perturbations do not affect the measurement values. For the special case for which x^o is a feasible solution for the process graph G and δx is a feasible solution G_1, where $G_1 \subset G$ and $\delta x \to 0$, G_1 is a perturbation subgraph of G_o at x^o, if the measurements on G cannot distinguish x^o from $x^o + \delta x$.

Properties of biconnected components, feasible unmeasurable perturbations and perturbation subgraphs are developed and summarized by Stanley and Mah (1977, 1981b). By applying these properties systematically to the process network and its various sub-networks, it is possible to construct extremely efficient algorithms which will classify the observability and redundancy of individual variables in a process network in only $O(m)$ time, where m is the number of arcs of the process graph.

DETECTION AND ELIMINATION OF GROSS ERRORS

In our treatment of data enhancement thus far only normally distributed measurement errors with zero means and known variances are assumed to be present in the data. In practice the raw process data may also contain gross errors which are caused by non-random events such as leaks, depositions and inadequate accounting of departures from steady-state operations as well as measurement biases and malfunctioning instruments. In comparison with the random errors there should normally be a very small number of gross errors present in any given set of data. Nonetheless their presence invalidates the statistical basis of reconciliation procedures. When a set of process data is subject to constrained least-squares reconciliation, a high penalty is imposed on making any single large correction, because the corrections are squared when they appear in the objective function. Thus, a gross error present in one measurement will cause a series of "small corrections" in other measurements. The distribution of the gross error has two serious disadvantages: it makes the data as a whole inaccurate, and it obscures the source of the gross error, providing no information to guide instrument repair or other corrective actions.

The distinction between gross and random errors was first pointed out in a perceptive paper by Ripps (1965) who proposed to identify and eliminate gross errors by selectively discarding redundant measurements. He gave no explicit criterion, but suggested that the value of the objective function might be used to guide the selection. The ratio of balance residual at a node to the measured average flow through the node was proposed as an empirical measure by Vaclavek (1974a) and Vaclavek and Vosolsobe (1975a).

Nogita (1972) was the first to introduce a statistical test of significance. He proposed the sum of adjustments δ_j weighted by the reciprocals of population standard deviations σ_j of measurements

$$Z = \sum_{j=1}^{N} (\delta_j/\sigma_j) \qquad (14)$$

as a statistical test function. It can be readily shown that Z is normally distributed with a zero mean and variance σ_z^2 which is readily computed from the measurement error statistics. The normalized random variable Z/σ_z may thus be tested for Type I and Type II errors using a cumulative normal distribution table (Hald, 1952). Nogita also outlined a serial elimination procedure for discarding suspected measurements.

Mah et al. (1976) proposed an alternative test function which is based on the residual of each constraining equations,

$$\mathbf{z} = \mathbf{H}^{-1} \mathbf{A} \mathbf{x} \qquad (15)$$

where

$$\mathbf{H}_{ij} = \begin{cases} (\mathbf{A} \mathbf{Q} \mathbf{A}^T)_{ij}^{1/2} & i = j \\ 0, & i \neq j \end{cases} \qquad (16)$$

Each element of **z** is also normally distributed with mean of zero and variance of unity. In comparison with Nogita's test function (14), the new test function (15) offers two distinct advantages. First, it can be evaluated directly without any prior reconciliation calculations. It takes less time to compute, and it is also more sensitive to gross errors, since after reconciliation, a gross error is "smeared" over all the estimates and it becomes difficult to trace its effect. Second, Nogita uses a single statistic for the entire process, while a separate statistic is used at each node in the new test function. This difference is crucial for large problems because the effect of a single gross error may be swamped when combined with smaller errors under Nogita's procedure. Moreover, since Nogita takes linear combinations of all the residuals, gross errors can pass unnoticed because of error cancellations.

Independent investigation by Almasy and Sztano (1974, 1975) led to a related test function for affine constraints $\mathbf{A} \hat{\mathbf{x}} - \mathbf{b} = \mathbf{0}$,

$$t^2 = (\mathbf{A} \mathbf{x} - \mathbf{b})^T (\mathbf{A} \mathbf{Q} \mathbf{A}^T)^{-1} (\mathbf{A} \mathbf{x} - \mathbf{b}) \qquad (17)$$

where t^2 obeys chi-square distribution with p degrees of freedom, p being the number of affine constraints. More recently the same test function is proposed by Stephanopoulos and Romagnoli (1980a), and essentially the same test function has also been applied to gross error detection in reactor data by Madron et al. (1977).

It should be apparent from the above discussion that gross errors can only be checked for redundant measurements and that, whenever nonlinear constraints are involved, the statistics is based on the linearized relationship.

The statistical tests described above would only alert us to the possible existence of one or more gross errors either around a unit (Mah et al. 1976) or in the entire process network (Nogita 1972, Almasy and Sztano 1975). Further procesing is required to isolate and identify the error source.

For measurements containing only a single gross error Almasy and Sztano (1975) devised a procedure which would pick out the erroneous measurement. According to these authors, for the procedure to be successful the magnitude ratio of the gross error to the dispersion of the random error must not be "too small." No rigorous results on confidence levels are available for this procedure.

For one-component (mass or enthalpy) flow network Mah et al. (1976) showed that a complete set of 8 graph-theoretical rules may be developed based on certain simplifications (e.g. no accidental cancellation of gross errors). By applying the gross error detection test repeatedly to node clusters of increasing size these rules permit a scheme to be constructed for systematically identifying leaks or faulty flow measurements. Investigation by Stephanopoulos and Romagnoli (Stephanopoulos 1980b) is under way to develop a gross error identification scheme for multi-component flow networks by combining the above type of structural considerations with an analysis based on sequential deletion of measurements.

MEASUREMENT PLACEMENT

The analysis up to this point has not brought into consideration data enhancement from the viewpoint of process requirements. We shall now turn to the relationship between these requirements and the selection or placement of measurements in a process network, and to indicate how the analytical and computational techniques developed previously may be brought to bear on this problem.

First of all, process requirements may be specified in terms of locally or globally observable and locally or globally redundant variables. Such a specification would, of course, have immediate implications on the measurement placement. But the mapping need not be one-to-one. In fact it is likely to be one-to-many (as well as many-to-one) if the observable and the redundant subsets are at all extensive. Some mappings will be excluded because of measurement infeasibility. An optimal selection can then be made among the remainders on the basis of economic considerations.

It should be pointed out that the optimal measurement placement problem described above has not yet been satisfactorily tackled, although several attempts have been made in this direction. The approaches used may be broadly classified as direct and indirect methods. Two direct methods are attributed to Vaclavek and coworkers by Hlavacek (1977). The first method (Vaclavek et al. 1972b, 1975a) makes use of classification of various variable sets, and obtains a feasible measurement set by logical disjunction of various classified and required sets. A partial description of this method was given by Vaclavek and Loucka (1976). A similiar approach was outlined for 3 special problems by Ramagnoli and Stephanopoulos (1980a). The second method apparently makes use of the powers of an adjacency matrix. Except for a brief paragraph in Hlavacek's review (1977) no description of this method is available in open literature. None of these methods give consideration to economic optimization.

In the indirect approach a prior selection of measurement set is made on the basis of safety, experience and other heuristic considerations. The selection is then tested to see if it meets the process requirements. If not, a measurement is added according to some predetermined rules. A "lowest cost first" evolutionary strategy along these lines was in fact proposed by Stanley (1977) starting with all measurements required by process controllers. An indirect method making use of "reduction of the balance scheme" (Vaclavek and Vosolsobe 1975a) was also mentioned by Hlavacek (1977). In the specific context of feedback control Mellefont and Sargent (1977, 1978) presented an implicit enumeration algorithm for selecting the optimal measurement subsets for linear stochastic systems during the control interval. The relationship between measurement selection and plant control was further discussed by Morari (1981) at this conference.

The second aspect of process requirements is the specification of accuracies. Since this problem was reviewed at some length by Hlavacek (1977), we shall not dwell on it except to point out that linearized treatment is used in all known attempts in this direction.

CLOSING REMARKS

Widespread industrial interest in the design and analysis of performance monitoring systems is attested by continual and sustained industrial publications over a period of almost 20 years (Kuehn and Davidson 1961,

Swenker 1964, Ripps 1965, Hoffman and Muller 1966, Smith et al. 1969, Umeda et al. 1971, Meharg 1972, Nogita 1972, Murthy 1973, 1974, Clair et al. 1975, Duyfjes and Swenker 1976). After much initial attention to the computational procedures for data reconciliation the emphasis of recent research has gravitated towards the more central problems of variable and measurement classification, gross error detection and identification, and measurement placement. Significant progress has been made in the last 5 years towards a sound theoretical basis for treating the first two problems. Several algorithms have also been developed for network type of constraints. Further work is required to evaluate and generalize these algorithms. A good start has been made on the problem of measurement placement, but more definitive work remains to be done in this area. There is also a need to integrate many of the specific gains on different aspects of performance monitoring systems into an overall framework that could be of direct utility to industry. Finally, the interface of data treatment with process control and other performance monitoring functions offers much scope for exploration and creative research.

Past progress in this area was hampered by the lack of critical evaluation and consolidation of technical literature. The problems discussed in this paper have been repeatedly rediscovered by different authors. Publications seem poorly reviewed and uneven in standards. While there are some excellent papers on this subject, others give very superficial treatment and contain little substance in theory or practice. Earlier progress was also hampered by inaccessible literature, some proprietary and others in less familiar languages and journals. However, as the subject gains greater attention and following, these difficulties should diminish with time. There is therefore ground for optimism in the long run.

REFERENCES

Aho, A. V., J. E. Hopcroft and J. D. Ullman, *The Design and Analysis of Computer Algorithms*, Addison Wesley, Reading, Mass. 1974.

Almasy, G. A., "Dynamic Models Satisfying Balance Equations," CHEMPLANT 80, EFCE Symposium, Heviz-Hungary, September 1980.

Almasy, G. A., T. Sztano, P. Csillag, P. G. Veress, J. Holderith, 3rd Congress CHISA, Marianske' Lazne, 1969.

Almasy G. A. and T. Sztano, Problems of Control and Information Theory, 4(1):57, 1975.

Almasy, G. A. and T. Sztano, "Empirical Models Satisfying Balance Equations," EFCE Congress, Paris, March 1978.

Clair, R., F. Lebourgeosis, J. P. Caujolle, G. Bornard, "Measurement Data Set Improvement Through Material Balance Calculation—Application to an Industrial Chemical Plant," IFAC Congress, Boston, 1975.

Deo, N., *Graph Theory With Applications to Engineering and Computer Science*, Prentice-Hall, 1974.
Duyfjes, G., and A. G. Swenker, "Smoothing of Redundant Data—A Useful Tool in Plant Operation," DSM Symp. on The Use of Computers for the Operation of Production Plants in the Process Industries, Florence Sept. 23–24, 1976.
Fisher, D. G., "Systems Software for Process Control: Today's State of the Art," presented at this conference.
Gertler, J. and G. A. Almasy, Automatica 9:79, 1973.
Hald, A., *Statistical Theory with Engineering Applications*, Wiley, New York, 1952.
Hlavacek, V., Comp. Chem. Engng. 1:75, 1977.
Hoffman, R. and R. Muller, Messen-Steuern-Regeln, 9:233, 1966.
Kafarov, V. V., V. L. Perov and V. P. Meshalkin, Teoreticheskie Osnovy Khimicheskoi Tekhnologii, (Theoretical Foundations of Chemical Engineering), 4(5):745, 1970a.
Kafarov, V. V., V. L. Perov and V. P. Meshalkin, Teoreticheskie Osnovy Khimicheskoi Tekhnologii, 4(6):898, 1970b.
Korach, M., L. Hasko, Teoreticheskie Osnovy Khimicheskoi Tekhnologii, 2(3):346, 1968.
Kuehn, D. R. and H. Davidson, Computer Control, Chem. Eng. Prog. 57(6):44, 1961.
Madron, Frantisek, Vladimir Veverka and Vojtech Vanecek, AIChE J. 23:482, 1977.
Mah, R. S. H., G. M. Stanley and D. M. Downing, Ind. Eng. Chem. Proc. Des. Dev. 15:175, 1976.
Mathiesen, N. L., Automatica, 10:431 1974.
Meharg, E. B., "How to Use Data Adjustment," 7th Annual Conference on the Use of Digital Computers in Process Control, Louisiana State University, February 9–11, 1972.
Mellefont, D. J. and R. W. H. Sargent, Int. J. Contr., 26:595 1977.
Mellefont, D. J. and R. W. H. Sargent, Ind. Eng. Chem. Proc. Des. Dev. 17:549 1978.
Morari, M., "Integrated Plant Control. A Solution at Hand or a Research Topic for the Next Decade?", presented at this conference.
Murthy, A. K. S., Ind. Eng. Chem. Proc. Des. Dev. 12:246, 1973.
Murthy, A. K. S., Ind. Eng. Chem. Proc. Des. Dev. 13:347, 1974.
Nogita, S., Ind., Eng. Chem. Proc. Des. Dev. 11:197, 1972.
Ripps. D. L., Chem. Eng. Prog. Symp. Ser. No. 55 61:8, 1965.
Romagnoli, J. A. and G. Stephanopoulos, Chem. Eng. Sci., 35:1067, 1980a.
Stephanopoulos, G., J. A. Romagnoli, "A General Approach to Classify Operational Parameters and Rectify Measurements Errors for Complex Chemical Processes," ASC Symp. Series 124:153, 1980a.
Smith, R. A., R. L. Indiveri, W. M. Byrne, "Material Balancing Process Plants by Network Analysis," NPRA Computer Conference, National Petroleum Refiners Assoc., Nov. 1969.
Stadtherr, M. A., W. A. Gifford, and L. E. Scriven, Chem. Engng. Sci. 29:1025, 1974.
Stanley, G. M., Ph.D. Thesis, Northwestern University, Evanston, Il., 1977.
Stanley, G M., and R. S. H. Mah, AIChE 23:642, 1977.
Stanley, G. M., and R. S. H. Mah, Chem. Eng. Sci., 36:259 1981a.
Stanley, G. M. and R. S. H. Mah, "Observability and Redundancy Classification in Process Networks," Chem. Eng. Sci., accepted for publication 1981b.
Stephanopoulos, G., Personal Communication 1980b.
Swenker, A. G., Acta IMEKO, Budapest, Proc. 3rd Int. Measurement Conf. 1, 1964
Sztano, T. and G. A. Almasy, "Measurement Error Checking in Linear Stochastic Systems," IFAC Symposium on Stochastic Control, Budapest, September 1974.
Tarjan, R., SIAM Comput. 1:146, 1972.
Umeda, T., M. Nishio and S. Komatsu, Ind. Eng. Chem. Proc. Des. Dev. 10:236, 1971.

Vaclavek, V., M. Kubicek, V. Hlavacek, M. Marek, Collection Czech. Chem. Commun. 33:3653, 1968.
Vaclavek, V., Collection Czech. Chem. Commun. 34:364, 1969a.
Vaclavek, V., Chem. Eng. Sci., 24:947, 1969b.
Vaclavek, V., Collection Czech. Chem. Commun. 34:2662, 1969c.
Vaclavek, V., Chem. Eng. Sci. 27:1669, 1972a.
Vaclavek, V., Z. Bilek and J. Karasiewicz, Verfahrenstechnik 6:415, 1972b.
Vaclavek, V., Scientific Papers of the Prague Institute of Chemical Technology, K9:75, 1974a.
Vaclavek, V., Chem. Eng. Sci. 29:2307, 1974b.
Vaclavek, V. and J. Vosolsobe, Proc. Symp. Comp. Des. Erection Chem. Plants, Karlovy Vary, Czechoslovakia, 1975a.
Vaclavek, V., M. Kubicek and M. Loucka, Theoret. Osnovy Chim. Technologii, 9:415, 1975b.
Vaclavek, V. and M. Loucka, Chem. Eng. Sci. 31:1199, 1976.

CONTROL SYSTEMS FOR COMPLETE PLANTS — THE INDUSTRIAL VIEW

Irven R. Rinard
Halcon Research & Development Corporation
2 Park Avenue, New York, NY 10016

I would like to begin with some slides which depict what might be called the view from the control room floor (Slide 1). Now most of the time in most control rooms, the situation does not get this hectic. But now and then it does. After all, chemical plants are multivariable and as such are subject to the multivariable version of Murphy's Law (Slide 2). My point here is that there is a lot of territory between the analysis and tuning of individual control loops and the moment-to-moment control of an entire plant as seen by the Shift Supervisor.

What is this territory? One place to start is by asking what is control system. Of what is it comprised? First, one can take the narrow view (Slide 3). This is the definition usually encountered in the textbooks. We have some measurements available of some of the process variables (hopefully the right ones) and we have some actuators, primarily control valves. Process control engineering has traditionally been viewed as the determination of control law most appropriate to juggling the valve positions to achieve specified values of some of the process variables. It is the activity that has received the lion's share of the attention in the past.

But in reality one has to take a broader view of what constitutes a control system (Slide 4). In addition to the traditional control system, there are the operators. If they are experienced and well-trained, they will compensate for many of the shortcomings of the traditional control system. If not, they may become part of the control problem.

The View From The Control Room Floor

The Product Is Off Spec. Cause Unknown.
The High Level Alarm on The Rerun Tank Has Just Gone On.
The Reactor Temperature Control Valve Has Been Sticking And Is Getting Worse.
There Are Two Level Meters On The Bottom Of The Crude Product Column.
    One Shows The Level Going Up.
    The Other Shows It Going Down.
All Of The Column Overhead Condensers Are Air-Cooled
    A Severe Thunderstorm Is Approaching.
It Is 2:00 A.M. On Sunday Morning.
    The Area Superintendent Has Gone Fishing.
    The Plant Manager Is In Paris.
You Are The Shift Supervisor, What Do You Do?

Slide 1

Murphy's Law Multivariable Version

If Two Or More Things Can Go Wrong,
They Will,
And Usually At The Same Time.

Slide 2

What Do We Mean By A Control System?
The Narrow View

A. The Measurement And Sensing Instruments,
B. The Control Valves And Other Actuators, And
C. The Control And Display Equipement.

Slide 3

While we tend to think of the control valves and similar actuators as the final control elements, these are generally only part of the final control elements. For instance, if a distillation column is required to remove a certain component overhead, and an upset causes an increase in the amount of this component in the feed, merely opening the reflux valve will not handle the problem if the column is not correctly sized. The column must be sized based on the expected upset condition it must handle and not the steady-state value shown on the flowsheet. Back in the era when chemical plants were routinely oversized as a matter of course, this excess capacity for controllability was usually available without any special attention. But in these days of tighter design, the process designer and the control engineer need to consult before the fact, not after, a point which Jim Douglas has already addressed.

Not every piece of equipment is required by the steady-state flowsheet. Some are required by transient situations created by emergency actions—quench and purge systems, vent and flare systems, etc. The sizing of these items of equipment must be based on transient considerations.

And, in addition to the moment-to-moment control system, there is an emergency interlock system whose purpose is to shut down or drastically modify plant operations should any of a number of abnormal situations occur.

The plant analytical lab is generally the final arbitrer of product quality. It also serves to provide calibration checks for the on-line analyzer systems. Also, it may be the only source for key composition measure-

What Do We Mean By A Control System?
The Broader View

A. The Measurement And Sensing Instruments,
B. The Control Valves And Other Actuators,
C. The Control And Display Equipment,
D. The Operators,
E. Process Equipment Items Functioning As Final Control Elements,
F. Auxilary Equipment Required For Start-Up, Shut-Downs, Both Normal And Emergency, And Safety,
G. The Analytical Lab,
H. The Instrumentation Personnel,
I. Etc.

Slide 4

ments required to control the process. In this case it becomes a noticeable part of the overall plant control system.

The plant may have the finest available instrumentation, but if it is not properly installed, maintained and calibrated, it is next to or even worse than useless. Fault tree analysis shows that in many cases, instrumentation reliability is the dominating factor in plant safety and operability. Thus, the instrumentation personnel are a significant factor in the quality of control that can be achieved in a given plant.

If the pump does not pump, the control valve has no flow to control. Plant maintainance personnel also become a factor in control.

All this is to say that there are factors other than the mathematical and analytical which determine the quality control which a given process and plant will achieve. These are factors which must be considered by the control engineer.

I am encouraged by both the agenda and the content not only of this session but also of the entire conference. The thrust is considerably different than it was at its predecessor held only five years ago at Asilomar. Then the question seemed to be "Why aren't we making wider use of all of these marvelous control system methodologies?" The second time around we are more concerned with how computers and their software and people fit into the picture. I think all of this is the province of overall computer control and we should not lose sight of that fact.

I am glad to see that we are beginning to pay some real attention to structure of control systems rather than just the control compensation of individual or closely related loops. I have commented before (Rinard 1975) that what passes in many quarters as process control is really unit operations control. Also, there was a long induction period between the pioneering effort of Page Buckley in 1964 and a systematic approach to the configuration of many variable systems (Govind and Power 1976, also a pioneering paper in its own way). The analytical-synthesis approach has been further developed by a couple of the leading lights of this session (Morari, Arkun and Stephanopolous 1980). However, their approach is quite abstract and, I am afraid, too formidable for most practicing control engineers to assimilate. What is needed is something a little more concrete, perhaps Ed Bristrol extending some of his previous thoughts on codifying idioms to entire plants (Bristrol 1980) or perhaps along the lines of Stainthorp and Valdman (1980) who, like Douglas, are attempting to handle the control system configuration partly from steady-state considerations.

One of the most important functions of a control system is to provide timely, unambiguous and understandable information to the operating personnel. Much is to be learned from Three Mile Island both in terms of the information presented during a crisis (too much is as bad as too little) and the need for well-trained operators (Michaelis 1980). Fortunately, there now seems to be some effort being devoted to the subject. There is a need to extract the proper information and conclusions from the plant alarm system as a guide to the operators (Andow 1980). There needs to be worked out strategies for dealing with component failures (redundant control valves, automatic control system reconfiguration, etc.). This problem will be more severe for coal processing systems than it is for the plants we are used to.

Also, I am glad to see a paper on this general topic by Dick Mah as part of this session. Dick has been concerned for some time with the problem of making sense of plant data. I can say from personal experience that it becomes very difficult to make intelligent operating decisions if you do not have a decent material balance for your current operations.

There are other considerations, some quite mundane, which may have a substantial effect on the quality of control. Control room layout and design is just as important as some of the critical control loops (Farmer 1980).

The final point I would like to make is that overall process control, and we only call it overall because in the past the term process control has been too restricted in scope, is a way of thinking as much as it is a collection of techniques. Control considerations must enter early on in the development and commercialization of any process and must continue throughout the entire history of the process. No longer can the control system be something tacked onto the process long after it has been designed and the equipment ordered. Nor must we forget that even after the plant is built and started up, there is the need to continue monitoring all facets of the control system including operator training and instrumentation maintenance. In this context, diligence is the price of optimality.

REFERENCES

Andow, P. K., "Real-Time Analysis of Process Plant Alarms Using a Mini-Computer," *Computers and Chemical Engineering*, Vol. 4, pp. 143–155 (1980).

Bristol, E. H., "Strategic Design: A Practical Chapter in a Textbook on Control," Proceedings JACC, Paper WA4-A (1978).

Buckley, P. S., *Techniques of Process Control*, Wiley, New York, (1964).

Farmer, E., "Design Tips for Modern Control Rooms," *Instruments & Control Systems*, pp. 41–43 (March 1980).

Govind, R., and Powers, G. J., "Control System Synthesis Strategies," AIChE National Meeting, Atlantic City, New Jersey (1976).

Michaelis, A. R., "Editorial-High Technology Accidents: Unpredictable and Inevitable," *Interdisciplinary Science Reviews*, Vol. 5, No. 2, pp. 79–82 (1980).

Morari, M., Arkun, Y., and Stephanopoulos, G., "Studies in the Synthesis of Control Structures for Chemical Processes," *AIChE Journal*, Vol. 26, No. 2, pp. 220–245 (1980).

Rinard, I. H., "Process Control of Highly Integrated Processes", *Conference Report—Innovative Design Techniques for Energy Efficient Processes,* National Science Foundation, pp. A48–A51, (1975).

"CONTROL SYSTEMS FOR COMPLETE PLANTS — THE INDUSTRIAL VIEW"

J. P. Shunta
Systems Consultant
E. I. du Pont de Nemours & Co.
Engineering Dept.
Wilmington Delaware 19898

A chemical plant 15 years ago was usually straightforward to instrument and control because unit operations, typically separated by storage tanks, operated autonomously. In today's plant, unit operations are highly integrated to reduce capital investment and recover heat energy. We use process-to-process heat exchangers, multi-effect distillation, and waste heat boilers that often make single variable feedback control inadequate.

The development of useful, multivariable control methods for the chemical process industries is crucial to keep up with the extra burdens being placed on control. So far, so-called modern control methods are not being implemented to any significant extent for a variety of well-known reasons; but progress is being made and the work must continue.

In the meantime, control practitioners try to bridge the gap by stretching classical, linear methods to the multivariable case. We have developed a "systems" approach to the design of control schemes which deals with all the interconnected unit operations as a single entity. The purpose of this paper is to discuss this approach and hopefully provide some insights for those involved in multivariable control development.

DEFINITION OF CONTROL GOALS

A clear understanding of the goals we are trying to achieve with the control system must precede any control strategy design. Basically, we require a process that will produce specification products at the required rates, for minimum cost, and be able to ride through upsets smoothly.

Product Quality

The only variables we normally have to control closely are product compositions. (Products also include waste streams which must be disposed of). Other variables such as levels, pressures, and flows are secondary and can be more loosely controlled to minimize process interactions or decrease the sensitivity to controller tuning.

Process Turndown

Processes are often required to run at variable rates which will affect process gains and time constants. The controls must be stable over the expected range of operation without having to reconfigure control schemes or do extensive controller retuning.

Minimum Energy Consumption

Minimum energy consumption usually implies running the process to just make specification product. Overpurification results in excess energy consumption. Therefore, we should make generous use of composition control loops.

Thermodynamics shows us that we minimize lost work by minimizing gradients in temperature, pressure, composition, etc. Recovering heat as in feed-effluent heat exchangers is one way process engineers try to achieve this. These designs result in greater process interactions.

Control Stability

Controls must work in harmony to achieve an overall dynamic balance as opposed to achieving the fastest response to a setpoint change or minimum variance, which usually cause interactions. Process interactions are one of our greatest concerns because of safety, their negative effect upon product quality, and because they are also a source of frustration for operating personnel.

Process Constraints

Exceeding process constraints can result in misoperation and damaged equipment. Special controls called overrides are often implemented to assure that the process operates within defined limits. Overrides do not shut down the process as interlocks do, but rather they take over a valve and limit its travel so that the process can operate against the constraint.

The remainder of the paper discusses an approach to control system design that tries to meet these goals.

SYSTEMS APPROACH TO CONTROL STRATEGY DESIGN

As mentioned earlier, it is futile to design controls around individual pieces of equipment without considering what happens upstream and downstream. Control engineers who modernize a control system in an existing plant are often frustrated with the problem of not being able to make broad enough changes to get the most improvement. They are constrained by what already exists. We do not have that problem on grass roots plants if we get involved early enough in the project so that equipment design can be approached with control principles in mind.

The following approach for control system design has evolved over many years of working on new projects. Many of the principles have been previously stated in papers by P. S. Buckley [1, 2,]. A fundamental precept in this approach is that the control engineer and process engineer collaborate on the control system design so that piping and equipment designs are amenable to good control.

Understand the Process Requirements

Before control system design can begin, the control engineer must determine certain things about the process.
- The main flow through the plant must be determined first; then secondary flows and recycles. The main flow sets the production rate around which all the other controls are laid out.
- The location of storage tanks is important because they provide a natural division of equipment. Equipment between storage tanks is considered the "system" for which the controls are designed.
- The product streams, including waste streams, must be identified along with the desired purity specification. The location of on-line process analyzers must be specified if they are to be used for composition control.
- Turndown requirements affect process gains, time constants, and transport lags which dictate controller tuning. It is not unusual to analyze process dynamics at maximum and minimum rates to make sure the control system will be satisfactory across the expected range of operation.

- Complex equipment configurations and expected control problem areas need to be identified for special studies.
- The availability of measurements affects control strategy design. Troublesome measurements need to be identified so that proper measurement instruments can be obtained. In some cases, we must calculate a variable from measurements of other variables.

Material Balance Controls

Having determined the main flow through the process (the one that sets production rate), the inventory or material balance controls are designed. These controls move the main flow from raw material storage through the process to final product storage. They consist of a combination of flow and level controls. All the other controls depend on how the material balance controls are executed.

An important consideration for liquid level control is the speed of response. If a manipulated flow goes into storage, the level can be controlled tightly; the level controller can have a fast speed of response. If, however, the flow feeds a process vessel like a distillation column, we do not want a fast speed of response. The liquid inventory is sized to accommodate slow, gradual level control to filter out upsets rather than propagating them. Whenever economically feasible, we design the equipment to minimize interactions rather than placing an additional burden on control.

Another consideration for level control when more than one inflow or outflow is present, is the capacity of a variable to handle upsets. For example, bottoms flow or vapor boilup can control base level in a distillation column. We prefer bottoms flow; but, if it is only one-third or less of the boilup, boilup is preferable. This is not always a very controllable loop, depending on whether or not a severe inverse response occurs. It can happen that when boilup increases, the liquid on the trays momentarily froths up and more liquid dumps over the weir. This increases base level and confuses the controller. Enough surge capacity is provided so boilup can be changed gradually to minimize the problem.

Pressure controls regulate inventory in the vapor space. These controls are selected for the particular operating pressure level. For example, a vacuum column pressure may be regulated by a vacuum pump or steam jet, while a pressure column may be controlled by a vent valve. It is important to keep pressures fairly constant, particularly in distillation columns, so that temperature measurements are more indicative of composition, and the heat balance is not upset.

Product Quality Controls

Product quality controls are important not only to get salable product but to minimize energy consumption. It is generally true that minimum energy consumption occurs when all products just meet product specifications. If automatic control is not provided, operators will use excess energy to keep them on the safe side when upsets occur.

It is well known that multiple composition controls on distillation columns cause interactions. Decoupling methods have been the focus of much research in the past 10–15 years but a generally satisfactory solution is still not available. Mathematical methods have been developed that work in certain circumstances and not in others. One of the problems is the nonlinear behaviour of distillation columns which demands that the algorithm adapt to changing process parameters. Without automatic decoupling, we detune the least important composition controller to minimize interaction.

The proper pairing of manipulative and controlled variables has been the subject of many papers. In our experience, there is no substitute for actual plant data to determine which variables go together. We usually do not have that luxury in plant design. In some cases we do not have the freedom to choose anyway because the material balance controls have already dictated the pair. If, for example, we want to control top composition in a high reflux ratio column, we will use distillate flow because reflux has already been selected to control reflux tank level.

Ideally, product composition measurements are made with an online composition analyzer. Many times, a suitable analyzer cannot be found so we deduce composition some other way such as by temperature in a distillation column. The sensitivity of temperature to composition must be determined so that the temperature transmitter is located properly. Temperature is pressure-compensated if pressure is expected to vary. A differential vapor pressure transmitter is another measurement technique that has been used.

Inverse response can appear when bottoms composition is controlled by vapor boilup. In order to drive lights from the base, boilup is increased, but in so doing, liquid from the trays momentarily froths up and drops into the base, further increasing the lights. Composition cannot be controlled too tightly if this occurs. We have a computer program that predicts inverse response in a column.

Feedforward control enhances control of composition. The simplest form of feedforward control is flow ratio control such as steam/feed and reflux/feed ratio control. The controls begin to compensate for a feed rate upset before an error is generated in temperature or composition. Compensation is required on the feed signal to match the dynamics of the equipment. We normally use a first order model of the dynamics to keep the controls straightforward.

A feedforward controller on feed composition requires a feed composition measurement which we normally do not have. Sometimes we attempt to correlate some other measurable characteristic such as pH or density to composition.

Secondary Controls

Controls for secondary flows and temperatures are determined after all the more critical controls are defined. Examples are heat exchanger temperature controls and recycle flow controls.

Constraint Controls

There are certain equipment and operational constraints that impose boundaries on the range of control. For example, boilup in a distillation column is limited by conditions of flooding, weeping, degradation of material in the base, etc. So, in addition to the normal controls, constraint controls or overrides are superimposed on the control system. This is really a form of multivariable control. A steam valve, for example, may be manipulated to control steam rate, column pressure drop, base temperature, base level or column pressure depending on which variable is at its allowable limit. Two recent papers illustrate the concept [1, 3].

Startup Controls

Startup logic and controls are the last type to be designed. These ramp open steam valves on reboilers, and transfer from total reflux to normal operation on columns. Overrides make sure that the proper conditions exist before startup is initiated such as a minimum liquid level in the column base.

Process Modelling and Dynamic Studies

Dynamic studies are done in critical control situations to verify that the proposed design will be adequate. Dynamic models range from simple, low order linear ones to high order, nonlinear models that are solved numerically on the computer. Simulations compare alternative control strategies and investigate the process under various types of upsets. These approaches are really necessary when the process is so highly nonlinear or interactive that it is impossible to predict the behaviour by any simpler means.

On-line models predict unmeasurable variables for feedback control. In one plant, an on-line model predicts the product composition from a reactor which is used in a feedback loop to reset an exit temperature controller.

FUTURE NEEDS IN CONTROL

Process models are an important part of advanced control. We need them to predict what is happening in the plant, and for closed loop control. Models can predict operating parameters like inverse response, flooding, and weeping in distillation columns. Such information could then sound an alarm or be applied to override controllers. So far, not much use is make of models for these purposes.

Advanced control methods are needed to handle nonlinearities, variable process parameters, and interactions, and be simple enough for plant people to understand. Otherwise they will not be used. Some override controls have been abandoned over a period of a few years not because they were unnecessary, but because turnovers in plant personnel led to loss of understanding. Technical people are especially subject to turnovers.

Measurements have been and still are a weak link in the control chain. Development of accurate, versatile measuring devices is woefully slow. We would welcome more interest among the academicians in this area.

REFERENCES

[1] Buckley, P. S., "Status of Distillation Control System Design," ISA Chempid Meeting, Houston Texas, (May, 1978).
[2] Buckley, P. S., "A Systems Approach to Process Design," AIChE All-Day Symposium, Princeton, NJ, (October, 1978).
[3] Buckley, P. S., "Multivariable Control in the Process Industries," Purdue University Conference, (May, 1975).

THE EFFECT ON CONTROL SYSTEMS OF VIEWING THE PLANT AS A SYSTEM OF INTEGRATED UNITS

T. E. Marlin
Exxon Research and Engineering
Florham Park, N. J.

Integration of plant equipment is essential in satisfying today's demands for reduced energy consumption and maximum utilization of expensive feedstocks. The standard industrial approach when evaluating complete plant control systems is to decompose the plant into a system of integrated units. Each separate unit consists of a limited number of unit operations like distillation columns, reactors, and heat exchangers. The process dynamics and control of each unit are presumed well enough understood to allow analysis of the integrated units.

Integration among units occurs through many processing configurations which are summarized in Figures 1 to 6. The first two categories—sequential and recycle processing—are thoroughly investigated in the current control literature and the third—process heat integration—is now being analyzed although definitive results have not been achieved. The remaining three categories—utility energy distribution, fuel distribution, and product blending—receive little attention in theoretical papers but present challenging control problems. The influence of these six configurations in even a simple plant can result in strong interactions (Figure 7).

Two examples serve to illustrate the differences between those categories commonly investigated and those not. The example in Figure 8 demonstrates a typical research problem, a heat integrated distillation

Figure 1. Sequential Processing.

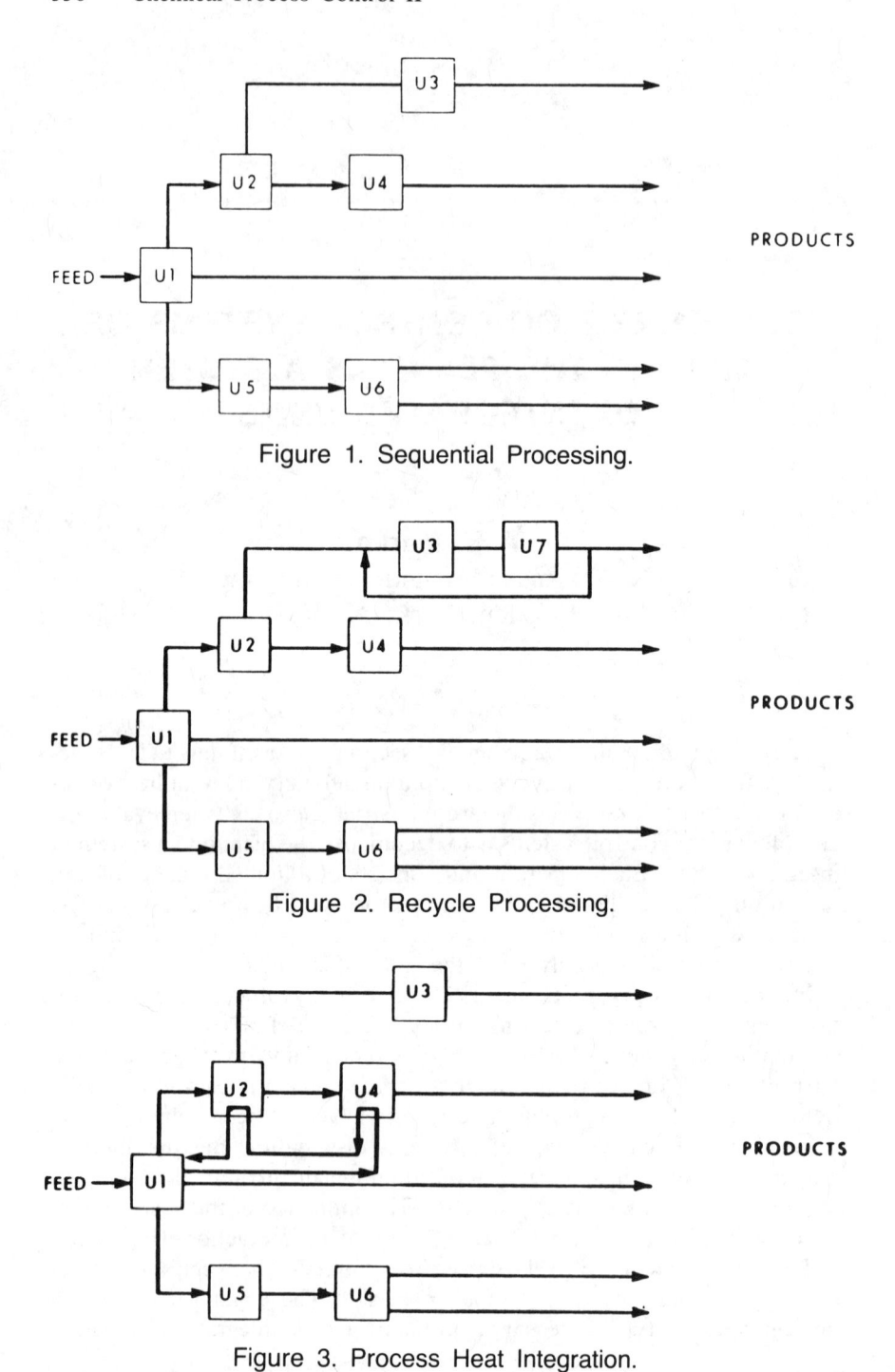

Figure 2. Recycle Processing.

Figure 3. Process Heat Integration.

T. E. Marlin 557

Figure 4. Utility Energy Distribution.

Figure 5. Fuel Distribution.

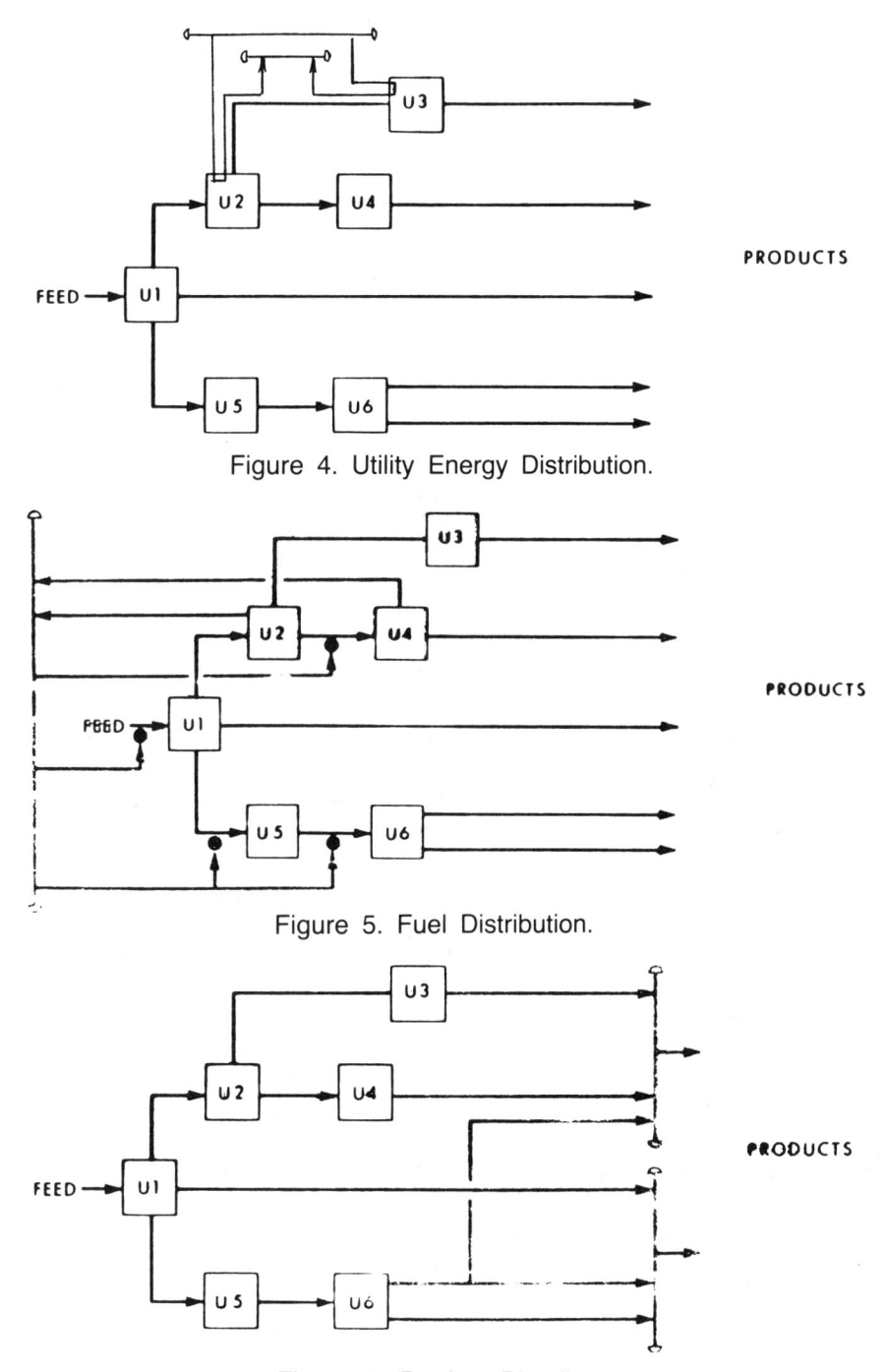

Figure 6. Product Blending.

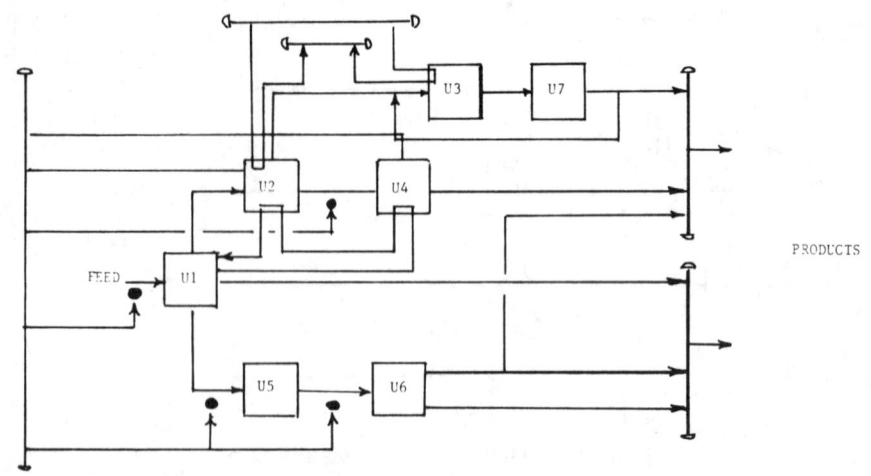

Figure 7. Fully Integrated Plant.

tower. The example in Figure 9 demonstrates the higher degree of integration in utility energy distribution. The correct amount of refrigeration supplied to each unit depends on the status of all other integrated units, the refrigeration compressor, and the steam system. Steam systems, refrigeration systems, hot oil belts, and fuel distribution systems are the most commonly encountered complex integrating configurations. Such systems are characterized by a wide scope of integrations, strong interaction, and the need to satisfy unit energy requirements rapidly.

Several plant design methods are commonly used to attenuate interunit upsets (Figure 9 and Table 1). These methods usually increase capital costs and, through reduced integration and subsequent increased utility consumption, increase operating costs.

Based on this brief review, several conclusions are drawn concerning additional technology needed to allow practitioners to realize fully the potential of unit integration. Better design techniques are needed to allow analysis of many dynamic responses without detailed simulation and to facilitate simulation when required (Table 2). Advanced control technology is needed for multivariable control systems and for optimizing integrated units (Table 3). Hardware and software requirements center around the need for an integrated system capable of controlling entire plants (Table 4).

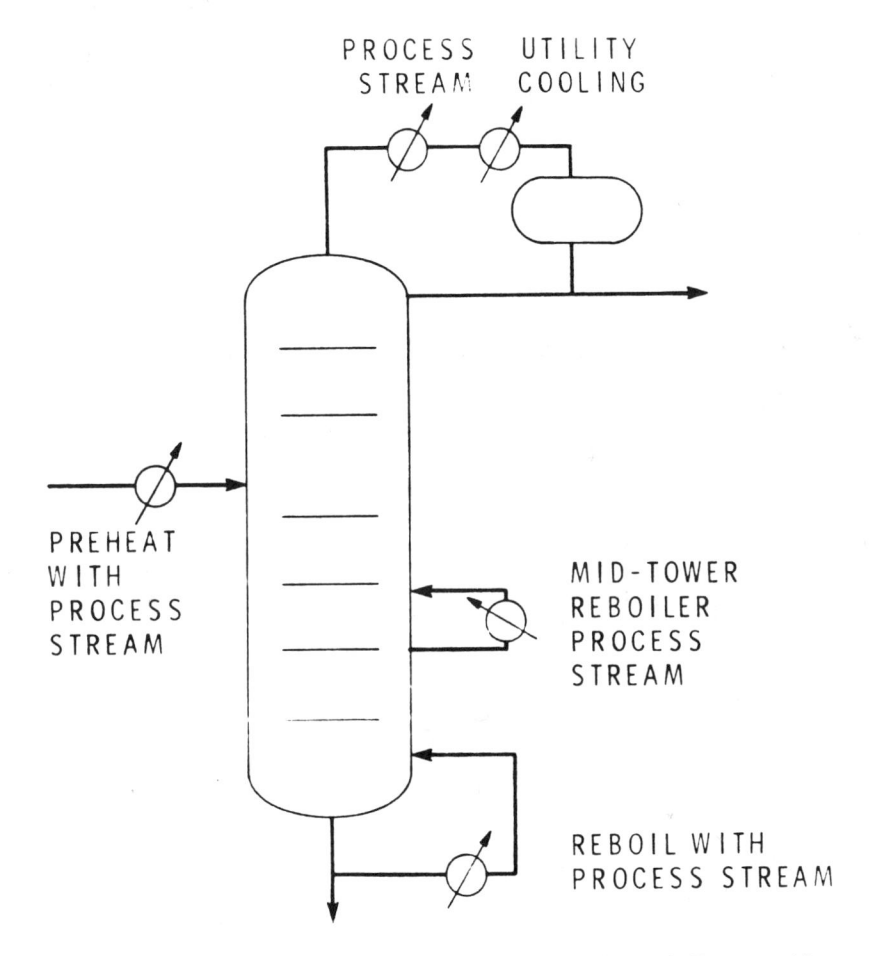

Figure 8. Example of Distillation Tower Inegrated through Process Heat Integration.

Finally, the crucial component in successful computer control, as in all endeavors, is the creative insight supplied by skilled, motivated individuals. I hope that, as a result of this conference, we take steps to attract more students to the flourishing field of process control.

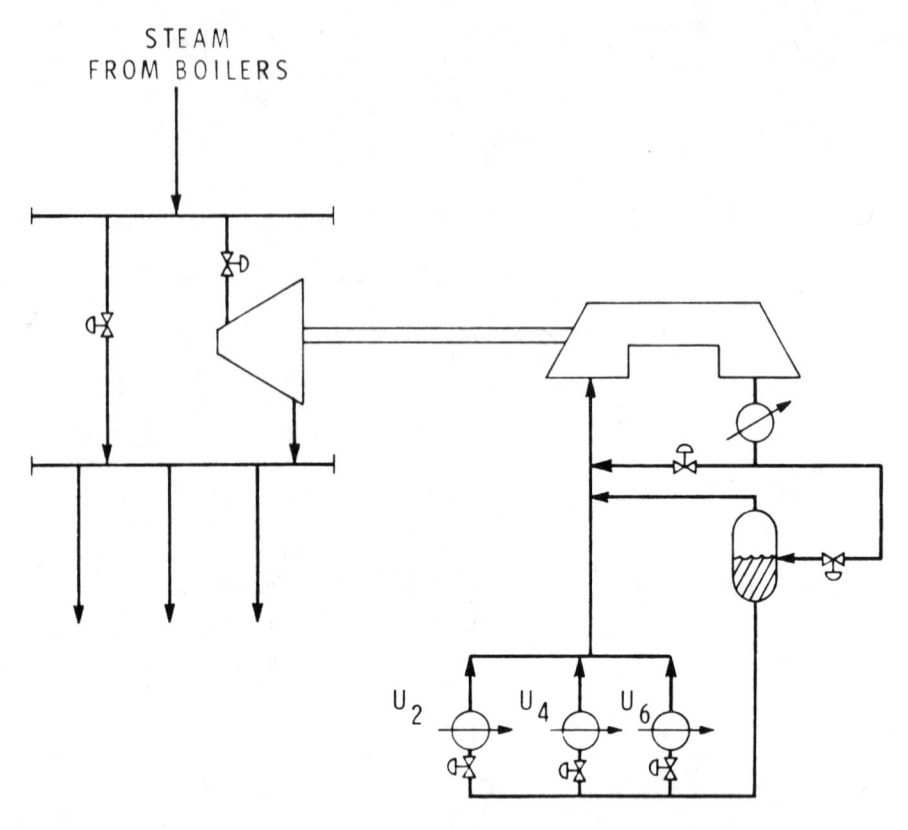

Figure 9. Example of Process Units U_2, U_4, and U_6 integrated through Refrigeration System.

Table 1. Design Variables Used to Attenuate Interaction.

- Liquid Holdup In Tanks And Drums
- Vapor Holdup In Lines And Vessels
- "Trim" Sources Of Heat And Cooling
- Alternate Fuel Sources

Table 2. Plant and Control Design Needs.

- Better User Oriented Software To Support Dynamic Simulations
- Accurate, Flexible Modules Modelling Dynamic Response Of Common Unit Operations
- Analysis Techniques For Dynamic Response, Sensitivity Analysis, And Control Design Not Requiring Detailed Simulation (Similar To Eigen- Values, Relative Gain)

Table 3. Control Technology Needs.

- Multivariable Control For Plants With Hard Constraints And Widely Differing Dynamics Among Variables
- Flexible Strategies Which Respond To Various Objectives And Constraints (Perhaps By Effectively Changing Structure)
- Improved Analysis Methods For Optimizing Integrated Process Units

Table 4. Control Hardware/Software Needs.

- Access To Information From Entire Plant For Integrated Computer Control System Rather Than Isolated Control Computers (Serving Units, Utilities, Tanks, Laboratories, Etc.)
- Support Software For Various Applications In One System (Continuous Control, Batch, Optimization)
- Much Greater Data Processing Capacity

Distributed Computer Process Control

THE STRUCTURING OF DISTRIBUTED INTELLIGENCE COMPUTER CONTROL SYSTEMS

Charles W. Rose
Department of Computer Engineering and Science
Case Institute of Technology
Case Western Reserve University
Cleveland, Ohio

Although several distributed intelligence computer control systems have been introduced commercially, the processes for creating them are largely *ad hoc*. Data Acquisition and control applications place stringent constraints upon the topological and logical structuring of the systems in the areas of modularity, reliability, and performance. Contention-based architectures, although new, have particular promise.

INTRODUCTION

Over the past four to five years, the confluence of theoretical developments in decentralized control, the maturation of powerful, low cost microprocessors, and the availability of reliable communications technologies has resulted in a great interest in distributed intelligence computer control in both universities [1, 2, 3] and industry [4, 5]. The process of developing these loosely-coupled multiprocessor architectures is still *ad*

566   Chemical Process Control II

hoc and, in practice, most available systems are deficient, particularly in their approaches to reliability [6, 7].

In this paper, we shall identify the particular requirements that data acquisition and control applications place upon a distributed intelligence architecture and then examine their implications upon the topological and logical structure of a system. Finally, we shall discuss a particular class of architectures, contention-based busses, which appear to hold significant promise.

REQUIREMENTS

Figure 1 represents a stylized, hierarchical distributed computer control system. While the requirements that the control application places upon the architecture generally hold at each level of the hierarchy, the amount of information communicated between the supervisory level and higher levels and the frequency of updating is clearly lower than at the

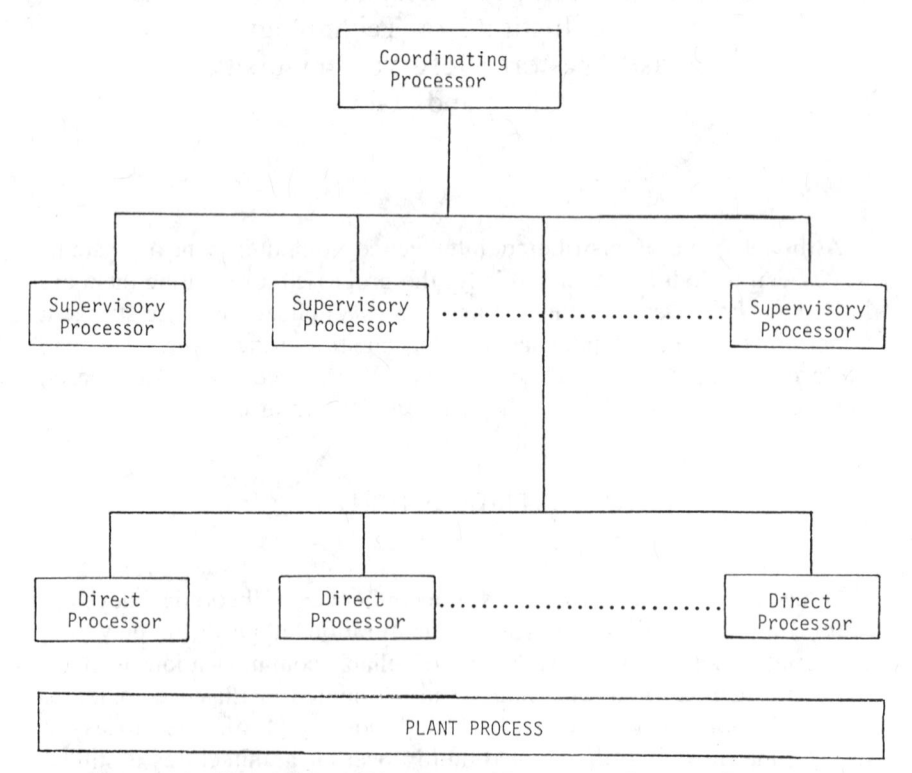

Figure 1. Hierarchical Distributed Computer Control System

direct and supervisory control levels. Furthermore, since the lower levels can usually maintain functions for a reasonable time in the absence of higher level control, reliability and availability constraints are less demanding at higher levels.

For these reasons, we shall concentrate our efforts on the requirements imposed at the direct and supervisory control levels.

Reliability

Clearly, a prime requirement on any computer control system is reliability. What is meant by reliability and how best to achieve it, however, is not as easily stated. System vendors frequently refer to mechanical (hardware) failure rates in terms of Mean Time Between Failures or system availability [8], but these figures lose significance as more of the system is realized in software. Users often have their own peculiar measures—not more than N variables out of control, or not more than M loops non-communicating, for example.

System architects tend to employ a combination of fault avoidance (high reliability parts), fault tolerance, and fault containment techniques in a rather *ad hoc* way to achieve reliability. The result is that commercially available systems tend to have idiosyncratic reliability deficiencies because systematic approaches to reliability design are not used [6]. In fact, most vendors routinely recommend that a second, redundant system be used to achieve required reliability.

The entire problem of designing reliable, distributed intelligence hardware/software systems requires a great deal of study. It is clear, however, that any suitable architecture must support graceful degradation. That is, its mechanism for detecting and containing failures must be sufficiently well-developed that single point failures will not flunk the entire system and that loss of function and/or performance resulting from a given failure will be minimized. These characteristics are sometimes referred to as failure effect and failure reconfiguration, respectively [9].

Response Time and Performance

Computer control systems are, by their very nature, real time, and for a given application, their response time requirements are well-defined. In a uniprocessor, centralized architecture, the processor speed, interrupt response time, operating system overhead, swapping channel speed, frequency of task execution, etc. are all determinates of the response time of the complete control system. The methods for achieving a given level of performance are well understood.

In distributed intelligence systems, the issues become more complicated. Intertask communication is subject to interprocessor communications delays. Synchronization among tasks can no longer be easily accomplished using a single real time clock interrupt. Interrupts themselves frequently must be ultimately handled by tasks which are not resident in the processor which initially receives the stimulus, *eg.* alarm conditions. At the same time, the software structure within a given processor should be simpler since much of the complexity of multiprogrammed, memory managed operating systems can be avoided.

In summary, real time performance requirements dictate that distributed intelligence architectures have minimum internode communications delays, have provision for interrupts for high priority messages, and have some means for global synchronization.

Modularity

One of the potential advantages of distributed intelligence systems is that functionality and performance can be incrementally improved by adding processing and sensor/activator nodes. Computer control applications require that these additions be accomplished *inexpensively wherever* they are needed. Inexpensively implies that the hardware interface and cabling costs should be low and that software modifications costs should be minimized. Wherever implies that elements should be easily added at their points of application in the plant, refinery, or airframe, etc. These characteristics are referred to as Cost and Place Modularity, respectively [9], and are extremely important for control-oriented architectures.

TOPOLOGICAL CONSIDERATIONS

A wide variety of topologies for multicomputer architectures has been either implemented or suggested. Anderson and Jensen [9] suggested a taxonomy, shown in Figure 2, relating these architectures in terms of 1) whether the paths are direct or switched; 2) if switched, whether the paths are shared or dedicated.

The requirements of distributed intelligence data acquisition and control imply that switched systems, particularly these employing store and forwarding are not acceptable. The delays imposed by the intermediate hops tend to be unacceptable, and thus broadcast messages cannot be used for synchronization purposes. Furthermore, the complexity of switched architectures tends to lower reliability unless redundant switches and paths are added at additional cost.

Direct topologies avoid these problems, but several of them fall prey to other requirements of the control application. For example, the direct dedicated-complete interconnection topology (DDC), in which a dedi-

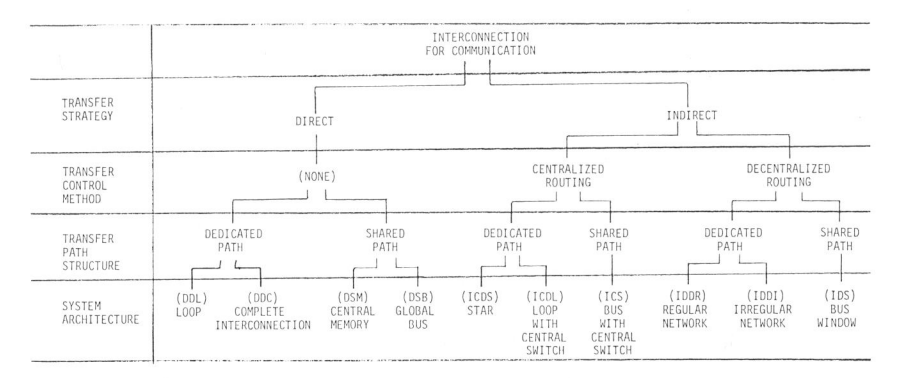

Figure 2. Taxonomy of Interconnection Topologies

cated path exists between every pair of direct-shared-central memory topology (DSM) cannot be geographically distributed.

The DDL topology or loop, on the other hand, may be distributed and has good cost and place modularity. A sensor/PE may be placed anywhere on the loop with a simple interface and communicate with every other node. Since messages are passed from node to node successively, the failure of a single node or path between two nodes can bring down the entire ring. Thus, for unidirectional, active repeater loops, the failure effect and failure reconfiguration attributes are poor. Passive repeater systems, in which each node simply looks at each message as it goes by to determine whether it is addressed to that node, are less vulnerable to PE failure, and redundant path-loop collapsing mechanisms may be added to improve the graceful degradation of loop. These additions increase the cost, of course.

The logical complexity is quite low. This topology is subject to bottlenecks if the message traffic between adjacent or nearly adjacent nodes on the loop is high for sustained periods.

The primary advantage of this topology is the cost (since most loops are bit or byte serial) and the high modularity. For passive repeater loops, a PE failure can be masked by load sharing among the remaining elements if process or task addresses are soft; that is, if they are not bound to one processor but are kept in a table and checked by the communications software in each PE.

The Direct shared bus, DSB, or global bus, is very attractive for distributed intelligence data acquisition and control systems. Access to the bus is shared by the PE's by some allocation scheme and messages are sent directly to the destination node.

Cost and place modularity of the communications bus itself is poor in that the total bandwidth is fixed and increasing it would require extensive changes to PE interfaces and perhaps to the communications medium itself. However, for data acquisition and control systems, acceptable bandwidth (to avoid bottlenecks) can be provided, if at all, at design time.

Failure effect for PE's is good, provided that the failure does not hang up the bus. If soft process addresses are used, the dropped load may be picked up by other processors.

As with DDL topologies the failure characteristics of the bus itself are poor. Replication may be costly depending upon the width of the bus.

The primary advantages of this topology for data acquisition are the cost and place modularity of PE's and PE failure characteristics.

It is clear, therefore, that of the DX topologies, DDL and DSB are the principle candidates for data acquisition and control systems because of cost and place modularity for PE's, low overall cost arising from little logical complexity, and good fault reconfiguration characteristics for PEs (this is only true for passive repeater loops, however) in the presence of soft process addresses.

Their disadvantage lies in the vulnerability of the topologies to communication path failures. Redundancy of paths can mitigate this disadvantage, particularly in bit or byte serial systems.

Indicative of the suitability of DDL and DSB topologies for distributed data acquisition and control systems is the fact that of those systems commercially available or proposed as standards, all utilize one of the two topologies.

For example, the Camac standard [10] developed for nuclear instrumentation is a DSB system, and a DDL serial CAMAC standard [11] has been proposed.

FOXBORO's SPECTRUM [12] and Honeywell's line sharing system, TDC 2000, [13] are both DSB's.

LOGICAL CONSIDERATIONS

The logical structuring of a distributed intelligence architecture is a critical factor in determining its performance, flexibility, and reliability and maintainability. The logical structure of a generalized distributed system (network) is shown in Figure 3a [14]. In this section we shall examine the implications of data acquisition and control applications upon the elements of this structure.

At the outset, it should be noted that proper structuring of a network requires that, in any pair of nodes, the corresponding protocol layers

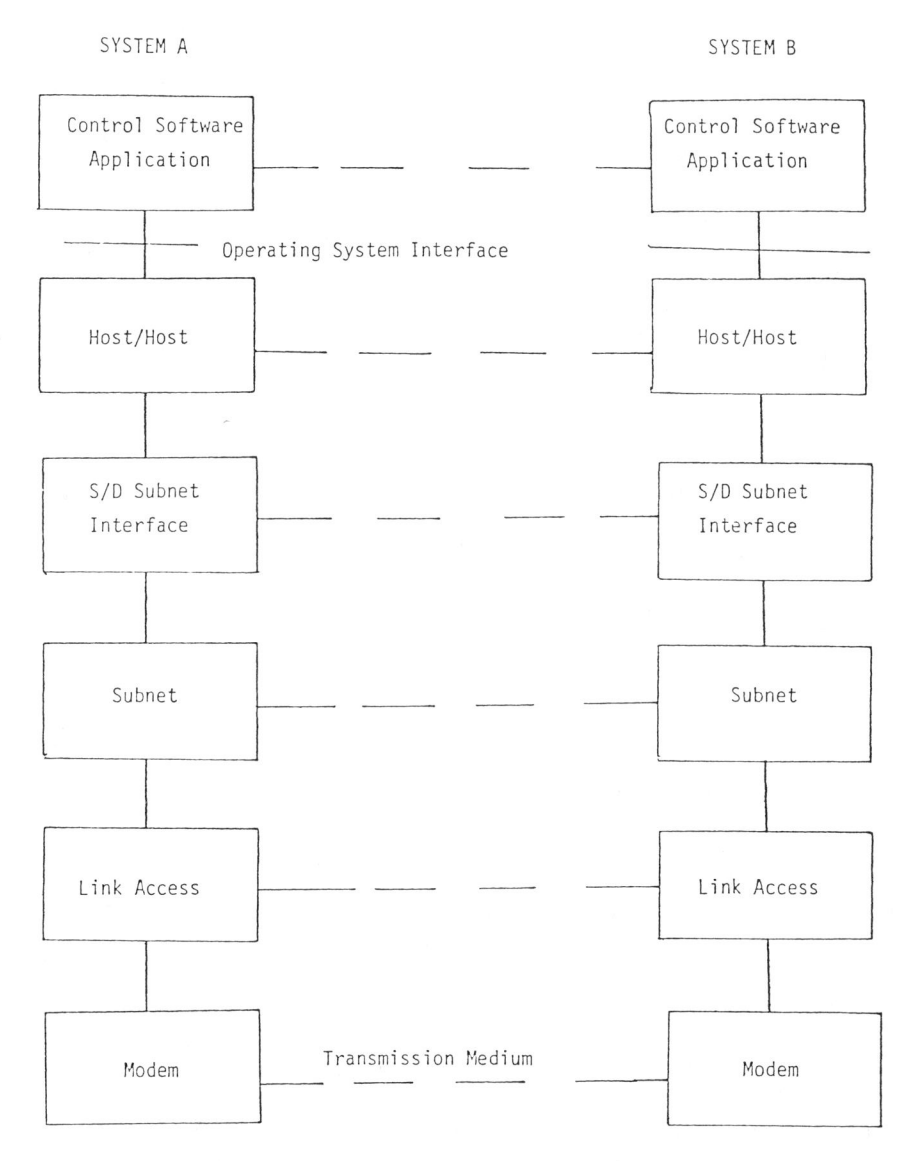

Figure 3a. Logical Structure of Computer Network

communicate only with each other. The importance of this constraint will be seen later.

At the highest layer are the APPLICATIONS TASKS. In control applications these would include operator communications, supervisory

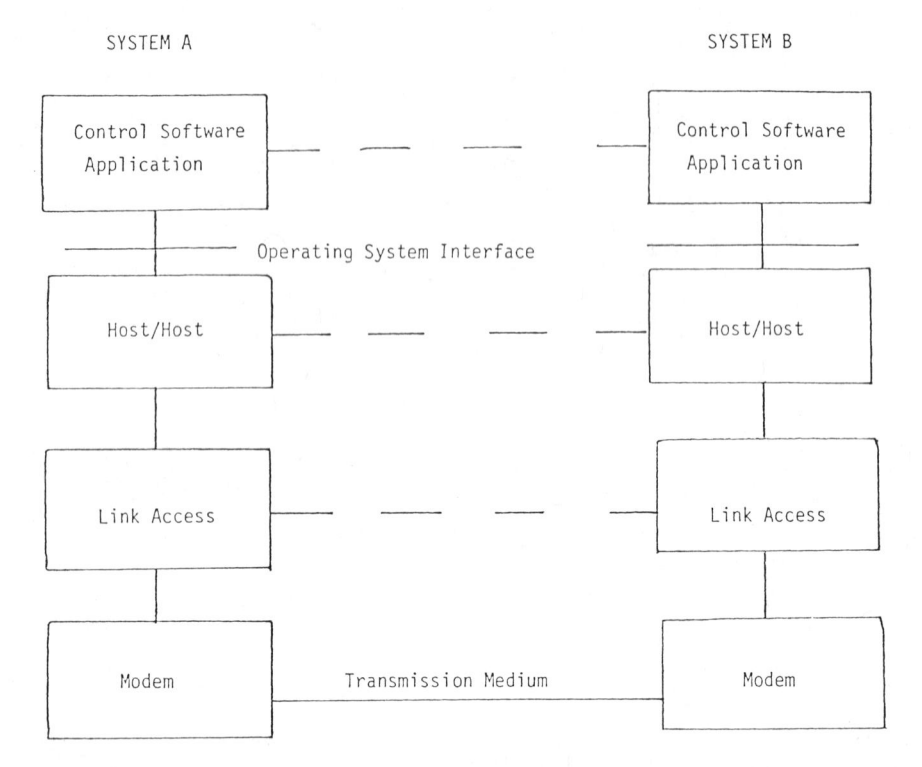

Figure 3b. Logical Structure of Directly Connected Network

control module, data acquisition tasks, etc. The intent is that they should communicate as defined by the applications without *a priori* knowledge of network structure beneath them. Ideally, the intertask communication interface at the operating system level should be the same whether the target task is co-resident or remote. This will allow dynamic reconfiguration in the event of failure and will allow the tasks to be distributed across a varying number of PE's as performance requirements dictate without reprogramming the applications tasks.

Moving now to the bottom of the structure, the TRANSMISSION MEDIUM is the physical path over which messages are sent. For control applications, the medium is typically twisted pair wire or coaxial cable. Coaxial cable is used in most commercial systems because of its relatively high noise immunity and high bandwidth capacity. As demand increases and prices drop, fiber optics will become an increasingly important transmission medium. Presently, taps which will allow signals to be "bled off" and inserted have not been commercially perfected, and the cost of fiber and interfaces is about twice as expensive as coaxial cable for comparable capacity.

The MODEMS connect the digital signal to a form appropriate for transmission, either baseband or modulated. Modulating the signal increases the cost of the MODEM, but permits multiplexing several channels of data on the same physical channel, and, as we shall see later, this allows the transmission of synchronizing messages, etc.

The lowest layer of protocol is at the LINK ACCESS level. The function of this protocol is to transfer information across a transmission link sequentially and without error. Link access protocols manage flow control—the rate of insertion of messages onto the link and the mechanism for insertion—and error detection and correction which is usually accomplished by a redundancy check in the message with retransmission and timeout on errors. Link access protocols are well-developed [15] and fall into three main classes—byte or character-oriented, bit-oriented, and count-oriented. Byte-oriented protocols are suited to asynchronous transmission schemes; IBM's Binary Synchronous Control Protocol [16] is a typical example. Bit-oriented protocols are suited to synchronous communications and are fast becoming the *defacto* standard for most network applications; IBM's Synchronous Data Link Control Protocol [17], the virtually identical ISO High Level Data Link Control Procedures [18], and the ANSI Advanced Data Communications Control Protocol are prime examples. Finally, Digital Equipment Corporation's Digital Data Communications Message Protocol [19] is count-oriented and was developed for use in DECNET with both asynchronous and synchronous transmission schemes. Figure 4 illustrates the general format of byte-, bit-, and count-oriented messages.

In virtually all new systems introduced since 1975, logical messages between tasks, which may be of arbitrary length, are broken into fixed length packets, each with its own source and destination information so that it can be routed independently through the network. This technique is called packet switching [20], and, by appropriately selecting the packet sizes, an acceptable trade-off between network response time and throughout can be achieved.

In control applications, short interrupt and operator communications messages will be interspersed with much longer messages. In order to achieve acceptable response time for these short, high priority messages, packet switching should be used, and the link access protocol drivers (in either hardware or software) must be able to insert high priority packets onto the link ahead of those of lower priority.

The COMMUNICATIONS SUBNET in a network is responsible for routing packets to their ultimate destination through switching nodes. The subnet protocol provides the necessary information to these nodes. We

574   Chemical Process Control II

| Control Char | Header | Control Char | Information | Control Char | Frame Check Redundancy |

| Address | Sequencing Control | Control Fcns |

Byte or Character-Oriented Protocol

| Frame Flag | Header | Information | Frame Check Redundancy | Frame Flag |

| Address | Sequencing Control | Control Fcns |

Bit-Oriented Protocol

| Control Char | Header | Header Check Redundancy | Information | Information Check Redundancy |

| Address | Sequencing Control | Control Fcns | Info Char Count |

Count-Oriented Protocol

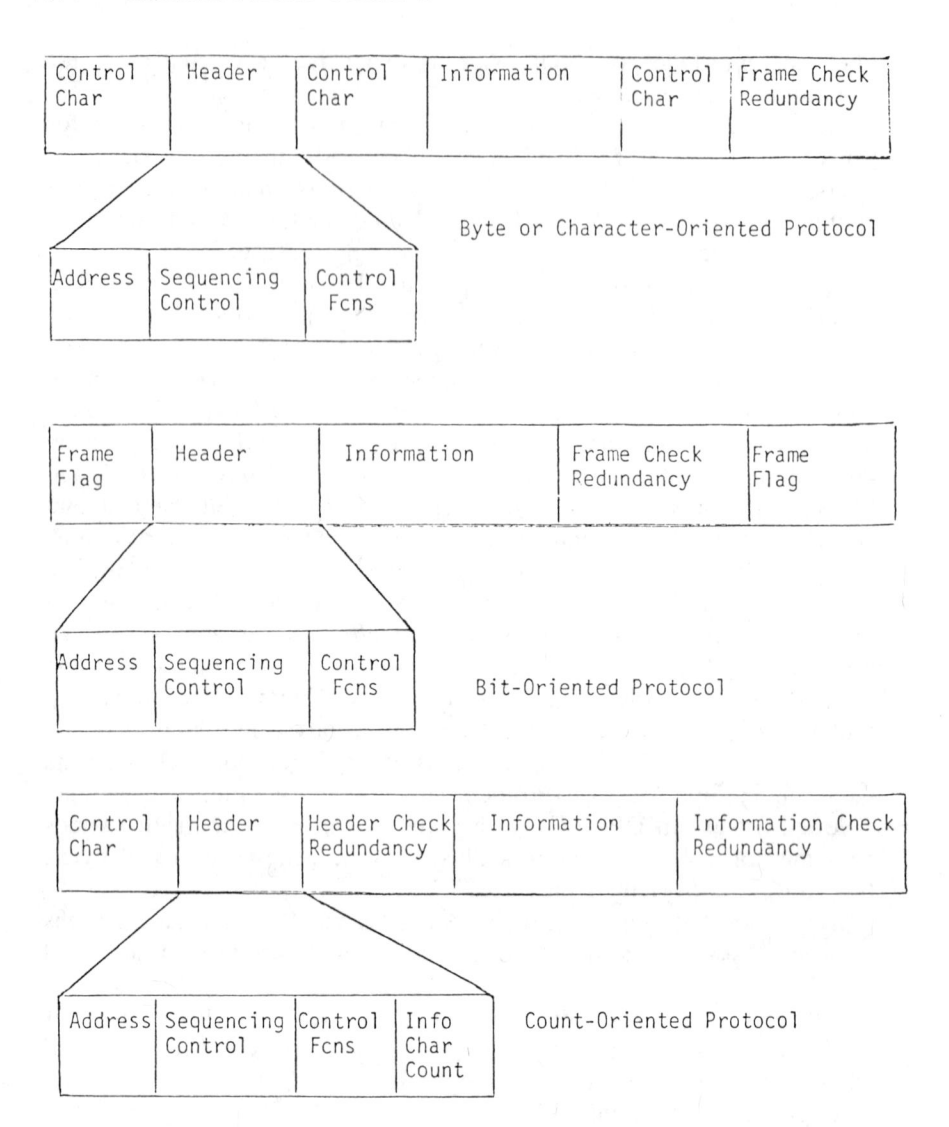

Figure 4. General Formats of Link Access Level Protocol Messages

have already determined that switched, or indirect, topologies are not acceptable for our purposes. Thus, in distributed architectures for computer control, the subnet is degenerate, and the logical structure is as shown in Figure 3b.

The most important protocol layer, and unfortunately, the least well-understood, is the HOST to HOST or task to task protocol. For our purposes we will define a task to be any addressable entity in a PE, that is, a program, a device, or a file. Tasks tend to communicate in one or both of two modes—session mode in which a number of messages which are related pass beween them over time and are sequenced, and single message mode where single, unrelated messages are occasionally sent. The latter mode is frequently referred to as Datagram mode in which no sequentiality of message delivery is guaranteed. In some cases, delivery is not guaranteed. Thus Datagram node is in many ways analogous to the post office system.

Session-oriented communications, on the other hand, are analogous to the services provided by the telephone system. To accomplish this, the host-to-host protocol driver accepts requests from a task to create and maintain a logical channel between it and another task. Should that task not be coresident, the host-host protocol will establish, with its counterpart at the target PE, a logical channel or pipe over which the dialogue will take place. Of course, if the destination task is unwilling or unable to enter into dialogue with the source task, the channel will not be established and the source will be notified.

The existence of a logical channel for each dialogue guarantees that messages will be delivered sequentially and that flow control can be maintained for each dialogue. That is, that messages will not arrive for which the destination has no buffers. One might ask why the link access level flow control will not suffice. The answer is that each task may have several dialogues active at the same time and in a given PE, several tasks may have active dialogues with tasks in the same target PE. Were the flow control mechanisms not independent, a problem in any single dialogue could cause delay or abortion of the others, thus violating the graceful degradation requirement.

There are no standards for host-host protocols, and there are no accepted measures for evaluating them. Both ARPANET [21] and DECNET [22] have developed host-host protocols, but they are quite different in detail, and CCITT has prepared a host-host protocol interface as one level of its recommended standard interfacing terminal equipment to packet node communications facilities X.25 [23]. Two of the major difficulties in host-host protocol development in general are that, 1) if the network is to be transparent to the tasks, the interface must be integrated into the operating system which requires costly software design and parallel maintenance of a standard product, and 2) if the PE's are heterogenous, addressable entities will differ between PE's.

Given today's state of knowledge regarding host-host protocols, a computer control system vendor shoud be able to design one for a homogenous, distributed intelligence architecture with custom operating systems software which has all of the necessary features to support the application. As of now, however, no commercially available system supports a well-developed host-host capability.

In summary, proper logical structuring of a network requires that the structure and protocol at any interface be transparent above and below. For example, the link access protocol structure should be unknown to the host-host protocol except for the calling interface, and the content and structure of a host-host message should be transparent to the link access protocol. In other words, host-host messages are imbedded in the data portions of link access packets as shown in Figure 5. In this way, protocols can be modified on one level without impacting other layers, reliability is improved and maximum flexibility is maintained.

CONTENTION-BASED ARCHITECTURES

One aspect of the topological and logical design of distributed intelligence systems which dramatically impacts performance, reliability and modularity is the method by which access to the DSB or DDL is arbitrated. Traditionally, in computer control architectures, access is arbitrated by a centralized bus controller as in SPECTRUM [12] and TDC 2000 [13] or by passing a token around the ring as in CAMAC [11].

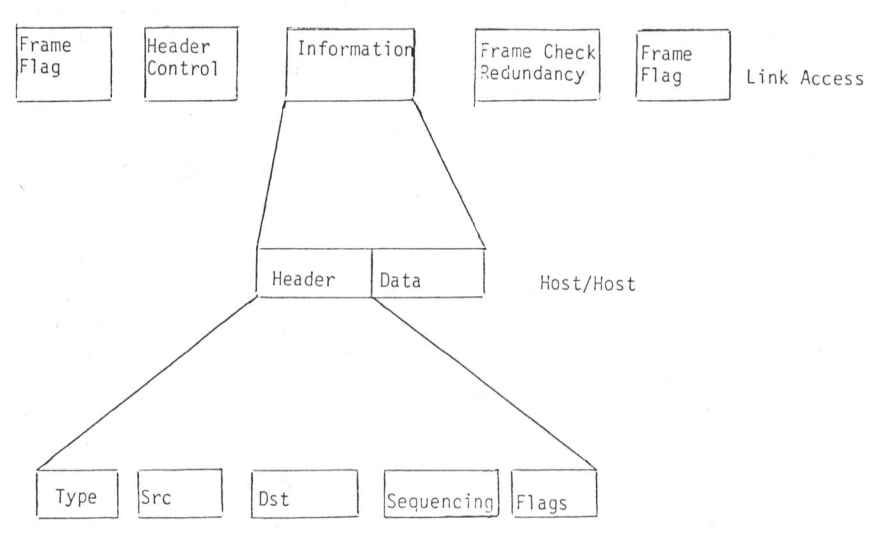

Figure 5. Imbedded Protocol Messages

Recently, another access scheme based upon packet switched radio technology developed by DARPA in the ALOHA project [24] has been exploited commercially in local area computer communication networks. This scheme appears to have many very attractive characteristics for distributed intelligence computer control.

Briefly, a DSB topology is usually used in which there is no centralized arbitrator. In the original pure ALOHA scheme, if the link access protocol control module (LAM) in a PE has a packet to transmit, it simply does so. If it does not receive an acknowledge, it waits a random time and retransmits again. Obviously, collisions among packets will occur, and many packets will be garbled, but the random time outs guarantee that sooner or later, every packet will be transmitted to its destination. As traffic increases, however, so will collisions, and the efficiency of the system degrades badly—the maximum line utilization drops to about 18% [25].

More modern systems avoid many collisions by requiring each LAM to check the bus for the presence of carrier before inserting a packet. There are several variations to this scheme, but essentially, a LAM never inserts a packet when it can detect that the bus is busy. With this variation, called Carrier Sense Multiple Access (CSMA), line utilization of about 90% can be achieved [25].

Since the bus has a non-zero propogation delay, it is still possible for collisions to occur when two PE's detect no carrier and transmit within the propogation delay. A further enhancement, collision detect or listen while talking, allows a LAM to truncate a packet if it detects a collision or transmission error while sending a packet. This technology, CSMA/CD, represents the state of the art in contention DSB's and is embodied commercially in Ethernet [26], Mitrebus, built by Mitre Corporation and Localnet, built by Sytek Corporation. The line utilization in these systems approaches 98% [27], and by allowing high priority insertions in the link access protocol, response time for interrupt messages is good.

The DSB in these systems is coaxial cable with transmission rates ranging from .5 to 10 Mbits/second. Ethernet is a baseband system while Localnet is modulated. A particularly attractive feature of a modulated system is that by using frequency division multiplexing techniques, secondary channels which broadcast synchronizing messages and reliability data can be added at incremental cost.

Since access control is distributed, the bus is not vulnerable to a single failure in an access control unit which could flunk the entire bus. Failure of a LAM will result only in the loss of one PE unless the LAM fails in "babbler mode," that is, unless the LAM fails by continually inserting carrier onto the bus. This condition can be contained by timers in each LAM which will disable it if its carrier persists longer than appropriate or by using a secondary channel to disable the offender. If packet addressing is to destination task rather than to PE, load sharing reconfiguration in the event of a PE or LAM failure can be achieved easily. Cost and place modularity are quite good as for any DSB.

While contention-based architectures have yet to be exploited commercially for data acquisition and control applications, they are currently being evaluated by a number of vendors.

CONCLUSION

The particular requirements of data acquisition and control applications impose modularity, reliability and performance constraints upon the topological and logical design of distributed intelligence architectures. DDL and DSB topologies are suited to these applications, and a structured approach to protocol design can meet the requirements logically. Contention-based architectures are new and particularly attractive candidates.

The development of systematic approaches to reliability design and standardization of host-host protocols are required before the design of these systems becomes routine.

REFERENCES

[1] Rose, C. W. and Schoeffler, J. D., "Distribution of Intelligence and Input/Output in Data Acquisition Systems," Proceedings of the International Telemetering Conference, Vol. uXII, pp. 705–720 (1976).

[2] Buchner, Marc and Schaeffer, Donald, "Task Allocation in a Distributed Computer Control System," presented at the 18th IEEE Conference on Decision and Control, Florida, (1979).

[3] Schoeffler, J. D., "Analysis of Performance of Distributed Data Acquisition and Control Systems," Proceedings of the Joint Automatic Control Conference, Vol. 11, p. FA4B, (1980).

[4] Keyes, M. A., "Distributed Control . . . Relevance and Ramification for Utility and Process Applications," Proceedings of the International Telemetering Conference, Vol. XII, pp. 705–720 (1978).

[5] Bristol, E. H., "The Organizational Requirements of the Process Control Distributed Systems," Proceedings of the Joint Automatic Control Conference, Vol. 3, pp. 27–36 (1978).

[6] Rose, C. W. and Dulik, R. P., "Systematic Reliability Design for Computer Control Systems," Proceedings of the Joint Automatic Control Conference, Vol. II, p. FA4A, (1980).

[7] "A Fault Tolerance Assessment of DAIS," USAF Avionics Laboratory Report AFAL-TR-79-1007, (1979).
[8] Randell, B., et al, "Reliability Issues in Computing Systems Design," *Computer Surveys*, Vol. 10, No. 2, pp. 123–165 (1978).
[9] Anderson, G. A., and Jensen, E. D., "Computer interconnection structures," *Computer Surveys*, Vol. 7, No. 4, pp. 197–213 (1975).
[10] Costrell, L., "CAMAC-A Modular Instrumentation System for Data Handling; Revised Description and Specification," National Technical Information Service, TID-25875 (1972).
[11] AEC/NIM and ESONE, "CAMAC Serial System Organization, National Technical Information Service, TID-26488, (1973); "Addendum and Errata" (1975).
[12] Foxboro Company, "3CRA Series Foxnet Communications Subsystem," TI 821–002, (1979).
[13] Bibbero, R. J., *Microprocessors in Instruments and Control*, Wiley, New York, New York (1977).
[14] McQuillan, J. M. and Ceif, V. G., *A Practical View of Communications Protocols*, IEEE Computer Society, New York (1978).
[15] Pouzen, Louis and Zimmerman, Hubert, *Proceedings of the IEEE*, 66, (11), pp. 1346–1370 (1978).
[16] I.B.M., "General Information-Binary Synchronous Communication," Systems Reference Library Manual No. GA27–3004, IBM Corporation (1970).
[17] I.B.M., "Synchronous Data Link Control-General Information," No. GA27–3093, IBM Corporation (1974).
[18] International Organization for Standardization, "Data Communication-High-level Data Link Control Procedures-Frame Structure," ISO 3309–1976 (E) (1976).
[19] D. E. C., "Specifications for: DDCMP Digital Data Communications Message Protocol," Edition 3, Digital Equipment Corporation, Maynard, Massachusetts (1974).
[20] Roberts, Larry, *Proceedings of the IEEE*, 66 (11), 1307–1313 (1978).
[21] McKenzie, A. A., "Host/Host Protocol for the ARPA Network," Network Information Center #8246 (1972).
[22] D. E. C., "Specification for: NSP Network Services Protocol," Version 2, Digital Equipment Corproation, Maynard, Massachusetts (1976).
[23] C. C. I. T. T., "Interface Between Data Terminal Equipement (DTE) and Data Circuit-Terminating Equipment (DCE) for Terminals Operating in the Packet Mode on Public Data Networks," Provisional Recommendation X.25, CCITT, Geneva, Switzerland (1976).
[24] Abrahamson, N., "The ALOHA System-Another Alternative for Complete Communications," AFIPS-FJCC Vol. 37, pp. 281–295 (1970).
[25] Heitmeyer, C. L., Kullback, J. H. and Shore, J. E., "A Survey of Packet Switching Techniques for Broadcast Media," NRL Report 8035, Naval Research Laboratory, Washington, D.C. (1976).
[26] Metcalfe, R. M. and Boggs, D. R., *Communications of the ACM* 19 (7) pp. 395–404 (1976).
[27] Herr, D. E., "The Effects of Random Noise and Packet Truncation on the Throughput of CSMA Networks," M. S. Thesis, Case Western Reserve University, Cleveland, Ohio (1978).

AN APPROACH TO DIGITAL PROCESS CONTROL

M. Masak
Chevron Research Company
Richmond, California

INTRODUCTION

Industrial automatic process control is at a very exiciting stage of its evolution. As with most technologies, there are a number of problems crying for a solution and there are a number of economic opportunities awaiting successful exploitation. In contrast to other technologies, however, a revolution is occurring in the related field of microelectronics and computer technology that promises to help solve these problems and thereby capitalize on the opportunities.

Ever since industrial computers first became available, people have recognized their economic potential for process control. Latour [1] describes these incentive areas as industry views them at the present time. Very simply, they can be placed in the following four categories:

a. Energy Savings
b. Product Yield Improvement
c. Plant Throughput Increase
d. Upset Reduction

Note, incidentally, the absence of manpower savings.

Although process control computers have been available for 20+ years, their application in many instances has not been an unqualified success. Projects were overambitious, combining unreliable equipment, primitive error-ridden software, and inexperienced people. As we see them, the most critical *technical* problems in process control are now:

1. *Deficiencies in the Operator Interface*—It is absolutely essential that operators have a consistent, meaningful, and easy-to-use operator interface to the process. This is particularly true for off-normal operating conditions during a plant problem. This need is becoming even more pressing because operators today appear to be younger and less experienced than in earlier decades and they are given the responsibility for operating more complex plants which are built to closer tolerances. Three Mile Island provides a compelling illustration [2] that many problems exist in the man-machine interface area.

Encl. — Tables 1 and 2 (RE 808809
  and RE 808810)
  Figures 1 and 2 (RE 808811 and RE 808812)
  Figures 3–5 (PM 801963-PM 801965)
  Figures 6 and 7 (RE 808813 and RE 808814)
  Figure 8 (PM 801979)
  Figures 9–11 (RE 808815-RE 808817)

2. *Maintainability of Sophisticated Control Systems*—Economics favor larger process plants operated very close to equipment limits [3]. This usually requires sophisticated control systems. Conventional pneumatic or analog electronic equipment becomes unwieldy and consequently too unreliable to accommodate these requirements. Similarly, many commercially available programmable process control software systems are very cumbersome to use and difficult to maintain after initial installation in these kinds of applications.

3. *Absence of Auto-logic*—Closely related to the above two problems is the fact that sophisticated control systems need more states than MANUAL/AUTOMATIC to cope with off-normal operating conditions. This is a topic that is only recently beginning to receive some attention in the open literature [4, 5].

The purpose of this paper is to describe the control system which we in Socal are using with considerable success. It is an example of a system which employs the latest microelectronic and computer technology to solve these above problems.

OVERALL SYSTEM CONSIDERATIONS

Within the past few years, most instrument manufacturers have announced microprocessor-based digital control equipment products with cathode ray tube (CRT) displays. In our opinion, none of these products has the capability to replace a fully programmable computer control system performing sophisticated, advanced control; and, in particular, they offer a poor operator interface when advanced control is attempted. Cutler [6], in a recent letter to the editor of Intech, agrees with this opinion.

However, the programmable computer by itself, with the necessary input-output equipment, does not meet all the control system requirements. Although its onstream factors are in the 99%+ range, some backup instrumentation is necessary to control the plant during those periods when the computer is down and to assist in plant startups.

For backup instrumentation, there are now two options: (1) conventional, panel-mounted, single-loop controllers or (2) CRT-based, digital instrumentation referred to above. The high costs of interconnection cabling and control room space all favor the CRT-based option for new plants of even modest size. For example, a study [7] by Foster-Wheeler for a plant with 150 control loops, which is not large by today's standards, showed that the conventional panel instrumentation installation was approximately 10% higher than the CRT approach. This incentive would increase markedly with a larger plant size.

To help solve the system maintainability problems, it is most desirable to separate the minute-by-minute control computer from the longer term plant monitoring and information computer. The control system should be secure, reasonably simple, and sufficiently comprehensive that software changes after initial startup be infrequent. These factors call for a "small" minicomputer with limited or no moving parts but one in which control system software changes can be made readily when necessary. The plant monitoring and information system, on the other hand, is for the use of operations, maintenance, and technical service people to gather large quantities of data about the process and its equipment. It is used to analyze the cause for upsets and to maintain historical plant data. It is characterized by large data files and software to enable a variety of people frequent access both to the real time and the historical data files. Such a software system, with acceptable response times to users, tends to be incompatible with secure process control. As a result, our Company has separated these two functions into separate mainframes with radically different software systems. This paper will not discuss the information systems.

Finally, because of the relative complexity of a modern control system and the shortage of highly trained people, it is encumbent on any organization to standardize on a few specific products. This avoids solving the same systems problems again and again in different locations, allows easier and more effective transfer of staff from one location to

584   Chemical Process Control II

another, and enables a central corporate engineering facility the maximum effectiveness in helping different locations with their applications.

All of these factors, summarized in Table 1, have caused Socal to develop the system described in this paper.

OVERVIEW OF THE CONTROL SYSTEM

Figure 1 shows a simplified layout of our "typical" computer control and monitoring system together with the backup instrumentation. Parts of this overall system are working in different configurations and locations in over a dozen applications, while four total systems are presently under construction and due for installation in the next 6-24 months.

For reasons discussed above, the overall system includes two types of computers: one for control and the other for information and monitoring. The two computers are totally different and are not intended for redundancy. They both access real time data through the instrumentation system communication bus.

The instrumentation system has three roles:

1. It provides basic automatic control when the control computer is unavailable. In other words, it provides backup capability to allow the process to remain safely in operation.
2. It acts as the process signal input-output system between the plant and both the control and the monitoring computers. During normal operation, the controllers in the instrument racks are in the computer mode, and the control computer sends signals via these controllers directly to the valve. The reason for this direct digital control (DDC) mode is discussed in a subsequent section.
3. It provides trend data for the operator which is used in conjunction with the control page in the integrated operator console.

Table 1. Important Factors Guiding The Evolution of Process Control System.

---

1. Sophisticated Control Requires Programmable Computer System
2. Backup Instrumentation is Necessary, Should Be Fully Compatible with Control Computer
3. Split Plant Control and Monitoring/Information Systems for Increased Reliability
4. Standardize for Manpower Effectiveness

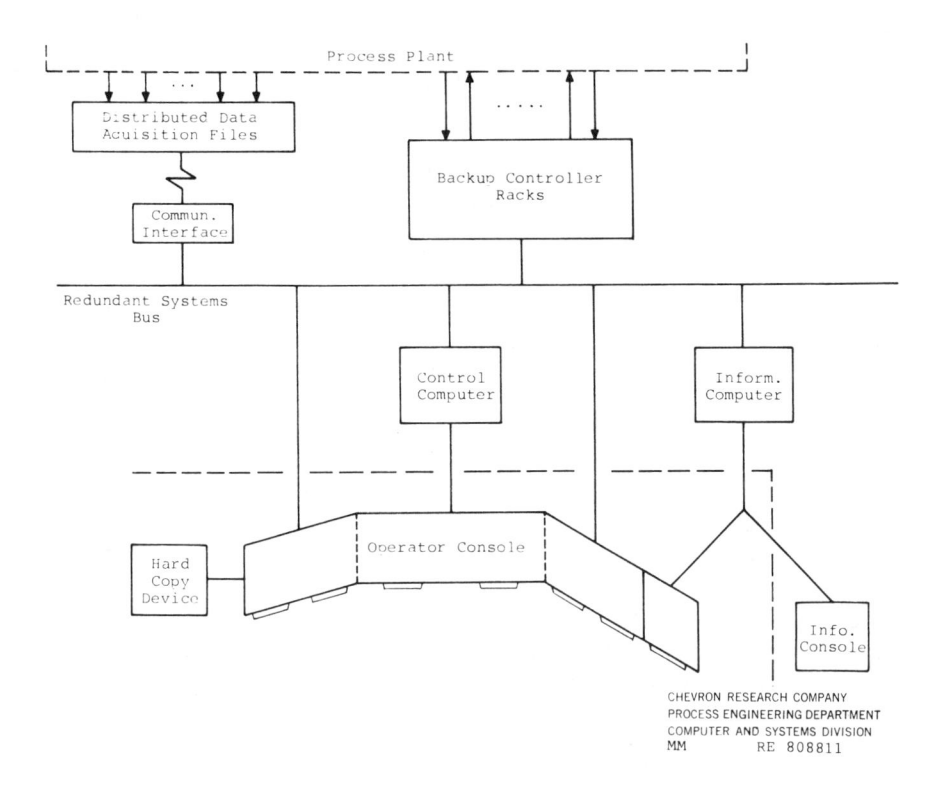

Figure 1. Simplified Layout of A"Typical" Control and Information System.

OPERATOR INTERFACE

The operator runs the process plant from a console, as illustrated in Figure 2. The console includes two types of CRT's with associated function buttons from three different devices; namely, the instrumentation system, the control computer, and the monitoring and information computer. Above the CRT's are a small number of hard-wired alarm annunciators.

The instrumentation CRT's are tied directly to the system bus, the main data communication channel between the backup controllers and the computers. As such, these CRT's can be used to run the plant via the backup control system. However, during normal operation, they are used to display trend data. The particular system we have chosen, the Taylor MOD-III, provides high resolution phosphorescent CRT's which allow

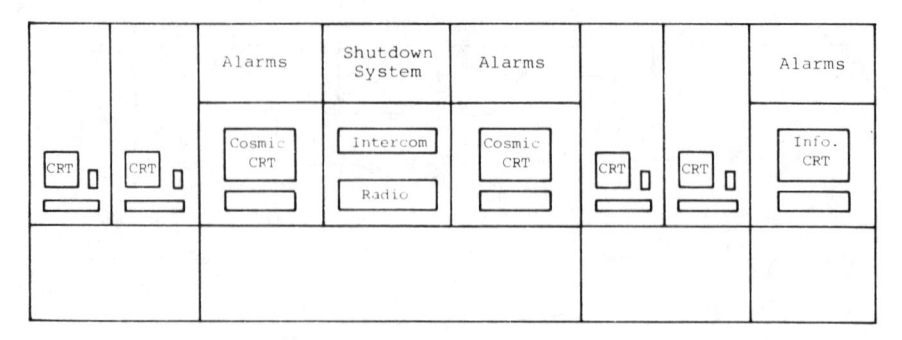

Figure 2. Simplified Layout of A"Typical" operator Console.

simultaneous display of up to 12 trends over two time periods or three selected trends over eight time periods. These displays, illustrated in Figures 3 and 4, can be called up with a single button push. The trend display quality was one of the major reasons why this particular backup instrumentation system was selected above others currently available. It is noteworthy that the operator has no conventional paper trend charts available at all; hard copies of the trends on any given page are available on a copier device. Further detailed information is available from the manufacturer.

Primary operator interaction with the control system is via "control pages." These are tabular displays, illustrated on Figures 5 and 6, through which operators can:

1. Turn loops on and off automatic.
2. Change set points in automatic mode.
3. Change valve outputs in manual mode.
4. Observe set points, measurements, calculated values, outputs, control modes, and alarms.
5. View the status of complex, multivariable control procedures and change their modes if necessary.

The control page on the instrumentation system CRT is a subset of the control page on the control computer CRT. The former shows only the lowest levels of a cascade while the computer control page displays the entire hierarchy, almost all of which is resident only in the control

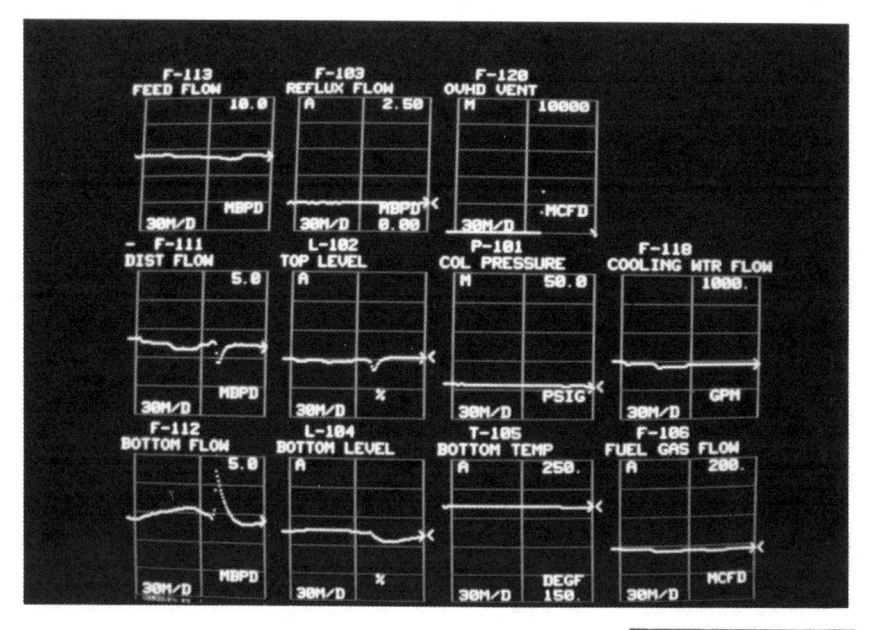

Figure 3. Multitrend Operator Display, Variables Selected and Preconfigured.

computer. Figure 7 illustrates this concept in a simple case. Note that this control scheme includes only a two-level cascade, and the computer page shows the higher level strategies on lines 10, 20, and 31.

As a result, when all equipment is functioning, the operator interacts through the computer control page and views trends on the instrumentation CRT. He can also observe and interact through graphic displays which show complete plant sections, as shown in Figure 8. When there is a control computer failure, he falls back to a more primitive automatic control system, interacting with the instrumentation system CRT.

As noted, the control page displays are tabular. This design has evolved, at the insistence of operators, because of their need for vast quantities of simultaneous data. It has also evolved because of a desire to employ function buttons rather than generalized typewriter keyboards to

588   Chemical Process Control II

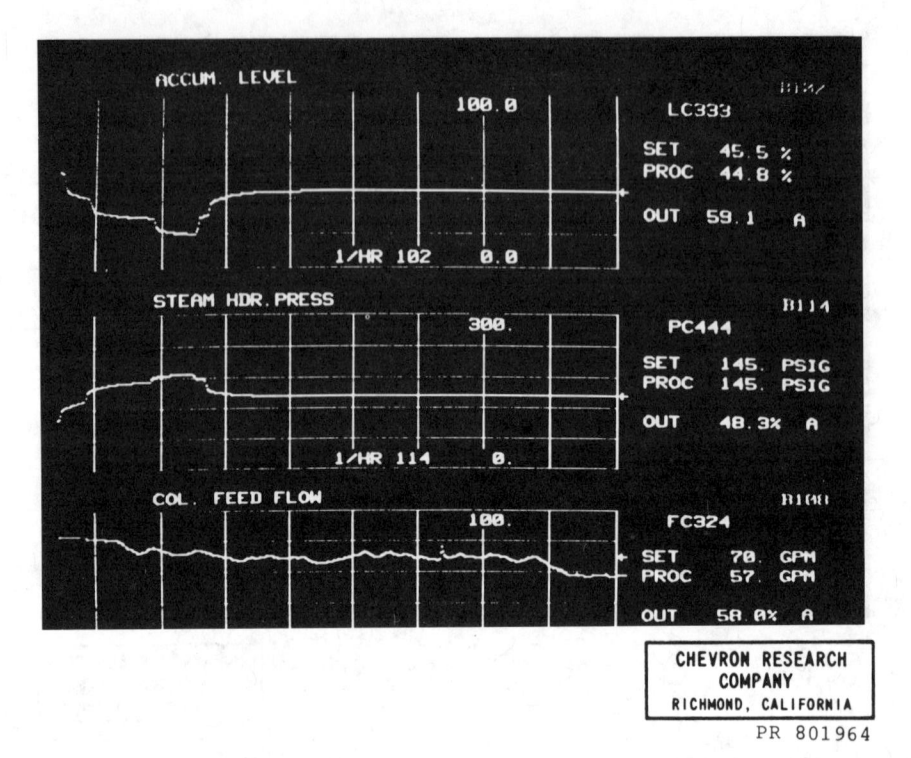

Figure 4. Loop Display of Any Arbitrary Three Loops, Trend History Four Times That of Multitrend.

access and change displays. Regular tabular displays facilitate the function button design.

The function button layout for the control computer CRT is illustrated in Figure 9. The displays are organized by plant sections and types of display. The left two groups of buttons identify a plant section and the type of display required. A maximum of two button pushes is needed to call a particular type of display for a plant section. The right two groups allow interaction with the display shown. All interaction involves moving a cursor on the display and then entering a number at the position selected.

The types of displays available include the control pages already discussed as well as limits data, instrument data, graphics, and computer trend pages. These latter trend pages are intended for displays of

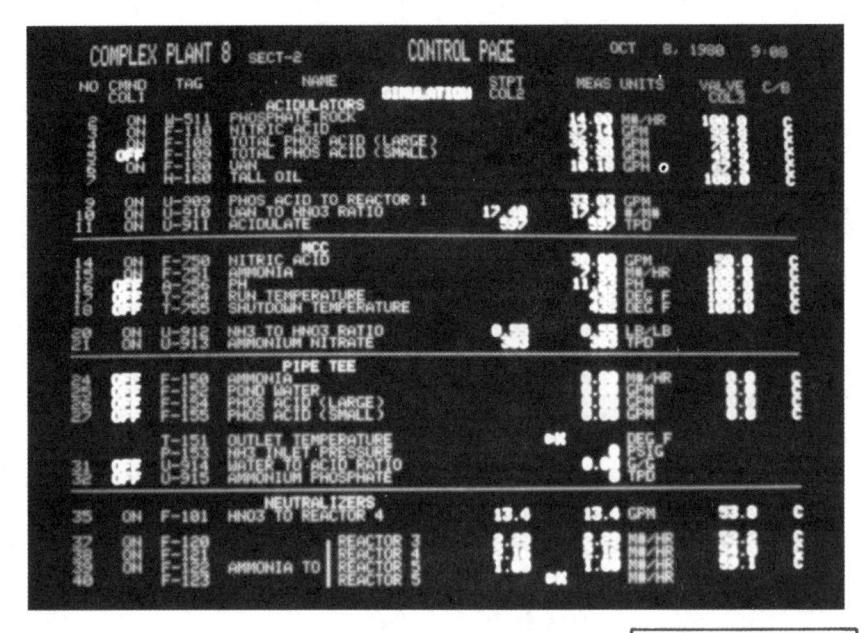

Figure 5. Typical Control Page, Full Color.

calculated variable trends, the information for which is unavailable to the backup instrumentation system.

Further details of this operator interface system are left as a subject of a future paper.

CONTROL COMPUTER HARDWARE AND SOFTWARE

Table 2 provides a hardware parts list for a typical control computer. The core and semiconductor memory contain the main control data files and most frequently used displays and control programs. The one million word drums are necessary primarily because of the comprehensive display and utility systems, while the floppy disk system is used only for system cold-starts.

```
4CU C1190 STABILIZER  CP-10                                JULY  8, 1980  10:37
                                                           LSFO EAST CRT COPY
NO CMD             NAME              TAG       MEAS UNITS      STPT       VALVE  C/B
   COL1                                                        COL2       COL3
 1 FAIL |CALC BEST C1190 FD TEMP    C592  ▷X        DEG F
 2   ON ↓C1190 FD TEMP              T85       240.3 DEG F     238.0   ▷V  99.9    C

 5 FAIL |MIN V1190 VENTGAS
 6   ON ↓C1190 OH PRESS             P91        84.7 PSIG       85.0       79.1    C

 8   ON |C1190 OH C5'S FREDICT      C594        0.4 PCT    ▷S   0.5
     ON |C1190 VAPOR LINE TEMP      T91       110.3 DEG F
10   ON ↓C1190 REFLUX RATE          F90         6.6 MBPD                  30.1    C

12   ON |V1190 HC LEVEL             L91        57.0 PCT        54.5
13   ON ↓C1190 C3/C4 PROD RATE      F94        3.91 MBPD                  37.5    C

15 COND |C1190 L/V                  C597       0.69 BL/BL
16      |C1160 HIGH RVP     O/RIDE  C596            RVP
17 FAIL |C1190 MIN  REFLUX  O/RIDE  F900R ▷X        MBPD
18 FAIL |C1190 LOADING      O/RIDE  C642  ▷X        PCT
19  OFF |C1190 TRAY 2 TEMP  O/RIDE  T90       265.6 DEG F
20   ON ↓STEAM TO E1197A/B          F91        25.2 MLB/HR     25.3       33.1    C

22   ON |C1190 BTM LEVEL            L90        60.0 PCT        60.0
23   ON ↓C1190 BTMS                 F80-1      28.8 MBPD                  46.4    C

--------------------------------------------------------------------------------------
       C1190  FEED          |     C1190  TEMPS      |
                            |                       | C1190 A90 C5S      6.6 PCT
C1190 TOT FD      32.6 MBD  |                       | C1190 OH  C5S     5.00 PCT
LPD GRUB           2.3 MBD  | RFX TEMP       63 F   |
TOT GRUB           2.3 MBD  | BTMS TEMP     306 F   |
                            |                       | C1190 OFF GAS      0.6 MFD
--------------------------------------------------------------------------------------
```

CHEVRON RESEARCH COMPANY
PROCESS ENGINEERING DEPARTMENT
COMPUTER AND SYSTEMS DIVISION
MM RE 808813

Figure 6. Another Example of Control Page, Hard Copy Directly Off CRT.

Chevron employs in-house developed proprietary process control software, COSMIC (Command Oriented Software for Modern Industrial Control). This software system runs on a Taylor 1010 computer which is built around a SPERRY-UNIVAC V-77 mainframe. Figure 10 illustrates the interrelationships within COSMIC.

COSMIC consists of assembly language subroutines for specific tasks involving data acquisition, point processing, and display processing. High level language procedures are used for the main process control software. The high level language is a multitask version of BASIC, executed via an interpreter.

The total software is, therefore, an optimized collection of high speed, special purpose assembly language routines which are rarely changed and slower speed, highly flexible BASIC procedures readily adaptable to varying process requirements. The latter procedures fall into two categories:

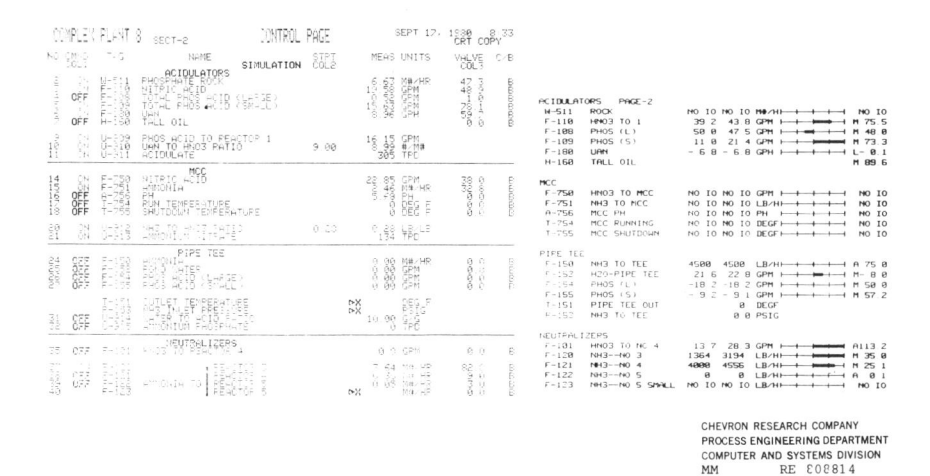

Figure 7. "Cosmic" And Instrumentation Control Pages, Side By Side, As Available To Operator.

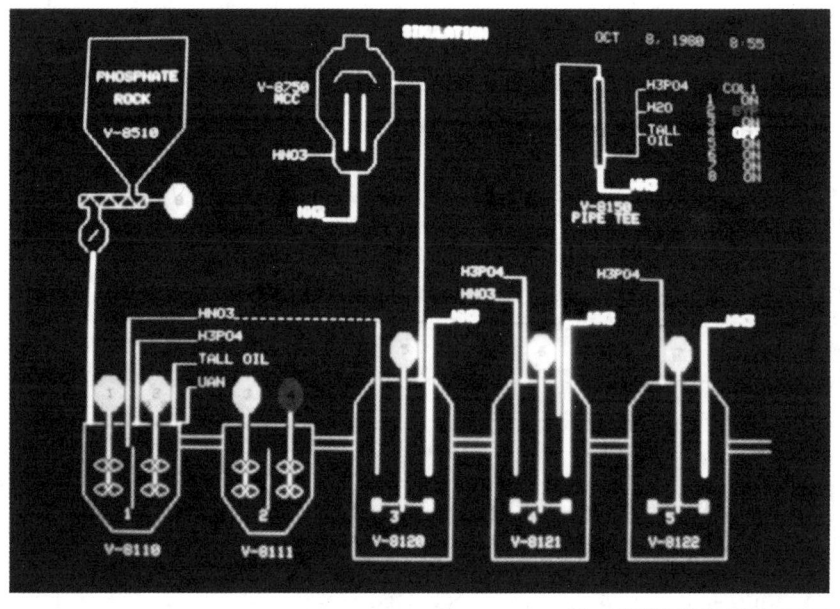

Figure 8. Typical Graphic Display, Available On "Cosmic" Console.

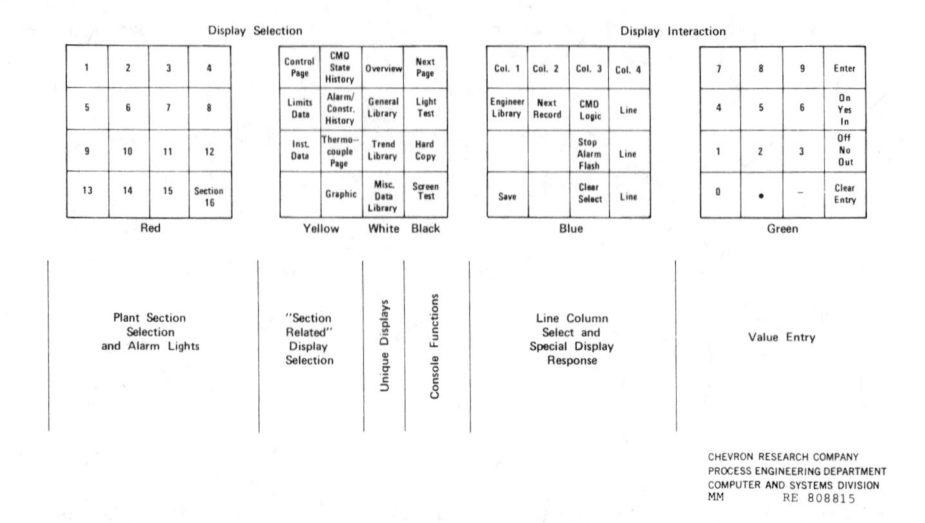

Figure 9. Function Buttons Integrated Within "Cosmic".

Table 2. Hardware Parts List Typical Control Computer.

- Central Processing Unit with 320K 16-Bit Memory, Accelerator Firmware
- Real Time Clock, Memory Parity, Power Fail/Restart
- 1 Million Work Drum
- Dual Floppy Discs
- Three 9-In. Color CRT's with Function Buttons
- One Black and White CRT
- Decwriter III
- Hard Copy Recorder, Interfaced to CRT's

1. COSMIC procedures, designed and supported by Chevron Research, which form the backbone of the table-driven process control software package, relatively standard for all installations.
2. Supervisory control procedures, designed by various project teams, which accomplish the necessary control functions for the specific process plant.

Because the COSMIC system is table-driven, the displays and most simpler control loops are configured simply by filling in data tables, and

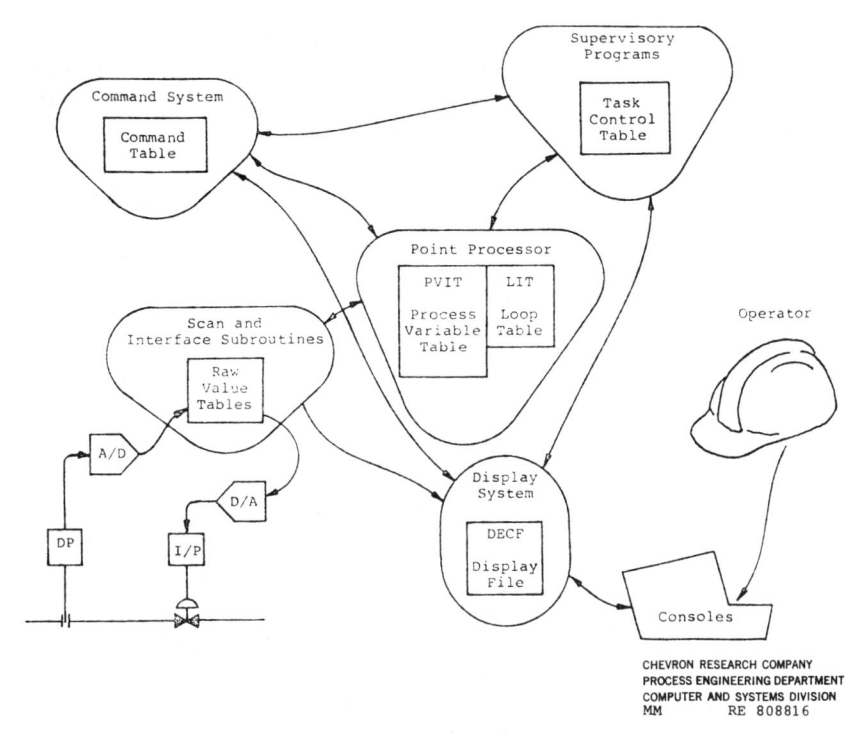

Figure 10. Simplified Schematic of "Cosmic" Software Layout.

invoking the appropriate COSMIC procedures for control system operation.

The main control system function is accomplished with entities called "commands." These "commands" are generalized controllers which have mutiple states such as ON, OFF, FAIL, CONDITIONAL, etc. A command may be anything from a simple proportional controller to a very complicated inferential control calculation involving multiple procedures with several process measurements.

The state of a command is established by the command executive subsystem. A combination of preprogrammed and user-programmed events can trigger a command evaluation. These events may include operator actions, timers, limit checks, and so forth. When an event occurs, the command evaluator calls a state transition table to see whether the state of the command should be changed and what it should be changed to. Although there is nothing inherent in the command evaluation system limiting it to this, the common use of the different command states are accepted by convention; so that FAIL means the actual failure of a critical piece of control equipment such as a measurement, CONDITIONAL means that the control loop cannot function because of some reason outside its own domain, and so forth.

In effect, the command executive can be thought of as a generalized mode switching logic or auto-logic system which is superimposed on the control system to ensure the internal consistency of control process in the face of different external events. The operator display are integrated with the command executive so that the operator always sees the correct current state of the entire control system.

PROCESS CONTROL

The process control strategies employed in our control systems are all organized in the form of a hierarchy of multiple cascade loops, as illustrated in a simplified way in Figure 11. This particular hierarchy of loops would be displayed to the operator as a group of three commands. The operator has the option to close any or all of this cascade, depending on circumstances. The command evaluator checks to ensure that control system integrity is always maintained and that the display always shows the operator the correct state and structure of the control system. For example, in contrast to some commercially available systems, the display will never show the outside, higher level control loops to be operating if an inner slave loop is in manual!

The process control which we employ is almost invariably DDC, meaning that the final signal which is sent to the valve has been calculated by the process control computer. There are two main reasons for this:

1. It provides the easiest method for dynamically changing the control system configuration in the event that some part of the process becomes limiting. For example, when a process has two valves, it has two degrees of freedom, each of which can be employed to accomplish an objective. Now, if one valve, because of other circumstances, becomes wide open, a degree of freedom is lost. Consequently, the control system uses the remaining valve (or degree of freedom) to continue accomplishing the more important of the two objectives.
2. Our philosophy is that the operator should be able to view the entire state and structure of a control system through one "window." He should not have to look at two rather different displays to see whether a complex cascade is all connected and working properly.

Objective 10

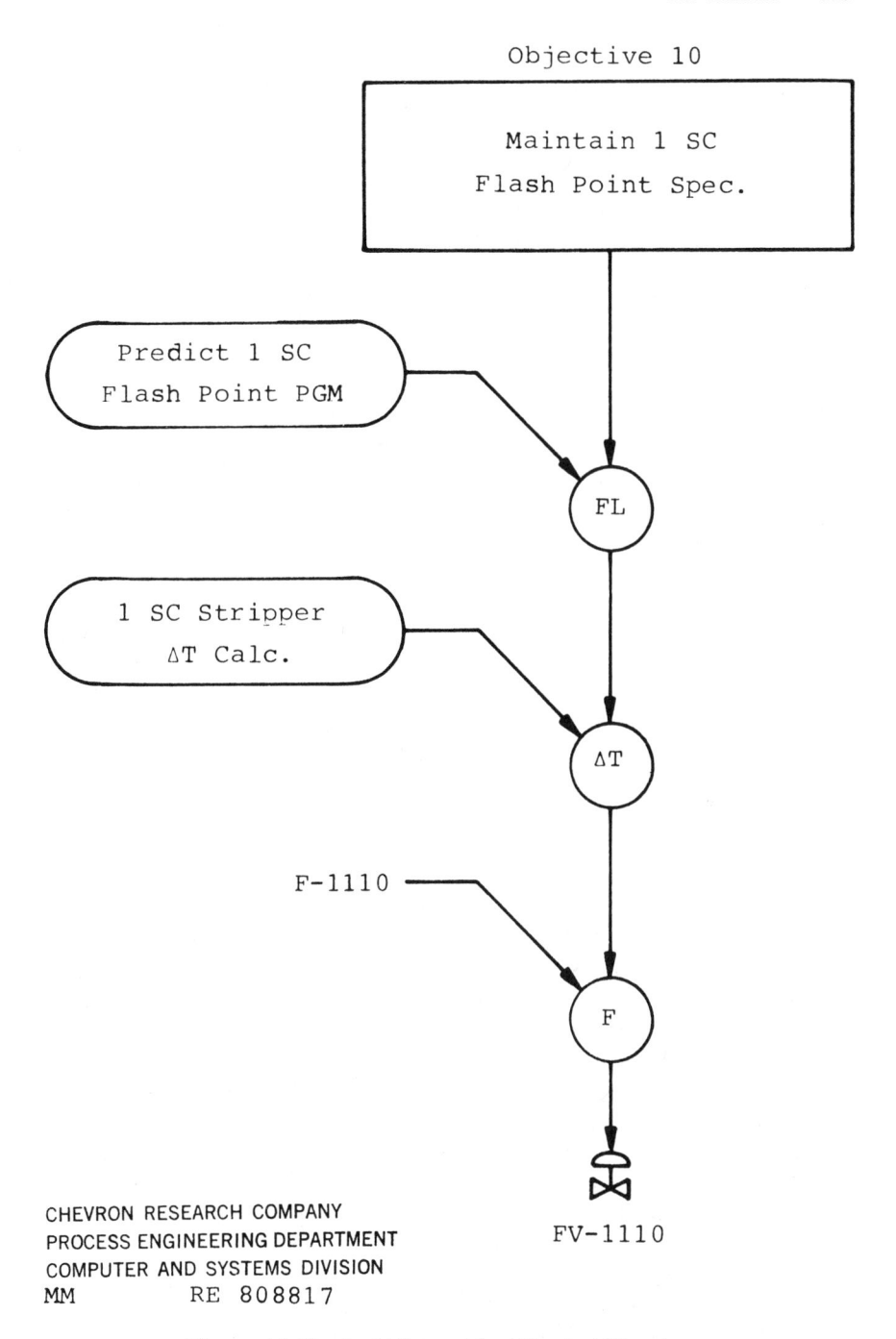

```
┌─────────────────────┐
│  Maintain 1 SC      │
│  Flash Point Spec.  │
└─────────────────────┘
```

Predict 1 SC
Flash Point PGM

(FL)

1 SC Stripper
ΔT Calc.

(ΔT)

F-1110

(F)

FV-1110

CHEVRON RESEARCH COMPANY
PROCESS ENGINEERING DEPARTMENT
COMPUTER AND SYSTEMS DIVISION
MM   RE 808817

Figure 11. Typical Hierarchical Control Structure.

The operator's window is the control computer's CRT, with the display system running under COSMIC. Because of this display requirement, the control computer has to have frequent updates of measurements, outputs, and statuses. This display processing requirement places a significant load on the computer; the addition of the lower level control calculations, DDC, is minimum, adding less than 4% to the computer load.

The upper level control strategies which we have employed with great success fall into the following general categories:

Inferential Control—Here the objective is to control a process variable to a given target. However, the direct measurement of that variable is either too expensive or not technically feasible. Consequently, the computer is employed to infer that process variable from a number of other measurements using standard process engineering correlations to arrive at the desired result [8]. Some examples are octane control, distillation cut point control, and percent desulfurization.

Constraint Control—Invariably, process economics dictate that some part of the process be pushed against some equipment limit, be it flooding in a column, wide open control valve, or whatever. Our highest level strategies normally check a number of constraints; and if none are violated, some control valve is slightly opened or closed and the check repeated. In this manner, the process is driven against the desired limit. It is noted that we do not employ online linear (or nonlinear) programming; rather, it is assumed that the engineers can always analyze the process situation sufficiently and arrive at a control algorithm which drives the process to the desired limit. Examples of this type of strategy are combustion control, feed maximization, and distillation column pressure minimization.

Measurement Redundancy—Although not strictly a "strategy," we employ redundant measurements in a variety of situations to ensure maximum control system effectiveness and up-time. This redundancy manifests itself in several ways such as using several thermocouples for the same temperature measurement; checking level measurements with mass balance calculations; using laboratory samples to update process inferences; and in combustion control, using at least two different types of analyzers and a process measurement to ensure that an analyzer failure does not cause a catastrophe.

Adaptive Control—We have attempted adaptive control on some DDC loops. There is an incentive to keep a control loop well tuned regardless of the position of the manual bypass valves. Although we have had some limited success, a great deal of work is still required. We do not feel that we can justify further research and development in this field.

CONCLUSION

The system described here is the current version of a rapidly evolving control system employed by one company. It offers one solution to the operator interface, maintainability, and advanced control deficiency problems described in the introduction. It is the product of many people, both engineers and operators, working in a variety of locations on different processes. Undoubtedly it will continue to evolve as new problems present themselves and new opportunities are perceived. It does represent the current, 1981, state-of-the-art in automatic process control.

REFERENCES

[1] Latour, P. R., *Hydrocarbon Processing,* 58(6):73 (1979).
[2] Livingston, W. L., *Control Engineering,* 27(5):86 (1980).
[3] Asbjornsen, O. A., "The Economical, Technical, Intellectual, and Social Gap Between Theory and Practice in Process Control" in *AIChE Symposium Series, Chemical Process Control,* A. S. Foss and M. M. Denn (ed.), 72(159):80 (1976).
[4] Wilhelm, R. G., Jr., *IEEE TransAuto Control,* AC-24 (1), 27 (1979).
[5] Bristol, E. H., *Chem. Engrg Progress,* 76(11):84 (1980).
[6] Cutler, C. R., *Intech,* 27(10):40 (1980).
[7] Foster-Wheeler Corporation Study No. 11-53096, Livingston, New Jersey (1976).
[8] Joseph, B. and Brosilow, C. B., *AIChE Journ,* 24(3):485 (1978).

IMPORTANT DESIGN ISSUES FOR PROSPECTIVE USER OF DISTRIBUTED CONTROL

Remsi R. Messare
Exxon Research and Engineering
Florham Park, N. J.

This presentation will address aspects of engineering and system design with currently used control systems, where distributed control is performed at the basic regulatory (instrument) level.

The impact on engineering effort and the design complexity of aspects such as remote installations of instruments, desired integration of man-machine interface and allocation of control functions in computer based systems will be briefly reviewed. Also, desired design improvements in the areas of reliability, data transfer security, system diagnostics and environmental hardening will be discussed. Tables 1 through 5 highlight important facets of the above issues.

Table 1.

- Distribution of Control Functions
- Operator Interface
- Instrument System Configuration And Wiring
- Desired Design Improvements

TABLE 2.

- Division of Process Control Between Instruments And Process Control Computer Depends On:
  - Control Capabilities of Instruments
  - Ease of Implementation
  - Back-up Control (Computer Outages)
  - Cost

TABLE 3.

- Elimination of Traditional Instrument Panel Makes Design of Operator CRT Station Critical
  - Layout
  - Degree of Redundancy
  - Integration of Hardwired Components
- Instrument Console Data Base Definition Requires Increased Owner's Involvement
  - Tag Names And Descriptor Libraries
  - Display Arrangement
  - CRT Alarming
- Degree of Integration Between Instrument CRT's And Computer CRT's Depends On
  - Host Computer
  - Design Features of Computer/Instrument Interface
  - Owner's Involvement

TABLE 4.

- Remote Instrument Installation From Control Center
  - Requires Early Decision
  - Impacts On Control Center Sizing, Field Wiring And Design of Field Instrument Shelters
  - Requires Significant Owner's Involvement (Operations, Maintenance) During Early Design Phase
- Instrument Configuration And Data Base Definition
  - Requires Understanding of Instrument Features, Modularity and Failure Modes
  - Depends On Allocation of Control Functions Between Instruments And Process Control Computer
- Integration of Data Acquisition Subsystems Into Instrument System
  - Impacts On Operator Interface Design
  - Requires Early Decision And Cost Comparison Studies

TABLE 5.

- Reliability
  - Increased Data Transfer Security
  - Protection And Monitoring of Instrument Data Base
  - Easy Recovery From Multiple Output Failures
  - No Undetected Failure Modes Affecting Integrity of Inputs/Outputs
  - Self Checking For Failures At The Device Level
  - Increased MTBF
- Maintainability
  - Built-In Diagnostics Down To The I/O Level
  - Card Replacement With Power On
  - System Diagnostics Not To Penalize Operator Interface (Separate Maintainance Console)
  - Service At Card Level (Replacement)
- Operability
  - Streamlined Operator Interface
  - Improved CRT Trending
  - Intelligent Use of Color
  - Increased Console Display Capacity
- Environmental Hardening
  - Div 2 Design For Control And Data Acquisition Devices
  - Resistance To Field Exposure of Corrosive Gasses
  - Operating Temperature Range Same As Field Transmitters
- Other
  - Input Conditioning And Validity Checks At The Device Level
  - Adoption of Proway As Standard Digital Tranmission Subsystem

COMMENTS: PANEL SESSION ON "DISTRIBUTED COMPUTER CONTROL"

Duncan A. Mellichamp
University of California, Santa Barbara

By definition, perhaps, the applications in this area which interest us involve networks of computers. Some of these computers may be larger ones, intended for optimization and management-level decision making; many will be smaller, i.e. microcomputers functioning at the control loop level. One fact of life which has to be dealt with at the present time is that commercial real-time networks are both poorly categorized by vendors and poorly understood by users. Much of this is the result of recent rapid growth of the area and a lack of definitions which are commonly agreed on. As a result, each application must be considered individually; each design is unique or nearly unique.

PROBLEMS AND ISSUES

1. Specification of a system, i.e. the design process, is difficult: Where should hardware (processors, memories, communication links, etc.) be located? Are these elements designed for efficient network operation? Are software and firmware truly distributed? Transparent to the user? Where should data bases be located? How much redundancy should be involved?
2. Generally, the real-time performance of a distributed control system cannot be evaluated until it is built: We need to be able to predict

operating properties in advance or to be able to simulate alternative designs.
3. System reliability usually is unknown until after considerable operating experience is obtained.
4. Modifications to the original system or expansions should be easy to make but often are not.
   Result: Distributed computer process control systems often are
   - designed by ad hoc (inefficient) methods,
   - subject to large risk of inadequate performance, even failure.

PROMISES

Distributed systems possess a number of important potential advantages:

1. The system can be optimally configured, i.e. the size and location of processors and other hardware can be determined by process and information flow requirements rather than constrained by computer limitations; software and firmware can be located where calculations need to be made; communications requirements can be reduced.
2. The system can be modularly designed so as to be easily expandable.
3. The run-time characteristics can be optimally designed: Response to alarm situations can be fast; much potential computing power can be placed only where it is required; the system can be friendly to the operator; data bases can be much more accessible to the process engineer; soft failure can be a built-in feature. On this last point, failure at any one point in the network should mean only that a small subset of network operations is temporarily lost. For example, at the local-level computers (control elements), several loops may need to be left on manual; at the supervisory-level computer, no optimization of set points may be possible (however, local control operations at previously-determined set points continues); at the management level, management may have to make decisions temporarily without all the facts (Is this anything new?).

A VIEW OF DISTRIBUTED SYSTEMS IN PROCESS CONTROL

E. Bristol
Foxboro Company

In the Conference we have seen expressed a view of distributed computation that I would call the centralized computer view emphasizing multiprocessing of computing elements not assigned to any process computation and the distributed data base. This view achieves reliability by spreading computation over whichever elements are functional at the time, and unity by relying on centralized data conventions to manage the distributed data base.

I would like to speak to a different view of distributed control (to some extent implementable within the above framework), more consistent with process control's practical history and more oriented toward a distinctive process control view of computation. That view takes control systems as being made up of functionally specialized boxes that I call self-aware boxes. This view achieves reliability by minimizing the number for functions in a box (as well as by in box redundancy) and unity through basing function and functional interbox communications on explicitly process control oriented concepts. Each box has its own functionally specialized language totally controlled by the box. This view like the previous one uniquely benefits from cheap microprocessor oriented computation and suggests its own exciting and distinctive future developments.

There are many advantages to this approach, but time only permits discussion of one. By investing in careful engineering of an integrated general modular approach to all of the process control problems: control, man process interface, sequencing, alarming and diagnosis, trending, and supervisory or process management functions one can drastically cut per application engineering costs in two ways. First many of the basic functions can then be implemented *better* as basic functions not requiring user programming. Second often the control design implies the man-process design and other functions. Thus, whereas these functions are redundantly programmed in the general purpose language, they need not be in the specialized one.

Such an approach requires that each of the functions be handled with a well defined philosophy. In many areas it is possible to pick philosophies which are easily translated to each other. Thus, logical control functions are translated from relay ladder diagrams to logic diagrams. Other philosophies are not so mutually consistent. Thus, a modular control system based on classical (and neoclassical) control functions such as PID's, selectors, decouplers, and feedforwards with parallel man process interface concepts such as single loop displays and multiloop displays would tend to make a state space oriented design a second class citizen particularly if that design were to be developed in terms of its own man process interface view.

For these reasons, at least at a commercial level, it is not sufficient in a general way to say that there are a diversity of control techniques each with favorable characteristics. It is necessary to bring the methods into sharper comparison not necessarily to reject some, but to develop each completely into an attitude to deal with the larger design and operational issues and to make the translations between philosophies clearer. Then we can concentrate on minimum engineering cost application designs. The question is: How do you want to talk about control? Because you will talk that way not only to people but to machines. And in the very long run the forces for standarization at this level commercially are not too different (at least in kind) than the forces for 4–20 ma or 3–15 pounds.

Conference Summary

A THEME PROBLEM FOR CHEMICAL PROCESS CONTROL

Morton M. Denn
Department of Chemical Engineering
University of Delaware
Newark, Delaware 19711

In our summary of the first Engineering Foundation Conference on Chemical Process Control in 1976 [1], Alan Foss and I made the following recommendation to the National Science Foundation:

> "There is a need for a set of standardized processes which can be used by researchers to test advanced control theories, in terms of both effectiveness of control and utility as design tools. One mechanism for providing such processes is to develop a set of simulation models for processes whose dynamics are well understood and to make the models generally available to researchers through a computer network . . . substantial industrial-university cooperation would be required. The simulation model should be available to the user only in a form in which he can make changes comparable to those that he could make on the real process, and he should be able to obtain only noise-distorted output."

At that time, we hoped that we might be able to gain access to existing industrial simulations, with attention focused on the fluid cat cracker (FCC). The participants at that conference had two goals in mind. First, we believed that groups developing procedures for multivariable control system design should have the opportunity to make comparisons on a common basis, using a system that was typical of the expected applications. Second, such a simulation could ultimately be useful for educational purposes. Unfortunately, nothing happened; perhaps no group or individual was sufficiently motivated to undertake the task of coordinating and developing such a project. Subsequently, and independently, such a project was undertaken by a group of electrical engineers using a turbofan engine as a theme problem [2]. While this problem, in retrospect, had many drawbacks (the system was linear, deterministic, and contained no disturbances), it is generally felt to have been a useful exercise—some would say competition—in control system design methodology, and the results appear to have had an impact on the specific technology as well.

The question of a chemical process theme problem was raised again early in this Conference, and then discussed informally throughout. The passage of time has diminished interest in the FCC as a candidate, in part because the order of existing reactor models is perhaps too low to be of current interest, in part because some studies (our own, for example) suggest that there may be pathologies in a control system designed for this system by modern methods that would not lead to useful generalization. In the opening remarks on the subject, the ammonia plant was offered as a viable candidate which appeared to satisfy the several requirements of a theme problem that would attract the attention and interest of a wide group of academic and industrial control investigators: It is reasonable to focus on a full plant, because this provides the opportunity for unit interactions, and it is consistent with current interest in distributed control. The process contains both reaction and separation (but, unfortunately, not distillation); reactor configurations are published, and the system chemistry and thermodynamics are well-established and available in the open literature. There are both overall process and local unit operation control problems, which should interact. Finally, an effective control scheme, designed by heuristic methods, is known to exist, thus providing a base against which to compare control schemes designed using advanced methods.

The general response to the proposal for a model plant was well-received by participants, both industrial and academic, and included offers of cooperation if the program were undertaken. Those commenting generally supported the idea that the model system should comprise an entire plant. A number of participants felt that, given the interest in distillation control, the model process should consist of, or at least include, one or more distillation columns. Several other specific processes or process units were suggested, but on the whole the problems tended to be too specific for the theme purposes. The most viable alternative to the ammonia process that was suggested was the methanol process, since it is generally similar in structure and does contain distillation. The additional complexities associated with the methanol reactor relative to the ammonia reactor, together with the rapidly evolving nature of methanol technology, appear to make the process less suitable for our purposes.

There was considerable discussion, both informally and during the final Conference session, regarding user accessibility to the plant model. There was general agreement that the basic model should be nonlinear and should contain upsets in the form of both continuous disturbances and isolated events. Many people felt that the idea first proposed in 1976, of allowing the user only to make measurements, remained valid, and to require him to construct his own model if he wished; a possible compromise in this regard would be to provide the basic physical-chemical information that was used. Others felt that the user should be provided with a plant model in order to ensure that all designs were carried out on a common basis, at least with regard to the model; some felt that a linear model, perhaps slightly inaccurate with regard to the true linearization of the plant dynamic model, should be made available. It seems apparent that some form of model must be made available in order to avoid the necessity of requiring all users to repeat the modeling effort, which is expected to require a number of man-years. The most reasonable position to me is to provide the full nonlinear model in a deterministic form, and to allow each user to utilize that model in any way that he feels appropriate. The actual running model on the computer network would not only include disturbances, but might also use slightly altered parameter values from those published. (Indeed, the suggestion was made that there be a systematic change in some parameter values with time.) Several participants from outside North America expressed concern that they be able to access the model, and it is clear that this problem must be resolved; access through a network could be a problem even for North American users, since long contact times may be required, and costs could become excessive. One possible resolution is to provide object codes on selected standard computers, so that most participants could use their own facilities.

There was sufficient interest in the development of the plant model that a group did appear to be forming to undertake the project. The final focus during informal discussions was on the mechamism for funding and for ensuring adequate input by individuals from industry who have relevant expertise with regard to the specific plant ultimately chosen, both its design and control.

REFERENCES

[1] Foss, A. S., and Denn, M. M., in "Chemical Process Control," AIChE Symposium Series, 72, No. 159 (1976), p. 232.

[2] Sain, M. K., Peczkowski, J. L., and Melsa, J. L., "Alternatives for Linear Multivariable Control with Turbofan Engine Theme Problem," Nat. Engg. Consortium, Inc., Chicago, 1978.

SYSTEM SOFTWARE FOR PROCESS CONTROL
A Summary of Conference Presentations and Discussion.

D. Grant Fisher
Department of Chemical Engineering
University of Alberta
Edmonton, Alberta, Canada. T6G 2G6

SYSTEM SOFTWARE FOR PROCESS CONTROL

The primary consensus from the Monday morning presentations and other comments made throughout the conference, is that software can, and should, be "engineered." Software engineering is a new field but one that can make significant contributions towards reducing total costs, improving management control of software projects and improving software reliability, maintenance, useability, etc.

Grant Fisher from the University of Alberta opened the conference with a wide ranging talk that served as an introduction and overview of computer control systems for industrial processes. Process control systems have grown from the "bottom-up" and simultaneously from the "top-down", i.e. control computers have evolved from direct digital control (DDC) to supervisory systems and at the same time management systems have moved from corporate financial planning to providing targets for individual production units. Unfortunately the software design philosophy and hardware used for implementation were not always compatible and there are now problems as these two approaches "meet in the middle." In the future, system design should include all levels from a single control loop to corporate planning. It was suggested that business (management) systems are a much stronger driving force in the computer market than process control systems and hence greater attention should be paid to the use of commercial business products rather than trying to develop special systems for process control. This appears to be particularly true in the area of man-machine interfaces (cf. dynamic, custom, colored graphic displays) plus network wide communication and database management. Other points raised were: (1) the difference between a geographically distributed network of micro and minicomputers versus a distributed, multi-processor architecture that appears to the application engineer as a single entity. (The former is a current trend but the latter may be more appropriate), (2) the major economic gains are being made at the supervisory level (constraint control, energy and material optimization etc.) rather than at the DDC level, (3) the rapid rate of change in technology and the turnover in computer-oriented personnel should be included in system planning, (4) database management, communication and man-machine interfaces are key points to focus on in current systems, (5) there is a need for higher level application support and programming languages.

Lalive d'Epinany from Brown Boveri Research Centre emphasized one of the key concepts for good software design and implementation—a "layered design" with clearly defined functions for each level and a carefully specified means of interfacing between levels. This approach decomposes the total problem into more manageable components and formally separates the work of the application and cotrol specialists from that of the operating system and hardware specialists. For example, the degree of separation that is possible between a "global computer system" and the application related programs that run on it was demonstrated at Brown Boveri by running a multiprocessor system and then plugging or unplugging processors on-line. Adding processors increased system throughput. Removing processors (cf. failures) reduced speed, but the programs still ran. Obviously the programs were independent of the number, type and network interconnection of the hardware processors.

One of the essential "levels" in any software for industrial control is the "application level" which is often written by the user's process control engineering group. The less these people have to know about the details of hardware operation, communication protocols, etc. and the higher the level of support for program development (eg. compilers, program libraries, etc.) then the more they can concentrate on process control. Bob Wilhelm of Accuray Corp. gave specific examples of how application-oriented, system software can simplify the job of the application programmer and provide a simple systematic basis for handling problems associated with such common functions as the changing of a process control device from operable to non-operable status. For example, a "subscription list" can be associated with the "state" of a particular device or task. If the state changes them everyone on the subscription list is notified and can take appropriate action, eg. if a supervisory program is notified that a measurement instrument has failed then it could perhaps replace the direct reading by an estimate based on other data. The protocol of communication is invisible to the user so he does not have to worry about network architecture or in which node. the supervisory program is located.

Bill Vaughan from Honeywell compared the features of different programming languages and showed how they affected such important areas as program reliability, ease of implementation and maintenance, error recovery, user control of system operation, etc. In a nutshell, FORTRAN is good, PASCAL is better, and ADA (the new language being developed under the auspices of the U.S. Department of Defence for embedded computer systems) looks excellent. Refer to his paper for a brief introduction and history of ADA and a helpful comparison and evaluation of different high level languages.

The bottom line is always "money." The figures differ from project to project, but for large applications such as energy management for an entire chemical plant, coordination of control at the precell and management levels throughout an entire refinery, etc. a typical breakdown of costs over the total lifetime of the software is: 25–30% design and coding, 20–25% testing and integration and 50% for "maintenance." One author estimated maintenance costs as 10% of development costs per year. (Over ten years this confirms that development costs equal maintenance costs). Note that "maintenance" includes fixing bugs, making program modifications to reflect process or operating changes, and minor enhancements. Most people are surprised that a software project that required 20 man-years in the development phase needs 2 people full time for maintenance.

Software engineering can help, and should be considered as important as process engineering and control engineering.

HUMAN FACTORS IN PROCESS CONTROL

John E. Rijnsdorp

The example of description of an Indian Elephant by a committee of blind men was used quite appropriately a number of times during the conference, perhaps most effectively in the session of Distillation Column Control.

Process Control Engineers usually come into contact with the lower limbs of the elephant (i.e., the four limbs rooted in common sense).

1. When in our work we meet a man/machine interface problem, we like to have quick answers to a number of straightforward questions such as: What is a good keyboard for a V.D.U. (Visual Display Unit)? How can trends be combined with process values? How shall I use color? What is better: a flow graph, a bar graph, or a table?
2. Those active in reliability and safety issues would like to know how to set up the alarm system and how to protect process control from operator errors.
3. There is also an interest in operator behavior. What does the operator look at? What does the operator ignore?
4. Engineers responsible for system implementation would like to know how to train plant people. In particular, they like to have information for designing trainers and simulators.

Human Factors Engineering can give some answers to such questions, based on a long series of experimental results. However, more appropriate answers, relevant for modern computerized automation technology, require additional research work. When the elephant extends its trunk towards us that stands for cooperation between Human Factors and Process Control Engineers, in view of appropriate problem formulation and feedback of practical experience. In this way, the brains behind the upper end of the trunk can be put to work for us. As the upper parts of the elephant are rather far above the ground and the viewing angle is unfavorable, not many people can see what they represent. Therefore, Magne Fjeld has hired four donkeys in order to set up an audible display. It is well known that audible displays have a more forceful impact than visual displays.

The first of Magne's donkeys has made a plea for a thorough study of human error. The second one for deep involvement in task analyzing and its implications for training. The third one is for a sincere application of interdisciplinary Systems Engineering oriented design. The fourth one is for harmonizing work organization and automation system design. So far the acceptability of these seemingly far flung ideas is rather doubtful. Still, I would like to invite everbody to ponder the following questions. Your responses, if any, could possibly be an item for the next conference.

Are you willing to let control and information system design be influenced by consideration of human error idiosyncrasies? Is there a remote possibility for you to invoice your colleagues and your boss to follow an integrated Human Factor and System oriented design approach? If so, would you also appoint for this? Can you find the time and the assistance for obtaining a task analysis of the future uses of the system you are designing? Can you arrange for appropriate training of future users in particular today's operators? Do you see opportunities for the future user to adapt the system to his/her needs and wishes? Can you promote a human factor evaluation of your system under actual operating conditions?

ADVANCED STRATEGIES FOR PROCESS CONTROL AND ESTIMATION

J. H. Seinfeld

Significant progress in the development of advanced control algorithms has taken place since the first Engineering Foundation Conference on Process Control. Although the industrial rate of adoption of such algorithms has not kept pace with that of the developments themselves, there is evidence that industry is beginning to recognize the power and usefulness of the available theory. In Japan, for example, Dr. Hashimoto has reported that 63 percent of the companies he surveyed are greatly interested in advanced control technology and 83 percent of the actual implementations of advanced control theory were successful. U.S. industries appear to lag somewhat behind those of Japan in the adoption of advanced strategies. Nevertheless, the explosion in semiconductor technology is removing computational limitations to implementation of multivariable control theory.

The only new advanced control technique discussed at the meeting was Model Algorithmic Control (Mehra), a promising method that places certain demands on the process operator. It is clear that this method deserves study, in particular careful comparison with other techniques. Although algorithm research and development are continually needed, this aspect of process control no longer appears to deserve to occupy the position of prominence it has for many years. It would perhaps not be overly optimistic to assert that algorithms currently exist in the literature to solve somewhere over 95 percent of all process control problems. Certain exceptions, such as emergency or event control and the graceful inclusion of state and control constraints into multivariable control algorithms, still exist and are ripe for continued attention, as are the special control problems arising in very nonlinear systems. The major needs in the area of advanced control technology now seem to lie at the interface between design and control. Traditionally, process control has been largely a cleanup operation. It is now evident that the distinction between process engineers and control engineers should be less sharp. Control engineers must enter the project at the design phase. Questions such as:

1. What is the control objective?
2. What variables can be controlled?
3. What manipulated variables do we choose?

need to be addressed at the design stage.

The intervention of control system design at the process design stage underlies the coming shift of process control from a purely defensive field to an offensive, or synthetic, field. Problems such as integrated process operation and information reliability improvement loom important as new areas of endeavor. In short, there was a strong consensus at the Conference that process control will experience a shift from algorithm development to synthesis.

There was considerable discussion concerning the most desirable level of process detail at which to test new algorithms and methods, the single unit or the entire plant. The single unit (distillation tower, reactor) is simpler to handle and is amenable to generalization since the principles are common to most such units. The whole plant is, of course, our business and the control of the entire plant is really the bottom line. Effort on control system design at the plant level is clearly needed and was strongly endorsed by the attendees.

A few final comments reflecting the opinions of the reviewer can be made on the subject of advanced control strategies. The period during which university researchers could afford to focus almost exclusively on advanced algorithm developments appears to be drawing to a close. Continued university research on advanced control and the control/design interface will benefit greatly from university/industry interaction. Such interaction, in the form of summer employment for students and faculty, the use of industrial process equipment for control studies and theses suggested by industrial problems is a source of fresh ideas and invaluable experience for the student and professor. A concrete proposal can be offered. University researchers in process control should attempt to contact industrial laboratories to learn about current and anticipated control problems. Proposals for joint research can be prepared, proposals that include provision for significant feedback on project goals. On the other hand, managers of industrial process control groups should attempt to obtain funds for and encourage such joint research. (The cost of these joint projects is not large, basically modest equipment and student salary support). Experience in other areas of chemical engineering has borne out the advantages to both university research and industrial development of joint academic/industry research programs.

DISTILLATION CONTROL/ ENERGY MANAGEMENT

J. Patrick Kennedy
Oil Systems, Inc.

Distillation is the reverse of the irreversible mixing of two pure streams and thus extracts its thermodynamic penalty, to the tune of 3% of the total U.S. energy usage. There were four papers presented upon this subject that set the position of distillation control of this time. Page Buckley presented several examples of heat integration of distillation columns that have been constructed at DuPont. This is quite significant because, due to a variety of practical reasons, heat integration has not grown nearly at the rate that steady state economics indicate is reasonable. Indeed, there were problems that were reported, but solved, to put these projects on-line and the experience base was widened. This was followed by a discussion by Carroll Ryskamp of the methods for avoiding complex decoupling schemes by proper design. Carroll used the physics of the process to descibe the reasons for several observed control problems. A very sound technique for avoiding the coupling in dual composition control was presented–called combined reflux and distillate control. Tom McAvoy followed these papers with the results from a study on the problems of extrapolating dynamic behavior from low purity distillations to high purity distillation. These results utilized a simulation of a binary distillation but agreed with observed phenomena for multicomponent applications. The section was rounded out by a survey by Kurt Waller. Although entitled University Research, Kurt included references to several industrial research projects that were appropriate.

The energy management section was begun by George Quentin from EPRI. He described the control analysis of the Gasification/Combined

Cycle (GCC) power plant project of EPRI. This plant is likely the new technology that will efficiently convert coal to electrical power. Since the project design and engineering has been continuously expedited, it was appropriate that controllability be studied at the inception by experiment and study. Ron Simpkins presented a planned system for energy management in DuPont. This system uses multi-processor equipment based upon earlier process control computers built by DuPont. The final paper was by Mr. Dick Hanson describing an energy management system built by Taylor Instrument. This system is currently being implemented on a series of boilers in a paper mill.

SUMMARIZING REMARKS ON THE SEA ISLAND DISCUSSION ON PLANT CONTROL

A. S. Foss
University of California, Berkeley

I believe that I shall not be far from wrong when I state that the topic of plant control was for all participants the most stimulating and exciting topic treated at this conference. Everyone, academics and industrial practitioners, are seeing that new effort on the many problems posed by the control of integrated plants is absolutely necessary. In this growing appreciation of the importance of plant control, the profession is experiencing a wide-scale convergence of academic and industrial recognition that this is a central and pressing area for work. This concurrence was evident in many of the discussions at this conference, not just this particular session.

A general concurrence was evident on two matters.
- Process design and control system design must be coordinated.

    An alertness must be maintained (somehow) in both of these design activities to all eventualities in process operation.
- A broad scope is needed.

    The entire plant, of course, must be considered, but most importantly, one must consider the different aspects of process operation. Several of these were treated by the speakers and panelists in this section.
    - Data reconciliation (Dick Mah).
        - The need for resilience in process and control system (Manfred Morari).
    - Process operability as influenced by process and control system design (Jim Douglas).

- The very substantial influence of ever-present abnormal process conditions on one's approach to plant control (Irv Rinard).
- An overall process approach centered on product quality (Joe Shunta).
- The influence of energy integration, interrelation of process with utility energy and fuel distribution, and product blending (Tom Marlin).

In all of these discussions, the predominant concern was the maintenance and improvement of plant operability. My impression of the work so far reported on the matter of plant operability is that there has been an overwhelming emphasis on the *design* of control systems claimed to promote process operability. This is most evident in Manfred Morari's definition of the plant control problem. The definition comprised these five decisions at the design stage:

1. Formulating the control objectives.
2. Selection of variables to be controlled.
3. Selection of variables to be measured.
4. Selection of variables to be manipulated.
5. Selection of the control links.

In this definition, it seems to me that there is a least one "forgotten" constituent. That I would describe as

6. Insuring of operability in the *operating* plant.

Designing for improved operability is certainly to be commended, but because there are always overlooked effects at the design stage and unexpected operating conditions, something more has to be done during operations to insure that the plant-of-the-hour weathers the storm.

One participant told of being challenged by his boss to invent a "box" that, upon being interrogated with any problem-of-the-hour, would provide process operators with directives to guide the process safely past the approaching shoals. Such an omnibox is what is needed to handle item 6 of the amended plant control problem and might include features of the following type.

- Relative stability of the process is to be maintained.
  This matter is a recurring problem because of:
    - Production rate changes.
    - Equipment failures and operator blunders.
    - Abnormal conditions.
    - Process optimization drivers.
- A means to assess process stability.
- Data reconciliation.
- Priorities switching.

Product quality as a control criterion under normal conditions might be relaxed and replaced under abnormal conditions by process stability.

I make one further observation about the heightened interest among university researchers and industrial practitioners in plant control. The industrial segment has always had to face plant operability problems. University researchers have seldom met with that challenge; rather, they have concentrated heavily on the design aspects of process control. Possibly this difference in concentration has been the source of differences in points of view and directions of these two groups. But now the unmistakable confluence of opinion regarding the importance of the plant control problem promises a marshalling of a great wealth of human resources. Circumstances are auspicious for significant advances in all of chemical process control.

DISTRIBUTED COMPUTER PROCESS CONTROL

Duncan A. Mellichamp
University of California, Santa Barbara

The session on distributed computer process control consisted of two papers and a panel discussion following.

"The structuring of distributed intelligence computer control system" by C. W. Rose was an excellent tutorial paper on the concepts of distributed digital computing. The author dealt with distributed computing systems the way they someday will be built and not as most commercial systems presently are designed. In this regard, a comment raised from the floor during the later panel discussion should be noted, viz. that many of the distributed systems available today are actually instrumentation systems rather than distributed computer control systems. The author dealt substantially with the latter case, i.e. with distributed intelligence whose actual location in the network should be transparent to the user. A description of the requirements for such systems was given, emphasizing reliability, response time and performance, and modularity. Following a discussion of network topologies, the author concluded with an extended introduction to digital communication techniques and protocols.

"An Approach to Digital Process Control" by M. Masak described an industrial computer control system with a geographically-distributed, i.e. remote front-end data acquisition sub-system and dual processors. One of these processors is substantially dedicated to process control functions while the other is used for information handling and displays. The system has been adopted by Chevron for general application within the company; four new systems presently are due for installation in the coming two-year period with a dozen systems already installed.

The distribution of computing functions within the system is a traditional one with important, time-critical duties assigned to the control processor. The author discussed the application of inferential control, constraint control, measurement redundancy, and adaptive control, all briefly. One unusual feature of the system was the use of process control software (COSMIC) which was developed in-house. At least one questioner indicated slight surprise at the relatively low total of manpower effort required to develop this system and to replicate it.

The two papers presented in this session, plus the panel discussion and several papers presented earlier on real-time applications, furnish a good brief introduction to the field. Nevertheless, it was clear that a proper session on distributed computing will have to wait until commercial systems just coming on the market have arrived and users have had the opportunity to evaluate them in comparison to traditional computer network configurations.

Author Index

Belanger, Pierre R.
Review of Some Adaptive Control Schemes for Process Control, 269

Bristol, Edgar
Important Design Issues and Recommendation for Prospective Users of Distributed Control—III, 605

Buckley, Page S.
Control of Heat-Integrated Columns, 347

Denn, M.
Industrial Theme Problems in Process Control, 609

d'Epinay, Th. Lalive
Principles in Structuring Software for Process Control, 33

Douglas, James M.
Process Operability and Control of Preliminary Designs, 497

Edgar, Thomas F.
New Results and the Status of Computer-Aided Process Control Systems Design in North America, 213

Field, Magne
On Closing "The Gap"—With Modern Control Theory that Works, 229

Fisher, D. Grant
Systems Software, 613
Systems Software for Process Control: Today's State of the Art, 3

Foss, Alan S.
Control Systems for Integrated Plants, 625

Gilles, E. Dieter
Application of Modelling, Estimation and Control to Chemical Processes, 311

Hanson, Richard A.
Plant Utility Management, 449

Hashimoto, Tori
New Results and the Status of Computer-Aided Process Control Systems Design in Japan, 147

Kanous, Lawrence E.
Training and Selection of Operators in Complex Process Control Systems, 137

Kennedy, J. Patrick
Distillation column Control/Energy Management, 623

Livingston, William L.
The Systems Engineering Context of Process Control and Man-Machine System Design, 117

Mah, Richard S.H.
Design and Analysis of Process Performance: Monitoring Systems, 525

Marlin, Thomas E.
Control Systems for Complete Plants—The Industrial View—III, 555

Masak, Martin
An Approach to Digital Process Control, 581

McAvoy, Thomas J.
On the Difference Between Distillation Column Control Problems, 377

Mehra, Raman K.
Model Algorithmic Control (MAC): Review and Recent Developments, with Ramine Rouhani, 287

Mellichamp, Duncan A.
Distributed Computer Process Control, 629
Important Design Issues and Recommendation for Prospective Users of Distributed Control—II, 603

Messare, Ramsi R.
Important Design Issues and Recommendation for Prospective Users of Distributed Control—I, 599

Morari, Manfred
Integrated Plant Control: A Solution at Hand or a Research Topic for the Next Decade?, 467

Quentin, George H.
Control Analysis of Gasification/Combined Cycle (GCC) Power Plants, 415

Ray, W. Harmon
New Approaches to the Dynamics of Nonlinear Systems with Implicationsfor Process and Control System Design, 245

Rijnsdorp, John E.
Human Factors, 617
Important Problems and Challenges in Human Factors and Man-Machine Engineering for Process Control Systems, 93

Rinard, Irven
Control Systems for Complete Plants—The Industrial View—I, 541

Rose, Charles W.
The Structuring of Distributed Intelligence Computer Control Systems, 565

Rouhani, Ramine
See Raman K. Mehra, 287

Ryskamp, Carroll J.
Explicit vs. Implicit Decoupling in Distillation Control, 361

Seinfeld, John H.
Advanced Strategies for Process Control and Estimation, 619

Senders, John W.
Human Error and Human Reliability in Process Control, 111

Shunta, Joseph P.
Control Systems for Complete Plants—The Industrial View—II, 547

Simpkins, C. Ronald
General Industrial Energy Management System—Functionality and Architecture, 433

Tysso, Arne
New Results and the Status of Computer-Aided Process Control Systems in Europe, 187

Vaughan, William C.M.
Application of the ADA Language to Process Control, 71

Waller, Kurt V.
University Research on Dual Composition Control of Distillation: A Review, 395

Wilhelm, Robert G., Jr.
Operational Needs in Software for Complex Process Control Systems, 51

Subject Index

Abnormalities; Chemical plants; Control systems; Control theory; Design improvements; Design practices; Industrial plants; Quality control
Control Systems for Integrated Plants, Alan S. Foss, 625

Actuators; Algorithms; Computers; Constraints; Design practices; Emergencies; Model studies; Optimization; Plants (industrial); Process control; Selection
Model Algorithmic Control (MAC): Review and Recent Developments, Raman K. Mehra and Ramine Rouhani, 287

Adaptive systems; Control systems; Convergence; Identification; Industries; Paper industry; Process control; Processing; Regulators
Review of Some Adaptive Control Schemes for Process Control, Pierre R. Belanger, 269

Alarm systems; Chemical plants; Control equipment; Control systems; Display devices; Laboratories; Measuring instruments; Operator performance; Valves
Control Systems for Complete Plants—The Industrial View—I, Irven Rinard, 541

Alarm systems; Computer languages; Cost analysis; Distributed processing; Man machine systems; Modules; Process control; Reliability
Important Design Issues and Recommendation for Prospective Users of Distributed Control—III, Edgar Bristol, 605

Algorithms; Computer programs; Computers; Control; Modules; Nodes; Operating criteria; Paper industry; Process control
Operational Needs in Software for Complex Process Control Systems, Robert G. Wilhelm, Jr., 51

Algorithms; Computers; Constraints; Design practices; Emergencies; Model studies; Optimization; Plants (industrial); Process control; Selection; Actuators
Model Algorithmic Control (MAC): Review and Recent Developments, Raman K. Mehra and Ramine Rouhani, 287

Algorithms; Constraints; Control systems; Control theory; Emergencies; Estimation; Operator performance; Process control; Research
Advanced Strategies for Process Control and Estimation, John H. Seinfeld, 619

Application methods; Computer applications; Computer programs; Layered systems; Mathematical models; Nodes; Process control; Sequential analysis; Systems engineering
Principles in Structuring Software for Process Control, Th. Lalive d'Epinay, 33

Automatic control; Computer programs; Control systems; Digital systems; Interfaces; Maintainability; Operator performance; Process control
An Approach to Digital Process Control, Martin Masak, 581

Benefits; Computerized simulation; Control systems; Economic impact; Gasification; Loading; Performance; Powerplants; Transient response
Control Analysis of Gasification/Combined Cycle (GCC) Power Plants, George H. Quentin, 415

Blending; Computer programs; Computers; Control systems; Distillation equipment; Fuels; Heat balance; Power supplies; Sequential analysis
Control Systems for Complete Plants—The Industrial View—III, Thomas E. Marlin, 555

Boilers; Chemical industry; Control systems; Costs; Energy conservation; Energy conversion; Energy methods; Management systems; Petroleum industry; Powerhouses
General Industrial Energy Management System—Functionality and Architecture, C. Ronald Simpkins, 433

Boilers; Combustion control; Environmental factors; Header pipes; Management; Optimization; Plants (industrial); Pressure responses; Remote sensing; Steam generators; Utilities
Plant Utility Management, Richard A. Hanson, 449

Boilers; Control systems; Control theory; Distillation; Distillation equipment; Energy conversion; Energy methods; Gasification; Paper mills; Powerplants
Distillation column Control/Energy Management, J. Patrick Kennedy, 623

Chemical industry; Chemical reactors; Control; Distillation equipment; Estimation; Mathematical models; Model studies; Polymerization; Systems analysis
Application of Modelling, Estimation and Control to Chemical Processes, E. Dieter Gilles, 311

Chemical industry; Compositions; Computer controlled signals; Control systems; Distillation; Experimental data; Feedback control; Process control
University Research on Dual Composition Control of Distillation: A Review, Kurt V. Waller, 395

Chemical industry; Computer applications; Computerized design; Costs; Emission control; Energy conservation; Environmental quality regulations; Process control; Safety standards; Systems engineering
New Results and the Status of Computer-Aided Process Control Systems Design in Japan, Tori Hashimoto, 147

Chemical industry; Control systems; Costs; Energy conservation; Energy conversion; Energy methods; Management systems; Petroleum industry; Powerhouses; Boilers
General Industrial Energy Management System—Functionality and Architecture, C. Ronald Simpkins, 433

Chemical industry; Industries; Model studies; National Science Foundation; Process control; Recommendations; Simulation models; Universities
Industrial Theme Problems in Process Control, M. Denn, 609

Chemical plants; Chemical reactions; Communication systems; Computer programs; Management systems; Process control; Systems analysis
  Systems Software for Process Control: Today's State of the Art, D. Grant Fisher, 3

Chemical plants; Constraints; Control systems; Energy conservation; Materials control; Model studies; Process control; Quality control; Systems analysis
  Control Systems for Complete Plants—The Industrial View—II, Joseph P. Shunta, 547

Chemical plants; Control equipment; Control systems; Display devices; Laboratories; Measuring instruments; Operator performance; Valves; Alarm systems
  Control Systems for Complete Plants—The Industrial View—I, Irven Rinard, 541

Chemical plants; Control systems; Control theory; Design improvements; Design practices; Industrial plants; Quality control; Abnormalities
  Control Systems for Integrated Plants, Alan S. Foss, 625

Chemical plants; Control systems; Design; Operations; Preliminary investigations; Process charting; Process control; Steady state; Variations
  Process Operability and Control of Preliminary Designs, James M. Douglas, 497

Chemical reactions; Communication systems; Computer programs; Management systems; Process control; Systems analysis; Chemical plants
  Systems Software for Process Control: Today's State of the Art, D. Grant Fisher, 3

Chemical reactors; Control; Distillation equipment; Estimation; Mathematical models; Model studies; Polymerization; Systems analysis; Chemical industry
  Application of Modelling, Estimation and Control to Chemical Processes, E. Dieter Gilles, 311

Chemical reactors; Dynamic characteristics; Dynamics; Input output routines; Nonlinear systems; Nonlinear systems; Oscillations; Parametric equations; Process control; Sensitivity analysis; Systems engineering
  New Approaches to the Dynamics of Nonlinear Systems with Implicationsfor Process and Control System Design, W. Harmon Ray, 245

Classification; Data analysis; Error analysis; Measurement; Monitor routines; Performance; Process charting; Process control; Research
  Design and Analysis of Process Performance: Monitoring Systems, Richard S.H. Mah, 525

Columns (process engineering); Compositions; Control systems; Digital simulation; Distillation equipment; Material balance; Purity; Towers
  On the Difference Between Distillation Column Control Problems, Thomas J. McAvoy, 377

Combustion control; Environmental factors; Header pipes; Management; Optimization; Plants (industrial); Pressure responses; Remote sensing; Steam generators; Utilities; Boilers
  Plant Utility Management, Richard A. Hanson, 449

Communication systems; Computer applications; Computer languages; Computer programs; Computer systems programs; Error analysis; Optimization; Process control; Reliability
  Application of the ADA Language to Process Control, William C.M. Vaughan, 71

Communication systems; Computer programs; Management systems; Process control; Systems analysis; Chemical plants; Chemical reactions
  Systems Software for Process Control: Today's State of the Art, D. Grant Fisher, 3

Compositions; Computer controlled signals; Control systems; Distillation; Experimental data; Feedback control; Process control; Chemical industry
  University Research on Dual Composition Control of Distillation: A Review, Kurt V. Waller, 395

Compositions; Control systems; Digital simulation; Distillation equipment; Material balance; Purity; Towers; Columns (process engineering)
  On the Difference Between Distillation Column Control Problems, Thomas J. McAvoy, 377

Computer applications; Computerized design; Computer programming; Computers; Europe; Identification; Industrial plants; Models; Process control; Simulation
  New Results and the Status of Computer-Aided Process Control Systems in Europe, Arne Tysso, 187

Computer applications; Computerized design; Computer programming; Estimates; Multivariate analysis; Process control; Research; Systems engineering
  New Results and the Status of Computer-Aided Process Control Systems Design in North America, Thomas F. Edgar, 213

Computer applications; Computerized design; Costs; Emission control; Energy conservation; Environmental quality regulations; Process control; Safety standards; Systems engineering; Chemical industry
  New Results and the Status of Computer-Aided Process Control Systems Design in Japan, Tori Hashimoto, 147

Computer applications; Computer languages; Computer programs; Computer systems programs; Error analysis; Optimization; Process control; Reliability; Communication systems
  Application of the ADA Language to Process Control, William C.M. Vaughan, 71

Computer applications; Computer programs; Layered systems; Mathematical models; Nodes; Process control; Sequential analysis; Systems engineering; Application methods
  Principles in Structuring Software for Process Control, Th. Lalive d'Epinay, 33

Computer applications; Design improvements; Design practices; Design standards; Distributed processing; Instrumentation; Interfaces; Maintainability; Operator performance; Reliability
  Important Design Issues and Recommendation for Prospective Users of Distributed Control—I, Ramsi R. Messare, 599

Computer controlled signals; Computer languages; Computer programming; Control equipment; Digital systems; Industrial plants; Management systems; Process control; Systems analysis
  Systems Software, D. Grant Fisher, 613

Computer controlled signals; Computer programming; Constraints; Control systems; Control theory; Digital systems; Distributed processing; Intelligence; Process control
  Distributed Computer Process Control, Duncan A. Mellichamp, 629

Computer controlled signals; Computers; Decision making; Distributed processing; Networks; Optimization; Reliability; Systems analysis
  Important Design Issues and Recommendation for Prospective Users of Distributed Control—II, Duncan A. Mellichamp, 603

Computer controlled signals; Control systems; Data acquisition; Distributed processing; Intelligence; Performance; Reliability; Topology
  The Structuring of Distributed Intelligence Computer Control Systems, Charles W. Rose, 565

Computer controlled signals; Control systems; Distillation; Experimental data; Feedback control; Process control; Chemical industry; Compositions
  University Research on Dual Composition Control of Distillation: A Review, Kurt V. Waller, 395

Computerized design; Computer programming; Computers; Europe; Identification; Industrial plants; Models; Process control; Simulation; Computer applications
  New Results and the Status of Computer-Aided Process Control Systems in Europe, Arne Tysso, 187

Computerized design; Computer programming; Estimates; Multivariate analysis; Process control; Research; Systems engineering; Computer applications
  New Results and the Status of Computer-Aided Process Control Systems Design in North America, Thomas F. Edgar, 213

Computerized design; Costs; Emission control; Energy conservation; Environmental quality regulations; Process control; Safety standards; Systems engineering; Chemical industry; Computer applications
  New Results and the Status of Computer-Aided Process Control Systems Design in Japan, Tori Hashimoto, 147

Computerized simulation; Control systems; Economic impact; Gasification; Loading; Performance; Powerplants; Transient response; Benefits
  Control Analysis of Gasification/Combined Cycle (GCC) Power Plants, George H. Quentin, 415

Computer languages; Computer programming; Control equipment; Digital systems; Industrial plants; Management systems; Process control; Systems analysis; Computer controlled signals
  Systems Software, D. Grant Fisher, 613

Computer languages; Computer programs; Computer systems programs; Error analysis; Optimization; Process control; Reliability; Communication systems; Computer applications
  Application of the ADA Language to Process Control, William C.M. Vaughan, 71

635

Computer languages; Cost analysis; Distributed processing; Man machine systems; Modules; Process control; Reliability; Alarm systems
  Important Design Issues and Recommendation for Prospective Users of Distributed Control—III, Edgar Bristol, 605

Computer programming; Computers; Europe; Identification; Industrial plants; Models; Process control; Simulation; Computer applications; Computerized design
  New Results and the Status of Computer-Aided Process Control Systems in Europe, Arne Tysso, 187

Computer programming; Constraints; Control systems; Control theory; Digital systems; Distributed processing; Intelligence; Process control; Computer controlled signals
  Distributed Computer Process Control, Duncan A. Mellichamp, 629

Computer programming; Control equipment; Digital systems; Industrial plants; Management systems; Process control; Systems analysis; Computer controlled signals; Computer languages
  Systems Software, D. Grant Fisher, 613

Computer programming; Estimates; Multivariate analysis; Process control; Research; Systems engineering; Computer applications; Computerized design
  New Results and the Status of Computer-Aided Process Control Systems Design in North America, Thomas F. Edgar, 213

Computer programs; Computers; Control; Modules; Nodes; Operating criteria; Paper industry; Process control; Algorithms
  Operational Needs in Software for Complex Process Control Systems, Robert G. Wilhelm, Jr., 51

Computer programs; Computers; Control systems; Distillation equipment; Fuels; Heat balance; Power supplies; Sequential analysis; Blending
  Control Systems for Complete Plants—The Industrial View—III, Thomas E. Marlin, 555

Computer programs; Computer systems programs; Error analysis; Optimization; Process control; Reliability; Communication systems; Computer applications; Computer languages
  Application of the ADA Language to Process Control, William C.M. Vaughan, 71

Computer programs; Control systems; Digital systems; Interfaces; Maintainability; Operator performance; Process control; Automatic control
  An Approach to Digital Process Control, Martin Masak, 581

Computer programs; Layered systems; Mathematical models; Nodes; Process control; Sequential analysis; Systems engineering; Application methods; Computer applications
  Principles in Structuring Software for Process Control, Th. Lalive d'Epinay, 33

Computer programs; Management systems; Process control; Systems analysis; Chemical plants; Chemical reactions; Communication systems
  Systems Software for Process Control: Today's State of the Art, D. Grant Fisher, 3

Computers; Constraints; Design practices;
Emergencies; Model studies; Optimization; Plants
(industrial); Process control; Selection; Actuators;
Algorithms
   Model Algorithmic Control (MAC): Review and
   Recent Developments, Raman K. Mehra and
   Ramine Rouhani, 287

Computers; Control; Modules; Nodes; Operating
criteria; Paper industry; Process control;
Algorithms; Computer programs
   Operational Needs in Software for Complex
   Process Control Systems, Robert G. Wilhelm,
   Jr., 51

Computers; Control systems; Distillation equipment;
Fuels; Heat balance; Power supplies; Sequential
analysis; Blending; Computer programs
   Control Systems for Complete Plants—The
   Industrial View—III, Thomas E. Marlin, 555

Computers; Decision making; Distributed processing;
Networks; Optimization; Reliability; Systems
analysis; Computer controlled signals
   Important Design Issues and Recommendation for
   Prospective Users of Distributed Control—II,
   Duncan A. Mellichamp, 603

Computers; Europe; Identification; Industrial plants;
Models; Process control; Simulation; Computer
applications; Computerized design; Computer
programming
   New Results and the Status of Computer-Aided
   Process Control Systems in Europe, Arne Tysso,
   187

Computer systems hardware; Computer systems
programs; Control systems; Design improvements;
Design practices; Human factors; Man machine
systems; Process control; Systems engineering
   The Systems Engineering Context of Process
   Control and Man-Machine System Design,
   William L. Livingston, 117

Computer systems programs; Control systems;
Design improvements; Design practices; Human
factors; Man machine systems; Process control;
Systems engineering; Computer systems hardware
   The Systems Engineering Context of Process
   Control and Man-Machine System Design,
   William L. Livingston, 117

Computer systems programs; Error analysis;
Optimization; Process control; Reliability;
Communication systems; Computer applications;
Computer languages; Computer programs
   Application of the ADA Language to Process
   Control, William C.M. Vaughan, 71

Condensers; Control; Distillation; Distillation
equipment; Energy conservation; Heat; Heat
recovery; Process control; Vapor pressure
   Control of Heat-Integrated Columns, Page S.
   Buckley, 347

Constraints; Control systems; Control theory;
Digital systems; Distributed processing;
Intelligence; Process control; Computer controlled
signals; Computer programming
   Distributed Computer Process Control, Duncan
   A. Mellichamp, 629

Constraints; Control systems; Control theory;
Emergencies; Estimation; Operator performance;
Process control; Research; Algorithms
   Advanced Strategies for Process Control and
   Estimation, John H. Seinfeld, 619

Constraints; Control systems; Control theory;
Gaussian quadrature; Industrial engineering; Linear
systems; Quadratic forms; Systems engineering
   On Closing "The Gap"—With Modern Control
   Theory that Works, Magne Field, 229

Constraints; Control systems; Energy conservation;
Materials control; Model studies; Process control;
Quality control; Systems analysis; Chemical plants
   Control Systems for Complete Plants—The
   Industrial View—II, Joseph P. Shunta, 547

Constraints; Design practices; Emergencies; Model
studies; Optimization; Plants (industrial); Process
control; Selection; Actuators; Algorithms;
Computers
   Model Algorithmic Control (MAC): Review and
   Recent Developments, Raman K. Mehra and
   Ramine Rouhani, 287

Control; Controller characteristics; Plants
(industrial); Process control; Quality control;
Research; Sensitivity; Systems analysis; Tuning
   Integrated Plant Control: A Solution at Hand or a
   Research Topic for the Next Decade?, Manfred
   Morari, 467

Control; Couplings; Distillation; Nonlinear systems;
Process control; Responses; Stability
   Explicit vs. Implicit Decoupling in Distillation
   Control, Carroll J. Ryskamp, 361

Control; Distillation; Distillation equipment; Energy
conservation; Heat; Heat recovery; Process
control; Vapor pressure; Condensers
   Control of Heat-Integrated Columns, Page S.
   Buckley, 347

Control; Distillation equipment; Estimation;
Mathematical models; Model studies;
Polymerization; Systems analysis; Chemical
industry; Chemical reactors
   Application of Modelling, Estimation and Control
   to Chemical Processes, E. Dieter Gilles, 311

Control; Modules; Nodes; Operating criteria; Paper
industry; Process control; Algorithms; Computer
programs; Computers
   Operational Needs in Software for Complex
   Process Control Systems, Robert G. Wilhelm,
   Jr., 51

Control equipment; Control systems; Display
devices; Laboratories; Measuring instruments;
Operator performance; Valves; Alarm systems;
Chemical plants
   Control Systems for Complete Plants—The
   Industrial View—I, Irven Rinard, 541

Control equipment; Digital systems; Industrial
plants; Management systems; Process control;
Systems analysis; Computer controlled signals;
Computer languages; Computer programming
   Systems Software, D. Grant Fisher, 613

Controller characteristics; Plants (industrial);
Process control; Quality control; Research;
Sensitivity; Systems analysis; Tuning; Control
   Integrated Plant Control: A Solution at Hand or a
   Research Topic for the Next Decade?, Manfred
   Morari, 467

Control systems; Control theory; Design
improvements; Design practices; Industrial plants;
Quality control; Abnormalities; Chemical plants
   Control Systems for Integrated Plants, Alan S.
   Foss, 625

Control systems; Control theory; Digital systems; Distributed processing; Intelligence; Process control; Computer controlled signals; Computer programming; Constraints
  Distributed Computer Process Control, Duncan A. Mellichamp, 629

Control systems; Control theory; Distillation; Distillation equipment; Energy conversion; Energy methods; Gasification; Paper mills; Powerplants; Boilers
  Distillation column Control/Energy Management, J. Patrick Kennedy, 623

Control systems; Control theory; Emergencies; Estimation; Operator performance; Process control; Research; Algorithms; Constraints
  Advanced Strategies for Process Control and Estimation, John H. Seinfeld, 619

Control systems; Control theory; Gaussian quadrature; Industrial engineering; Linear systems; Quadratic forms; Systems engineering; Constraints
  On Closing "The Gap"—With Modern Control Theory that Works, Magne Field, 229

Control systems; Convergence; Identification; Industries; Paper industry; Process control; Processing; Regulators; Adaptive systems
  Review of Some Adaptive Control Schemes for Process Control, Pierre R. Belanger, 269

Control systems; Costs; Energy conservation; Energy conversion; Energy methods; Management systems; Petroleum industry; Powerhouses; Boilers; Chemical industry
  General Industrial Energy Management System—Functionality and Architecture, C. Ronald Simpkins, 433

Control systems; Data acquisition; Distributed processing; Intelligence; Performance; Reliability; Topology; Computer controlled signals
  The Structuring of Distributed Intelligence Computer Control Systems, Charles W. Rose, 565

Control systems; Design; Operations; Preliminary investigations; Process charting; Process control; Steady state; Variations; Chemical plants
  Process Operability and Control of Preliminary Designs, James M. Douglas, 497

Control systems; Design improvements; Design practices; Human factors; Man machine systems; Process control; Systems engineering; Computer systems hardware; Computer systems programs
  The Systems Engineering Context of Process Control and Man-Machine System Design, William L. Livingston, 117

Control systems; Digital simulation; Distillation equipment; Material balance; Purity; Towers; Columns (process engineering); Compositions
  On the Difference Between Distillation Column Control Problems, Thomas J. McAvoy, 377

Control systems; Digital systems; Interfaces; Maintainability; Operator performance; Process control; Automatic control; Computer programs
  An Approach to Digital Process Control, Martin Masak, 581

Control systems; Display devices; Laboratories; Measuring instruments; Operator performance; Valves; Alarm systems; Chemical plants; Control equipment
  Control Systems for Complete Plants—The Industrial View—I, Irven Rinard, 541

Control systems; Distillation; Experimental data; Feedback control; Process control; Chemical industry; Compositions; Computer controlled signals
  University Research on Dual Composition Control of Distillation: A Review, Kurt V. Waller, 395

Control systems; Distillation equipment; Fuels; Heat balance; Power supplies; Sequential analysis; Blending; Computer programs; Computers
  Control Systems for Complete Plants—The Industrial View—III, Thomas E. Marlin, 555

Control systems; Economic impact; Gasification; Loading; Performance; Powerplants; Transient response; Benefits; Computerized simulation
  Control Analysis of Gasification/Combined Cycle (GCC) Power Plants, George H. Quentin, 415

Control systems; Energy conservation; Materials control; Model studies; Process control; Quality control; Systems analysis; Chemical plants; Constraints
  Control Systems for Complete Plants—The Industrial View—II, Joseph P. Shunta, 547

Control theory; Design improvements; Design practices; Industrial plants; Quality control; Abnormalities; Chemical plants; Control systems
  Control Systems for Integrated Plants, Alan S. Foss, 625

Control theory; Digital systems; Distributed processing; Intelligence; Process control; Computer controlled signals; Computer programming; Constraints; Control systems
  Distributed Computer Process Control, Duncan A. Mellichamp, 629

Control theory; Distillation; Distillation equipment; Energy conversion; Energy methods; Gasification; Paper mills; Powerplants; Boilers; Control systems
  Distillation column Control/Energy Management, J. Patrick Kennedy, 623

Control theory; Emergencies; Estimation; Operator performance; Process control; Research; Algorithms; Constraints; Control systems
  Advanced Strategies for Process Control and Estimation, John H. Seinfeld, 619

Control theory; Gaussian quadrature; Industrial engineering; Linear systems; Quadratic forms; Systems engineering; Constraints; Control systems
  On Closing "The Gap"—With Modern Control Theory that Works, Magne Field, 229

Convergence; Identification; Industries; Paper industry; Process control; Processing; Regulators; Adaptive systems; Control systems
  Review of Some Adaptive Control Schemes for Process Control, Pierre R. Belanger, 269

Cost analysis; Distributed processing; Man machine systems; Modules; Process control; Reliability; Alarm systems; Computer languages
  Important Design Issues and Recommendation for Prospective Users of Distributed Control—III, Edgar Bristol, 605

Costs; Emission control; Energy conservation; Environmental quality regulations; Process control; Safety standards; Systems engineering; Chemical industry; Computer applications; Computerized design
  New Results and the Status of Computer-Aided Process Control Systems Design in Japan, Tori Hashimoto, 147

Costs; Energy conservation; Energy conversion; Energy methods; Management systems; Petroleum industry; Powerhouses; Boilers; Chemical industry; Control systems
  General Industrial Energy Management System—Functionality and Architecture, C. Ronald Simpkins, 433

Couplings; Distillation; Nonlinear systems; Process control; Responses; Stability; Control
  Explicit vs. Implicit Decoupling in Distillation Control, Carroll J. Ryskamp, 361

Data acquisition; Distributed processing; Intelligence; Performance; Reliability; Topology; Computer controlled signals; Control systems
  The Structuring of Distributed Intelligence Computer Control Systems, Charles W. Rose, 565

Data analysis; Error analysis; Measurement; Monitor routines; Performance; Process charting; Process control; Research; Classification
  Design and Analysis of Process Performance: Monitoring Systems, Richard S.H. Mah, 525

Decision making; Distributed processing; Networks; Optimization; Reliability; Systems analysis; Computer controlled signals; Computers
  Important Design Issues and Recommendation for Prospective Users of Distributed Control—II, Duncan A. Mellichamp, 603

Design; Operations; Preliminary investigations; Process charting; Process control; Steady state; Variations; Chemical plants; Control systems
  Process Operability and Control of Preliminary Designs, James M. Douglas, 497

Design improvements; Design practices; Design standards; Distributed processing; Instrumentation; Interfaces; Maintainability; Operator performance; Reliability; Computer applications
  Important Design Issues and Recommendation for Prospective Users of Distributed Control—I, Ramsi R. Messare, 599

Design improvements; Design practices; Errors; Human factors; Human resources; Man machine systems; Performance evaluation; Process control; Reliability; Systems engineering
  Human Error and Human Reliability in Process Control, John W. Senders, 111

Design improvements; Design practices; Human factors; Man machine systems; Process control; Systems engineering; Computer systems hardware; Computer systems programs; Control systems
  The Systems Engineering Context of Process Control and Man-Machine System Design, William L. Livingston, 117

Design improvements; Design practices; Industrial plants; Quality control; Abnormalities; Chemical plants; Control systems; Control theory
  Control Systems for Integrated Plants, Alan S. Foss, 625

Design practices; Design standards; Distributed processing; Instrumentation; Interfaces; Maintainability; Operator performance; Reliability; Computer applications; Design improvements
  Important Design Issues and Recommendation for Prospective Users of Distributed Control—I, Ramsi R. Messare, 599

Design practices; Emergencies; Model studies; Optimization; Plants (industrial); Process control; Selection; Actuators; Algorithms; Computers; Constraints
  Model Algorithmic Control (MAC): Review and Recent Developments, Raman K. Mehra and Ramine Rouhani, 287

Design practices; Errors; Human factors; Human resources; Man machine systems; Performance evaluation; Process control; Reliability; Systems engineering; Design improvements
  Human Error and Human Reliability in Process Control, John W. Senders, 111

Design practices; Human factors; Man machine systems; Process control; Systems engineering; Computer systems hardware; Computer systems programs; Control systems; Design improvements
  The Systems Engineering Context of Process Control and Man-Machine System Design, William L. Livingston, 117

Design practices; Industrial plants; Quality control; Abnormalities; Chemical plants; Control systems; Control theory; Design improvements
  Control Systems for Integrated Plants, Alan S. Foss, 625

Design standards; Distributed processing; Instrumentation; Interfaces; Maintainability; Operator performance; Reliability; Computer applications; Design improvements; Design practices
  Important Design Issues and Recommendation for Prospective Users of Distributed Control—I, Ramsi R. Messare, 599

Digital simulation; Distillation equipment; Material balance; Purity; Towers; Columns (process engineering); Compositions; Control systems
  On the Difference Between Distillation Column Control Problems, Thomas J. McAvoy, 377

Digital systems; Distributed processing; Intelligence; Process control; Computer controlled signals; Computer programming; Constraints; Control systems; Control theory
  Distributed Computer Process Control, Duncan A. Mellichamp, 629

Digital systems; Industrial plants; Management systems; Process control; Systems analysis; Computer controlled signals; Computer languages; Computer programming; Control equipment
  Systems Software, D. Grant Fisher, 613

Digital systems; Interfaces; Maintainability; Operator performance; Process control; Automatic control; Computer programs; Control systems
  An Approach to Digital Process Control, Martin Masak, 581

Display devices; Laboratories; Measuring instruments; Operator performance; Valves; Alarm systems; Chemical plants; Control equipment; Control systems
  Control Systems for Complete Plants—The Industrial View—I, Irven Rinard, 541

Distillation; Distillation equipment; Energy conservation; Heat; Heat recovery; Process control; Vapor pressure; Condensers; Control
Control of Heat-Integrated Columns, Page S. Buckley, 347

Distillation; Distillation equipment; Energy conversion; Energy methods; Gasification; Paper mills; Powerplants; Boilers; Control systems; Control theory
Distillation column Control/Energy Management, J. Patrick Kennedy, 623

Distillation; Experimental data; Feedback control; Process control; Chemical industry; Compositions; Computer controlled signals; Control systems
University Research on Dual Composition Control of Distillation: A Review, Kurt V. Waller, 395

Distillation; Nonlinear systems; Process control; Responses; Stability; Control; Couplings
Explicit vs. Implicit Decoupling in Distillation Control, Carroll J. Ryskamp, 361

Distillation equipment; Energy conservation; Heat; Heat recovery; Process control; Vapor pressure; Condensers; Control; Distillation
Control of Heat-Integrated Columns, Page S. Buckley, 347

Distillation equipment; Energy conversion; Energy methods; Gasification; Paper mills; Powerplants; Boilers; Control systems; Control theory; Distillation
Distillation column Control/Energy Management, J. Patrick Kennedy, 623

Distillation equipment; Estimation; Mathematical models; Model studies; Polymerization; Systems analysis; Chemical industry; Chemical reactors; Control
Application of Modelling, Estimation and Control to Chemical Processes, E. Dieter Gilles, 311

Distillation equipment; Fuels; Heat balance; Power supplies; Sequential analysis; Blending; Computer programs; Computers; Control systems
Control Systems for Complete Plants—The Industrial View—III, Thomas E. Marlin, 555

Distillation equipment; Material balance; Purity; Towers; Columns (process engineering); Compositions; Control systems; Digital simulation
On the Difference Between Distillation Column Control Problems, Thomas J. McAvoy, 377

Distributed processing; Instrumentation; Interfaces; Maintainability; Operator performance; Reliability; Computer applications; Design improvements; Design practices; Design standards
Important Design Issues and Recommendation for Prospective Users of Distributed Control—I, Ramsi R. Messare, 599

Distributed processing; Intelligence; Performance; Reliability; Topology; Computer controlled signals; Control systems; Data acquisition
The Structuring of Distributed Intelligence Computer Control Systems, Charles W. Rose, 565

Distributed processing; Intelligence; Process control; Computer controlled signals; Computer programming; Constraints; Control systems; Control theory; Digital systems
Distributed Computer Process Control, Duncan A. Mellichamp, 629

Distributed processing; Man machine systems; Modules; Process control; Reliability; Alarm systems; Computer languages; Cost analysis
Important Design Issues and Recommendation for Prospective Users of Distributed Control—III, Edgar Bristol, 605

Distributed processing; Networks; Optimization; Reliability; Systems analysis; Computer controlled signals; Computers; Decision making
Important Design Issues and Recommendation for Prospective Users of Distributed Control—II, Duncan A. Mellichamp, 603

Dynamic characteristics; Dynamics; Input output routines; Nonlinear systems; Nonlinear systems; Oscillations; Parametric equations; Process control; Sensitivity analysis; Systems engineering; Chemical reactors
New Approaches to the Dynamics of Nonlinear Systems with Implicationsfor Process and Control System Design, W. Harmon Ray, 245

Dynamics; Input output routines; Nonlinear systems; Nonlinear systems; Oscillations; Parametric equations; Process control; Sensitivity analysis; Systems engineering; Chemical reactors; Dynamic characteristics
New Approaches to the Dynamics of Nonlinear Systems with Implicationsfor Process and Control System Design, W. Harmon Ray, 245

Dynamic tests; Human factors engineering; Operators (personnel); Personnel development; Personnel management; Personnel selection; Process control; Simulation; Training
Training and Selection of Operators in Complex Process Control Systems, Lawrence E. Kanous, 137

Economic impact; Gasification; Loading; Performance; Powerplants; Transient response; Benefits; Computerized simulation; Control systems
Control Analysis of Gasification/Combined Cycle (GCC) Power Plants, George H. Quentin, 415

Emergencies; Estimation; Operator performance; Process control; Research; Algorithms; Constraints; Control systems; Control theory
Advanced Strategies for Process Control and Estimation, John H. Seinfeld, 619

Emergencies; Model studies; Optimization; Plants (industrial); Process control; Selection; Actuators; Algorithms; Computers; Constraints; Design practices
Model Algorithmic Control (MAC): Review and Recent Developments, Raman K. Mehra and Ramine Rouhani, 287

Emission control; Energy conservation; Environmental quality regulations; Process control; Safety standards; Systems engineering; Chemical industry; Computer applications; Computerized design; Costs
New Results and the Status of Computer-Aided Process Control Systems Design in Japan, Tori Hashimoto, 147

Energy conservation; Energy conversion; Energy methods; Management systems; Petroleum industry; Powerhouses; Boilers; Chemical industry; Control systems; Costs
General Industrial Energy Management System—Functionality and Architecture, C. Ronald Simpkins, 433

Energy conservation; Environmental quality
regulations; Process control; Safety standards;
Systems engineering; Chemical industry; Computer
applications; Computerized design; Costs; Emission
control
   New Results and the Status of Computer-Aided
   Process Control Systems Design in Japan, Tori
   Hashimoto, 147

Energy conservation; Heat; Heat recovery; Process
control; Vapor pressure; Condensers; Control;
Distillation; Distillation equipment
   Control of Heat-Integrated Columns, Page S.
   Buckley, 347

Energy conservation; Materials control; Model
studies; Process control; Quality control; Systems
analysis; Chemical plants; Constraints; Control
systems
   Control Systems for Complete Plants—The
   Industrial View—II, Joseph P. Shunta, 547

Energy conversion; Energy methods; Gasification;
Paper mills; Powerplants; Boilers; Control
systems; Control theory; Distillation; Distillation
equipment
   Distillation column Control/Energy Management,
   J. Patrick Kennedy, 623

Energy conversion; Energy methods; Management
systems; Petroleum industry; Powerhouses;
Boilers; Chemical industry; Control systems; Costs;
Energy conservation
   General Industrial Energy Management System—
   Functionality and Architecture, C. Ronald
   Simpkins, 433

Energy methods; Gasification; Paper mills;
Powerplants; Boilers; Control systems; Control
theory; Distillation; Distillation equipment; Energy
conversion
   Distillation column Control/Energy Management,
   J. Patrick Kennedy, 623

Energy methods; Management systems; Petroleum
industry; Powerhouses; Boilers; Chemical industry;
Control systems; Costs; Energy conservation;
Energy conversion
   General Industrial Energy Management System—
   Functionality and Architecture, C. Ronald
   Simpkins, 433

Environmental factors; Header pipes; Management;
Optimization; Plants (industrial); Pressure
responses; Remote sensing; Steam generators;
Utilities; Boilers; Combustion control
   Plant Utility Management, Richard A. Hanson,
   449

Environmental quality regulations; Process control;
Safety standards; Systems engineering; Chemical
industry; Computer applications; Computerized
design; Costs; Emission control; Energy
conservation
   New Results and the Status of Computer-Aided
   Process Control Systems Design in Japan, Tori
   Hashimoto, 147

Error analysis; Measurement; Monitor routines;
Performance; Process charting; Process control;
Research; Classification; Data analysis
   Design and Analysis of Process Performance:
   Monitoring Systems, Richard S.H. Mah, 525

Error analysis; Optimization; Process control;
Reliability; Communication systems; Computer
applications; Computer languages; Computer
programs; Computer systems programs
   Application of the ADA Language to Process
   Control, William C.M. Vaughan, 71

Errors; Human factors; Human resources; Man
machine systems; Performance evaluation; Process
control; Reliability; Systems engineering; Design
improvements; Design practices
   Human Error and Human Reliability in Process
   Control, John W. Senders, 111

Estimates; Multivariate analysis; Process control;
Research; Systems engineering; Computer
applications; Computerized design; Computer
programming
   New Results and the Status of Computer-Aided
   Process Control Systems Design in North
   America, Thomas F. Edgar, 213

Estimation; Mathematical models; Model studies;
Polymerization; Systems analysis; Chemical
industry; Chemical reactors; Control; Distillation
equipment
   Application of Modelling, Estimation and Control
   to Chemical Processes, E. Dieter Gilles, 311

Estimation; Operator performance; Process control;
Research; Algorithms; Constraints; Control
systems; Control theory; Emergencies
   Advanced Strategies for Process Control and
   Estimation, John H. Seinfeld, 619

Europe; Identification; Industrial plants; Models;
Process control; Simulation; Computer
applications; Computerized design; Computer
programming; Computers
   New Results and the Status of Computer-Aided
   Process Control Systems in Europe, Arne Tysso,
   187

Experimental data; Feedback control; Process
control; Chemical industry; Compositions;
Computer controlled signals; Control systems;
Distillation
   University Research on Dual Composition
   Control of Distillation: A Review, Kurt V.
   Waller, 395

Feedback control; Process control; Chemical
industry; Compositions; Computer controlled
signals; Control systems; Distillation;
Experimental data
   University Research on Dual Composition
   Control of Distillation: A Review, Kurt V.
   Waller, 395

Fuels; Heat balance; Power supplies; Sequential
analysis; Blending; Computer programs;
Computers; Control systems; Distillation
equipment
   Control Systems for Complete Plants—The
   Industrial View—III, Thomas E. Marlin, 555

Gasification; Loading; Performance; Powerplants;
Transient response; Benefits; Computerized
simulation; Control systems; Economic impact
   Control Analysis of Gasification/Combined Cycle
   (GCC) Power Plants, George H. Quentin, 415

Gasification; Paper mills; Powerplants; Boilers;
Control systems; Control theory; Distillation;
Distillation equipment; Energy conversion; Energy
methods
   Distillation column Control/Energy Management,
   J. Patrick Kennedy, 623

Gaussian quadrature; Industrial engineering; Linear systems; Quadratic forms; Systems engineering; Constraints; Control systems; Control theory
On Closing "The Gap"—With Modern Control Theory that Works, Magne Field, 229

Header pipes; Management; Optimization; Plants (industrial); Pressure responses; Remote sensing; Steam generators; Utilities; Boilers; Combustion control; Environmental factors
Plant Utility Management, Richard A. Hanson, 449

Heat; Heat recovery; Process control; Vapor pressure; Condensers; Control; Distillation; Distillation equipment; Energy conservation
Control of Heat-Integrated Columns, Page S. Buckley, 347

Heat balance; Power supplies; Sequential analysis; Blending; Computer programs; Computers; Control systems; Distillation equipment; Fuels
Control Systems for Complete Plants—The Industrial View—III, Thomas E. Marlin, 555

Heat recovery; Process control; Vapor pressure; Condensers; Control; Distillation; Distillation equipment; Energy conservation; Heat
Control of Heat-Integrated Columns, Page S. Buckley, 347

Human factors; Human resources; Man machine systems; Performance evaluation; Process control; Reliability; Systems engineering; Design improvements; Design practices; Errors
Human Error and Human Reliability in Process Control, John W. Senders, 111

Human factors; Man machine systems; Process control; Systems engineering; Computer systems hardware; Computer systems programs; Control systems; Design improvements; Design practices
The Systems Engineering Context of Process Control and Man-Machine System Design, William L. Livingston, 117

Human factors engineering; Interfaces; Man machine systems; Operator performance; Process control; Reliability; Training
Human Factors, John E. Rijnsdorp, 617

Human factors engineering; Man machine systems; Manual control; Multivariate analysis; Participative management; Process control; Quality control; Research and development; Training; Working conditions
Important Problems and Challenges in Human Factors and Man-Machine Engineering for Process Control Systems, John E. Rijnsdorp, 93

Human factors engineering; Operators (personnel); Personnel development; Personnel management; Personnel selection; Process control; Simulation; Training; Dynamic tests
Training and Selection of Operators in Complex Process Control Systems, Lawrence E. Kanous, 137

Human resources; Man machine systems; Performance evaluation; Process control; Reliability; Systems engineering; Design improvements; Design practices; Errors; Human factors
Human Error and Human Reliability in Process Control, John W. Senders, 111

641

Identification; Industrial plants; Models; Process control; Simulation; Computer applications; Computerized design; Computer programming; Computers; Europe
New Results and the Status of Computer-Aided Process Control Systems in Europe, Arne Tysso, 187

Identification; Industries; Paper industry; Process control; Processing; Regulators; Adaptive systems; Control systems; Convergence
Review of Some Adaptive Control Schemes for Process Control, Pierre R. Belanger, 269

Industrial engineering; Linear systems; Quadratic forms; Systems engineering; Constraints; Control systems; Control theory; Gaussian quadrature
On Closing "The Gap"—With Modern Control Theory that Works, Magne Field, 229

Industrial plants; Management systems; Process control; Systems analysis; Computer controlled signals; Computer languages; Computer programming; Control equipment; Digital systems
Systems Software, D. Grant Fisher, 613

Industrial plants; Models; Process control; Simulation; Computer applications; Computerized design; Computer programming; Computers; Europe; Identification
New Results and the Status of Computer-Aided Process Control Systems in Europe, Arne Tysso, 187

Industrial plants; Quality control; Abnormalities; Chemical plants; Control systems; Control theory; Design improvements; Design practices
Control Systems for Integrated Plants, Alan S. Foss, 625

Industries; Model studies; National Science Foundation; Process control; Recommendations; Simulation models; Universities; Chemical industry
Industrial Theme Problems in Process Control, M. Denn, 609

Industries; Paper industry; Process control; Processing; Regulators; Adaptive systems; Control systems; Convergence; Identification
Review of Some Adaptive Control Schemes for Process Control, Pierre R. Belanger, 269

Input output routines; Nonlinear systems; Nonlinear systems; Oscillations; Parametric equations; Process control; Sensitivity analysis; Systems engineering; Chemical reactors; Dynamic characteristics; Dynamics
New Approaches to the Dynamics of Nonlinear Systems with Implicationsfor Process and Control System Design, W. Harmon Ray, 245

Instrumentation; Interfaces; Maintainability; Operator performance; Reliability; Computer applications; Design improvements; Design practices; Design standards; Distributed processing
Important Design Issues and Recommendation for Prospective Users of Distributed Control—I, Ramsi R. Messare, 599

Intelligence; Performance; Reliability; Topology; Computer controlled signals; Control systems; Data acquisition; Distributed processing
The Structuring of Distributed Intelligence Computer Control Systems, Charles W. Rose, 565

Intelligence; Process control; Computer controlled signals; Computer programming; Constraints; Control systems; Control theory; Digital systems; Distributed processing
   Distributed Computer Process Control, Duncan A. Mellichamp, 629

Interfaces; Maintainability; Operator performance; Process control; Automatic control; Computer programs; Control systems; Digital systems
   An Approach to Digital Process Control, Martin Masak, 581

Interfaces; Maintainability; Operator performance; Reliability; Computer applications; Design improvements; Design practices; Design standards; Distributed processing; Instrumentation
   Important Design Issues and Recommendation for Prospective Users of Distributed Control—I, Ramsi R. Messare, 599

Interfaces; Man machine systems; Operator performance; Process control; Reliability; Training; Human factors engineering
   Human Factors, John E. Rijnsdorp, 617

Laboratories; Measuring instruments; Operator performance; Valves; Alarm systems; Chemical plants; Control equipment; Control systems; Display devices
   Control Systems for Complete Plants—The Industrial View—I, Irven Rinard, 541

Layered systems; Mathematical models; Nodes; Process control; Sequential analysis; Systems engineering; Application methods; Computer applications; Computer programs
   Principles in Structuring Software for Process Control, Th. Lalive d'Epinay, 33

Linear systems; Quadratic forms; Systems engineering; Constraints; Control systems; Control theory; Gaussian quadrature; Industrial engineering
   On Closing "The Gap"—With Modern Control Theory that Works, Magne Field, 229

Loading; Performance; Powerplants; Transient response; Benefits; Computerized simulation; Control systems; Economic impact; Gasification
   Control Analysis of Gasification/Combined Cycle (GCC) Power Plants, George H. Quentin, 415

Maintainability; Operator performance; Process control; Automatic control; Computer programs; Control systems; Digital systems; Interfaces
   An Approach to Digital Process Control, Martin Masak, 581

Maintainability; Operator performance; Reliability; Computer applications; Design improvements; Design practices; Design standards; Distributed processing; Instrumentation; Interfaces
   Important Design Issues and Recommendation for Prospective Users of Distributed Control—I, Ramsi R. Messare, 599

Management; Optimization; Plants (industrial); Pressure responses; Remote sensing; Steam generators; Utilities; Boilers; Combustion control; Environmental factors; Header pipes
   Plant Utility Management, Richard A. Hanson, 449

Management systems; Petroleum industry; Powerhouses; Boilers; Chemical industry; Control systems; Costs; Energy conservation; Energy conversion; Energy methods
   General Industrial Energy Management System—Functionality and Architecture, C. Ronald Simpkins, 433

Management systems; Process control; Systems analysis; Chemical plants; Chemical reactions; Communication systems; Computer programs
   Systems Software for Process Control: Today's State of the Art, D. Grant Fisher, 3

Management systems; Process control; Systems analysis; Computer controlled signals; Computer languages; Computer programming; Control equipment; Digital systems; Industrial plants
   Systems Software, D. Grant Fisher, 613

Man machine systems; Manual control; Multivariate analysis; Participative management; Process control; Quality control; Research and development; Training; Working conditions; Human factors engineering
   Important Problems and Challenges in Human Factors and Man-Machine Engineering for Process Control Systems, John E. Rijnsdorp, 93

Man machine systems; Modules; Process control; Reliability; Alarm systems; Computer languages; Cost analysis; Distributed processing
   Important Design Issues and Recommendation for Prospective Users of Distributed Control—III, Edgar Bristol, 605

Man machine systems; Operator performance; Process control; Reliability; Training; Human factors engineering; Interfaces
   Human Factors, John E. Rijnsdorp, 617

Man machine systems; Performance evaluation; Process control; Reliability; Systems engineering; Design improvements; Design practices; Errors; Human factors; Human resources
   Human Error and Human Reliability in Process Control, John W. Senders, 111

Man machine systems; Process control; Systems engineering; Computer systems hardware; Computer systems programs; Control systems; Design improvements; Design practices; Human factors
   The Systems Engineering Context of Process Control and Man-Machine System Design, William L. Livingston, 117

Manual control; Multivariate analysis; Participative management; Process control; Quality control; Research and development; Training; Working conditions; Human factors engineering; Man machine systems
   Important Problems and Challenges in Human Factors and Man-Machine Engineering for Process Control Systems, John E. Rijnsdorp, 93

Material balance; Purity; Towers; Columns (process engineering); Compositions; Control systems; Digital simulation; Distillation equipment
   On the Difference Between Distillation Column Control Problems, Thomas J. McAvoy, 377

Materials control; Model studies; Process control; Quality control; Systems analysis; Chemical plants; Constraints; Control systems; Energy conservation
   Control Systems for Complete Plants—The Industrial View—II, Joseph P. Shunta, 547

Mathematical models; Model studies;
Polymerization; Systems analysis; Chemical
industry; Chemical reactors; Control; Distillation
equipment; Estimation
    Application of Modelling, Estimation and Control
to Chemical Processes, E. Dieter Gilles, 311

Mathematical models; Nodes; Process control;
Sequential analysis; Systems engineering;
Application methods; Computer applications;
Computer programs; Layered systems
    Principles in Structuring Software for Process
Control, Th. Lalive d'Epinay, 33

Measurement; Monitor routines; Performance;
Process charting; Process control; Research;
Classification; Data analysis; Error analysis
    Design and Analysis of Process Performance:
Monitoring Systems, Richard S.H. Mah, 525

Measuring instruments; Operator performance;
Valves; Alarm systems; Chemical plants; Control
equipment; Control systems; Display devices;
Laboratories
    Control Systems for Complete Plants—The
Industrial View—I, Irven Rinard, 541

Models; Process control; Simulation; Computer
applications; Computerized design; Computer
programming; Computers; Europe; Identification;
Industrial plants
    New Results and the Status of Computer-Aided
Process Control Systems in Europe, Arne Tysso,
187

Model studies; National Science Foundation;
Process control; Recommendations; Simulation
models; Universities; Chemical industry; Industries
    Industrial Theme Problems in Process Control, M.
Denn, 609

Model studies; Optimization; Plants (industrial);
Process control; Selection; Actuators; Algorithms;
Computers; Constraints; Design practices;
Emergencies
    Model Algorithmic Control (MAC): Review and
Recent Developments, Raman K. Mehra and
Ramine Rouhani, 287

Model studies; Polymerization; Systems analysis;
Chemical industry; Chemical reactors; Control;
Distillation equipment; Estimation; Mathematical
models
    Application of Modelling, Estimation and Control
to Chemical Processes, E. Dieter Gilles, 311

Model studies; Process control; Quality control;
Systems analysis; Chemical plants; Constraints;
Control systems; Energy conservation; Materials
control
    Control Systems for Complete Plants—The
Industrial View—II, Joseph P. Shunta, 547

Modules; Nodes; Operating criteria; Paper industry;
Process control; Algorithms; Computer programs;
Computers; Control
    Operational Needs in Software for Complex
Process Control Systems, Robert G. Wilhelm,
Jr., 51

Modules; Process control; Reliability; Alarm
systems; Computer languages; Cost analysis;
Distributed processing; Man machine systems
    Important Design Issues and Recommendation for
Prospective Users of Distributed Control—III,
Edgar Bristol, 605

Monitor routines; Performance; Process charting;
Process control; Research; Classification; Data
analysis; Error analysis; Measurement
    Design and Analysis of Process Performance:
Monitoring Systems, Richard S.H. Mah, 525

Multivariate analysis; Participative management;
Process control; Quality control; Research and
development; Training; Working conditions;
Human factors engineering; Man machine systems;
Manual control
    Important Problems and Challenges in Human
Factors and Man-Machine Engineering for
Process Control Systems, John E. Rijnsdorp, 93

Multivariate analysis; Process control; Research;
Systems engineering; Computer applications;
Computerized design; Computer programming;
Estimates
    New Results and the Status of Computer-Aided
Process Control Systems Design in North
America, Thomas F. Edgar, 213

National Science Foundation; Process control;
Recommendations; Simulation models;
Universities; Chemical industry; Industries; Model
studies
    Industrial Theme Problems in Process Control, M.
Denn, 609

Networks; Optimization; Reliability; Systems
analysis; Computer controlled signals; Computers;
Decision making; Distributed processing
    Important Design Issues and Recommendation for
Prospective Users of Distributed Control—II,
Duncan A. Mellichamp, 603

Nodes; Operating criteria; Paper industry; Process
control; Algorithms; Computer programs;
Computers; Control; Modules
    Operational Needs in Software for Complex
Process Control Systems, Robert G. Wilhelm,
Jr., 51

Nodes; Process control; Sequential analysis; Systems
engineering; Application methods; Computer
applications; Computer programs; Layered
systems; Mathematical models
    Principles in Structuring Software for Process
Control, Th. Lalive d'Epinay, 33

Nonlinear systems; Nonlinear systems; Oscillations;
Parametric equations; Process control; Sensitivity
analysis; Systems engineering; Chemical reactors;
Dynamic characteristics; Dynamics; Input output
routines
    New Approaches to the Dynamics of Nonlinear
Systems with Implicationsfor Process and
Control System Design, W. Harmon Ray, 245

Nonlinear systems; Oscillations; Parametric
equations; Process control; Sensitivity analysis;
Systems engineering; Chemical reactors; Dynamic
characteristics; Dynamics; Input output routines;
Nonlinear systems
    New Approaches to the Dynamics of Nonlinear
Systems with Implicationsfor Process and
Control System Design, W. Harmon Ray, 245

Nonlinear systems; Process control; Responses;
Stability; Control; Couplings; Distillation
    Explicit vs. Implicit Decoupling in Distillation
Control, Carroll J. Ryskamp, 361

Operating criteria; Paper industry; Process control;
Algorithms; Computer programs; Computers;
Control; Modules; Nodes
　Operational Needs in Software for Complex
　Process Control Systems, Robert G. Wilhelm,
　Jr., 51

Operations; Preliminary investigations; Process
charting; Process control; Steady state; Variations;
Chemical plants; Control systems; Design
　Process Operability and Control of Preliminary
　Designs, James M. Douglas, 497

Operator performance; Process control; Automatic
control; Computer programs; Control systems;
Digital systems; Interfaces; Maintainability
　An Approach to Digital Process Control, Martin
　Masak, 581

Operator performance; Process control; Reliability;
Training; Human factors engineering; Interfaces;
Man machine systems
　Human Factors, John E. Rijnsdorp, 617

Operator performance; Process control; Research;
Algorithms; Constraints; Control systems; Control
theory; Emergencies; Estimation
　Advanced Strategies for Process Control and
　Estimation, John H. Seinfeld, 619

Operator performance; Reliability; Computer
applications; Design improvements; Design
practices; Design standards; Distributed
processing; Instrumentation; Interfaces;
Maintainability
　Important Design Issues and Recommendation for
　Prospective Users of Distributed Control—I,
　Ramsi R. Messare, 599

Operator performance; Valves; Alarm systems;
Chemical plants; Control equipment; Control
systems; Display devices; Laboratories; Measuring
instruments
　Control Systems for Complete Plants—The
　Industrial View—I, Irven Rinard, 541

Operators (personnel); Personnel development;
Personnel management; Personnel selection;
Process control; Simulation; Training; Dynamic
tests; Human factors engineering
　Training and Selection of Operators in Complex
　Process Control Systems, Lawrence E. Kanous,
　137

Optimization; Plants (industrial); Pressure
responses; Remote sensing; Steam generators;
Utilities; Boilers; Combustion control;
Environmental factors; Header pipes; Management
　Plant Utility Management, Richard A. Hanson,
　449

Optimization; Plants (industrial); Process control;
Selection; Actuators; Algorithms; Computers;
Constraints; Design practices; Emergencies; Model
studies
　Model Algorithmic Control (MAC): Review and
　Recent Developments, Raman K. Mehra and
　Ramine Rouhani, 287

Optimization; Process control; Reliability;
Communication systems; Computer applications;
Computer languages; Computer programs;
Computer systems programs; Error analysis
　Application of the ADA Language to Process
　Control, William C.M. Vaughan, 71

Optimization; Reliability; Systems analysis;
Computer controlled signals; Computers; Decision
making; Distributed processing; Networks
　Important Design Issues and Recommendation for
　Prospective Users of Distributed Control—II,
　Duncan A. Mellichamp, 603

Oscillations; Parametric equations; Process control;
Sensitivity analysis; Systems engineering;
Chemical reactors; Dynamic characteristics;
Dynamics; Input output routines; Nonlinear
systems; Nonlinear systems
　New Approaches to the Dynamics of Nonlinear
　Systems with Implicationsfor Process and
　Control System Design, W. Harmon Ray, 245

Paper industry; Process control; Algorithms;
Computer programs; Computers; Control; Modules;
Nodes; Operating criteria
　Operational Needs in Software for Complex
　Process Control Systems, Robert G. Wilhelm,
　Jr., 51

Paper industry; Process control; Processing;
Regulators; Adaptive systems; Control systems;
Convergence; Identification; Industries
　Review of Some Adaptive Control Schemes for
　Process Control, Pierre R. Belanger, 269

Paper mills; Powerplants; Boilers; Control systems;
Control theory; Distillation; Distillation
equipment; Energy conversion; Energy methods;
Gasification
　Distillation column Control/Energy Management,
　J. Patrick Kennedy, 623

Parametric equations; Process control; Sensitivity
analysis; Systems engineering; Chemical reactors;
Dynamic characteristics; Dynamics; Input output
routines; Nonlinear systems; Nonlinear systems;
Oscillations
　New Approaches to the Dynamics of Nonlinear
　Systems with Implicationsfor Process and
　Control System Design, W. Harmon Ray, 245

Participative management; Process control; Quality
control; Research and development; Training;
Working conditions; Human factors engineering;
Man machine systems; Manual control;
Multivariate analysis
　Important Problems and Challenges in Human
　Factors and Man-Machine Engineering for
　Process Control Systems, John E. Rijnsdorp, 93

Performance; Powerplants; Transient response;
Benefits; Computerized simulation; Control
systems; Economic impact; Gasification; Loading
　Control Analysis of Gasification/Combined Cycle
　(GCC) Power Plants, George H. Quentin, 415

Performance; Process charting; Process control;
Research; Classification; Data analysis; Error
analysis; Measurement; Monitor routines
　Design and Analysis of Process Performance:
　Monitoring Systems, Richard S.H. Mah, 525

Performance; Reliability; Topology; Computer
controlled signals; Control systems; Data
acquisition; Distributed processing; Intelligence
　The Structuring of Distributed Intelligence
　Computer Control Systems, Charles W. Rose,
　565

Performance evaluation; Process control; Reliability; Systems engineering; Design improvements; Design practices; Errors; Human factors; Human resources; Man machine systems
   Human Error and Human Reliability in Process Control, John W. Senders, 111

Personnel development; Personnel management; Personnel selection; Process control; Simulation; Training; Dynamic tests; Human factors engineering; Operators (personnel)
   Training and Selection of Operators in Complex Process Control Systems, Lawrence E. Kanous, 137

Personnel management; Personnel selection; Process control; Simulation; Training; Dynamic tests; Human factors engineering; Operators (personnel); Personnel development
   Training and Selection of Operators in Complex Process Control Systems, Lawrence E. Kanous, 137

Personnel selection; Process control; Simulation; Training; Dynamic tests; Human factors engineering; Operators (personnel); Personnel development; Personnel management
   Training and Selection of Operators in Complex Process Control Systems, Lawrence E. Kanous, 137

Petroleum industry; Powerhouses; Boilers; Chemical industry; Control systems; Costs; Energy conservation; Energy conversion; Energy methods; Management systems
   General Industrial Energy Management System—Functionality and Architecture, C. Ronald Simpkins, 433

Plants (industrial); Pressure responses; Remote sensing; Steam generators; Utilities; Boilers; Combustion control; Environmental factors; Header pipes; Management; Optimization
   Plant Utility Management, Richard A. Hanson, 449

Plants (industrial); Process control; Quality control; Research; Sensitivity; Systems analysis; Tuning; Control; Controller characteristics
   Integrated Plant Control: A Solution at Hand or a Research Topic for the Next Decade?, Manfred Morari, 467

Plants (industrial); Process control; Selection; Actuators; Algorithms; Computers; Constraints; Design practices; Emergencies; Model studies; Optimization
   Model Algorithmic Control (MAC): Review and Recent Developments, Raman K. Mehra and Ramine Rouhani, 287

Polymerization; Systems analysis; Chemical industry; Chemical reactors; Control; Distillation equipment; Estimation; Mathematical models; Model studies
   Application of Modelling, Estimation and Control to Chemical Processes, E. Dieter Gilles, 311

Powerhouses; Boilers; Chemical industry; Control systems; Costs; Energy conservation; Energy conversion; Energy methods; Management systems; Petroleum industry
   General Industrial Energy Management System—Functionality and Architecture, C. Ronald Simpkins, 433

Powerplants; Boilers; Control systems; Control theory; Distillation; Distillation equipment; Energy conversion; Energy methods; Gasification; Paper mills
   Distillation column Control/Energy Management, J. Patrick Kennedy, 623

Powerplants; Transient response; Benefits; Computerized simulation; Control systems; Economic impact; Gasification; Loading; Performance
   Control Analysis of Gasification/Combined Cycle (GCC) Power Plants, George H. Quentin, 415

Power supplies; Sequential analysis; Blending; Computer programs; Computers; Control systems; Distillation equipment; Fuels; Heat balance
   Control Systems for Complete Plants—The Industrial View—III, Thomas E. Marlin, 555

Preliminary investigations; Process charting; Process control; Steady state; Variations; Chemical plants; Control systems; Design; Operations
   Process Operability and Control of Preliminary Designs, James M. Douglas, 497

Pressure responses; Remote sensing; Steam generators; Utilities; Boilers; Combustion control; Environmental factors; Header pipes; Management; Optimization; Plants (industrial)
   Plant Utility Management, Richard A. Hanson, 449

Process charting; Process control; Research; Classification; Data analysis; Error analysis; Measurement; Monitor routines; Performance
   Design and Analysis of Process Performance: Monitoring Systems, Richard S.H. Mah, 525

Process charting; Process control; Steady state; Variations; Chemical plants; Control systems; Design; Operations; Preliminary investigations
   Process Operability and Control of Preliminary Designs, James M. Douglas, 497

Process control; Algorithms; Computer programs; Computers; Control; Modules; Nodes; Operating criteria; Paper industry
   Operational Needs in Software for Complex Process Control Systems, Robert G. Wilhelm, Jr., 51

Process control; Automatic control; Computer programs; Control systems; Digital systems; Interfaces; Maintainability; Operator performance
   An Approach to Digital Process Control, Martin Masak, 581

Process control; Chemical industry; Compositions; Computer controlled signals; Control systems; Distillation; Experimental data; Feedback control
   University Research on Dual Composition Control of Distillation: A Review, Kurt V. Waller, 395

Process control; Computer controlled signals; Computer programming; Constraints; Control systems; Control theory; Digital systems; Distributed processing; Intelligence
   Distributed Computer Process Control, Duncan A. Mellichamp, 629

Process control; Processing; Regulators; Adaptive systems; Control systems; Convergence; Identification; Industries; Paper industry
   Review of Some Adaptive Control Schemes for Process Control, Pierre R. Belanger, 269

Process control; Quality control; Research; Sensitivity; Systems analysis; Tuning; Control; Controller characteristics; Plants (industrial)
 Integrated Plant Control: A Solution at Hand or a Research Topic for the Next Decade?, Manfred Morari, 467

Process control; Quality control; Research and development; Training; Working conditions; Human factors engineering; Man machine systems; Manual control; Multivariate analysis; Participative management
 Important Problems and Challenges in Human Factors and Man-Machine Engineering for Process Control Systems, John E. Rijnsdorp, 93

Process control; Quality control; Systems analysis; Chemical plants; Constraints; Control systems; Energy conservation; Materials control; Model studies
 Control Systems for Complete Plants—The Industrial View—II, Joseph P. Shunta, 547

Process control; Recommendations; Simulation models; Universities; Chemical industry; Industries; Model studies; National Science Foundation
 Industrial Theme Problems in Process Control, M. Denn, 609

Process control; Reliability; Alarm systems; Computer languages; Cost analysis; Distributed processing; Man machine systems; Modules
 Important Design Issues and Recommendation for Prospective Users of Distributed Control—III, Edgar Bristol, 605

Process control; Reliability; Communication systems; Computer applications; Computer languages; Computer programs; Computer systems programs; Error analysis; Optimization
 Application of the ADA Language to Process Control, William C.M. Vaughan, 71

Process control; Reliability; Systems engineering; Design improvements; Design practices; Errors; Human factors; Human resources; Man machine systems; Performance evaluation
 Human Error and Human Reliability in Process Control, John W. Senders, 111

Process control; Reliability; Training; Human factors engineering; Interfaces; Man machine systems; Operator performance
 Human Factors, John E. Rijnsdorp, 617

Process control; Research; Algorithms; Constraints; Control systems; Control theory; Emergencies; Estimation; Operator performance
 Advanced Strategies for Process Control and Estimation, John H. Seinfeld, 619

Process control; Research; Classification; Data analysis; Error analysis; Measurement; Monitor routines; Performance; Process charting
 Design and Analysis of Process Performance: Monitoring Systems, Richard S.H. Mah, 525

Process control; Research; Systems engineering; Computer applications; Computerized design; Computer programming; Estimates; Multivariate analysis
 New Results and the Status of Computer-Aided Process Control Systems Design in North America, Thomas F. Edgar, 213

Process control; Responses; Stability; Control; Couplings; Distillation; Nonlinear systems
 Explicit vs. Implicit Decoupling in Distillation Control, Carroll J. Ryskamp, 361

Process control; Safety standards; Systems engineering; Chemical industry; Computer applications; Computerized design; Costs; Emission control; Energy conservation; Environmental quality regulations
 New Results and the Status of Computer-Aided Process Control Systems Design in Japan, Tori Hashimoto, 147

Process control; Selection; Actuators; Algorithms; Computers; Constraints; Design practices; Emergencies; Model studies; Optimization; Plants (industrial)
 Model Algorithmic Control (MAC): Review and Recent Developments, Raman K. Mehra and Ramine Rouhani, 287

Process control; Sensitivity analysis; Systems engineering; Chemical reactors; Dynamic characteristics; Dynamics; Input output routines; Nonlinear systems; Nonlinear systems; Oscillations; Parametric equations
 New Approaches to the Dynamics of Nonlinear Systems with Implicationsfor Process and Control System Design, W. Harmon Ray, 245

Process control; Sequential analysis; Systems engineering; Application methods; Computer applications; Computer programs; Layered systems; Mathematical models; Nodes
 Principles in Structuring Software for Process Control, Th. Lalive d'Epinay, 33

Process control; Simulation; Computer applications; Computerized design; Computer programming; Computers; Europe; Identification; Industrial plants; Models
 New Results and the Status of Computer-Aided Process Control Systems in Europe, Arne Tysso, 187

Process control; Simulation; Training; Dynamic tests; Human factors engineering; Operators (personnel); Personnel development; Personnel management; Personnel selection
 Training and Selection of Operators in Complex Process Control Systems, Lawrence E. Kanous, 137

Process control; Steady state; Variations; Chemical plants; Control systems; Design; Operations; Preliminary investigations; Process charting
 Process Operability and Control of Preliminary Designs, James M. Douglas, 497

Process control; Systems analysis; Chemical plants; Chemical reactions; Communication systems; Computer programs; Management systems
 Systems Software for Process Control: Today's State of the Art, D. Grant Fisher, 3

Process control; Systems analysis; Computer controlled signals; Computer languages; Computer programming; Control equipment; Digital systems; Industrial plants; Management systems
 Systems Software, D. Grant Fisher, 613

Process control; Systems engineering; Computer systems hardware; Computer systems programs; Control systems; Design improvements; Design practices; Human factors; Man machine systems
   The Systems Engineering Context of Process Control and Man-Machine System Design, William L. Livingston, 117

Process control; Vapor pressure; Condensers; Control; Distillation; Distillation equipment; Energy conservation; Heat; Heat recovery
   Control of Heat-Integrated Columns, Page S. Buckley, 347

Processing; Regulators; Adaptive systems; Control systems; Convergence; Identification; Industries; Paper industry; Process control
   Review of Some Adaptive Control Schemes for Process Control, Pierre R. Belanger, 269

Purity; Towers; Columns (process engineering); Compositions; Control systems; Digital simulation; Distillation equipment; Material balance
   On the Difference Between Distillation Column Control Problems, Thomas J. McAvoy, 377

Quadratic forms; Systems engineering; Constraints; Control systems; Control theory; Gaussian quadrature; Industrial engineering; Linear systems
   On Closing "The Gap"—With Modern Control Theory that Works, Magne Field, 229

Quality control; Abnormalities; Chemical plants; Control systems; Control theory; Design improvements; Design practices; Industrial plants
   Control Systems for Integrated Plants, Alan S. Foss, 625

Quality control; Research; Sensitivity; Systems analysis; Tuning; Control; Controller characteristics; Plants (industrial); Process control
   Integrated Plant Control: A Solution at Hand or a Research Topic for the Next Decade?, Manfred Morari, 467

Quality control; Research and development; Training; Working conditions; Human factors engineering; Man machine systems; Manual control; Multivariate analysis; Participative management; Process control
   Important Problems and Challenges in Human Factors and Man-Machine Engineering for Process Control Systems, John E. Rijnsdorp, 93

Quality control; Systems analysis; Chemical plants; Constraints; Control systems; Energy conservation; Materials control; Model studies; Process control
   Control Systems for Complete Plants—The Industrial View—II, Joseph P. Shunta, 547

Recommendations; Simulation models; Universities; Chemical industry; Industries; Model studies; National Science Foundation; Process control
   Industrial Theme Problems in Process Control, M. Denn, 609

Regulators; Adaptive systems; Control systems; Convergence; Identification; Industries; Paper industry; Process control; Processing
   Review of Some Adaptive Control Schemes for Process Control, Pierre R. Belanger, 269

Reliability; Alarm systems; Computer languages; Cost analysis; Distributed processing; Man machine systems; Modules; Process control
   Important Design Issues and Recommendation for Prospective Users of Distributed Control—III, Edgar Bristol, 605

647

Reliability; Communication systems; Computer applications; Computer languages; Computer programs; Computer systems programs; Error analysis; Optimization; Process control
   Application of the ADA Language to Process Control, William C.M. Vaughan, 71

Reliability; Computer applications; Design improvements; Design practices; Design standards; Distributed processing; Instrumentation; Interfaces; Maintainability; Operator performance
   Important Design Issues and Recommendation for Prospective Users of Distributed Control—I, Ramsi R. Messare, 599

Reliability; Systems analysis; Computer controlled signals; Computers; Decision making; Distributed processing; Networks; Optimization
   Important Design Issues and Recommendation for Prospective Users of Distributed Control—II, Duncan A. Mellichamp, 603

Reliability; Systems engineering; Design improvements; Design practices; Errors; Human factors; Human resources; Man machine systems; Performance evaluation; Process control
   Human Error and Human Reliability in Process Control, John W. Senders, 111

Reliability; Topology; Computer controlled signals; Control systems; Data acquisition; Distributed processing; Intelligence; Performance
   The Structuring of Distributed Intelligence Computer Control Systems, Charles W. Rose, 565

Reliability; Training; Human factors engineering; Interfaces; Man machine systems; Operator performance; Process control
   Human Factors, John E. Rijnsdorp, 617

Remote sensing; Steam generators; Utilities; Boilers; Combustion control; Environmental factors; Header pipes; Management; Optimization; Plants (industrial); Pressure responses
   Plant Utility Management, Richard A. Hanson, 449

Research; Algorithms; Constraints; Control systems; Control theory; Emergencies; Estimation; Operator performance; Process control
   Advanced Strategies for Process Control and Estimation, John H. Seinfeld, 619

Research; Classification; Data analysis; Error analysis; Measurement; Monitor routines; Performance; Process charting; Process control
   Design and Analysis of Process Performance: Monitoring Systems, Richard S.H. Mah, 525

Research; Sensitivity; Systems analysis; Tuning; Control; Controller characteristics; Plants (industrial); Process control; Quality control
   Integrated Plant Control: A Solution at Hand or a Research Topic for the Next Decade?, Manfred Morari, 467

Research; Systems engineering; Computer applications; Computerized design; Computer programming; Estimates; Multivariate analysis; Process control
   New Results and the Status of Computer-Aided Process Control Systems Design in North America, Thomas F. Edgar, 213

Research and development; Training; Working conditions; Human factors engineering; Man machine systems; Manual control; Multivariate analysis; Participative management; Process control; Quality control
　Important Problems and Challenges in Human Factors and Man-Machine Engineering for Process Control Systems, John E. Rijnsdorp, 93

Responses; Stability; Control; Couplings; Distillation; Nonlinear systems; Process control
　Explicit vs. Implicit Decoupling in Distillation Control, Carroll J. Ryskamp, 361

Safety standards; Systems engineering; Chemical industry; Computer applications; Computerized design; Costs; Emission control; Energy conservation; Environmental quality regulations; Process control
　New Results and the Status of Computer-Aided Process Control Systems Design in Japan, Tori Hashimoto, 147

Selection; Actuators; Algorithms; Computers; Constraints; Design practices; Emergencies; Model studies; Optimization; Plants (industrial); Process control
　Model Algorithmic Control (MAC): Review and Recent Developments, Raman K. Mehra and Ramine Rouhani, 287

Sensitivity; Systems analysis; Tuning; Control; Controller characteristics; Plants (industrial); Process control; Quality control; Research
　Integrated Plant Control: A Solution at Hand or a Research Topic for the Next Decade?, Manfred Morari, 467

Sensitivity analysis; Systems engineering; Chemical reactors; Dynamic characteristics; Dynamics; Input output routines; Nonlinear systems; Nonlinear systems; Oscillations; Parametric equations; Process control
　New Approaches to the Dynamics of Nonlinear Systems with Implicationsfor Process and Control System Design, W. Harmon Ray, 245

Sequential analysis; Blending; Computer programs; Computers; Control systems; Distillation equipment; Fuels; Heat balance; Power supplies
　Control Systems for Complete Plants—The Industrial View—III, Thomas E. Marlin, 555

Sequential analysis; Systems engineering; Application methods; Computer applications; Computer programs; Layered systems; Mathematical models; Nodes; Process control
　Principles in Structuring Software for Process Control, Th. Lalive d'Epinay, 33

Simulation; Computer applications; Computerized design; Computer programming; Computers; Europe; Identification; Industrial plants; Models; Process control
　New Results and the Status of Computer-Aided Process Control Systems in Europe, Arne Tysso, 187

Simulation; Training; Dynamic tests; Human factors engineering; Operators (personnel); Personnel development; Personnel management; Personnel selection; Process control
　Training and Selection of Operators in Complex Process Control Systems, Lawrence E. Kanous, 137

Simulation models; Universities; Chemical industry; Industries; Model studies; National Science Foundation; Process control; Recommendations
　Industrial Theme Problems in Process Control, M. Denn, 609

Stability; Control; Couplings; Distillation; Nonlinear systems; Process control; Responses
　Explicit vs. Implicit Decoupling in Distillation Control, Carroll J. Ryskamp, 361

Steady state; Variations; Chemical plants; Control systems; Design; Operations; Preliminary investigations; Process charting; Process control
　Process Operability and Control of Preliminary Designs, James M. Douglas, 497

Steam generators; Utilities; Boilers; Combustion control; Environmental factors; Header pipes; Management; Optimization; Plants (industrial); Pressure responses; Remote sensing
　Plant Utility Management, Richard A. Hanson, 449

Systems analysis; Chemical industry; Chemical reactors; Control; Distillation equipment; Estimation; Mathematical models; Model studies; Polymerization
　Application of Modelling, Estimation and Control to Chemical Processes, E. Dieter Gilles, 311

Systems analysis; Chemical plants; Chemical reactions; Communication systems; Computer programs; Management systems; Process control
　Systems Software for Process Control: Today's State of the Art, D. Grant Fisher, 3

Systems analysis; Chemical plants; Constraints; Control systems; Energy conservation; Materials control; Model studies; Process control; Quality control
　Control Systems for Complete Plants—The Industrial View—II, Joseph P. Shunta, 547

Systems analysis; Computer controlled signals; Computer languages; Computer programming; Control equipment; Digital systems; Industrial plants; Management systems; Process control
　Systems Software, D. Grant Fisher, 613

Systems analysis; Computer controlled signals; Computers; Decision making; Distributed processing; Networks; Optimization; Reliability
　Important Design Issues and Recommendation for Prospective Users of Distributed Control—II, Duncan A. Mellichamp, 603

Systems analysis; Tuning; Control; Controller characteristics; Plants (industrial); Process control; Quality control; Research; Sensitivity
　Integrated Plant Control: A Solution at Hand or a Research Topic for the Next Decade?, Manfred Morari, 467

Systems engineering; Application methods; Computer applications; Computer programs; Layered systems; Mathematical models; Nodes; Process control; Sequential analysis
　Principles in Structuring Software for Process Control, Th. Lalive d'Epinay, 33

Systems engineering; Chemical industry; Computer applications; Computerized design; Costs; Emission control; Energy conservation; Environmental quality regulations; Process control; Safety standards
  New Results and the Status of Computer-Aided Process Control Systems Design in Japan, Tori Hashimoto, 147

Systems engineering; Chemical reactors; Dynamic characteristics; Dynamics; Input output routines; Nonlinear systems; Nonlinear systems; Oscillations; Parametric equations; Process control; Sensitivity analysis
  New Approaches to the Dynamics of Nonlinear Systems with Implicationsfor Process and Control System Design, W. Harmon Ray, 245

Systems engineering; Computer applications; Computerized design; Computer programming; Estimates; Multivariate analysis; Process control; Research
  New Results and the Status of Computer-Aided Process Control Systems Design in North America, Thomas F. Edgar, 213

Systems engineering; Computer systems hardware; Computer systems programs; Control systems; Design improvements; Design practices; Human factors; Man machine systems; Process control
  The Systems Engineering Context of Process Control and Man-Machine System Design, William L. Livingston, 117

Systems engineering; Constraints; Control systems; Control theory; Gaussian quadrature; Industrial engineering; Linear systems; Quadratic forms
  On Closing "The Gap"—With Modern Control Theory that Works, Magne Field, 229

Systems engineering; Design improvements; Design practices; Errors; Human factors; Human resources; Man machine systems; Performance evaluation; Process control; Reliability
  Human Error and Human Reliability in Process Control, John W. Senders, 111

Topology; Computer controlled signals; Control systems; Data acquisition; Distributed processing; Intelligence; Performance; Reliability
  The Structuring of Distributed Intelligence Computer Control Systems, Charles W. Rose, 565

Towers; Columns (process engineering); Compositions; Control systems; Digital simulation; Distillation equipment; Material balance; Purity
  On the Difference Between Distillation Column Control Problems, Thomas J. McAvoy, 377

Training; Dynamic tests; Human factors engineering; Operators (personnel); Personnel development; Personnel management; Personnel selection; Process control; Simulation
  Training and Selection of Operators in Complex Process Control Systems, Lawrence E. Kanous, 137

Training; Human factors engineering; Interfaces; Man machine systems; Operator performance; Process control; Reliability
  Human Factors, John E. Rijnsdorp, 617

Training; Working conditions; Human factors engineering; Man machine systems; Manual control; Multivariate analysis; Participative management; Process control; Quality control; Research and development
  Important Problems and Challenges in Human Factors and Man-Machine Engineering for Process Control Systems, John E. Rijnsdorp, 93

Transient response; Benefits; Computerized simulation; Control systems; Economic impact; Gasification; Loading; Performance; Powerplants
  Control Analysis of Gasification/Combined Cycle (GCC) Power Plants, George H. Quentin, 415

Tuning; Control; Controller characteristics; Plants (industrial); Process control; Quality control; Research; Sensitivity; Systems analysis
  Integrated Plant Control: A Solution at Hand or a Research Topic for the Next Decade?, Manfred Morari, 467

Universities; Chemical industry; Industries; Model studies; National Science Foundation; Process control; Recommendations; Simulation models
  Industrial Theme Problems in Process Control, M. Denn, 609

Utilities; Boilers; Combustion control; Environmental factors; Header pipes; Management; Optimization; Plants (industrial); Pressure responses; Remote sensing; Steam generators
  Plant Utility Management, Richard A. Hanson, 449

Valves; Alarm systems; Chemical plants; Control equipment; Control systems; Display devices; Laboratories; Measuring instruments; Operator performance
  Control Systems for Complete Plants—The Industrial View—I, Irven Rinard, 541

Vapor pressure; Condensers; Control; Distillation; Distillation equipment; Energy conservation; Heat; Heat recovery; Process control
  Control of Heat-Integrated Columns, Page S. Buckley, 347

Variations; Chemical plants; Control systems; Design; Operations; Preliminary investigations; Process charting; Process control; Steady state
  Process Operability and Control of Preliminary Designs, James M. Douglas, 497

Working conditions; Human factors engineering; Man machine systems; Manual control; Multivariate analysis; Participative management; Process control; Quality control; Research and development; Training
  Important Problems and Challenges in Human Factors and Man-Machine Engineering for Process Control Systems, John E. Rijnsdorp, 93

DATE DUE

DEMCO 38-297